THEORIES, SITES, TOPOSES

Theories, Sites, Toposes

Relating and studying mathematical theories through topos-theoretic 'bridges'

OLIVIA CARAMELLO

Università degli Studi dell'Insubria – Como

OXFORD
UNIVERSITY PRESS

Great Clarendon Street, Oxford, OX2 6DP,
United Kingdom

Oxford University Press is a department of the University of Oxford.
It furthers the University's objective of excellence in research, scholarship,
and education by publishing worldwide. Oxford is a registered trade mark of
Oxford University Press in the UK and in certain other countries

© Olivia Caramello 2018

The moral rights of the author have been asserted

First Edition published in 2018

Impression: 3

All rights reserved. No part of this publication may be reproduced, stored in
a retrieval system, or transmitted, in any form or by any means, without the
prior permission in writing of Oxford University Press, or as expressly permitted
by law, by licence or under terms agreed with the appropriate reprographics
rights organization. Enquiries concerning reproduction outside the scope of the
above should be sent to the Rights Department, Oxford University Press, at the
address above

You must not circulate this work in any other form
and you must impose this same condition on any acquirer

Published in the United States of America by Oxford University Press
198 Madison Avenue, New York, NY 10016, United States of America

British Library Cataloguing in Publication Data

Data available

Library of Congress Control Number: 2016947978

ISBN 978–0–19–875891–4

Printed and bound by
CPI Group (UK) Ltd, Croydon, CR0 4YY

Links to third party websites are provided by Oxford in good faith and
for information only. Oxford disclaims any responsibility for the materials
contained in any third party website referenced in this work.

PREFACE

Scope of the book

The genesis of this book, which focuses on geometric theories and their classifying toposes, dates back to the author's Ph.D. thesis *The Duality between Grothendieck Toposes and Geometric Theories* [12] defended in 2009 at the University of Cambridge.

The idea of regarding Grothendieck toposes from the point of view of the structures that they classify dates back to A. Grothendieck and his student M. Hakim, who characterized in her book *Topos annelés et schémas relatifs* [48] four toposes arising in algebraic geometry, notably including the Zariski topos, as the classifiers of certain special kinds of rings. Later, Lawvere's work on the *Functorial Semantics of Algebraic Theories* [59] implicitly showed that all finite algebraic theories are classified by presheaf toposes. The introduction of geometric logic, that is, the logic that is preserved under inverse images of geometric functors, is due to the Montréal school of categorical logic and topos theory active in the seventies, more specifically to G. Reyes, A. Joyal and M. Makkai. Its importance is evidenced by the fact that every geometric theory admits a classifying topos and that, conversely, every Grothendieck topos is the classifying topos of some geometric theory. After the publication, in 1977, of the monograph *First Order Categorical Logic* [64] by Makkai and Reyes, the theory of classifying toposes, in spite of its promising beginnings, stood essentially undeveloped; very few papers on the subject appeared in the following years and, as a result, most mathematicians remained unaware of the existence and potential usefulness of this fundamental notion.

One of the aims of this book is to give new life to the theory of classifying toposes by addressing in a systematic way some of the central questions that have remained unanswered throughout the past years, such as:

- The problem of elucidating the structure of the collection of geometric theory-extensions of a given geometric theory, which we tackle in Chapters 3, 4 and 8;

- The problem of characterizing (syntactically and semantically) the class of geometric theories classified by a presheaf topos, which we treat in Chapter 6;

- The crucial meta-mathematical question of how to fruitfully apply the theory of classifying toposes to get 'concrete' insights on theories of natural mathematical interest, to which we propose an answer by means of the 'bridge technique' described in Chapter 2.

It is our hope that by the end of the book the reader will have appreciated that the field is far from being exhausted and that in fact there is still much room for theoretical developments as well as great potential for applications.

Pre-requisites and reading advice

The only pre-requisite for reading this book is a basic familiarity with the language of category theory. This can be achieved by reading any introductory text on the subject, for instance the classic but still excellent *Categories for the Working Mathematician* [62] by S. Mac Lane.

The intended readership of this book is therefore quite large: mathematicians, logicians and philosophers with some experience of categories, graduate students wishing to learn topos theory, etc.

Our treatment is essentially self-contained, the necessary topos-theoretic background being recalled in Chapter 1 and referred to at various points of the book. The development of the general theory is complemented by a variety of examples and applications in different areas of mathematics which illustrate its scope and potential (cf. Chapter 10). Of course, these are not meant to exhaust the possibilities of application of the methods developed in the book; rather, they are aimed at giving the reader a flavour of the variety and mathematical depth of the 'concrete' results that can be obtained by applying such techniques.

The chapters of the book should normally be read sequentially, each one being dependent on the previous ones (with the exception of Chapter 5, which only requires Chapter 1, and of Chapters 6 and 7, which do not require Chapters 3 and 4). Nonetheless, the reader who wishes to immediately jump to the applications described in Chapter 10 may profitably do so by pausing from time to time to read the theory referred to in a given section to complement his understanding.

Acknowledgements

As mentioned above, the genesis of this book dates back to my Ph.D. studies carried out at the University of Cambridge in the years 2006-2009. Thanks are therefore due to Trinity College, Cambridge (U.K.), for fully supporting my Ph.D. studies through a Prince of Wales Studentship, as well as to Jesus College, Cambridge (U.K.) for its support through a Research Fellowship. The one-year, post-doctoral stay at the De Giorgi Center of the Scuola Normale Superiore di Pisa (Italy) was also important in connection with the writing of this book, since it was in that context that the general systematization of the unifying methodology 'toposes as bridges' took place. Later, I have been able to count on the support of a two-month visiting position at the Max Planck Institute for Mathematics (Bonn, Germany), where a significant part of Chapter 5 was written, as well as of a one-year CARMIN post-doctoral position at IHÉS, during which I wrote, amongst other texts, the remaining parts of the book. Thanks are also due to the University of Paris 7 and the Università degli Studi di Milano, who hosted my Marie Curie fellowship (cofunded by the Istituto Nazionale di Alta Matematica "F. Severi"), and again to IHÉS as well as to the Università degli Studi dell'Insubria for employing me in the period during which the final revision of the book has taken place.

Several results described in this book have been presented at international conferences and invited talks at universities around the world; the list is too long to be reported here, but I would like to collectively thank the organizers of such events for giving me the opportunity to present my work to responsive and stimulating audiences.

Special thanks go to Laurent Lafforgue for his unwavering encouragement to write a book on my research and for his precious assistance during the final revision phase.

I am also grateful to Marta Bunge for reading and commenting on a preliminary version of the book, to the anonymous referees contacted by Oxford University Press and to Alain Connes, Anatole Khelif, Steve Vickers and Noson Yanofsky for their valuable remarks on results presented in this book.

Como
October 2017

Olivia Caramello

CONTENTS

Notation and terminology   1

Introduction   3

1 Topos-theoretic background   9
  1.1 Grothendieck toposes   9
    1.1.1 The notion of site   10
    1.1.2 Sheaves on a site   12
    1.1.3 Basic properties of categories of sheaves   14
    1.1.4 Geometric morphisms   17
    1.1.5 Diaconescu's equivalence   20
  1.2 First-order logic   22
    1.2.1 First-order theories   23
    1.2.2 Deduction systems for first-order logic   27
    1.2.3 Fragments of first-order logic   28
  1.3 Categorical semantics   29
    1.3.1 Classes of 'logical' categories   31
    1.3.2 Completions of 'logical' categories   34
    1.3.3 Models of first-order theories in categories   35
    1.3.4 Elementary toposes   39
    1.3.5 Toposes as mathematical universes   43
  1.4 Syntactic categories   45
    1.4.1 Definition   45
    1.4.2 Syntactic sites   48
    1.4.3 Models as functors   49
    1.4.4 Categories with 'logical structure' as syntactic categories   50
    1.4.5 Soundness and completeness   51

2 Classifying toposes and the 'bridge' technique   53
  2.1 Geometric logic and classifying toposes   53
    2.1.1 Geometric theories   53
    2.1.2 The notion of classifying topos   55
    2.1.3 Interpretations and geometric morphisms   60
    2.1.4 Classifying toposes for propositional theories   63
    2.1.5 Classifying toposes for cartesian theories   64

		2.1.6 Further examples	65
		2.1.7 A characterization theorem for universal models in classifying toposes	67
	2.2	Toposes as 'bridges'	69
		2.2.1 The 'bridge-building' technique	69
		2.2.2 *Decks of 'bridges'*: Morita equivalences	70
		2.2.3 *Arches of 'bridges'*: site characterizations	74
		2.2.4 Some simple examples	76
		2.2.5 A theory of 'structural translations'	80
3	**A duality theorem**		83
	3.1	Preliminary results	83
		3.1.1 A 2-dimensional Yoneda lemma	83
		3.1.2 An alternative view of Grothendieck topologies	84
		3.1.3 Generators for Grothendieck topologies	86
	3.2	Quotients and subtoposes	88
		3.2.1 The duality theorem	88
		3.2.2 The proof-theoretic interpretation	94
	3.3	A deduction theorem for geometric logic	105
4	**Lattices of theories**		107
	4.1	The lattice operations on Grothendieck topologies and quotients	107
		4.1.1 The lattice operations on Grothendieck topologies	108
		4.1.2 The lattice operations on theories	112
		4.1.3 The Heyting implication in $\mathfrak{Th}_\Sigma^\mathbb{T}$	115
	4.2	Transfer of notions from topos theory to logic	119
		4.2.1 Relativization of local operators	119
		4.2.2 Open, closed, quasi-closed subtoposes	126
		4.2.3 The Booleanization and DeMorganization of a geometric theory	130
		4.2.4 The dense-closed factorization of a geometric inclusion	132
		4.2.5 Skeletal inclusions	133
		4.2.6 The surjection-inclusion factorization	134
		4.2.7 Atoms	135
		4.2.8 Subtoposes with enough points	138
5	**Flat functors and classifying toposes**		139
	5.1	Preliminary results on indexed colimits in toposes	140
		5.1.1 Background on indexed categories	140
		5.1.2 $\mathcal{E}$-filtered indexed categories	144
		5.1.3 Indexation of internal diagrams	145
		5.1.4 Colimits and tensor products	146
		5.1.5 $\mathcal{E}$-final functors	150
		5.1.6 A characterization of $\mathcal{E}$-indexed colimits	157

		5.1.7 Explicit calculation of set-indexed colimits	169
	5.2	Extensions of flat functors	176
		5.2.1 General extensions	176
		5.2.2 Extensions along embeddings of categories	178
		5.2.3 Extensions from categories of set-based models to syntactic categories	181
		5.2.4 A general adjunction	187
	5.3	Yoneda representations of flat functors	194
		5.3.1 Cauchy completion of sites	195
6	**Theories of presheaf type: general criteria**		**197**
	6.1	Preliminary results	198
		6.1.1 A canonical form for Morita equivalences	198
		6.1.2 Universal models and definability	199
		6.1.3 A syntactic criterion for a theory to be of presheaf type	202
		6.1.4 Finitely presentable = finitely presented	204
		6.1.5 A syntactic description of the finitely presentable models	208
	6.2	Internal finite presentability	210
		6.2.1 Objects of homomorphisms	210
		6.2.2 Strong finite presentability	212
		6.2.3 Semantic $\mathcal{E}$-finite presentability	216
	6.3	Semantic criteria for a theory to be of presheaf type	217
		6.3.1 The characterization theorem	217
		6.3.2 Concrete reformulations	223
		6.3.3 Abstract reformulation	239
7	**Expansions and faithful interpretations**		**241**
	7.1	Expansions of geometric theories	241
		7.1.1 General theory	241
		7.1.2 Another criterion for a theory to be of presheaf type	246
		7.1.3 Expanding a geometric theory to a theory of presheaf type	247
		7.1.4 Presheaf-type expansions	249
	7.2	Faithful interpretations of theories of presheaf type	251
		7.2.1 General results	251
		7.2.2 Injectivizations of theories	256
		7.2.3 Finitely presentable and finitely generated models	258
		7.2.4 Further reformulations of condition (iii) of Theorem 6.3.1	261
		7.2.5 A criterion for injectivizations	268
8	**Quotients of a theory of presheaf type**		**273**
	8.1	Studying quotients through the associated Grothendieck topologies	274

		8.1.1	The notion of J-homogeneous model	274
		8.1.2	Axiomatizations for the J-homogeneous models	280
		8.1.3	Quotients with enough set-based models	284
		8.1.4	Coherent quotients and topologies of finite type	288
		8.1.5	An example	292
	8.2	Presheaf-type quotients		293
		8.2.1	Finality conditions	293
		8.2.2	Rigid topologies	295
		8.2.3	Finding theories classified by a given presheaf topos	299
9	**Examples of theories of presheaf type**			**305**
	9.1	Theories whose finitely presentable models are finite		305
	9.2	The theory of abstract circles		307
	9.3	The geometric theory of finite sets		310
	9.4	The theory of Diers fields		312
	9.5	The theory of algebraic extensions of a given field		317
	9.6	Groups with decidable equality		318
	9.7	Locally finite groups		320
	9.8	Vector spaces		321
	9.9	The theory of abelian ℓ-groups with strong unit		322
10	**Some applications**			**325**
	10.1	Restrictions of Morita equivalences		325
	10.2	A solution to the boundary problem for subtoposes		326
	10.3	Syntax-semantics 'bridges'		327
	10.4	Topos-theoretic Fraïssé theorem		331
	10.5	Maximal theories and Galois representations		340
	10.6	A characterization theorem for geometric logic		345
	10.7	The maximal spectrum of a commutative ring		346
	10.8	Compactness conditions for geometric theories		354
	Bibliography			359
	Index			363

NOTATION AND TERMINOLOGY

The terminology and notation used in this book is essentially standard and borrowed from [55], if not stated otherwise; some specific notations that we shall employ are as follows:
- The category of sets and functions between them will be denoted by **Set**.
- The classifying topos of a geometric theory $\mathbb{T}$ will be denoted by $\mathbf{Set}[\mathbb{T}]$ or $\mathcal{E}_{\mathbb{T}}$.
- $\gamma_{\mathcal{E}} : \mathcal{E} \to \mathbf{Set}$ will denote the unique (up to isomorphism) geometric morphism from a Grothendieck topos $\mathcal{E}$ to the category **Set** of sets.
- By a cartesian category (resp. cartesian functor) we mean a category with finite limits (resp. a finite-limit-preserving functor).
- Given a category $\mathcal{C}$, we shall often write $c \in \mathcal{C}$ to mean that c is an object of $\mathcal{C}$.
- The collection of objects (resp. arrows) of a category $\mathcal{C}$ will be denoted by $\mathrm{Ob}(\mathcal{C})$ (resp. $\mathrm{Arr}(\mathcal{C})$).
- The category of functors from a category $\mathcal{C}$ to a category $\mathcal{D}$ will be denoted by $[\mathcal{C}, \mathcal{D}]$.
- The terminal object of a category $\mathcal{C}$, if it exists, will be denoted by $1_{\mathcal{C}}$, or simply by 1.
- The initial object of a category $\mathcal{C}$, if it exists, will be denoted by $0_{\mathcal{C}}$, or simply by 0.
- The identity arrow on an object c in a category will be denoted by 1_c.
- By a subterminal object in a category $\mathcal{C}$ having a terminal object 1 we mean an object c such that the unique arrow $c \to 1$ in $\mathcal{C}$ is a monomorphism.
- In a category $\mathcal{C}$ with a terminal object 1, for any object c of $\mathcal{C}$, we denote by $!_c : c \to 1$ the unique arrow in $\mathcal{C}$ from c to 1.
- In a category with finite products, we shall denote by $<f_1, \ldots, f_n>$ the arrow $c \to a_1 \times \cdots \times a_n$ determined by the universal property of the product by a collection of arrows $f_i : c \to a_i$ (for $i = 1, \ldots, n$).
- Given an object c of a category $\mathcal{C}$ and a category $\mathcal{I}$, we shall denote by Δc the constant functor $\mathcal{I} \to \mathcal{C}$ with value c.
- We shall say that a category is small if it has only a set of objects and arrows, and that a category is essentially small if it is equivalent to a small category.
- The (meta-)2-category of (large) categories, functors and natural transformations between them will be denoted by $\mathfrak{CAT}$.

- The 2-category of small categories, functors and natural transformations between them will be denoted by $\mathfrak{Cat}$.
- The 2-category of Grothendieck toposes, geometric morphisms and geometric transformations between them will be denoted by $\mathfrak{BTop}$ and, given two Grothendieck toposes $\mathcal{E}$ and $\mathcal{F}$, **Geom**$(\mathcal{E}, \mathcal{F})$ will denote the category of geometric morphisms $\mathcal{E} \to \mathcal{F}$ and geometric transformations between them.
- Given a Grothendieck topos $\mathcal{E}$ and a category $\mathcal{C}$, we will write **Flat**$(\mathcal{C}, \mathcal{E})$ for the category of flat functors $\mathcal{C} \to \mathcal{E}$ and natural transformations between them. Given a Grothendieck topology J on $\mathcal{C}$, **Flat**$_J(\mathcal{C}, \mathcal{E})$ will denote the full subcategory of J-continuous flat functors $\mathcal{C} \to \mathcal{E}$.
- Given a geometric theory $\mathbb{T}$, we shall denote by $\mathcal{C}_\mathbb{T}$ the geometric syntactic category of $\mathbb{T}$ and by $J_\mathbb{T}$ the canonical (Grothendieck) topology on $\mathcal{C}_\mathbb{T}$.
- Given a theory of presheaf type $\mathbb{T}$, we shall denote by f.p.$\mathbb{T}$-mod(**Set**) the category of (representatives of isomorphism classes of) finitely presentable models of $\mathbb{T}$ in **Set**.
- By a poset we mean a partially ordered set.
- Given a locally small category $\mathcal{C}$ and an object c of $\mathcal{C}$, the covariant (resp. contravariant) representable functor associated with c will be denoted by $\mathcal{C}(c, -)$ or $\text{Hom}_\mathcal{C}(c, -)$ (resp. by $\mathcal{C}(-, c)$ or $\text{Hom}_\mathcal{C}(-, c)$).
- A sieve generated by a single arrow f in a category will be denoted by (f).

INTRODUCTION

This book is devoted to a general study of geometric theories through the associated classifying toposes.

A central theme of the book is a duality, established in Chapter 3, between the subtoposes of the classifying topos of a geometric theory $\mathbb{T}$ over a signature Σ and the (syntactic-equivalence classes of) 'quotients' of the theory $\mathbb{T}$ (i.e. geometric theories obtained from $\mathbb{T}$ by adding geometric sequents over Σ); via this duality, each quotient of $\mathbb{T}$ corresponds to a subtopos which classifies it.

We make a detailed investigation of several different aspects of this duality, and derive a variety of applications. In fact, many concepts of elementary topos theory which apply to the lattice of subtoposes of a given topos can be reinterpreted, via this duality, in the context of geometric theories. It turns out that the resulting notions are of natural logical and mathematical interest. For instance, as we show in Chapter 4, the collection of (syntactic equivalence classes of) quotients of a given geometric theory has the structure of a lattice (in fact, of a Heyting algebra) with respect to a natural notion of ordering of theories. Open (resp. closed) subtoposes correspond via the duality to quotients obtained by adding sequents of the form $(\top \vdash_{[]} \phi)$ (resp. $(\phi \vdash_{[]} \bot)$), where ϕ is a geometric sentence over the signature of the theory. Also, the surjection-inclusion factorization of a geometric morphism has a natural semantic interpretation (cf. section 4.2.6), while subtoposes obtained by means of topos-theoretic constructions such as the Booleanization or DeMorganization of a topos have natural logical counterparts (cf. section 4.2.3) which often specialize, in the case of important mathematical theories, to quotients of genuine mathematical interest; for example, the Booleanization of the theory of linear orders is the theory of dense linear orders without endpoints (cf. section 10.4), the DeMorganization of the (coherent) theory of fields is the geometric theory of fields of finite characteristic in which every element is algebraic over the prime field (cf. Proposition 2.3 [29]) and the Booleanization of the (coherent) theory of fields is the theory of algebraically closed fields of finite characteristic in which every element is algebraic over the prime field (cf. Proposition 2.4 [29]).

A crucial aspect of the duality between quotients and subtoposes is that the notion of subtopos of a given topos is a topos-theoretic invariant which behaves naturally with respect to sites. Indeed, subtoposes of equivalent toposes correspond to each other bijectively, and for any small site $(\mathcal{C}, J)$, the subtoposes of the topos $\mathbf{Sh}(\mathcal{C}, J)$ correspond bijectively to the Grothendieck topologies on $\mathcal{C}$ which contain J. This is particularly relevant for our purposes since it allows

Theories, Sites, Toposes. Olivia Caramello.
© Olivia Caramello 2018. Published 2018 by Oxford University Press.

us to effectively transfer information between quotients of geometric theories classified by the same topos, as well as between different sites of definition of the classifying topos of a given quotient, according to the technique 'toposes as bridges' described in Chapter 2. In fact, this technique is profitably exploited in different parts of the book for obtaining insights on geometric theories through a comparison of different sites of definition of their classifying toposes; in particular, the various applications described in Chapter 10 all arise from an implementation of this methodology.

A particularly natural context of application of the above-mentioned duality is provided by the class of theories classified by a presheaf topos. Indeed, as any Grothendieck topos is a subtopos of a presheaf topos, so any geometric theory is a quotient of a theory classified by a presheaf topos. Moreover, if a theory $\mathbb{T}$ is classified by a topos $[\mathcal{C}^{\mathrm{op}}, \mathbf{Set}]$ then its quotients are in bijective correspondence with the Grothendieck topologies J on $\mathcal{C}$. Following [5], we say that a geometric theory is *of presheaf type* if it is classified by a presheaf topos.

As we show in section 6.1.1, a geometric theory $\mathbb{T}$ is of presheaf type if and only if it is classified by the topos $[\text{f.p.}\mathbb{T}\text{-mod}(\mathbf{Set}), \mathbf{Set}]$, where f.p.$\mathbb{T}$-mod($\mathbf{Set}$) is (a skeleton of) the full subcategory of $\mathbb{T}$-mod($\mathbf{Set}$) on the finitely presentable $\mathbb{T}$-models.

The subject of theories of presheaf type has a long history, starting with the book [48] by Hakim, which first introduced the point of view of classifying toposes in the context of the theory of commutative rings with unit and its quotients. The subsequent pioneering work [59] by Lawvere led to the discovery that any finitary algebraic theory is of presheaf type, classified by the topos of presheaves on the opposite of its category of finitely presentable models (cf. [56]). This result was later generalized to cartesian (or essentially algebraic) theories as well as to universal Horn theories (cf. [7]). At the same time, new examples of non-cartesian theories of presheaf type were discovered (cf. for instance [5] for a long, but by no means exhaustive, list of examples) and partial results in connection to the problem of characterizing the class of theories of presheaf type emerged; for instance, [51], [5] and [73] contain different sets of sufficient conditions for a theory to be of presheaf type.

Theories of presheaf type occupy a central role in topos theory for a number of reasons:

(i) Every small category $\mathcal{C}$ can be seen, up to Cauchy completion, as the category of finitely presentable models of a theory of presheaf type (namely, the theory of flat functors on $\mathcal{C}^{\mathrm{op}}$);

(ii) As every Grothendieck topos is a subtopos of some presheaf topos, so every geometric theory is a quotient of some theory of presheaf type (cf. the duality theorem of Chapter 3);

(iii) Every finitary algebraic theory (and more generally, any cartesian theory) is of presheaf type;

(iv) The class of theories of presheaf type contains, besides all cartesian theories, many other interesting mathematical theories pertaining to different fields of mathematics (for instance, the coherent theory of linear orders or the geometric theory of algebraic extensions of a given field – cf. Chapter 9);

(v) The 'bridge technique' of Chapter 2 can be fruitfully applied in the context of theories of presheaf type due to the fact that the classifying topos of any such theory admits (at least) two quite different representations, one of semantic nature (namely, as the category of set-valued functors on the category of finitely presentable models of the theory) and one of syntactic nature (namely, as the category of sheaves on the syntactic site of the theory).

It is therefore important to have effective criteria available for testing whether a theory is of presheaf type, as well as methods for generating new theories of presheaf type. We address these questions systematically in Chapters 6 and 7, considering in particular the stability of the presheaf-type condition with respect to natural operations or relations on theories. Old and new examples of theories of presheaf type are discussed in Chapter 9 in light of the theory developed in the previous chapters, while Chapter 8 is devoted to the important subject of quotients of theories of presheaf type.

In the final chapter, we present a selection of applications in different mathematical contexts of the methods developed in the book.

Overview of the book

Let us now describe the contents of the book in more detail.

In Chapter 1 we provide the topos-theoretic background necessary for understanding the contents of the book. The presentation is self-contained and only assumes a basic familiarity with the language of category theory. We start by reviewing the basic theory of Grothendieck toposes, including the fundamental equivalence between geometric morphisms and flat functors. We then present first-order logic and its interpretation in categories having 'enough' structure. Lastly, we discuss the key concept of syntactic category of a first-order theory, which is used in Chapter 2 for constructing classifying toposes of geometric theories.

Chapter 2 consists of two parts. In the first part we review the fundamental notion of classifying topos of a geometric theory and discuss the appropriate kinds of interpretations between theories which induce morphisms between the associate classifying toposes; the theoretical presentation is accompanied by a few concrete examples of classifying toposes of theories naturally arising in mathematics. We also establish a characterization theorem for universal models of geometric theories inside classifying toposes. In the second part we explain the general unifying technique 'toposes as bridges'. This technique, which allows us to extract 'concrete' information from the existence of different representations for the classifying topos of a geometric theory, is systematically exploited in the course of the book to establish theoretical results as well as applications.

The 'decks' of our topos-theoretic 'bridges' are normally given by Morita equivalences, while the 'arches' are given by site characterizations of topos-theoretic invariants.

In Chapter 3, after establishing some preliminary facts about syntactic categories and Grothendieck topologies, we state the duality theorem between quotients and subtoposes and provide two different proofs of it, one relying on the theory of classifying toposes and the other, of purely syntactic nature, based on a proof-theoretic interpretation of the notion of Grothendieck topology. More specifically, we establish a proof-theoretic equivalence between the classical system of geometric logic over a geometric theory $\mathbb{T}$ and a new proof system for sieves in the geometric syntactic category $\mathcal{C}_{\mathbb{T}}$ of $\mathbb{T}$ whose inference rules are given by the pullback stability and transitivity rules for Grothendieck topologies. This equivalence is interesting since the latter proof system turns out to be computationally better-behaved than the former; indeed, checking that a sieve belongs to the Grothendieck topology generated by a given family of sieves is often technically easier than showing that a geometric sequent is provable in a given theory. This equivalence of deduction systems can therefore be used for shedding light on axiomatization problems for geometric theories. We use it in particular for proving a deduction theorem for geometric logic, which we obtain by means of a calculation on Grothendieck topologies.

In Chapter 4, by using the duality theorem established in Chapter 3, we transfer many ideas and concepts of elementary topos theory to geometric logic, by passing through the theory of Grothendieck toposes. Specifically, we analyze notions such as the coHeyting algebra structure on the lattice of subtoposes of a given topos, open, closed, quasi-closed subtoposes, the dense-closed factorization of a geometric inclusion, coherent subtoposes, subtoposes with enough points, the surjection-inclusion factorization of a geometric morphism, skeletal inclusions, atoms in the lattice of subtoposes of a given topos, the Booleanization and DeMorganization of a topos. We also obtain, in section 4.1.1, explicit descriptions of the Heyting operation between Grothendieck topologies on a given category and of the Grothendieck topology generated by a given collection of sieves, while in section 4.2.1 we establish a number of results about the problem of 'relativizing' a local operator with respect to a given subtopos.

Chapter 5 is devoted to flat functors in relation to classifying toposes. Section 5.1 contains some general results about colimits of internal diagrams in toposes, section 5.2 consists in the development of a general theory of extensions of flat functors along geometric morphisms of toposes, and section 5.3 discusses a way of representing flat functors by using a suitable internalized version of the Yoneda lemma. These general results will be instrumental for establishing in Chapter 6 our main characterization theorem for theories of presheaf type.

In Chapter 6 we carry out a systematic investigation of the class of theories classified by a presheaf topos, obtaining in particular a characterization theorem providing necessary and sufficient conditions for a theory to be of presheaf type, expressed in terms of the models of the theory in arbitrary Grothendieck toposes. This theorem, whose general statement is quite abstract, admits several ramifications and simpler corollaries which can be effectively applied in practice to test whether a given theory is of presheaf type as well as for generating new examples of theories of presheaf type, also through appropriate 'modifications' of a given geometric theory. All the partial results and recognition criteria previously obtained on the subject are naturally recovered from this general result; moreover, the constructive nature of the characterization theorem allows us to replace the requirements that the theory should have enough set-based models in the sufficient criteria of [5] and [51] with explicit semantic conditions which can be directly verified without having to invoke any form of the axiom of choice.

In Chapter 7 we introduce the concept of expansion of a geometric theory and develop some basic theory about it, including a logical treatment of the hyperconnected-localic factorization of a geometric morphism. We then investigate the preservation, by 'faithful interpretations' of theories, of each of the conditions in the characterization theorem for theories of presheaf type, obtaining results of the form 'under appropriate conditions, a geometric theory in which a theory of presheaf type faithfully interprets is again of presheaf type'.

In Chapter 8, we present a semantic perspective on classifying toposes based on the notion of theory of presheaf type. Specifically, given a theory $\mathbb{T}$ classified by a presheaf topos $[\mathcal{C}^{\mathrm{op}}, \mathbf{Set}]$, where $\mathcal{C}$ is (a skeleton of) the opposite of the category of finitely presentable $\mathbb{T}$-models, and a subtopos $\mathbf{Sh}(\mathcal{C}, J)$ of $[\mathcal{C}^{\mathrm{op}}, \mathbf{Set}]$, we achieve an entirely semantic description (in terms of the Grothendieck topology J) of the quotient $\mathbb{T}'$ of $\mathbb{T}$ corresponding to $\mathbf{Sh}(\mathcal{C}, J)$ via the duality theorem of Chapter 3; in fact, we introduce the notion of J-homogeneous $\mathbb{T}$-model and show that, up to equivalence, the models of $\mathbb{T}'$ (in any Grothendieck topos) can be characterized among the models of $\mathbb{T}$ as the J-homogeneous ones. If $\mathcal{C}$ satisfies the right Ore condition and J is the atomic topology on it, the notion of J-homogeneous model is seen to specialize essentially to the notion of homogeneous model in model theory; this fact plays a crucial role in the topos-theoretic interpretation of Fraïssé's construction in model theory presented in section 10.4. We then unify this semantic approach to classifying toposes with the syntactic perspective provided by the duality theorem of Chapter 3; more specifically, we give results describing the classifying topos of a quotient of a theory of presheaf type and provide explicit axiomatizations for the J-homogeneous models. We also identify some sufficient conditions for a quotient of a theory of presheaf type to be again of presheaf type.

In Chapter 9 we discuss some classical, as well as new, examples of theories of presheaf type from the perspective of the theory developed in the previous chapters. We revisit in particular well-known examples of theories of presheaf type, such as the theory of intervals and the geometric theory of finite sets, and introduce new ones, including the theory of algebraic extensions of a given field, the theory of locally finite groups, the theory of vector spaces with linear independence predicates and the theory of lattice-ordered abelian groups with strong unit.

Finally, Chapter 10 contains some selected applications of the theory developed in the book into different areas of mathematics. Specifically, the first section addresses the problem of restricting Morita equivalences to quotients of the two theories, the second contains a solution to a problem posed by F. W. Lawvere concerning the effect of the boundary operation on subtoposes on quotients classified by them, the third contains a number of results connecting syntactic properties of quotients of theories of presheaf type and 'geometric' properties of the Grothendieck topologies corresponding to them via the duality theorem of Chapter 3, the fourth presents a topos-theoretic interpretation of Fraïssé's construction in model theory leading to new results on countably categorical theories, the fifth describes a general theory of topological Galois representations, the sixth presents the solution to I. Moerdijk's long-standing problem of finding an intrinsic semantic characterization of geometric logic obtained in [15], the seventh is devoted to an analysis of the maximal spectrum of a commutative ring with unit from the point of view of the duality theorem of Chapter 3, and the eighth investigates compactness conditions for geometric theories allowing one to identify theories lying in smaller fragments of geometric logic.

1
TOPOS-THEORETIC BACKGROUND

In this chapter we review the classical topos-theoretic preliminaries necessary for understanding the contents of the book, for the benefit of readers who are not familiar with topos theory. We shall only assume an acquaintance with the basic notions of category theory.

Besides the original work [3], where the theory of toposes was first introduced, classical books on the subject, to which we shall systematically refer in the course of the chapter, are, in increasing order of technical sophistication, [46], [63] and [55]. We shall in particular follow section D1 of [55] in our presentation of the basic notions and results of first-order categorical logic, which are mainly due to [64], in sections 1.2, 1.3 and 1.4.

The structure of this introductory chapter is as follows.

In section 1.1 we introduce the notion of Grothendieck topos, review the basic properties of categories of sheaves and discuss the fundamental equivalence between geometric morphisms and flat functors.

In section 1.2 we introduce the concept of first-order theory and the various deductive systems for fragments of first-order logic that we shall consider in the course of the book.

In section 1.3 we review the way in which first-order theories can be soundly interpreted in categories having 'enough' structure; in particular, we explain the sense in which a Grothendieck topos (or, more generally, an elementary topos) can be regarded as a mathematical universe in which one can consider models of first-order theories.

In section 1.4 we present the fundamental concept of syntactic category of a first-order theory and illustrate the classification of models through structure-preserving functors defined on such categories. Lastly, we use syntactic categories for establishing the soundness and completeness of the categorical semantics defined in section 1.3.

1.1 Grothendieck toposes

Before introducing the notion of Grothendieck topos, we need to talk about Grothendieck topologies on categories. Indeed, Grothendieck toposes will be defined in section 1.1.2 as categories of sheaves on a small category with respect to a Grothendieck topology on it.

Theories, Sites, Toposes. Olivia Caramello.
© Olivia Caramello 2018. Published 2018 by Oxford University Press.

1.1.1 The notion of site

The notion of Grothendieck topology represents a categorical abstraction of the classical topological notion of covering of an open set of a topological space by a family of smaller open subsets.

Recall that a category is said to be *cartesian* if it has finite limits. A functor between two cartesian categories is said to be *cartesian* if it preserves finite limits.

Definition 1.1.1 (a) A *sieve* on an object c of a category $\mathcal{C}$ is a family of arrows in $\mathcal{C}$ with codomain c such that for any $f \in S$, $f \circ g \in S$ for any arrow g composable with f in $\mathcal{C}$.

(b) A sieve S is said to be *generated* by a family $\mathcal{F}$ of arrows contained in it if every arrow in S factors through an arrow in $\mathcal{F}$.

(c) A *Grothendieck topology* on a category $\mathcal{C}$ is an assignment J sending any object c of $\mathcal{C}$ to a collection $J(c)$ of sieves on $\mathcal{C}$ in such a way that the following properties are satisfied:

- (Maximality axiom) For any object c of $\mathcal{C}$, the maximal sieve $M_c := \{f \mid \cod(f) = c\}$ on c belongs to $J(c)$.
- (Pullback stability) For any arrow $f : d \to c$ in $\mathcal{C}$ and any sieve $S \in J(c)$, the sieve $f^*(S) = \{g : e \to d \mid f \circ g \in S\}$ belongs to $J(d)$.
- (Transitivity) For any sieve S in $\mathcal{C}$ on an object c and any $T \in J(c)$, if $f^*(S) \in J(\dom(f))$ for all $f \in T$ then $S \in J(c)$.

A sieve on an object c of $\mathcal{C}$ is said to be *J-covering* (or a *J-cover*), for a Grothendieck topology J on $\mathcal{C}$, if it belongs to $J(c)$.

(d) A *site* is a pair $(\mathcal{C}, J)$ consisting of a category $\mathcal{C}$ and a Grothendieck topology J on $\mathcal{C}$; it is said to be *small* (resp. *essentially small*) if the category $\mathcal{C}$ is small (resp. essentially small).

(e) Given two cartesian sites (i.e. sites whose underlying categories are cartesian) $(\mathcal{C}, J)$ and $(\mathcal{D}, K)$, a *morphism of sites* $F : (\mathcal{C}, J) \to (\mathcal{D}, K)$ is a cartesian functor $F : \mathcal{C} \to \mathcal{D}$ which is cover-preserving (in the sense that the image under F of every J-covering sieve generates a K-covering sieve).

Remark 1.1.2 One can define morphisms of general, i.e. non-necessarily cartesian, sites (cf. section 1 of Exposé I.iii [3]); however, this general notion is more involved and we shall not use it in the book, so we do not discuss it here.

The following notions will be important for illuminating the relationship between topological spaces and the toposes of sheaves on them.

Recall that a *lattice* is a partially ordered set in which every pair of elements a and b has an infimum $a \wedge b$ and a supremum $a \vee b$.

Definition 1.1.3 (a) A Heyting algebra is a lattice H with bottom and top elements (denoted respectively by 0 and 1) in which for any two elements $a, b \in H$ there exists an element $a \Rightarrow b$ satisfying the universal property that for any $c \in H$, $c \leq a \Rightarrow b$ if and only if $c \wedge a \leq b$. The pseudocomplement $\neg a$ of an element a of H is defined to be $a \Rightarrow 0$. A Heyting algebra is said to be *complete* if it possesses arbitrary suprema (equivalently, arbitrary infima).

(b) A *frame* F is a partially ordered set with arbitrary joins (and meets), in which the distributivity law of arbitrary joins with respect to finite meets holds:
$$\left(\bigvee_{i \in I} a_i\right) \wedge b = \bigvee_{i \in I} a_i \wedge b.$$
Since, by the special adjoint functor theorem, the functor $b \wedge - : F \to F$ has a right adjoint $b \Rightarrow - : F \to F$ if and only if it preserves arbitrary joins, a frame is the same thing as a complete Heyting algebra (cf. Remark C1.1.2 [55]).

The category **Frm** of frames has as objects frames and as arrows the frame homomorphisms between them (i.e. the maps which preserve finite meets and arbitrary joins).

(c) A *locale* is a frame, regarded as an object of the opposite of the category **Frm**.

The category **Loc** of locales is the opposite $\mathbf{Frm}^{op}$ of the category of frames.

For any topological space X, the lattice $\mathcal{O}(X)$ of its open sets is a locale. Conversely, with any locale F one can associate a topological space X_F, whose points are the frame homomorphism $F \to \{0,1\}$ and whose open sets are the subsets of frame homomorphisms $F \to \{0,1\}$ which send a given element $f \in F$ to 1. In fact, the assignments $X \to \mathcal{O}(X)$ and $F \to X_f$ lift to an adjunction between the category **Top** of topological spaces and continuous maps and the category $\mathbf{Loc} = \mathbf{Frm}^{op}$ of locales (cf. section C1.2 of [55]).

The following are notable examples of Grothendieck topologies:

Examples 1.1.4

(a) The *trivial topology* T on a category $\mathcal{C}$ is the Grothendieck topology on $\mathcal{C}$ whose only covering sieves are the maximal ones.

(b) The *dense* or *double negation topology* on a category $\mathcal{C}$ is the Grothendieck topology D on $\mathcal{C}$ whose covering sieves are exactly the stably non-empty ones (recall that a sieve S in $\mathcal{C}$ on an object c is said to be *stably non-empty* if $f^*(S) \neq \emptyset$ for all arrows f in $\mathcal{C}$ with codomain c).

(c) If $\mathcal{C}$ satisfies the *right Ore condition* (that is, the property that every pair of arrows with common codomain can be completed to a commutative square), the dense topology on $\mathcal{C}$ specializes to the *atomic topology* J_{at} on $\mathcal{C}$, whose covering sieves are exactly the non-empty ones.

(d) Given a topological space X, there is a natural topology $J_{\mathcal{O}(X)}$ on the category $\mathcal{O}(X)$ of its open sets (whose objects are the open sets U of X and whose arrows are the inclusions $V \subseteq U$ between them), whose covering sieves are precisely the ones generated by (small) *covering families* (i.e. families of inclusions $\{U_i \subseteq U \mid i \in I\}$ such that $\bigcup_{i \in I} U_i = U$).

(e) More generally, given a locale L, regarded as a preorder category, the *canonical topology* J_L on L has as covering sieves $S \in J_L(l)$ the sieves S generated by families $\{l_i \leq l \mid i \in I\}$ such that $\bigvee_{i \in I} l_i = l$.

1.1.2 Sheaves on a site

The introduction of the concept of Grothendieck topos stemmed from the observation that many important properties of topological spaces, such as compactness and connectedness, admit reformulations as categorically invariant properties of the associated categories of sheaves of sets; moreover, if a topological space X is sufficiently well-behaved (technically speaking, sober, cf. Remark 1.1.28), it can be recovered from the associated category $\mathbf{Sh}(X)$, as well as from the frame $\mathcal{O}(X)$, up to homeomorphism. Replacing topological spaces with the corresponding categories of sheaves on them presents the advantage that the latter are very rich in terms of categorical structure and can be studied by using a variety of different invariants naturally defined on them, notably including cohomology groups.

In studying the assignment $X \to \mathbf{Sh}(X)$ one immediately realizes that the definition of $\mathbf{Sh}(X)$ starting from X does not require the consideration of the points of X but only of the open sets of X and of the classical notion of covering for families of open subsets U_i of a given open set U (defined by saying that a given family $\{U_i \subseteq U \mid i \in I\}$ covers U if $\bigcup_{i \in I} U_i = U$). Now, the open sets of X can be organized in the preorder category $\mathcal{O}(X)$, while the above-mentioned notion of covering can be formulated as an additional *datum* on this category.

Abstracting these two fundamental ingredients in the construction of categories of sheaves led Grothendieck to introduce the notion of site recalled in section 1.1.1. Sheaves on a general site $(\mathcal{C}, J)$ are defined in a formally analogous way to sheaves of sets on a topological space, as follows.

Definition 1.1.5 Let $(\mathcal{C}, J)$ be a site.
(a) A *presheaf* on a category $\mathcal{C}$ is a functor $P : \mathcal{C}^{\mathrm{op}} \to \mathbf{Set}$.
(b) A *sheaf* on $(\mathcal{C}, J)$ (or J-*sheaf* on $\mathcal{C}$) is a presheaf $P : \mathcal{C}^{\mathrm{op}} \to \mathbf{Set}$ on $\mathcal{C}$ such that for every J-covering sieve $S \in J(c)$ and every family $\{x_f \in P(\mathrm{dom}(f)) \mid f \in S\}$ such that $P(g)(x_f) = x_{f \circ g}$ for any $f \in S$ and any arrow g in $\mathcal{C}$ composable with f, there exists a unique element $x \in P(c)$ such that $x_f = P(f)(x)$ for all $f \in S$.
(c) The category $\mathbf{Sh}(\mathcal{C}, J)$ of sheaves on $(\mathcal{C}, J)$ has as objects the sheaves on $(\mathcal{C}, J)$ and as arrows the natural transformations between them, regarded as functors $\mathcal{C}^{\mathrm{op}} \to \mathbf{Set}$.
(d) A *Grothendieck topos* is a category which is equivalent to a category $\mathbf{Sh}(\mathcal{C}, J)$ of sheaves on a small site $(\mathcal{C}, J)$.
(e) A *site of definition* for a Grothendieck topos $\mathcal{E}$ is a small site $(\mathcal{C}, J)$ such that $\mathcal{E} \simeq \mathbf{Sh}(\mathcal{C}, J)$.
(f) A family of arrows $\mathcal{F}$ with common codomain e in a category $\mathcal{E}$ is said to be *epimorphic* if for any arrows $f, g : e \to a$ in $\mathcal{E}$, $f \circ h = g \circ h$ for all $h \in \mathcal{F}$ implies $f = g$.
(g) A family $\mathcal{C}$ of objects of a category $\mathcal{E}$ is said to be *separating* for $\mathcal{E}$ if for any object e of $\mathcal{E}$, the arrows from objects of $\mathcal{C}$ to e form an epimorphic family.

(h) A Grothendieck topology J on a locally small category $\mathcal{C}$ is said to be *subcanonical* if every representable functor $\mathrm{Hom}_{\mathcal{C}}(-,c) : \mathcal{C}^{\mathrm{op}} \to \mathbf{Set}$ is a J-sheaf.

Notice that for any subcanonical topology J on $\mathcal{C}$, the Yoneda embedding yields a full embedding of $\mathcal{C}$ into $\mathbf{Sh}(\mathcal{C}, J)$.

(i) A *subterminal object* of a Grothendieck topos $\mathcal{E}$ is an object e of $\mathcal{E}$ such that, denoting by 1 'the' terminal object of $\mathcal{E}$, the unique arrow $e \to 1$ in $\mathcal{E}$ is a monomorphism.

Examples 1.1.6

(a) The category **Set** of sets and functions between them is a Grothendieck topos. It can be represented as the category of sheaves on the one-point topological space.

(b) More generally, given a small category $\mathcal{C}$ with the trivial topology T on it, the T-sheaves on $\mathcal{C}$ are clearly just the presheaves on $\mathcal{C}$; the category $\mathbf{Sh}(\mathcal{C}, T)$ thus coincides with the category $[\mathcal{C}^{\mathrm{op}}, \mathbf{Set}]$ of functors $\mathcal{C}^{\mathrm{op}} \to \mathbf{Set}$ and natural transformations between them.

(c) For any topological space X, the topos $\mathbf{Sh}(\mathcal{O}_X, J_{\mathcal{O}(X)})$, where $J_{\mathcal{O}(X)}$ is the Grothendieck topology defined in Example 1.1.4(d), coincides with the usual category $\mathbf{Sh}(X)$ of sheaves of sets on the space X.

(d) For any locale L, the topos $\mathbf{Sh}(L, J_L)$ is often simply denoted by $\mathbf{Sh}(L)$. The locale L can be recovered from it, up to isomorphism, as the frame $\mathrm{Sub}_{\mathbf{Sh}(L, J_L)}(1)$ of subterminal objects of $\mathbf{Sh}(L, J_L)$. A topos is said to be *localic* if it is (up to equivalence) of the form $\mathbf{Sh}(L)$ for a locale L.

More generally, for any *localic groupoid* $\mathcal{G}$ (i.e. groupoid internal to the category of locales, in the sense of section 1.3.3), there exists a Grothendieck topos $\mathbf{Sh}(\mathcal{G})$ of sheaves on it. By a well-known theorem of Joyal and Tierney (cf. [57]), every Grothendieck topos can be represented in this form.

(e) Given a topological group G (with a prodiscrete topology), the category $\mathbf{Cont}(G)$ whose objects are the continuous (left) actions of G on discrete sets and whose arrows are the equivariant maps between them is a Grothendieck topos. In fact, $\mathbf{Cont}(G) \simeq \mathbf{Sh}(\mathcal{C}, J_{\mathrm{at}})$, where $\mathcal{C}$ is the full subcategory of $\mathbf{Cont}(G)$ on the non-empty transitive actions and J_{at} is the atomic topology on it.

More generally, for any *topological groupoid* $(d, c : G \to X)$ (i.e. groupoid internal to the category of topological spaces, in the sense of section 1.3.3), the category $\mathbf{Sh}_G(X)$ of G-equivariant sheaves on X is a Grothendieck topos. Butz and Moerdijk have shown in [11] that every Grothendieck topos with enough points can be represented in this form.

Definition 1.1.7 Let $(\mathcal{C}, J)$ be a site. A full subcategory $\mathcal{D}$ of $\mathcal{C}$ is said to be *J-dense* if for every object c of $\mathcal{C}$ the sieve generated by the family of arrows to c from objects in $\mathcal{D}$ is J-covering.

Given such a subcategory $\mathcal{D}$, the *Grothendieck topology* $J|_\mathcal{D}$ on $\mathcal{D}$ induced by J is defined as follows: for any sieve S in $\mathcal{D}$ on an object d, $S \in J|_\mathcal{D}(d)$ if and only if $\overline{S} \in J(d)$, where $\overline{S}$ is the sieve in $\mathcal{C}$ generated by the arrows in S.

A site of definition for a given topos can be seen as a sort of presentation of it by generators and relations (one can think of the objects of the category underlying the site as being the generators, and of the arrows and covering sieves as being the relations). This idea has for instance been exploited in section 8 of [16] for building various kinds of ordered structures presented by generators and relations.

As a group can have many different presentations, so a Grothendieck topos can have many different sites of definition, as for instance shown by the following theorem of [3] (cf. Theorem 4.1 in the Exposé I.iii):

Theorem 1.1.8 (Comparison Lemma). *Let $(\mathcal{C}, J)$ be a site such that $\mathcal{C}$ is locally small, and let $\mathcal{D}$ be a small full subcategory of $\mathcal{C}$ which is J-dense. Then the categories* $\mathbf{Sh}(\mathcal{C}, J)$ *and* $\mathbf{Sh}(\mathcal{D}, J|_\mathcal{D})$ *are equivalent.*

There are more refined versions of the Comparison Lemma in the literature; however, we shall not be concerned with them in this book. For a proof of the version reported above, we refer the reader to pp. 547-8 of [55].

Another important source of different sites of definition for a given topos is provided by its separating sets of objects.

Definition 1.1.9 The *canonical topology* $J_\mathcal{E}$ on a Grothendieck topos $\mathcal{E}$ is the Grothendieck topology on $\mathcal{E}$ whose covering sieves are exactly the sieves generated by small epimorphic families.

Every separating set of objects $\mathcal{C}$ of a Grothendieck topos $\mathcal{E}$, regarded as a full subcategory of $\mathcal{E}$, is $J_\mathcal{E}$-dense and yields a representation $\mathcal{E} \simeq \mathbf{Sh}(\mathcal{C}, J_\mathcal{E}|_\mathcal{C})$ (cf. the proof of Theorem C2.2.8 [55]).

In spite of the existence of multiple representations for a given topos, there exists an intrinsic characterization, first obtained by Giraud, of the notion of Grothendieck topos: a locally small category is a Grothendieck topos if and only if it is a ∞-pretopos with a separating set of objects (cf. Definition 1.3.17(b) below for the definition of an ∞-pretopos). We shall not make use of this characterization in the book; rather, we shall systematically exploit the existence of different sites of definition for a given topos (cf. section 2.2).

1.1.3 Basic properties of categories of sheaves

Grothendieck toposes are very rich in terms of categorical structure: they possess all (small) limits and colimits, exponentials and a subobject classifier (as defined below in this section).

Recall that a *limit* (or *projective limit*, in the terminology of [3]) of a functor $D : \mathcal{I} \to \mathcal{E}$ defined on a small category $\mathcal{I}$ is a universal cone over D, i.e. it is an object e of $\mathcal{E}$ together with arrows $\lambda_i : e \to D(i)$ indexed by the objects of $\mathcal{I}$ such that, for any arrow $f : i \to j$ in $\mathcal{I}$, $D(f) \circ \lambda_i = \lambda_j$, which is universal with this property. The notion of *colimit* (or *inductive limit*) is dual to that of limit; see, for instance, pp. 10-23 of [63] for more details. All the limits and colimits considered in this book will be *small*, i.e. indexed over small categories; so, we generically speak of (arbitrary) (co)limits, we actually mean *small (co)limits*.

An important property of functors which admit a right (resp. a left) adjoint, which will be exploited in the sequel, is the fact that they preserve all the colimits (resp. all the limits) which exist in the domain category.

For any small site $(\mathcal{C}, J)$ the inclusion $\mathbf{Sh}(\mathcal{C}, J) \hookrightarrow [\mathcal{C}^{\mathrm{op}}, \mathbf{Set}]$ has a left adjoint $a_J : [\mathcal{C}^{\mathrm{op}}, \mathbf{Set}] \to \mathbf{Sh}(\mathcal{C}, J)$, called the *associated sheaf functor*, which preserves finite limits. Limits in $\mathbf{Sh}(\mathcal{C}, J)$ are computed pointwise as in **Set** (they are created by the inclusion $\mathbf{Sh}(\mathcal{C}, J) \hookrightarrow [\mathcal{C}^{\mathrm{op}}, \mathbf{Set}]$) while colimits in $\mathbf{Sh}(\mathcal{C}, J)$ can be calculated by taking the image under a_J of the colimit of the given diagram in $[\mathcal{C}^{\mathrm{op}}, \mathbf{Set}]$.

To understand how exponentials are defined in general categories of sheaves, let us recall their construction in the base topos of sets.

For any two sets X and Y, we can always form the set Y^X of functions $X \to Y$. This set enjoys the following (universal) property in the category **Set** of sets: the familiar bijection

$$\mathrm{Hom}_{\mathbf{Set}}(Z, Y^X) \cong \mathrm{Hom}_{\mathbf{Set}}(Z \times X, Y)$$

is natural in both Y and Z and hence gives rise to an adjunction between the functor $- \times X : \mathbf{Set} \to \mathbf{Set}$ (left adjoint) and the functor $(-)^X : \mathbf{Set} \to \mathbf{Set}$ (right adjoint).

In a general category $\mathcal{C}$ with finite products, one defines the *exponential* A^B of two objects A and B by the universal property that for any object C of $\mathcal{C}$ the arrows $C \to A^B$ are in natural bijective correspondence with the arrows $C \times B \to A$. In a category $\mathbf{Sh}(\mathcal{C}, J)$ of sheaves on a small site $(\mathcal{C}, J)$ exponentials are computed as in $[\mathcal{C}^{\mathrm{op}}, \mathbf{Set}]$; specifically, an immediate application of the Yoneda lemma yields $P^Q(c) = \mathrm{Nat}(yc \times Q, P)$ for all objects c of $\mathcal{C}$, where yc denotes the representable functor associated with the object c and Nat denotes the set of natural transformations.

Definition 1.1.10 A category $\mathcal{C}$ is said to be *cartesian closed* if it has finite products and exponentials for each pair of objects.

Note that a Heyting algebra can be described as a lattice with bottom and top elements which is cartesian closed when regarded as a preorder category with products.

Let us now recall the notion of *subobject classifier*. A *subobject* of an object c in a category $\mathcal{C}$ is an equivalence class of monomorphisms with codomain c modulo the equivalence relation which identifies two such monomorphisms when they factor one through the other. In the category **Set**, subobjects of a set X correspond bijectively to subsets of X. The notion of subobject thus represents a natural categorical generalization of that of subset in set theory.

In the category **Set**, the subsets S of a given set X can be identified with their characteristic functions $\chi_S : X \to \{0,1\}$; in fact, denoting by $\top : \{*\} = 1_{\mathbf{Set}} \to \{0,1\}$ the function which sends $*$ to 1, we have a pullback square

$$\begin{array}{ccc} S & \xrightarrow{!} & \{*\} \\ {\scriptstyle i}\downarrow & & \downarrow{\scriptstyle \top} \\ X & \xrightarrow{\chi_S} & \{0,1\}, \end{array}$$

where $i : S \to X$ is the inclusion and $! : S \to \{*\}$ is the unique arrow in **Set** from S to the terminal object $1_{\mathbf{Set}} = \{*\}$.

This motivates the following

Definition 1.1.11 In a category $\mathcal{C}$ with finite limits a *subobject classifier* is a monomorphism $\top : 1_{\mathcal{C}} \to \Omega$, where $1_{\mathcal{C}}$ is 'the' terminal object of $\mathcal{C}$, such that for every monomorphism $m : a' \to a$ there is a unique arrow $\chi_m : a \to \Omega$, called the *classifying arrow* of m, such that we have a pullback square

$$\begin{array}{ccc} a' & \xrightarrow{!} & 1_{\mathcal{C}} \\ {\scriptstyle m}\downarrow & & \downarrow{\scriptstyle \top} \\ a & \xrightarrow{\chi_m} & \Omega. \end{array}$$

Equivalently, if $\mathcal{C}$ is small, a subobject classifier is a representing object for the functor $\mathcal{C}^{\mathrm{op}} \to \mathbf{Set}$ sending an object a to the poset $\mathrm{Sub}_{\mathcal{C}}(a)$ of subobjects of a and an arrow $f : a \to b$ to the pullback map $f^* : \mathrm{Sub}_{\mathcal{C}}(b) \to \mathrm{Sub}_{\mathcal{C}}(a)$.

Note that, for any object a of $\mathcal{C}$, we have an arrow $\in_a : a \times \Omega^a \to \Omega$ generalizing the belonging relation $\in$ of set theory, given by the transpose of the identity arrow on Ω^a.

Summarizing, we have the following result:

Theorem 1.1.12 *Let $(\mathcal{C}, J)$ be a small site.*

(i) *The inclusion* $\mathbf{Sh}(\mathcal{C}, J) \hookrightarrow [\mathcal{C}^{\mathrm{op}}, \mathbf{Set}]$ *has a left adjoint* $a_J : [\mathcal{C}^{\mathrm{op}}, \mathbf{Set}] \to \mathbf{Sh}(\mathcal{C}, J)$, *called the* associated sheaf functor, *which preserves finite limits.*

(ii) *The category* $\mathbf{Sh}(\mathcal{C}, J)$ *has all (small) limits, which are preserved by the inclusion functor* $\mathbf{Sh}(\mathcal{C}, J) \hookrightarrow [\mathcal{C}^{\mathrm{op}}, \mathbf{Set}]$; *in particular, (small) limits are computed pointwise and the terminal object* $1_{\mathbf{Sh}(\mathcal{C}, J)}$ *of* $\mathbf{Sh}(\mathcal{C}, J)$ *is the functor* $T : \mathcal{C}^{\mathrm{op}} \to \mathbf{Set}$ *sending each object* $c \in \mathcal{C}$ *to the singleton* $\{*\}$.

(iii) *The associated sheaf functor* $a_J : [\mathcal{C}^{\mathrm{op}}, \mathbf{Set}] \to \mathbf{Sh}(\mathcal{C}, J)$ *preserves (small) colimits; in particular,* $\mathbf{Sh}(\mathcal{C}, J)$ *has all (small) colimits.*

(iv) *The category* $\mathbf{Sh}(\mathcal{C}, J)$ *has exponentials, which are constructed as in the topos* $[\mathcal{C}^{\mathrm{op}}, \mathbf{Set}]$.

(v) *The category* $\mathbf{Sh}(\mathcal{C}, J)$ *has a subobject classifier.*

Subobject classifiers in Grothendieck toposes are constructed as follows:

- Given a site $(\mathcal{C}, J)$ and a sieve S in $\mathcal{C}$ on an object c, we say that S is *J-closed* if, for any arrow $f : d \to c$, $f^*(S) \in J(d)$ implies $f \in S$.
- For a small site $(\mathcal{C}, J)$, let us define $\Omega_J : \mathcal{C}^{\mathrm{op}} \to \mathbf{Set}$ as follows:

$$\Omega_J(c) = \{R \mid R \text{ is a } J\text{-closed sieve on } c\} \text{ (for an object } c \in \mathcal{C});$$

$$\Omega_J(f) = f^*(-) \text{ (for an arrow } f \text{ in } \mathcal{C}),$$

where $f^*(-)$ denotes the operation of pullback of sieves in $\mathcal{C}$ along f. Then the arrow $\top : 1_{\mathbf{Sh}(\mathcal{C}, J)} \to \Omega_J$ defined by

$$\top(*)(c) = M_c \text{ (the maximal sieve on } c\text{) for each } c \in \mathcal{C}$$

is a subobject classifier for $\mathbf{Sh}(\mathcal{C}, J)$.

- The classifying arrow $\chi_{A'} : A \to \Omega_J$ of a subobject $A' \subseteq A$ in $\mathbf{Sh}(\mathcal{C}, J)$ is given by

$$\chi_{A'}(c)(x) = \{f : d \to c \mid A(f)(x) \in A'(d)\}$$

for any $c \in \mathcal{C}$ and $x \in A(c)$.

For a detailed proof of this theorem the reader is referred to sections III.5-7 of [63].

The following result describes the main properties of subobject lattices in a Grothendieck topos:

Theorem 1.1.13

(i) *For any Grothendieck topos $\mathcal{E}$ and any object a of $\mathcal{E}$, the poset $\mathrm{Sub}_\mathcal{E}(a)$ of all subobjects of a in $\mathcal{E}$ is a complete Heyting algebra.*

(ii) *For any arrow $f : a \to b$ in a Grothendieck topos $\mathcal{E}$, the pullback functor $f^* : \mathrm{Sub}_\mathcal{E}(b) \to \mathrm{Sub}_\mathcal{E}(a)$ has both a left adjoint $\exists_f : \mathrm{Sub}_\mathcal{E}(a) \to \mathrm{Sub}_\mathcal{E}(b)$ and a right adjoint $\forall_f : \mathrm{Sub}_\mathcal{E}(a) \to \mathrm{Sub}_\mathcal{E}(b)$.*

A detailed proof of this theorem can be found in section III.8 of [63].

1.1.4 Geometric morphisms

There are two natural types of morphisms to consider between toposes: *geometric morphisms* and *logical functors*. The former preserve the 'geometric structure' of toposes, the latter the 'elementary logical' one (cf. section 1.3.4). The latter are the natural class of morphisms to consider when one is interested in regarding an elementary topos as the syntactic category of a higher-order intuitionistic type theory (cf. section 1.3.4), while the former are the appropriate kind of morphisms to consider when toposes are regarded as generalized spaces or as classifying toposes of geometric theories (cf. section 2.1). Accordingly, we will focus in this book on the former class.

Definition 1.1.14 (a) A *geometric morphism* $f : \mathcal{F} \to \mathcal{E}$ of toposes is a pair of adjoint functors $f_* : \mathcal{F} \to \mathcal{E}$ and $f^* : \mathcal{E} \to \mathcal{F}$, respectively called the *direct* *image* and the *inverse image* of f, such that the left adjoint f^* preserves finite limits (notice that f^* always preserves colimits as it has a right adjoint, while f_* always preserves limits).

(b) A *geometric transformation* $\alpha : f \to g$ between two geometric morphisms $f, g : \mathcal{F} \to \mathcal{E}$ is a natural transformation $f^* \to g^*$ (equivalently, a natural transformation $g_* \to f_*$ – cf. the remarks following Definition A4.1.1 [55]).

We shall denote by $\mathfrak{Btop}$ the 2-category of Grothendieck toposes, geometric morphisms and geometric transformations between them. Given two Grothendieck toposes $\mathcal{E}$ and $\mathcal{F}$, we shall denote by **Geom**$(\mathcal{E}, \mathcal{F})$ the category of geometric morphisms $\mathcal{E} \to \mathcal{F}$ and geometric transformations between them.

Recall that adjoints to a given functor, if they exist, are uniquely determined up to isomorphism. By the special adjoint functor theorem, giving a geometric morphism $\mathcal{F} \to \mathcal{E}$ is thus equivalent to giving a cocontinuous (i.e. colimit-preserving) cartesian functor $\mathcal{E} \to \mathcal{F}$.

Detailed proofs of the facts stated in the following list of examples can be found in sections VII.1-2 and VII.10 of [63].

Examples 1.1.15
(a) Any continuous map $f : X \to Y$ of topological spaces induces a geometric morphism **Sh**$(f) :$ **Sh**$(X) \to$ **Sh**(Y) whose direct image **Sh**$(f)_*$ is given by **Sh**$(f)_*(P)(V) = P(f^{-1}(V))$ (for each $V \in \mathcal{O}(Y)$).

(b) More generally, any arrow $f : L \to L'$ in **Loc** canonically induces a geometric morphism **Sh**$(L) \to$ **Sh**(L'); conversely, every geometric morphism **Sh**$(L) \to$ **Sh**(L') is, up to isomorphism, of this form. This defines a full and faithful embedding of the (2-)category **Loc** into the (2-)category $\mathfrak{Btop}$.

(c) Even more generally, any morphism of sites $f : (\mathcal{C}, J) \to (\mathcal{D}, K)$ induces a geometric morphism **Sh**$(f) :$ **Sh**$(\mathcal{D}, K) \to$ **Sh**$(\mathcal{C}, J)$, whose direct image is the functor $- \circ f^{\mathrm{op}} :$ **Sh**$(\mathcal{D}, K) \to$ **Sh**$(\mathcal{C}, J)$.

(d) Any functor $f : \mathcal{C} \to \mathcal{D}$ between small categories induces a geometric morphism $[f, \mathbf{Set}] : [\mathcal{C}, \mathbf{Set}] \to [\mathcal{D}, \mathbf{Set}]$, whose inverse image is given by $- \circ f : [\mathcal{D}, \mathbf{Set}] \to [\mathcal{C}, \mathbf{Set}]$. This functor has both a left and a right adjoint, respectively given by the left and right Kan extensions along the functor f. A geometric morphism whose inverse image has a left adjoint is said to be *essential*. If $\mathcal{D}$ is Cauchy-complete (i.e. all idempotents split in it) then every essential geometric morphism $[\mathcal{C}, \mathbf{Set}] \to [\mathcal{D}, \mathbf{Set}]$ is, up to isomorphism, of the form $[f, \mathbf{Set}]$ for a functor $f : \mathcal{C} \to \mathcal{D}$ (cf. Lemma A4.1.5 [55]).

(e) For any small site $(\mathcal{C}, J)$, the canonical inclusion functor **Sh**$(\mathcal{C}, J) \hookrightarrow [\mathcal{C}^{\mathrm{op}}, \mathbf{Set}]$ is the direct image of a geometric morphism **Sh**$(\mathcal{C}, J) \hookrightarrow [\mathcal{C}^{\mathrm{op}}, \mathbf{Set}]$ whose inverse image $a_J : [\mathcal{C}^{\mathrm{op}}, \mathbf{Set}] \to$ **Sh**$(\mathcal{C}, J)$ is the associated sheaf functor.

(f) For any (Grothendieck) topos $\mathcal{E}$ and any arrow $f : F \to E$ in $\mathcal{E}$, the pullback functor $f^* : \mathcal{E}/E \to \mathcal{E}/F$ between the slice toposes $\mathcal{E}/E$ and $\mathcal{E}/F$ is the inverse image of a geometric morphism $\mathcal{E}/F \to \mathcal{E}/E$.

(g) Every Grothendieck topos $\mathcal{E}$ admits a unique (up to isomorphism) geometric morphism $\gamma_{\mathcal{E}} : \mathcal{E} \to \mathbf{Set}$. The direct image of $\gamma_{\mathcal{E}}$ is the *global sections functor*

$\mathrm{Hom}_{\mathcal{E}}(1_{\mathcal{E}}, -) : \mathcal{E} \to \mathbf{Set}$, while the inverse image functor $\gamma_{\mathcal{E}}^{*}$ is given by $S \to \coprod_{s \in S} 1_{\mathcal{E}}$.

Definition 1.1.16 (a) A geometric morphism $f : \mathcal{F} \to \mathcal{E}$ is said to be a *surjection* if its inverse image $f^* : \mathcal{E} \to \mathcal{F}$ is faithful (equivalently, conservative).

(b) A geometric morphism $f : \mathcal{F} \to \mathcal{E}$ is said to be an *inclusion* if its direct image functor $f_* : \mathcal{F} \to \mathcal{E}$ is full and faithful.

Remark 1.1.17 As we shall see in section 1.3.4 (cf. Theorem 1.3.35), the *subtoposes* of $\mathcal{E}$, i.e. the (isomorphism classes of) geometric inclusions with codomain $\mathcal{E}$, correspond precisely to the local operators on $\mathcal{E}$.

Theorem 1.1.18 (cf. Theorems VII.4.6 and VII.4.8 [63]). *Every geometric morphism of (Grothendieck) toposes can be factored, in a unique way up to commuting isomorphisms, as a surjection followed by an inclusion.*

Definition 1.1.19 By a *point* of a topos $\mathcal{E}$, we mean a geometric morphism $\mathbf{Set} \to \mathcal{E}$.

Examples 1.1.20
(a) For any essentially small site $(\mathcal{C}, J)$, the points of the topos $\mathbf{Sh}(\mathcal{C}, J)$ can be identified with the J-continuous flat functors $\mathcal{C} \to \mathbf{Set}$ (cf. Theorem 1.1.34).
(b) For any locale L, the points of the topos $\mathbf{Sh}(L)$ correspond precisely to the frame homomorphisms $L \to \{0, 1\}$.
(c) For any small category $\mathcal{C}$ and any object c of $\mathcal{C}$ we have a point $\mathbf{Set} \to [\mathcal{C}^{\mathrm{op}}, \mathbf{Set}]$ of the topos $[\mathcal{C}^{\mathrm{op}}, \mathbf{Set}]$ whose inverse image is the evaluation functor at c.

Proposition 1.1.21 *Any collection of points P of a Grothendieck topos $\mathcal{E}$ indexed by a set X via a function $\xi : X \to P$ can be identified with a geometric morphism $\tilde{\xi} : [X, \mathbf{Set}] \to \mathcal{E}$ (where X is regarded as a discrete category).*

Definition 1.1.22 (a) Let $\mathcal{E}$ be a topos and P a collection of points of $\mathcal{E}$. We say that P is *separating* for $\mathcal{E}$ if the points in P are jointly surjective, i.e. if the inverse image functors of the geometric morphisms in P jointly reflect isomorphisms (equivalently, if P is indexed by a set X via a function $\xi : X \to P$ as in Proposition 1.1.21, if the geometric morphism $\tilde{\xi} : [X, \mathbf{Set}] \to \mathcal{E}$ is surjective).

(b) A topos is said to *have enough points* if the collection of all the points of $\mathcal{E}$ is separating for $\mathcal{E}$.

The following classical construction provides a way of endowing a given set of points of a topos with a natural topology:

Definition 1.1.23 Let $\xi : X \to P$ be an indexing of a set P of points of a Grothendieck topos $\mathcal{E}$ by a set X. The *subterminal topology* $\tau_\xi^\mathcal{E}$ on X is the image of the function $\phi_\mathcal{E} : \mathrm{Sub}_\mathcal{E}(1) \to \mathcal{P}(X)$ given by

$$\phi_\mathcal{E}(u) = \{x \in X \mid \xi(x)^*(u) \cong 1_\mathbf{Set}\}.$$

We denote the space X endowed with the topology $\tau_\xi^\mathcal{E}$ by $X_{\tau_\xi^\mathcal{E}}$.

The following proposition follows at once from the definition of $\tau_\xi^\mathcal{E}$:

Proposition 1.1.24 *If P is a separating set of points for $\mathcal{E}$ then the frame $\mathcal{O}(X_{\tau_\xi^\mathcal{E}})$ of open sets of $X_{\tau_\xi^\mathcal{E}}$ is isomorphic to $\mathrm{Sub}_\mathcal{E}(1)$ (via $\phi_\mathcal{E}$).*

This result was used in [16] and [26] to build spectra for various kinds of partially ordered structures and, combined with other results, Stone-type and Priestley-type dualities for them.

Remark 1.1.25 As observed in Remark C2.3.21 [55], for any small site $(\mathcal{C}, J)$, the subterminal objects of the topos $\mathbf{Sh}(\mathcal{C}, J)$ can be identified with the *J-ideals* on $\mathcal{C}$, i.e. with the collections I of objects of $\mathcal{C}$ with the property that for any arrow $f : c \to d$ in $\mathcal{C}$, $d \in I$ implies $c \in I$, and for any J-covering sieve S on an object c, if $\mathrm{dom}(f) \in I$ for all $f \in S$ then $c \in I$.

Definition 1.1.26 A topological space is said to be *sober* if every irreducible closed subset is the closure of a unique point.

Example 1.1.27 Affine algebraic varieties (with the Zariski topology) are sober spaces.

Remark 1.1.28 The sober topological spaces are exactly the topological spaces X such that the canonical map $X \to X_{\tau_\xi^{\mathbf{Sh}(X)}}$ is a homeomorphism; in other words, they are the spaces of points of localic toposes (up to homeomorphism).

1.1.5 Diaconescu's equivalence

The following theorem, already proved in [3], is of fundamental importance in topos theory and is commonly referred to in the literature as Diaconescu's equivalence in honour of R. Diaconescu who proved its relativization to an arbitrary base topos. Before stating it, we need to introduce the equally fundamental notion of flat functor.

Recall that the *category of elements* $\int P$ of a presheaf $P : \mathcal{C}^{\mathrm{op}} \to \mathbf{Set}$ has as objects the pairs (c, x) where c is an object of $\mathcal{C}$ and x is an element of $P(c)$ and as arrows $(c, x) \to (d, y)$ the arrows $f : c \to d$ in $\mathcal{C}$ such that $P(f)(y) = x$.

The following proposition exhibits a very general hom-tensor adjunction:

Proposition 1.1.29 *Let $\mathcal{C}$ be a small category and $\mathcal{E}$ a locally small cocomplete category. Then, for any functor $A : \mathcal{C} \to \mathcal{E}$, the functor $R_A : \mathcal{E} \to [\mathcal{C}^{\mathrm{op}}, \mathbf{Set}]$ defined for each $e \in \mathcal{E}$ and $c \in \mathcal{C}$ by*

$$R_A(e)(c) = \mathrm{Hom}_\mathcal{E}(A(c), e)$$

has a left adjoint $- \otimes_\mathcal{C} A : [\mathcal{C}^{\mathrm{op}}, \mathbf{Set}] \to \mathcal{E}$.

Sketch of proof The left adjoint $- \otimes_\mathcal{C} A$ sends a presheaf $P : \mathcal{C}^{\mathrm{op}} \to \mathcal{E}$ to the 'generalized tensor product' $P \otimes_\mathcal{C} A = \mathrm{colim}(A \circ \pi_P)$, where $\pi_P : \int P \to \mathcal{C}$ is the canonical projection to $\mathcal{C}$ from the category of elements $\int P$ of P. For more details, see the proof of Theorem I.5.2 [63]. □

Definition 1.1.30 (a) A functor $A : \mathcal{C} \to \mathcal{E}$ from a small category $\mathcal{C}$ to a Grothendieck topos $\mathcal{E}$ is said to be *flat* if the functor $- \otimes_\mathcal{C} A : [\mathcal{C}^{\mathrm{op}}, \mathbf{Set}] \to \mathcal{E}$ preserves finite limits.
(b) The full subcategory of $[\mathcal{C}, \mathcal{E}]$ on the flat functors will be denoted by $\mathbf{Flat}(\mathcal{C}, \mathcal{E})$.

Theorem 1.1.31 *Let $\mathcal{C}$ be a small category and $\mathcal{E}$ be a Grothendieck topos. Then we have an equivalence of categories*

$$\mathbf{Geom}(\mathcal{E}, [\mathcal{C}^{\mathrm{op}}, \mathbf{Set}]) \simeq \mathbf{Flat}(\mathcal{C}, \mathcal{E})$$

natural in $\mathcal{E}$, which sends

- *a flat functor $A : \mathcal{C} \to \mathcal{E}$ to the geometric morphism $\mathcal{E} \to [\mathcal{C}^{\mathrm{op}}, \mathbf{Set}]$ given by the functors R_A and $- \otimes_\mathcal{C} A$, and*
- *a geometric morphism $f : \mathcal{E} \to [\mathcal{C}^{\mathrm{op}}, \mathbf{Set}]$ to the flat functor given by the composite $f^* \circ y$ of $f^* : [\mathcal{C}^{\mathrm{op}}, \mathbf{Set}] \to \mathcal{E}$ with the Yoneda embedding $y : \mathcal{C} \to [\mathcal{C}^{\mathrm{op}}, \mathbf{Set}]$.*

Sketch of proof By definition of a flat functor, the functor $- \otimes_\mathcal{C} A$ is the inverse image functor of a geometric morphism $\mathcal{E} \to [\mathcal{C}^{\mathrm{op}}, \mathbf{Set}]$. Conversely, given a geometric morphism $f : \mathcal{E} \to [\mathcal{C}^{\mathrm{op}}, \mathbf{Set}]$, the composite of f^* with the Yoneda embedding $y : \mathcal{C} \to [\mathcal{C}^{\mathrm{op}}, \mathbf{Set}]$ yields a functor which is flat since $f^* \cong - \otimes_\mathcal{C} (f^* \circ y)$ (as the two functors take the same values on representables and both preserve colimits). The fact that $(- \otimes_\mathcal{C} A) \circ y \cong A$ follows immediately from the definition of $- \otimes_\mathcal{C} A$. For more details, see the proof of Corollary I.5.4 [63]. □

Recall that a small category $\mathcal{C}$ is said to be *filtered* if it is non-empty, for any objects $c, d \in \mathcal{C}$ there exist an object $e \in \mathcal{C}$ and two arrows $f : c \to e$ and $g : d \to e$ and for any parallel arrows $f, g : a \to b$ in $\mathcal{C}$ there exists an arrow $h : b \to c$ in $\mathcal{C}$ such that $h \circ f = h \circ g$.

For a detailed proof of the facts stated in the following proposition we refer the reader to sections VII.7 and VII.9 of [63].

Proposition 1.1.32

(i) *For any small category $\mathcal{C}$, a functor $F : \mathcal{C}^{\mathrm{op}} \to \mathbf{Set}$ is flat if and only if it is a filtered colimit of representables, equivalently if and only if its category of elements $\int F$ is filtered. More generally, a functor $F : \mathcal{C}^{\mathrm{op}} \to \mathcal{E}$ with values in a Grothendieck topos $\mathcal{E}$ is flat if and only if it is filtering i.e. satisfies the following conditions:*

   (a) *For any object E of $\mathcal{E}$ there exist an epimorphic family $\{e_i : E_i \to E \mid i \in I\}$ in $\mathcal{E}$ and for each $i \in I$ an object b_i of $\mathcal{C}$ and a generalized element $E_i \to F(b_i)$ in $\mathcal{E}$.*

(b) *For any two objects c and d in $\mathcal{C}$ and any generalized element $<x,y>:$
$E \to F(c) \times F(d)$ in $\mathcal{E}$ there is an epimorphic family $\{e_i : E_i \to E \mid i \in I\}$ in $\mathcal{E}$ and for each $i \in I$ an object b_i of $\mathcal{C}$ with arrows $u_i : c \to b_i$ and $v_i : d \to b_i$ in $\mathcal{C}$ and a generalized element $z_i : E_i \to F(b_i)$ in $\mathcal{E}$ such that $<F(u_i), F(v_i)> \circ z_i = \, <x,y> \circ e_i$ for all $i \in I$.*

(c) *For any two parallel arrows $u,v : d \to c$ in $\mathcal{C}$ and any generalized element $x : E \to F(c)$ in $\mathcal{E}$ for which $F(u) \circ x = F(v) \circ x$, there is an epimorphic family $\{e_i : E_i \to E \mid i \in I\}$ in $\mathcal{E}$ and for each $i \in I$ an arrow $w_i : c \to b_i$ and a generalized element $y_i : E_i \to F(b_i)$ such that $w_i \circ u = w_i \circ v$ and $F(w_i) \circ y_i = x \circ e_i$ for all $i \in I$.*

(ii) *Let $\mathcal{C}$ be a category with finite limits and $\mathcal{E}$ a Grothendieck topos. Then a functor $\mathcal{C} \to \mathcal{E}$ is flat if and only if it preserves finite limits.*

Definition 1.1.33 *Let $\mathcal{E}$ be a Grothendieck topos and $(\mathcal{C}, J)$ a site. A functor $F : \mathcal{C} \to \mathcal{E}$ is said to be J-continuous if it sends J-covering sieves to epimorphic families.*

The full subcategory of **Flat**$(\mathcal{C}, \mathcal{E})$ on the J-continuous flat functors will be denoted by **Flat**$_J(\mathcal{C}, \mathcal{E})$.

Theorem 1.1.34 *For any essentially small site $(\mathcal{C}, J)$ and Grothendieck topos $\mathcal{E}$, the above equivalence between geometric morphisms and flat functors restricts to an equivalence of categories*

$$\mathbf{Geom}(\mathcal{E}, \mathbf{Sh}(\mathcal{C}, J)) \simeq \mathbf{Flat}_J(\mathcal{C}, \mathcal{E})$$

natural in $\mathcal{E}$.

Sketch of proof Appeal to Theorem 1.1.31
- identifying the geometric morphisms $\mathcal{E} \to \mathbf{Sh}(\mathcal{C}, J)$ with the geometric morphisms $\mathcal{E} \to [\mathcal{C}^{\mathrm{op}}, \mathbf{Set}]$ which factor through the canonical geometric inclusion $\mathbf{Sh}(\mathcal{C}, J) \hookrightarrow [\mathcal{C}^{\mathrm{op}}, \mathbf{Set}]$, and
- using the characterization of such morphisms as the geometric morphisms $f : \mathcal{E} \to [\mathcal{C}^{\mathrm{op}}, \mathbf{Set}]$ such that the composite $f^* \circ y$ of the inverse image functor f^* of f with the Yoneda embedding $y : \mathcal{C} \to [\mathcal{C}^{\mathrm{op}}, \mathbf{Set}]$ sends J-covering sieves to epimorphic families in $\mathcal{E}$.

For more details, see section VII.7 of [63]. □

1.2 First-order logic

In logic, *first-order languages* are a wide class of formal languages used for describing mathematical structures. The attribute 'first-order' means that all the universal and existential quantifications occurring in the axioms of a theory in such a language concern only individuals rather than collections of individuals (or other higher-order constructions on individuals). For instance, the property of a group to be abelian can be expressed in a first-order way in the language of groups by the formula $(\forall x)(\forall y)(x+y = y+x)$, while the property of the ordered set $\mathbb{R}$ of real numbers that every bounded subset of it admits a supremum is not expressible in a first-order way (in the language of ordered sets) since it involves a quantification over subsets of the given structure rather than over elements of it (the formula defining it is a second-order one).

A first-order language contains *sorts*, which are meant to represent different *kinds* of individuals, *terms*, which denote individuals, and *formulae*, which make assertions about the individuals. Compound formulae are formed by using various logical operators, that is, either connectives (such as $\wedge$, $\vee$, $\Rightarrow$ etc.) or quantifiers ($\exists$ and $\forall$). For example, as we shall see in Example 1.2.7, it is natural to axiomatize the notion of (small) category by using a language with two sorts, one for the objects and one for the arrows.

It is well known, at least since the work of A. Tarski, that first-order languages can always be interpreted in the context of (a given model of) set theory (sorts are interpreted as sets, function symbols as functions and relation symbols as subsets). We will see in section 1.3 that one can meaningfully interpret them also in a general category, provided that the latter possesses enough categorical structure: sorts will be interpreted as *objects* of the given category, terms as *arrows* and formulae as *subobjects*, in a way which reflects their logical structure.

1.2.1 First-order theories

Definition 1.2.1 A first-order *signature* Σ consists of:

(a) A set Σ-Sort of *sorts*.

(b) A set Σ-Fun of *function symbols* f, each of which equipped with a finite list of source sorts and a single output sort: we write

$$f : A_1 \cdots A_n \to B$$

to indicate that f has source sorts $A_1, \ldots, A_n$ and output sort B. If $n = 0$, f is called a *constant* of sort B. The *arity* of f is the number n of its source sorts.

(c) A set Σ-Rel of *relation symbols* R, each of which equipped with a finite list of sorts: we write

$$R \rightarrowtail A_1 \cdots A_n$$

to indicate that the list of sorts associated with R is $A_1, \ldots, A_n$. The *arity* of R is the number n of sorts associated with it.

We shall say that a signature Σ is *finite* if it has only a finite number of sorts, function and relation symbols.

For each sort A of a signature Σ we assume to have at our disposal a stock of variables of sort A, used to denote individuals of type A.

A variable is said to be *free* in a given formula if it appears non-quantified in it, and *bound* otherwise. For example, in the formula $(\exists x)(x + y = 0)$ the variable y is free while the variable x is not. Note that a formula makes assertions only about the non-quantified variables occurring in it. For this reason, formulae without free variables are called *sentences*, as their validity does not depend on the value of any variable.

Starting from variables, terms are built up by repeated 'applications' of function symbols to them, as follows:

Definition 1.2.2 Let Σ be a signature. The collection of *terms* over Σ is defined inductively, together with the specification for each term of its sort (we write $t : A$ or t^A to denote that t is a term of sort A), by means of the following clauses:

(a) Any variable x^A of sort A is a term of sort A.
(b) For any list of terms $t_1 : A_1, \ldots, t_n : A_n$ and any function symbol $f : A_1 \cdots A_n \to B$, $f(t_1, \ldots, t_n)$ is a term of sort B.

Starting from terms, one can build classes F of formulae inductively by means of the following formation rules, which simultaneously define the set $\mathrm{FV}(\phi)$ of free variables of any formula ϕ in F:

(i) *Relations*: $R(t_1, \ldots, t_n)$ is in F for any terms $t_1 : A_1, \ldots, t_n : A_n$ and any relation symbol $R \rightarrowtail A_1 \cdots A_n$; $\mathrm{FV}(R(t_1, \ldots, t_n))$ is the set of variables occurring in some t_i (for $i = 1, \ldots, n$).
(ii) *Equality*: $(s = t)$ is in F for any terms s and t of the same sort; $\mathrm{FV}(s = t)$ is the set of variables occurring in s or t (or both).
(iii) *Truth*: $\top$ is in F; $\mathrm{FV}(\top) = \emptyset$.
(iv) *Binary conjunction*: $(\phi \wedge \psi)$ is in F whenever ϕ and ψ are in F; $\mathrm{FV}(\phi \wedge \psi) = \mathrm{FV}(\phi) \cup \mathrm{FV}(\psi)$.
(v) *Falsity*: $\bot$ is in F; $\mathrm{FV}(\top) = \emptyset$.
(vi) *Binary disjunction*: $(\phi \vee \psi)$ is in F whenever ϕ and ψ are in F; $\mathrm{FV}(\phi \vee \psi) = \mathrm{FV}(\phi) \cup \mathrm{FV}(\psi)$.
(vii) *Implication*: $(\phi \Rightarrow \psi)$ is in F whenever ϕ and ψ are in F; $\mathrm{FV}(\phi \Rightarrow \psi) = \mathrm{FV}(\phi) \cup \mathrm{FV}(\psi)$.
(viii) *Negation*: $\neg \phi$ is in F whenever ϕ is in F; $\mathrm{FV}(\neg \phi) = \mathrm{FV}(\phi)$.
(ix) *Existential quantification*: $(\exists x)\phi$ is in F whenever ϕ is in F and x is a free variable in ϕ; $\mathrm{FV}((\exists x)\phi) = \mathrm{FV}(\phi) \setminus \{x\}$.
(x) *Universal quantification*: $(\forall x)\phi$ is in F whenever ϕ is in F and x is a free variable in ϕ; $\mathrm{FV}((\forall x)\phi) = \mathrm{FV}(\phi) \setminus \{x\}$.
(xi) *Infinitary disjunction*: $\bigvee_{i \in I} \phi_i$ is in F whenever I is a set, ϕ_i is in F for each $i \in I$ and $\mathrm{FV}(\bigvee_{i \in I} \phi_i) := \bigcup_{i \in I} \mathrm{FV}(\phi_i)$ is finite.
(xii) *Infinitary conjunction*: $\bigwedge_{i \in I} \phi_i$ is in F whenever I is a set, ϕ_i is in F for each $i \in I$ and $\mathrm{FV}(\bigwedge_{i \in I} \phi_i) := \bigcup_{i \in I} \mathrm{FV}(\phi_i)$ is finite.

We can think of $\top$ as an empty conjunction and of $\bot$ as an empty disjunction (cf. the deduction rules for conjunction and disjunction in section 1.2.2 below).

A *context* is a finite list $\vec{x} = (x_1, \ldots, x_n)$ of distinct variables (the empty context, for $n = 0$, is allowed and indicated by $[]$). We will often consider *formulae-in-context*, that is, formulae ϕ equipped with a context $\vec{x}$ such that all the free variables of ϕ occur in $\vec{x}$; we shall write either $\phi(\vec{x})$ or $\{\vec{x}.\phi\}$ for a formula ϕ in a context $\vec{x}$. The *canonical context* of a given formula is the context consisting of all the variables which appear freely in the formula.

Definition 1.2.3 In relation to the above-mentioned formation rules:
(a) The set of *atomic formulae* over Σ is the smallest set closed under *Relations* and *Equality*.
(b) The set of *Horn formulae* over Σ is the smallest set containing the class of atomic formulae and closed under *Truth* and *Binary conjunction*.
(c) The set of *regular formulae* over Σ is the smallest set containing the class of atomic formulae and closed under *Truth, Binary conjunction* and *Existential quantification*.
(d) The set of *coherent formulae* over Σ is the smallest set containing the class of atomic formulae and closed under *Truth, Binary conjunction, Existential quantification, False* and *Binary disjunction*.
(e) The set of *first-order formulae* over Σ is the smallest set closed under all the above-mentioned formation rules except for the infinitary ones.
(f) The *class* of *geometric formulae* over Σ is the smallest class containing the class of atomic formulae and closed under *Truth, Binary conjunction, Existential quantification, False* and *Infinitary disjunction*.
(g) The *class* of *infinitary first-order formulae* over Σ is the smallest class closed under all the above-mentioned formation rules.

Definition 1.2.4 (a) By a *sequent* over a signature Σ we mean an expression of the form $(\phi \vdash_{\vec{x}} \psi)$, where ϕ and ψ are formulae over Σ in a context $\vec{x}$.
(b) We say that a sequent $(\phi \vdash_{\vec{x}} \psi)$ is Horn (resp. regular, coherent, ...) if both ϕ and ψ are Horn (resp. regular, coherent, ...) formulae.

The intended meaning of a sequent $(\phi \vdash_{\vec{x}} \psi)$ is that ψ is a logical consequence of ϕ in the context $\vec{x}$, i.e. that for any 'values' taken by the variables in $\vec{x}$, if ϕ is satisfied by them then ψ is satisfied by them as well.

Notice that sequents are not necessary in the context of full first-order logic; indeed, any sequent $(\phi \vdash_{\vec{x}} \psi)$ has the same intended meaning as the formula $(\forall \vec{x})(\phi \Rightarrow \psi)$.

Definition 1.2.5 (a) By a (first-order) *theory* over a signature Σ, we mean a set $\mathbb{T}$ of sequents over Σ, whose elements are called the (non-logical) *axioms* of $\mathbb{T}$.
(b) We say that $\mathbb{T}$ is an *algebraic theory* if its signature Σ has no relation symbols (apart from the equality relations on its sorts) and its axioms are all of the form $(\top \vdash_{\vec{x}} \phi)$ where ϕ is an atomic formula of the form $(s = t)$ and $\vec{x}$ its canonical context.

(c) We say that $\mathbb{T}$ is a *Horn* (resp. *regular, coherent, geometric*) theory if all the sequents in $\mathbb{T}$ are Horn (resp. regular, coherent, geometric).

(d) We say that $\mathbb{T}$ is a *universal Horn* theory if its axioms are all of the form $(\phi \vdash_{\vec{x}} \psi)$, where ϕ is a finite (possibly empty, i.e. equal to $\top$) conjunction of atomic formulae and ψ is an atomic formula or the formula $\bot$.

(e) We say that a regular theory $\mathbb{T}$ is *cartesian* if its axioms can be well-ordered in such a way that each axiom is cartesian relative to the subtheory consisting of all the axioms preceding it in the ordering, in the sense that all the existential quantifications which appear in the axiom are provably unique relative to that subtheory (cf. section 1.2.2 for the notion of provability in fragments of first-order logic). A *cartesian formula* relative to a cartesian theory $\mathbb{T}$ over a signature Σ ($\mathbb{T}$-*cartesian* formula, for short) is a formula built up from atomic formulae over Σ by only using binary conjunctions (possibly including the empty conjunction, i.e. $\top$) and $\mathbb{T}$-provably unique existential quantifications. We say a sequent $(\phi \vdash_{\vec{x}} \psi)$ is cartesian relative to a cartesian theory $\mathbb{T}$ over a signature Σ ($\mathbb{T}$-cartesian for short) if both ϕ and ψ are cartesian formulae over Σ (relative to $\mathbb{T}$).

(f) We say that $\mathbb{T}$ is a *propositional theory* if its signature has no sorts, i.e. it only consists of 0-ary relation symbols.

(g) We say that $\mathbb{T}$ is the *empty theory* over Σ if it does not contain any (non-logical) axioms.

Remark 1.2.6 The study of the model theory of algebraic theories has been initiated in the context of universal algebra and developed in a categorical setting since the pioneering work [59] by W. Lawvere introducing the functorial semantics of algebraic theories.

Example 1.2.7 An important example of a cartesian theory is given by the *theory* $\mathbb{C}$ *of small categories*. The language of $\mathbb{C}$ consists of two sorts O and A, respectively for objects and arrows, two function symbols $\text{dom}, \text{cod} : A \to O$ formalizing domain and codomain, a function symbol $1 : O \to A$ formalizing the map assigning to any object the identity arrow on it and a ternary predicate C formalizing the composition of arrows in the sense that $C(f, g, h)$ if and only if $h = f \circ g$ (notice that we have to use a relation symbol rather than a function symbol for formalizing composition since the latter is not everywhere defined). Over this signature, the axioms of $\mathbb{C}$ are the obvious ones. In particular, we have an axiom $(\text{dom}(f) = \text{cod}(g) \vdash_{f,g} (\exists h) C(f, g, h))$ expressing the existence of the composite of two arrows such that the codomain of the first coincides with the domain of the second, which is cartesian relative to the sequent $(C(f, g, h) \wedge C(f, g, h') \vdash_{f,g,h,h'} h = h')$ expressing the functionality of the predicate C.

Definition 1.2.8 Let $\mathbb{T}$ be a geometric theory over a signature Σ. The *cartesianization* (or *finite-limit part*) $\mathbb{T}_c$ of $\mathbb{T}$ is the cartesian theory over Σ consisting of all the $\mathbb{T}$-cartesian sequents over Σ which are provable in $\mathbb{T}$.

1.2.2 Deduction systems for first-order logic

With each of the fragments of first-order logic introduced above, we can naturally associate a *deduction system*, in the same spirit as in classical first-order logic. Such systems will be formulated as *sequent calculi*, that is, they will consist of *inference rules* enabling one to derive a sequent from a collection of other sequents; we shall write

$$\frac{\Gamma}{\sigma}$$

to mean that the sequent σ can be inferred from a collection of sequents Γ. A double line between two sequents will mean that each of the sequents can be inferred from the other.

Given the axioms and inference rules below, the notion of *proof* (or *derivation*) is the usual one: a chain of inference rules whose premises are the axioms in the system and whose conclusion is the given sequent. Allowing the axioms of a theory $\mathbb{T}$ to be taken as premises yields the notion of *proof* (or *derivation*) *relative to a theory* $\mathbb{T}$.

Consider the following rules:

- The *rules for finite conjunction* are the axioms

$$(\phi \vdash_{\vec{x}} \top) \quad ((\phi \wedge \psi) \vdash_{\vec{x}} \phi) \quad ((\phi \wedge \psi) \vdash_{\vec{x}} \psi)$$

  and the rule

$$\frac{(\phi \vdash_{\vec{x}} \psi)(\phi \vdash_{\vec{x}} \chi)}{(\phi \vdash_{\vec{x}} (\psi \wedge \chi))}.$$

- The *rules for finite disjunction* are the axioms

$$(\bot \vdash_{\vec{x}} \phi) \quad (\phi \vdash_{\vec{x}} (\phi \vee \psi)) \quad (\psi \vdash_{\vec{x}} \phi \vee \psi)$$

  and the rule

$$\frac{(\phi \vdash_{\vec{x}} \chi)(\psi \vdash_{\vec{x}} \chi)}{((\phi \vee \psi) \vdash_{\vec{x}} \chi)}.$$

- The *rules for infinitary conjunction (resp. disjunction)* are the infinitary analogues of the rules for finite conjunction (resp. disjunction).
- The *rules for implication* consist of the double rule

$$\frac{(\phi \wedge \psi \vdash_{\vec{x}} \chi)}{(\psi \vdash_{\vec{x}} (\phi \Rightarrow \chi))}.$$

- The *rules for existential quantification* consist of the double rule

$$\frac{(\phi \vdash_{\vec{x},y} \psi)}{((\exists y)\phi \vdash_{\vec{x}} \psi)},$$

  where we assume that y is not free in ψ.

- The *rules for universal quantification* consist of the double rule

$$\frac{(\phi \vdash_{\vec{x},y} \psi)}{(\phi \vdash_{\vec{x}} (\forall y)\psi)}.$$

- The *distributive axiom* is

$$(\phi \wedge (\psi \vee \chi) \vdash_{\vec{x}} (\phi \wedge \psi) \vee (\phi \wedge \chi)).$$

- The *Frobenius axiom* is

$$(\phi \wedge (\exists y)\psi \vdash_{\vec{x}} (\exists y)(\phi \wedge \psi)),$$

where y is a variable not in the context $\vec{x}$.

- The *Law of Excluded Middle* is

$$(\top \vdash_{\vec{x}} \phi \vee \neg \phi).$$

Remark 1.2.9 Note that these rules, although having been conceived with a particular semantics in mind (namely, the one that we shall describe in section 1.3), are completely formal: they do not have any meaning by themselves. Indeed, a fundamental principle of modern logic is that of strictly separating syntax and semantics, distinguishing in particular the notion of *provability* in a given formal system (intended as the possibility of deriving an assertion from a given set of premises by repeatedly applying certain specified 'inference rules') from that of *validity* (or satisfaction) in a given structure. Of central importance is therefore the investigation of the connections between syntax and semantics, possibly leading to soundness and completeness theorems relating the notion of provability in a deductive system with the notion of validity in an appropriately chosen class of structures. Results of this kind for the fragments of logic that we shall consider will be discussed in section 1.4.5.

1.2.3 Fragments of first-order logic

Definition 1.2.10 In addition to the usual structural rules of sequent calculi (*Identity axiom*, *Equality rules*, *Substitution rule* and *Cut rule*, cf. Definition D1.3.1 [55] for more details), our deduction systems consist of the following rules:

Algebraic logic	No additional rule
Horn logic	Finite conjunction
Regular logic	Finite conjunction, existential quantification and Frobenius axiom
Coherent logic	Finite conjunction, finite disjunction, existential quantification, distributive axiom and Frobenius axiom
Geometric logic	Finite conjunction, infinitary disjunction, existential quantification, 'infinitary' distributive axiom, Frobenius axiom
Intuitionistic first-order logic	All the finitary rules except for the law of excluded middle
Classical first-order logic	All the finitary rules

Definition 1.2.11 We say a sequent σ is *provable* (or *derivable*) in an algebraic (regular, coherent, ...) theory $\mathbb{T}$ ($\mathbb{T}$-*provable*, for short) if there exists a derivation of σ relative to $\mathbb{T}$, in the appropriate fragment of first-order logic.

Remark 1.2.12 Whilst there is a *class* of geometric formulae over a given signature, there is only a *set* of provable-equivalence (with respect to geometric logic) classes of geometric formulae in a given context (cf. the proof of Lemma D1.4.10(iv) [55]). So one can define a geometric theory over a given signature Σ also by giving a class, rather than a set, of geometric sequents over Σ; indeed, we are assured that there exists a set of geometric sequents that can act as representatives of the sequents in the given class in the sense that each of these sequents is provably equivalent in geometric logic to one of the sequents in the set. We shall specify geometric theories in this way at many points of the book.

In geometric logic, intuitionistic and classical provability of geometric sequents coincide, as shown by the following result:

Theorem 1.2.13 (cf. Proposition D3.1.16 [55]). *If a geometric sequent σ is derivable from the axioms of a geometric theory $\mathbb{T}$ using 'classical geometric logic' (i.e. the rules of geometric logic plus the Law of Excluded Middle), then σ is also derivable without using the Law of Excluded Middle.*

Remark 1.2.14 Since a given theory can in general be regarded as belonging to more than one of the fragments of geometric logic considered above, it is natural to wonder whether the notion of provability of sequents over the signature of the theory does not depend on the fragment. A positive answer to this question will be obtained in section 2.2 as an application of the 'bridge' technique.

1.3 Categorical semantics

Generalizing the classical Tarskian definition of satisfaction of first-order formulae in ordinary set-valued structures, one can naturally obtain, given a signature Σ, a notion of Σ-structure in a category with finite products, and define, according to the categorical structure present on the category, a notion of interpretation of an appropriate fragment of first-order logic into it.

Specifically, we shall introduce various classes of 'logical' categories, each of them providing a semantics for a corresponding fragment of first-order logic:

Categories with finite products	Algebraic logic
Cartesian categories	Cartesian logic
Regular categories	Regular logic
Coherent categories	Coherent logic
Geometric categories	Geometric logic
Heyting categories	First-order intuitionistic logic
Boolean coherent categories	First-order classical logic

The process by which these classes of categories are defined is quite canonical: first one looks at the set-theoretic structure needed to interpret the connectives and quantifiers occurring in the given fragment, then one proceeds to characterize it in categorical terms, and finally one uses the resulting categorical structure for interpreting the given fragment of logic. For example, one immediately realizes that in set theory the conjunction $\phi \wedge \psi$ of two formulae-in-context ϕ and ψ is interpreted as the intersection of the interpretations of ϕ and ψ. Now, intersections of subsets can be characterized in categorical terms as pullbacks of subobjects; this ensures that, in any category $\mathcal{C}$ with finite limits, it is possible to give a natural meaning to the conjunction of two formulae which are interpretable in $\mathcal{C}$.

Let us start defining the notion of Σ-structure in a category with finite products.

Definition 1.3.1 Let $\mathcal{C}$ be a category with finite products and Σ a signature. A Σ-*structure* M in $\mathcal{C}$ consists of:

(a) An object MA of $\mathcal{C}$ for each sort A in Σ-Sort.
(b) An arrow $Mf : MA_1 \times \cdots \times MA_n \to MB$ in $\mathcal{C}$ for each function symbol $f : A_1 \cdots A_n \to B$ in Σ-Fun.
(c) A subobject $MR \rightarrowtail MA_1 \times \cdots \times MA_n$ in $\mathcal{C}$ for each relation symbol $R \rightarrowtail A_1 \cdots A_n$ in Σ-Rel.

Notice that if $n = 0$ then the cartesian product $MA_1 \times \cdots \times MA_n$ is equal to the terminal object $1_\mathcal{C}$ of $\mathcal{C}$.

Definition 1.3.2 A Σ-*structure homomorphism* $h : M \to N$ between two Σ-structures M and N in $\mathcal{C}$ is a collection of arrows $h_A : MA \to NA$ in $\mathcal{C}$ indexed by the sorts of Σ such that:

(a) For each function symbol $f : A_1 \cdots A_n \to B$ in Σ-Fun, the diagram

$$\begin{array}{ccc} MA_1 \times \cdots \times MA_n & \xrightarrow{Mf} & MB \\ \downarrow{h_{A_1} \times \cdots \times h_{A_n}} & & \downarrow{h_B} \\ NA_1 \times \cdots \times NA_n & \xrightarrow{Nf} & NB \end{array}$$

commutes.

(b) For each relation symbol $R \rightarrowtail A_1 \cdots A_n$ in Σ-Rel, there is a commutative diagram in $\mathcal{C}$ of the form

$$\begin{array}{ccc} MR & \rightarrowtail & MA_1 \times \cdots \times MA_n \\ \downarrow & & \downarrow {\scriptstyle h_{A_1} \times \cdots \times h_{A_n}} \\ NR & \rightarrowtail & NA_1 \times \cdots \times NA_n. \end{array}$$

Given a category $\mathcal{C}$ with finite products, Σ-structures in $\mathcal{C}$ and Σ-homomorphisms between them form a category, denoted by Σ-str$(\mathcal{C})$. Identities and composition in Σ-str$(\mathcal{C})$ are defined componentwise from those in $\mathcal{C}$.

Remark 1.3.3 If $\mathcal{C}$ and $\mathcal{D}$ are two categories with finite products then every functor $F : \mathcal{C} \to \mathcal{D}$ which preserves finite products and monomorphisms induces a functor Σ-str$(F) : \Sigma$-str$(\mathcal{C}) \to \Sigma$-str$(\mathcal{D})$ in the obvious way.

1.3.1 Classes of 'logical' categories

In this section we shall introduce classes of categories in which the fragments of first-order logic considered in section 1.2.3 can be naturally interpreted.

In any category $\mathcal{C}$ with pullbacks, pullbacks of monomorphisms are again monomorphisms; thus, for any arrow $f : a \to b$ in $\mathcal{C}$, we have a *pullback functor*

$$f^* : \mathrm{Sub}_{\mathcal{C}}(b) \to \mathrm{Sub}_{\mathcal{C}}(a).$$

Recall that by a *finite limit* in a category $\mathcal{C}$ we mean a limit of a functor $F : \mathcal{J} \to \mathcal{C}$ where $\mathcal{J}$ is a *finite category* (i.e. a category with only a finite number of objects and arrows).

Definition 1.3.4 A *cartesian* category is any category with finite limits.

As we shall see below, in cartesian categories we can interpret atomic formulae as well as finite conjunctions of them; in fact, conjunctions will be interpreted as pullbacks (i.e. intersections) of subobjects.

Definition 1.3.5 (a) Given two monomorphisms $m_1 : a_1 \rightarrowtail c$ and $m_2 : a_2 \rightarrowtail c$ of an object c in a category $\mathcal{C}$, we say that m_1 factors through m_2 if there is a (necessarily unique) arrow $r : a_1 \to a_2$ in $\mathcal{C}$ such that $m_2 \circ r = m_1$. Note that this defines a preorder relation $\leq$ on the collection $\mathrm{Sub}_{\mathcal{C}}(c)$ of subobjects of a given object c.

(b) We say that a cartesian category $\mathcal{C}$ has *images* if for any morphism f of $\mathcal{C}$ there exists a subobject $\mathrm{Im}(f)$ of its codomain, called the *image of f*, which is the least (with respect to the preorder $\leq$) subobject of $\mathrm{cod}(f)$ through which f factors.

(c) A *regular category* is a cartesian category which has images that are stable under pullback.

Proposition 1.3.6 *Given an arrow $f : a \to b$ in a regular category $\mathcal{C}$, the pullback functor $f^* : \mathrm{Sub}_\mathcal{C}(b) \to \mathrm{Sub}_\mathcal{C}(a)$ has a left adjoint $\exists_f : \mathrm{Sub}_\mathcal{C}(a) \to \mathrm{Sub}_\mathcal{C}(b)$, which assigns to a subobject $m : c \rightarrowtail a$ the image of the composite arrow $f \circ m$.*

In a regular category, every arrow $f : a \to b$ factors uniquely through its image $\mathrm{Im}(f) \rightarrowtail b$ as the composite $a \to \mathrm{Im}(f) \to b$ of $\mathrm{Im}(f) \rightarrowtail b$ with an arrow $c(f) : a \to \mathrm{Im}(f)$; arrows of the form $c(f)$ for some f are called *covers*. In fact, every arrow in a regular category can be factored uniquely as a cover followed by a monomorphism, and covers are precisely the arrows g such that $\mathrm{Im}(g) = 1_{\mathrm{cod}(g)}$.

As we shall see in section 1.3.3, in regular categories we can interpret formulae built up from atomic formulae by using finite conjunctions and existential quantifications; in fact, the existential quantifications will be interpreted as images of certain arrows.

Definition 1.3.7 A *coherent category* is a regular category $\mathcal{C}$ in which every subobject lattice has finite unions which are stable under pullback.

As we shall see in section 1.3.3, in coherent categories we can interpret formulae built up from atomic formulae by using finite conjunctions, existential quantifications and finite disjunctions; in fact, finite disjunctions will be interpreted as finite unions of subobjects.

Note in passing that, if coproducts exist, a union of subobjects of an object c may be constructed as the image of the induced arrow from the coproduct of the domains of such subobjects to c.

Definition 1.3.8 (a) A (large) category $\mathcal{C}$ is said to be *well-powered* if each of the preorders $\mathrm{Sub}_\mathcal{C}(a)$, $a \in \mathcal{C}$, is equivalent to a small category.
(b) A *geometric category* is a well-powered regular category whose subobject lattices have arbitrary unions which are stable under pullback.

As we shall see in section 1.3.3, in geometric categories we can interpret formulae built up from atomic formulae by using finite conjunctions, existential quantifications, and infinitary disjunctions; indeed, disjunctions will be interpreted as unions of subobjects.

To understand how to categorically interpret quantifiers, let us analyse their interpretations in the category **Set**.

Let X and Y be two sets. For any given subset $S \subseteq X \times Y$, we can consider the sets

$$\forall_p S := \{y \in Y \mid \text{for all } x \in X, (x, y) \in S\} \text{ and}$$

$$\exists_p S := \{y \in Y \mid \text{there exists } x \in X, (x, y) \in S\}.$$

The projection map $p : X \times Y \to Y$ induces a map at the level of powersets $p^* = p^{-1} : \mathcal{P}(Y) \to \mathcal{P}(X \times Y)$. If we regard these powersets as poset categories (where the order is given by the inclusion relation) then this map becomes a functor. Also, the assignments $S \to \forall_p S$ and $S \to \exists_p S$ yield functors $\forall_p, \exists_p : \mathcal{P}(X \times Y) \to \mathcal{P}(Y)$.

Proposition 1.3.9 *The functors $\exists_p$ and $\forall_p$ are respectively left and right adjoints to the functor $p^* : \mathcal{P}(Y) \to \mathcal{P}(X \times Y)$ which sends each subset $T \subseteq Y$ to its inverse image p^*T under p.*

Of course, the proposition generalizes to the case of an arbitrary function in place of the projection p. This motivates the following

Definition 1.3.10 A *Heyting category* is a coherent category $\mathcal{C}$ such that for any arrow $f : a \to b$ in $\mathcal{C}$, the pullback functor $f^* : \mathrm{Sub}_\mathcal{C}(b) \to \mathrm{Sub}_\mathcal{C}(a)$ has a right adjoint $\forall_f : \mathrm{Sub}_\mathcal{C}(a) \to \mathrm{Sub}_\mathcal{C}(b)$ (in addition to its left adjoint $\exists_f : \mathrm{Sub}_\mathcal{C}(a) \to \mathrm{Sub}_\mathcal{C}(b)$).

Proposition 1.3.11 (Lemma A1.4.13 [55]). *Let $a_1 \rightarrowtail a$ and $a_2 \rightarrowtail a$ be subobjects in a Heyting category. Then there exists a largest subobject $(a_1 \Rightarrow a_2) \rightarrowtail a$ such that $(a_1 \Rightarrow a_2) \cap a_1 \leq a_2$. Moreover, the binary operation on subobjects thus defined is stable under pullback.*

In particular, all the subobject lattices in a Heyting category are Heyting algebras.

The *Heyting negation* (also called *pseudocomplement*) $\neg a_1$ of a subobject $a_1 \rightarrowtail a$ is defined as $a_1 \Rightarrow 0_a$ (where 0_a is the zero subobject of a), and is characterized by the property of being the largest subobject of a disjoint from $a_1 \rightarrowtail a$. Notice that in general $a_1 \cup \neg a_1 \neq a$.

Proposition 1.3.11 ensures that in a Heyting category we may interpret full finitary first-order logic. In particular, the negation $\neg \phi(\vec{x})$ of a first-order formula-in-context $\phi(\vec{x})$ is interpreted as the Heyting pseudocomplement $\neg[[\vec{x}.\phi]]$. Note that, while it is always the case that $a \wedge \neg a = 0$ (in any Heyting algebra), $a \vee \neg a$ is in general different from 1, whence the Law of Excluded Middle is not sound with respect to general Heyting categories; indeed, the logic of Heyting categories is intuitionistic, not classical.

Definition 1.3.12 A coherent category $\mathcal{C}$ is said to be *Boolean* if every subobject $m : a \rightarrowtail c$ in $\mathcal{C}$ is complemented, in the sense that there exists a unique subobject n of c such that $m \cup n = 1_c$ and $m \cap n = 0_c$.

Since every Boolean algebra is a Heyting algebra, we have the following result:

Proposition 1.3.13 *Every Boolean coherent category is a Heyting category.*

Proposition 1.3.14 (cf. Lemma A1.4.18 [55]). *Every geometric category is a Heyting category.*

Proposition 1.3.15 *Every Grothendieck topos is a geometric category.*

Sketch of proof Well-poweredness immediately follows from the fact that by Giraud's theorem every Grothendieck topos has a separating set of objects, while the other properties easily follow from Theorems 1.1.12 and 1.1.13. □

Thus every Grothendieck topos is a Heyting category. As we shall see in section 1.3.5, it is true more generally that every elementary topos is a Heyting category.

1.3.1.1 The internal language

Given a small category $\mathcal{C}$ with finite products one can define a first-order signature $\Sigma_{\mathcal{C}}$, called the *internal language* (or the *canonical signature*) of $\mathcal{C}$, for reasoning about $\mathcal{C}$ in a set-theoretic fashion.

Definition 1.3.16 The *internal language* $\Sigma_{\mathcal{C}}$ of $\mathcal{C}$ consists of

- one sort $\ulcorner A\urcorner$ for each object A of $\mathcal{C}$,
- one function symbol $\ulcorner f\urcorner : \ulcorner A_1\urcorner \times \cdots \times \ulcorner A_n\urcorner \to \ulcorner B\urcorner$ for each arrow $f : A_1 \times \cdots \times A_n \to B$, and
- one relation symbol $\ulcorner R\urcorner \rightarrowtail \ulcorner A_1\urcorner \cdots \ulcorner A_n\urcorner$ for each subobject $R \rightarrowtail A_1 \times \cdots \times A_n$ in $\mathcal{C}$.

We will sometimes write x^A instead of $x^{\ulcorner A\urcorner}$ to mean that x is a variable of sort $\ulcorner A\urcorner$.

In the sequel, we shall also informally consider the internal languages of non-small categories such as toposes, in situations where the size issues are not relevant.

There is a canonical $\Sigma_{\mathcal{C}}$-structure $\mathcal{S}_{\mathcal{C}}$ in $\mathcal{C}$ called the *tautological $\Sigma_{\mathcal{C}}$-structure*, which assigns A to $\ulcorner A\urcorner$, f to $\ulcorner f\urcorner$ and R to $\ulcorner R\urcorner$.

The usefulness of these notions lies in the fact that properties of $\mathcal{C}$ or constructions in it can often be formulated in terms of satisfaction of certain formulae over $\Sigma_{\mathcal{C}}$ in the tautological structure $\mathcal{S}_{\mathcal{C}}$. The internal language can thus be used for proving things about $\mathcal{C}$. Indeed, for any objects $A_1, \ldots, A_n$ of $\mathcal{C}$ and any first-order formula $\phi(\vec{x})$ over $\Sigma_{\mathcal{C}}$, where $\vec{x} = (x_1^{\ulcorner A_1\urcorner}, \ldots, x_n^{\ulcorner A_n\urcorner})$, the set-theoretic expression $\{\vec{x} \in A_1 \times \cdots \times A_n \mid \phi(\vec{x})\}$ can be given a meaning, namely the interpretation of the formula $\phi(\vec{x})$ in the $\Sigma_{\mathcal{C}}$-structure $\mathcal{S}_{\mathcal{C}}$.

We shall see basic examples of the use of the internal language of a topos in section 1.3.5.

It is also often convenient to use the reduced language $\Sigma_{\mathcal{C}}^{\mathrm{f}}$ obtained from $\Sigma_{\mathcal{C}}$ by removing all the relation symbols.

1.3.2 Completions of 'logical' categories

It is important for many purposes to be able to complete the 'logical' categories that we have considered in section 1.3.1 with respect to certain kinds of colimits that they lack.

As far as it concerns regular categories, we can 'complete' them to regular categories in which quotients by equivalence relations always exist, by formally adding them. Such a construction is called *effectivization* and is characterized by the following universal property: for any regular category $\mathcal{C}$ there exists an effective regular category (i.e. a regular category in which every equivalence relation occurs as the kernel pair of some morphism) **Eff**$(\mathcal{C})$ with a full and faithful functor $i_{\text{Eff}} : \mathcal{C} \to \mathbf{Eff}(\mathcal{C})$ such that for any effective regular category $\mathcal{D}$ the regular functors $\mathbf{Eff}(\mathcal{C}) \to \mathcal{D}$ correspond, naturally in $\mathcal{D}$, to the regular functors $\mathcal{C} \to \mathcal{D}$ (via composition with i_{Eff}). For a detailed description of this construction the reader is referred to the proof of Corollary A3.3.10 [55].

Coherent (resp. geometric) categories do not possess in general finite (resp. arbitrary) coproducts, but it is possible to 'complete' them to coherent (resp. geometric) categories in which finite (resp. arbitrary) disjoint coproducts of objects exist. Such a construction is called *positivization* (resp. *infinitary positivization*) and is characterized by the following universal property:

- For any coherent category $\mathcal{C}$ there exists a positive coherent category (i.e. a coherent category having disjoint finite coproducts) $\mathbf{Pos}(\mathcal{C})$ with a full and faithful functor $i_{\text{Pos}} : \mathcal{C} \to \mathbf{Pos}(\mathcal{C})$ such that for any positive category $\mathcal{D}$ the coherent functors $\mathbf{Pos}(\mathcal{C}) \to \mathcal{D}$ correspond, naturally in $\mathcal{D}$, to the coherent functors $\mathcal{C} \to \mathcal{D}$ (via composition with i_{Pos}).
- For any geometric category $\mathcal{C}$ there exists an ∞-positive coherent category (i.e. a geometric category which has disjoint arbitrary set-indexed coproducts) $\infty\text{-}\mathbf{Pos}(\mathcal{C})$ with a full and faithful functor $i_{\infty\text{-}\text{Pos}} : \mathcal{C} \to \infty\text{-}\mathbf{Pos}(\mathcal{C})$ such that for any ∞-positive category $\mathcal{D}$, the geometric functors $\infty\text{-}\mathbf{Pos}(\mathcal{C}) \to \mathcal{D}$ correspond, naturally in $\mathcal{D}$, to the geometric functors $\mathcal{C} \to \mathcal{D}$ (by composition with $i_{\infty\text{-}\text{Pos}}$).

More details about these constructions can be found at pp. 34-35 of [55].

Definition 1.3.17 (a) A *pretopos* is a positive and effective coherent category.
(b) An ∞-*pretopos* is an ∞-positive and effective geometric category.

Remarks 1.3.18 (a) For any coherent category $\mathcal{C}$, the category $\mathbf{Eff}(\mathbf{Pos}(\mathcal{C}))$, with the canonical embedding $i_{\text{Pr}} : \mathcal{C} \hookrightarrow \mathbf{Eff}(\mathbf{Pos}(\mathcal{C}))$, satisfies the universal property of the *pretopos completion* $\mathcal{P}_\mathcal{C}$ of $\mathcal{C}$, i.e. $\mathbf{Eff}(\mathbf{Pos}(\mathcal{C}))$ is a pretopos such that for any pretopos $\mathcal{D}$ the coherent functors $\mathcal{P}_\mathcal{C} \to \mathcal{D}$ correspond precisely to the coherent functors $\mathcal{C} \to \mathcal{D}$ (via composition with i_{Pr}).
(b) For any geometric category $\mathcal{C}$, the category $\mathbf{Eff}(\infty\text{-}\mathbf{Pos}(\mathcal{C}))$, with the canonical embedding $i_{\infty\text{-}\text{Pr}} : \mathcal{C} \hookrightarrow \mathbf{Eff}(\infty\text{-}\mathbf{Pos}(\mathcal{C}))$, satisfies the universal property of the ∞-*pretopos completion* $\mathcal{P}_\mathcal{C}^\infty$ of $\mathcal{C}$, i.e. $\mathbf{Eff}(\infty\text{-}\mathbf{Pos}(\mathcal{C}))$ is a ∞-pretopos such that for any ∞-pretopos $\mathcal{D}$ the geometric functors $\mathcal{P}_\mathcal{C}^\infty \to \mathcal{D}$ correspond precisely to the geometric functors $\mathcal{C} \to \mathcal{D}$ (via composition with $i_{\infty\text{-}\text{Pr}}$).

1.3.3 Models of first-order theories in categories

In this section we shall define the notion of model of a theory belonging to a fragment of first-order logic in a category in which it is interpretable.

First-order terms over a given signature can be interpreted in any category with finite products.

Definition 1.3.19 Let M be a Σ-structure in a category $\mathcal{C}$ with finite products. Given a term-in-context $\{\vec{x} \, . \, t\}$ over Σ (where $\vec{x} = (x_1^{A_1}, \ldots, x_n^{A_n})$ and $t : B$, say), we inductively define an arrow

$$[[\vec{x}.t]]_M : MA_1 \times \cdots \times MA_n \to MB$$

in $\mathcal{C}$ as follows:

- If t is a variable then it is necessarily x_i for a unique $i \in \{1, \ldots, n\}$, and we set $[[\vec{x}.t]]_M$ equal to the ith product projection

$$\pi_i : MA_1 \times \cdots \times MA_n \to MA_i.$$

- If t is $f(t_1, \ldots, t_m)$ (where $t_i : C_i$, say), then we set $[[\vec{x}.t]]_M$ equal to the composite

$$MA_1 \times \cdots \times MA_n \xrightarrow{<[[\vec{x}.t_1]]_M, \ldots, [[\vec{x}.t_m]]_M>} MC_1 \times \cdots \times MC_m \xrightarrow{Mf} MB$$

In order to interpret first-order formulae in a given category $\mathcal{C}$, we need to have a certain amount of categorical structure present on $\mathcal{C}$ so to give a meaning to the logical connectives and quantifiers which appear in the formulae. For example, to interpret finitary conjunctions, one needs to be able to form pullbacks, while to interpret disjunctions one needs to be able to take unions of subobjects, etc. In fact, the larger the fragment of logic is, the larger the amount of categorical structure is required to interpret it.

Let M be a Σ-structure in a category $\mathcal{C}$ with finite limits. A formula-in-context $\{\vec{x}.\phi\}$ over Σ (where $\vec{x} = x_1, \ldots, x_n$ and $x_i : A_i$, say) will be interpreted as a subobject $[[\vec{x}.\phi]]_M \rightarrowtail MA_1 \times \ldots \times MA_n$ according to the following inductive rules:

- If $\phi(\vec{x})$ is $R(t_1, \ldots, t_m)$ where R is a relation symbol (of type $B_1, \ldots, B_m$, say), then $[[\vec{x}.\phi]]_M$ is the pullback

$$\begin{array}{ccc} [[\vec{x}.\phi]]_M & \longrightarrow & MR \\ \downarrow & & \downarrow \\ MA_1 \times \cdots \times MA_n & \xrightarrow{<[[\vec{x}.t_1]]_M, \ldots, [[\vec{x}.t_m]]_M>} & MB_1 \times \cdots \times MB_m. \end{array}$$

- If $\phi(\vec{x})$ is $(s = t)$, where s and t are terms of sort B, then $[[\vec{x}.\phi]]_M$ is the equalizer of $[[\vec{x}.s]]_M, [[\vec{x}.t]]_M : MA_1 \times \cdots \times MA_n \to MB$.
- If $\phi(\vec{x})$ is $\top$ then $[[\vec{x}.\phi]]_M$ is the top element of $\mathrm{Sub}_{\mathcal{C}}(MA_1 \times \cdots \times MA_n)$.
- If ϕ is $\psi \wedge \chi$ then $[[\vec{x}.\phi]]_M$ is the intersection (i.e. pullback)

$$\begin{array}{ccc} [[\vec{x}.\phi]]_M & \rightarrowtail & [[\vec{x}.\chi]]_M \\ \downarrow & & \downarrow \\ [[\vec{x}.\psi]]_M & \rightarrowtail & MA_1 \times \cdots \times MA_n. \end{array}$$

- If $\phi(\vec{x})$ is $\bot$ and $\mathcal{C}$ is a coherent category then $[[\vec{x}.\phi]]_M$ is the bottom element of $\mathrm{Sub}_{\mathcal{C}}(MA_1 \times \cdots \times MA_n)$.

- If ϕ is $\psi \vee \chi$ and $\mathcal{C}$ is a coherent category then $[[\vec{x}.\phi]]_M$ is the union of the subobjects $[[\vec{x}.\psi]]_M$ and $[[\vec{x}.\chi]]_M$.
- If ϕ is $\psi \Rightarrow \chi$ and $\mathcal{C}$ is a Heyting category then $[[\vec{x}.\phi]]_M$ is the implication $[[\vec{x}.\psi]]_M \Rightarrow [[\vec{x}.\chi]]_M$ in the Heyting algebra $\mathrm{Sub}_{\mathcal{C}}(MA_1 \times \cdots \times MA_n)$ (similarly, the negation $\neg\psi$ is interpreted as the pseudocomplement of $[[\vec{x}.\psi]]_M$).
- If ϕ is $(\exists y)\psi$ where y is of sort B and $\mathcal{C}$ is a regular category then $[[\vec{x}.\phi]]_M$ is the image of the composite

$$[[\vec{x},y.\psi]]_M \longrightarrow MA_1 \times \cdots \times MA_n \times MB \xrightarrow{\pi} MA_1 \times \cdots \times MA_n,$$

where π is the product projection on the first n factors.
- If ϕ is $(\forall y)\psi$ where y is of sort B and $\mathcal{C}$ is a Heyting category then $[[\vec{x}.\phi]]_M$ is $\forall_\pi([[\vec{x},y.\psi]]_M)$, where π is the same projection as above.
- If ϕ is $\bigvee_{i \in I} \phi_i$ and $\mathcal{C}$ is a geometric category then $[[\vec{x}.\phi]]_M$ is the union of the subobjects $[[\vec{x}.\phi_i]]_M$.
- If ϕ is $\bigwedge_{i \in I} \phi_i$ and $\mathcal{C}$ has arbitrary intersections of subobjects then $[[\vec{x}.\phi]]_M$ is the intersection of the subobjects $[[\vec{x}.\phi_i]]_M$.

Definition 1.3.20 Let M be a Σ-structure in a category $\mathcal{C}$ with finite products.
(a) A sequent $\sigma \equiv (\phi \vdash_{\vec{x}} \psi)$ over Σ which is interpretable in $\mathcal{C}$ is said to be *satisfied in M* (and one writes $M \vDash \sigma$) if $[[\vec{x}.\phi]]_M \leq [[\vec{x}.\psi]]_M$ in $\mathrm{Sub}_{\mathcal{C}}(MA_1 \times \cdots \times MA_n)$.
(b) If $\mathbb{T}$ is a theory over Σ whose axioms are interpretable in $\mathcal{C}$, M is said to be a *model* of $\mathbb{T}$ (or a $\mathbb{T}$-*model*) if all the axioms of $\mathbb{T}$ are satisfied in M.
(c) M is said to be a *conservative* model of a theory $\mathbb{T}$ over Σ (within a given fragment of first-order logic) if any sequent over Σ which is satisfied in M is provable in $\mathbb{T}$ and conversely.

The full subcategory of Σ-str$(\mathcal{C})$ whose objects are the models of $\mathbb{T}$ will be denoted by $\mathbb{T}$-mod$(\mathcal{C})$.

Remark 1.3.21 One can easily prove, by induction on the structure of ϕ, that for any geometric formula $\phi(\vec{x}) = \phi(x_1^{A_1}, \ldots, x_n^{A_n})$ over a signature Σ and any Σ-structure homomorphism $h : M \to N$, there is a commutative diagram in $\mathcal{C}$ of the form

$$\begin{array}{ccc} [[\vec{x}.\phi]]_M & \longrightarrow & MA_1 \times \cdots \times MA_n \\ \downarrow & & \downarrow h_{A_1} \times \cdots \times h_{A_n} \\ [[\vec{x}.\phi]]_N & \longrightarrow & NA_1 \times \cdots \times NA_n. \end{array}$$

This is not in general true if ϕ is a first-order non-geometric formula. This motivates the following

Definition 1.3.22 A homomorphism of models of a first-order theory $\mathbb{T}$ in a category $\mathcal{C}$ is said to be an *elementary morphism* if there is a commutative diagram as in Remark 1.3.21 for all the first-order formulae $\phi(\vec{x})$ over the signature of $\mathbb{T}$. The category of $\mathbb{T}$-models in $\mathcal{C}$ and elementary embeddings between them will be denoted by $\mathbb{T}\text{-mod}_e(\mathcal{C})$.

If $\mathcal{C}$ and $\mathcal{D}$ are categories with finite products then any functor $F : \mathcal{C} \to \mathcal{D}$ which preserves finite products and monomorphisms canonically induces a functor $\Sigma\text{-str}(F) : \Sigma\text{-str}(\mathcal{C}) \to \Sigma\text{-str}(\mathcal{D})$ in the obvious way.

We say that a functor $F : \mathcal{C} \to \mathcal{D}$ between two cartesian (resp. regular, coherent, geometric, Heyting) categories is *cartesian* (resp. *regular*, *coherent*, *geometric*, *Heyting*) if it preserves finite limits (resp. finite limits and images, finite limits and images and finite unions of subobjects, finite limits and images and arbitrary unions of subobjects, finite limits and images and Heyting implications between subobjects).

Recall that a functor $F : \mathcal{C} \to \mathcal{D}$ is said to be *conservative* if it reflects isomorphisms (i.e. for any arrow $f : c \to c'$ in $\mathcal{C}$, if $F(f)$ is an isomorphism in $\mathcal{D}$ then f is an isomorphism in $\mathcal{C}$).

Lemma 1.3.23 (Lemma D1.2.13 [55]). *Let $F : \mathcal{C} \to \mathcal{D}$ be a cartesian (resp. regular, coherent, geometric, Heyting) functor between categories of the appropriate kind, M a Σ-structure in $\mathcal{C}$ and σ a sequent over Σ interpretable in $\mathcal{C}$. If $M \vDash \sigma$ in $\mathcal{C}$ then $F(M) \vDash \sigma$ in $\mathcal{D}$. The converse implication holds if F is conservative.*

Proof An easy induction shows that F preserves the interpretations of all formulae-in-context interpretable in the appropriate class of categories; from this the first assertion immediately follows. To prove the second, it suffices to note that a sequent $\sigma \equiv (\phi \vdash_{\vec{x}} \psi)$ is satisfied in M if and only if the canonical subobject $[[\vec{x}\,.\,\phi \wedge \psi]]_M \rightarrowtail [[\vec{x}\,.\,\phi]]_M$ is an isomorphism. $\square$

Lemma 1.3.23 implies the following result:

Theorem 1.3.24 *If $\mathbb{T}$ is a regular (resp. coherent, ...) theory over Σ, then for any regular (resp. coherent, ...) functor $F : \mathcal{C} \to \mathcal{D}$ the functor $\Sigma\text{-str}(F) : \Sigma\text{-str}(\mathcal{C}) \to \Sigma\text{-str}(\mathcal{D})$ defined above restricts to a functor $\mathbb{T}\text{-mod}(F) : \mathbb{T}\text{-mod}(\mathcal{C}) \to \mathbb{T}\text{-mod}(\mathcal{D})$.*

Remark 1.3.25 We shall see in section 1.4.5, by using the concept of syntactic category, that the categorical semantics defined above is *sound* and *complete*, in the sense that the provability of a sequent in a given theory belonging to a fragment of first-order logic is equivalent to its validity in all the models of the theory in categories in which the sequent is interpretable.

The following list of examples shows that the notion of model of a first-order theory in a category captures several important notions of genuine mathematical interest:

Examples 1.3.26

(a) A *topological group* can be seen as a model of the algebraic theory of groups in the category of topological spaces.

More generally, a *topological groupoid* can be seen as a model of the theory $\mathbb{G}$ of groupoids (obtained from the theory $\mathbb{C}$ of small categories considered in Example 1.2.7 by adding a unary function symbol Inv of sort A for formalizing the operation of inverse of an arrow and the obvious axioms for it) in the category of topological spaces.

(b) Similarly, an *algebraic* (resp. *Lie*) *group* is a model of the algebraic theory of groups in the category of algebraic varieties (resp. in the category of smooth manifolds).

(c) A *sheaf of rings* (more generally, a sheaf whose sections are models of a Horn theory $\mathbb{T}$) on a topological space X can be seen as a model of the theory of rings (resp. of the theory $\mathbb{T}$) in the topos $\mathbf{Sh}(X)$ of sheaves on X.

(d) A *sheaf of local rings* on a topological space X (i.e. a sheaf of rings on X whose stalks are all local rings) is precisely a model of the theory of local rings in the topos $\mathbf{Sh}(X)$.

(e) A *family of set-based models* of a geometric theory $\mathbb{T}$ indexed by a set I can be seen as a model of $\mathbb{T}$ in the functor category $[I, \mathbf{Set}]$.

Remark 1.3.27 An important aspect of categorical semantics is that a given piece of syntax can be soundly interpreted in a variety of independent contexts. The concrete results obtained in this way cannot in general be deduced from one another, but arise as different instances of a unique general result lying at the syntactic level. This can be useful in practice to avoid reproving the same results in different contexts, when it is possible to lift them to the syntactic level. For example, the fact that the category of algebraic groups and that of Lie groups have finite products are both immediate consequences of the syntactic property of the theory of groups to be algebraic.

1.3.4 Elementary toposes

The notion of elementary topos was introduced by W. Lawvere and M. Tierney in the late sixties as a generalization of the concept of Grothendieck topos in which it would still be possible to do some sort of abstract sheaf theory and consider models of arbitrary finitary first-order theories.

Definition 1.3.28 (a) An *elementary topos* is a cartesian closed category (i.e. a category with finite products in which exponentials exist for any pair of objects) with finite limits and a subobject classifier.

(b) A *logical functor* between elementary toposes is a functor preserving finite limits, exponentials and the subobject classifier.

As the name suggests, the notion of elementary topos can be formalized elementarily in the first-order language of categories (cf. Example 1.2.7).

By Theorem 1.1.12, every Grothendieck topos is an elementary topos. There are examples of elementary toposes that are not Grothendieck toposes, and which are useful in connection with the logical study of higher-order intuitionistic type theories and realizability. In fact, elementary toposes are exactly the syntactic categories, with respect to logical functors, of higher-order intuitionistic type theories (cf. section D4.3 of [55] for more details).

As shown by Giraud's theorem, an essential feature which distinguishes Grothendieck toposes among general elementary toposes is the fact that they admit separating sets of objects (equivalently, (small) sites of definition). The presence of sites allows one to effectively use geometric intuition when dealing with toposes, and to study first-order mathematical theories of a general specified form (technically speaking, geometric theories, cf. section 2.1.1) through their classifying toposes. On the other hand, when studying aspects of toposes arising from their elementary categorical structure, it is often possible, and even natural, to establish the relevant results in the setting elementary toposes. We shall see various instances of this phenomenon in the book.

The name 'topos' for this kind of categories is justified by the fact that it is possible to lift many natural notions and constructions which apply to Grothendieck toposes to this level of generality. For instance, as we shall see, the notion of Grothendieck topology J on a category $\mathcal{C}$ corresponds to an invariant notion defined at the level of the presheaf topos $[\mathcal{C}^{op}, \mathbf{Set}]$, namely to a *local operator* (also called *Lawvere-Tierney topology*) on it.

By an internal Heyting algebra in a topos $\mathcal{E}$ we mean a model of the algebraic theory of Heyting algebras in $\mathcal{E}$, in the sense of section 1.3.3, that is an object L of $\mathcal{E}$ with arrows $\wedge, \vee, \Rightarrow : L \times L \to L$ and $0, 1 : 1_\mathcal{E} \to L$ which make the diagrams expressing the identities used in the equational definition of a Heyting algebra commutative.

Theorem 1.3.29 *Any elementary topos $\mathcal{E}$ is a Heyting category and the subobject classifier Ω of $\mathcal{E}$ has the structure of an internal Heyting algebra, inducing by the Yoneda lemma a natural structure of Heyting algebra on each subobject lattice in $\mathcal{E}$.*

Sketch of proof Let $\mathcal{E}$ be an elementary topos. One can prove that $\mathcal{E}$ has all finite colimits (cf. section IV.5 of [63]). Using this, one can construct cover-mono factorizations for arrows which are stable under pullback, thereby showing that $\mathcal{E}$ is a regular category. The pullback functors between subobject lattices in $\mathcal{E}$ thus admit left adjoints. The fact that they also admit right adjoints follows from the existence of the cartesian closed structure (cf. the proof of Theorem I.9.4 [63]). The existence of these right adjoints ensures that finite unions of subobjects in $\mathcal{E}$, which exist in $\mathcal{E}$ since $\mathcal{E}$ has finite colimits as well as images, are stable under pullback. Given this, it is not hard to prove that Ω has the structure of an internal Heyting algebra. Specifically, one defines $0 : 1_\mathcal{E} \to \Omega$ to be the classifying arrow of the zero subobject $0 \rightarrowtail 1_\mathcal{E}$, $1 : 1_\mathcal{E} \to \Omega$ to be the arrow $\top$, $\wedge : \Omega \times \Omega \to \Omega$ to be the classifying arrow of the monomorphism $<\top, \top> : 1_\mathcal{E} \to \Omega \times \Omega$, $\Rightarrow : \Omega \times \Omega \to \Omega$ to be the classifying arrow of the equalizer of the arrows $\wedge$ and π_1 and $\vee : \Omega \times \Omega \to \Omega$ to be the classifying arrow of the union of the two subobjects $\pi_1^*(\top)$ and $\pi_2^*(\top)$, where π_1 and π_2 are the two canonical projections $\Omega \times \Omega \to \Omega$ (for more details, see the proof of Lemma A1.6.3 [55]).

It is also immediate to see that the Heyting algebra structure on the subobject lattices in $\mathcal{E}$ is induced by this internal structure via the Yoneda lemma. □

Definition 1.3.30 Let $\mathcal{E}$ be an elementary topos with subobject classifier $\top : 1 \to \Omega$. A *local operator* (or *Lawvere-Tierney topology*) on $\mathcal{E}$ is an arrow $j : \Omega \to \Omega$ in $\mathcal{E}$ such that the three diagrams

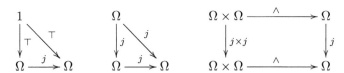

commute (where $\wedge : \Omega \times \Omega \to \Omega$ is the meet operation of the internal Heyting algebra Ω).

Interestingly, the notion of local operator admits several equivalent characterizations.

Definition 1.3.31 (a) A *closure operation* on subobjects in an elementary topos $\mathcal{E}$ is an operation c sending any subobject m in $\mathrm{Sub}_{\mathcal{E}}(A)$ to a subobject $c(m)$ in $\mathrm{Sub}_{\mathcal{E}}(A)$ in such a way that $m \leq c(m)$ and $c(c(m)) = c(m)$ for all m, where $\leq$ is the natural ordering between subobjects.
(b) A *universal closure operation* on an elementary topos $\mathcal{E}$ is a closure operation c on subobjects which commutes with pullback (in particular, intersection) of subobjects.

An important example of a universal closure operation on a topos $\mathcal{E}$ is the *double negation* one, which assigns to any subobject m in $\mathcal{E}$ the subobject $\neg\neg m$ (where the Heyting pseudocomplement $\neg$ is taken in the relevant subobject lattice).

Theorem 1.3.32 *For any elementary topos $\mathcal{E}$, there is a bijection between universal closure operations on $\mathcal{E}$ and local operators on $\mathcal{E}$.*

Sketch of proof The bijection of the theorem sends a universal closure operation c on $\mathcal{E}$ to the local operator $j_c : \Omega \to \Omega$ given by the classifying map of the subobject $c(1 \overset{\top}{\rightarrowtail} \Omega)$, and a local operator j to the closure operation c_j induced by composing classifying arrows with j. For more details, see the proof of Lemma A4.4.2 [55]. □

One can define an abstract notion of c-sheaf for a universal closure operation c on an elementary topos $\mathcal{E}$, as follows.

Definition 1.3.33 Let $\mathcal{E}$ be an elementary topos and c a universal closure operation on $\mathcal{E}$. An object A of $\mathcal{E}$ is said to be a *c-sheaf* if for any monomorphism

$m : B' \rightarrowtail B$ in $\mathcal{E}$ which is c-dense (i.e. such that $c(m) = 1_B$) and any arrow $f' : B' \to A$ in $\mathcal{E}$, there exists exactly one arrow $f : B \to A$ such that $f \circ m = f'$. The full subcategory of $\mathcal{E}$ on the c-sheaves will be denoted by $\mathbf{sh}_c(\mathcal{E})$.

Definition 1.3.34 A *reflector* L on a topos $\mathcal{E}$ is the left adjoint functor of a reflection (i.e. full embedding admitting a left adjoint) of a full subcategory of $\mathcal{E}$ into $\mathcal{E}$. A reflector is said to be *cartesian* if it preserves finite limits.

For a proof of the following result the reader is referred to sections A4.3 and A4.4 of [55].

Theorem 1.3.35 *For any local operator j on an elementary (resp. Grothendieck) topos $\mathcal{E}$, $\mathbf{sh}_{c_j}(\mathcal{E})$ is an elementary (resp. Grothendieck) topos, and the inclusion $\mathbf{sh}_{c_j}(\mathcal{E}) \hookrightarrow \mathcal{E}$ has a left adjoint $a_j : \mathcal{E} \to \mathbf{sh}_{c_j}(\mathcal{E})$ which preserves finite limits. In fact, local operators on $\mathcal{E}$ correspond bijectively to (isomorphism classes of) geometric inclusions to $\mathcal{E}$, and to (isomorphism classes of) cartesian reflectors on $\mathcal{E}$.*

Sketch of proof A closure operation c_j can be recovered from the associated cartesian reflector L_j by means of the pullbacks

$$\begin{array}{ccc} c_j(A') & \longrightarrow & L_j A' \\ {\scriptstyle c_j(m)} \downarrow & & \downarrow {\scriptstyle L_j m} \\ A & \xrightarrow{\eta_A} & L_j A, \end{array}$$

where $m : A' \rightarrowtail A$ is an arbitrary monomorphism in $\mathcal{E}$ and η is the unit of the reflection induced by L_j. $\square$

Theorem 1.3.36 *Let $\mathcal{C}$ be a small category. Then the Grothendieck topologies J on $\mathcal{C}$ correspond exactly to the local operators on the presheaf topos $[\mathcal{C}^{\mathrm{op}}, \mathbf{Set}]$. In fact, if J is the Grothendieck topology corresponding to a local operator j, an object of $[\mathcal{C}^{\mathrm{op}}, \mathbf{Set}]$ is a J-sheaf (in the sense of Definition 1.1.5(b)) if and only if it is a c_j-sheaf (in the sense of Definition 1.3.33).*

Sketch of proof The correspondence of the theorem sends a local operator $j : \Omega \to \Omega$ to the subobject $J \rightarrowtail \Omega$ which it classifies, that is to the Grothendieck topology J on $\mathcal{C}$ defined by

$$S \in J(c) \text{ if and only if } j(c)(S) = M_c.$$

In the other direction, it sends a Grothendieck topology J, regarded as a subobject $J \rightarrowtail \Omega$, to the arrow $j : \Omega \to \Omega$ that classifies it.

For more details, see section V.4 of [63]. $\square$

1.3.5 Toposes as mathematical universes

We have seen in section 1.3.1.1 that, given a category $\mathcal{C}$ with finite products, there is a first-order signature $\Sigma_\mathcal{C}$, called the internal language of $\mathcal{C}$, which can be profitably used for reasoning about $\mathcal{C}$ in a set-theoretic fashion, that is by using 'elements'.

If $\mathcal{C}$ is an elementary topos, we can extend the internal language by allowing the formation of formulae of the kind $\tau \in \Gamma$, where τ is a term of sort $\ulcorner A \urcorner$ and Γ is a term of sort $\ulcorner \Omega^A \urcorner$. Indeed, we may interpret such a formula as the subobject whose classifying arrow is the composite

$$W \xrightarrow{<\tau,\Gamma>} A \times \Omega^A \xrightarrow{\in_A} \Omega,$$

where W is the product of (the objects representing the) sorts of the variables occurring either in τ or in Γ (considered without repetitions), $<\tau, \Gamma>$ is the induced map to the product and $\in_A$ is the transpose of the identity arrow on Ω^A along the adjunction between $(-)^A$ and $A \times -$. Note that an object A of $\mathcal{C}$ gives rise to a constant term of type $\ulcorner \Omega^A \urcorner$. In fact, Ω^A behaves like the power set of A in $\mathcal{C}$.

We may also allow the formation of terms of type $t(w)$, where t is a term of type $\ulcorner B^A \urcorner$ and w is a term of type $\ulcorner A \urcorner$; indeed, we may interpret $t(w)$ as the composite $\mathrm{ev}_{A,B} \circ <t, w>$, where $\mathrm{ev}_{A,B}$ is the evaluation arrow $B^A \times A \to B$.

Thus in a topos we can also interpret all the common formulae that we use in set theory.

Note that, the logic of a topos being at least intuitionistic (cf. Theorems 1.3.29 and 1.4.15), for any topos $\mathcal{E}$, any formal proof involving first-order sequents over the signature $\Sigma_\mathcal{E}$ will be sound when interpreted in the tautological structure $S_\mathcal{E}$ provided that the law of excluded middle or any other non-constructive principles are not employed in it. In other words, the existence of the internal language of a topos and the corresponding tautological structure justifies the use of the standard set-theoretic intuition when working 'inside' a topos provided that one avoids invoking non-constructive principles.

An example of a reformulation of basic properties of sets in the internal language of a topos is provided by the following proposition.

Proposition 1.3.37 (cf. Lemma D1.3.11 [55]). *Let $\mathcal{E}$ be a topos. Then*

(i) *An arrow $f : A \to A$ of $\mathcal{E}$ is the identity on A if and only if $(\top \vdash_x \ulcorner f \urcorner(x) = x)$ holds in $S_\mathcal{E}$.*

(ii) *An arrow $f : A \to C$ of $\mathcal{E}$ is the composite of two arrows $g : A \to B$ and $h : B \to C$ if and only if $(\top \vdash_x \ulcorner f \urcorner(x) = \ulcorner h \urcorner(\ulcorner g \urcorner(x)))$ holds in $S_\mathcal{E}$.*

(iii) *An arrow $f : A \to B$ of $\mathcal{E}$ is a monomorphism if and only if $(\ulcorner f \urcorner(x) = \ulcorner f \urcorner(x') \vdash_{x,x'} x = x')$ holds in $S_\mathcal{E}$.*

(iv) *An arrow $f : A \to B$ of $\mathcal{E}$ is an epimorphism if and only if $(\top \vdash_y (\exists x)(\ulcorner f \urcorner(x) = y))$ holds in $S_\mathcal{E}$.*

(v) *An object A of $\mathcal{E}$ is terminal if and only if $(\top \vdash (\exists x)\top)$ and $(\top \vdash_{x,x'} (x = x'))$ hold in $\mathcal{S}_\mathcal{E}$ (where x and x' are variables of sort $\ulcorner A \urcorner$).*

1.3.5.1 Validity in an elementary topos

Definition 1.3.38 We say that a first-order formula $\phi(\vec{x})$ over a signature Σ is *valid* in an elementary topos $\mathcal{E}$ if for every Σ-structure M in $\mathcal{E}$ the sequent $(\top \vdash_{\vec{x}} \phi)$ is satisfied in M.

Theorem 1.3.39 *Let Σ be a signature and $\phi(\vec{x})$ a first-order formula over Σ. Then $\phi(\vec{x})$ is provable in intuitionistic (finitary) first-order logic if and only if it is valid in every elementary topos.*

Sketch of proof The soundness part follows from Theorems 1.3.29 and 1.4.15. The completeness part follows from the existence of canonical Kripke models and the fact that, given a poset P and a Kripke model $\mathcal{U}$ on P, there is a model $\mathcal{U}^*$ in the topos $[P, \mathbf{Set}]$ such that the first-order sequents valid in $\mathcal{U}$ are exactly those which are valid in $\mathcal{U}^*$. For more details, see section 8.4 of [46]. □

An elementary topos can thus be considered as a 'mathematical universe' in which one can do mathematics similarly to how one does it in the classical context of sets, with the only proviso being that one must argue constructively. This constructivity requirement should not be merely seen as a limitation, but rather as an opportunity for a more refined development of mathematics allowing one to discriminate between different notions that 'collapse' to equivalent ones in classical logic.

1.3.5.2 Kripke-Joyal semantics

Kripke-Joyal semantics represents the analogue for toposes of the usual Tarskian notion of satisfaction of a first-order formula by a tuple of elements of a set-theoretic structure in which the given formula is interpretable. It clearly makes no sense to speak of elements of a structure in a topos, but we can replace the classical notion of element of a set with that of *generalized element* of an object. A generalized element of an object B of a topos $\mathcal{E}$ is simply an arrow $\alpha : E \to B$ with codomain B. Note that, by the Yoneda lemma, an object is determined, up to isomorphism, by the collection of all its generalized elements (represented by the functor $\mathrm{Hom}_\mathcal{E}(-, B)$).

Definition 1.3.40 Let $\mathcal{E}$ be an elementary topos and $\mathcal{S}_\mathcal{E}$ the tautological $\Sigma_\mathcal{E}$-structure in $\mathcal{E}$. Given a first-order formula $\phi(x)$ in a variable x of sort $\ulcorner A \urcorner$ and a generalized element $\alpha : E \to A$ of A, we stipulate that

$$E \models \phi(\alpha) \text{ if and only if } \alpha \text{ factors through } [[x \,.\, \phi]]_{\mathcal{S}_\mathcal{E}} \rightarrowtail A.$$

Theorem 1.3.41 *Under the hypotheses of Definition 1.3.40, let $\alpha : E \to A$ be a generalized element of A and $\phi(x), \psi(x)$ formulae with a free variable x of sort $\ulcorner A \urcorner$. Then*

- $E \models (\phi \wedge \psi)(\alpha)$ if and only if $E \models \phi(\alpha)$ and $E \models \psi(\alpha)$.
- $E \models (\phi \vee \psi)(\alpha)$ if and only if there are jointly epimorphic arrows $p : E' \to E$ and $q : E'' \to E$ in $\mathcal{E}$ such that both $E' \models \phi(\alpha \circ p)$ and $E'' \models \psi(\alpha \circ q)$.
- $E \models (\phi \Rightarrow \psi)(\alpha)$ if and only if, for any arrow $p : E' \to E$ of $\mathcal{E}$ such that $E' \models \phi(\alpha \circ p)$, $E' \models \psi(\alpha \circ p)$.
- $E \models (\neg \phi)(\alpha)$ if and only if, whenever $p : E' \to E$ is an arrow of $\mathcal{E}$ such that $E' \models \phi(\alpha \circ p)$, $E' \cong 0_{\mathcal{E}}$.

If $\phi(x, y)$ has an additional free variable y of sort $\ulcorner B \urcorner$ then

- $E \models (\exists y)\phi(\alpha, y)$ if and only if there exist an epimorphism $p : E' \to E$ and a generalized element $\beta : E' \to B$ such that $E' \models \phi(\alpha \circ p, \beta)$.
- $E \models (\forall y)\phi(\alpha, y)$ if and only if, for every object E' of $\mathcal{E}$, every arrow $p : E' \to E$ of $\mathcal{E}$ and every generalized element $\beta : E' \to B$, $E' \models \phi(\alpha \circ p, \beta)$.

Of course, both Definition 1.3.40 and Theorem 1.3.41 can be generalized to the case of formulae with an arbitrary (finite) number of free variables. For a proof of Theorem 1.3.41 we refer the reader to section 6.6 of [9].

Remark 1.3.42 If $\mathcal{E}$ is a Grothendieck topos then infinitary formulae can also be interpreted in it. The infinitary version of the second point of the theorem reads as follows: $E \models (\bigvee_{i \in I} \phi_i)(\alpha)$ if and only if there is an epimorphic family of arrows $\{p_i : E_i \to E \mid i \in I\}$ in $\mathcal{E}$ such that $E_i \models \phi(\alpha \circ p_i)$ for all $i \in I$.

1.4 Syntactic categories

In this section we shall introduce the fundamental notion of *syntactic category* of a theory within a given fragment of first-order logic.

1.4.1 Definition

Let $\phi(\vec{x})$ and $\psi(\vec{y})$ be two formulae-in-context over a first-order signature, where $\vec{x} = (x_1, \ldots, x_n)$ and $\vec{y} = (y_1, \ldots, y_n)$ are contexts of the same type and length. We say that $\phi(\vec{x})$ and $\psi(\vec{y})$ are α-equivalent if $\psi(\vec{y})$ is obtained from $\phi(\vec{x})$ by (possibly renaming bound variables in ϕ and) replacing every free occurrence of x_i in ϕ by y_i. We write $\{\vec{x} \,.\, \phi\}$ for the α-equivalence class of the formula-in-context $\phi(\vec{x})$.

Definition 1.4.1 (a) Let $\mathbb{T}$ be a geometric theory over a signature Σ. The *syntactic category* $\mathcal{C}_{\mathbb{T}}$ of $\mathbb{T}$ has as objects the α-equivalence classes of geometric formulae-in-context $\{\vec{x} \,.\, \phi\}$ over Σ and as arrows $\{\vec{x} \,.\, \phi\} \to \{\vec{y} \,.\, \psi\}$ (where the contexts $\vec{x}$ and $\vec{y}$ are supposed to be disjoint without loss of generality) the $\mathbb{T}$-provable-equivalence classes $[\theta]$ of geometric formulae $\theta(\vec{x}, \vec{y})$ which are $\mathbb{T}$-provably functional, i.e. such that the sequents

$$(\phi \vdash_{\vec{x}} (\exists \vec{y})\theta),$$
$$(\theta \vdash_{\vec{x}, \vec{y}} \phi \wedge \psi), \text{ and}$$
$$((\theta \wedge \theta[\vec{z}/\vec{y}]) \vdash_{\vec{x}, \vec{y}, \vec{z}} (\vec{y} = \vec{z}))$$

are provable in $\mathbb{T}$.

The composite of two arrows

$$\{\vec{x}.\phi\} \xrightarrow{[\theta]} \{\vec{y}.\psi\} \xrightarrow{[\gamma]} \{\vec{z}.\chi\}$$

is defined as the $\mathbb{T}$-provable-equivalence class of the formula $(\exists \vec{y})(\theta \wedge \gamma)$. The identity arrow on an object $\{\vec{x}.\phi\}$ is the arrow

$$\{\vec{x}.\phi\} \xrightarrow{[\phi \wedge \vec{x}'=\vec{x}]} \{\vec{x}'.\phi[\vec{x}'/\vec{x}]\} \,.$$

(b) We define the *cartesian* (resp. *regular*, *coherent*, *first-order*) *syntactic category* $\mathcal{C}_{\mathbb{T}}^{\text{cart}}$ (resp. $\mathcal{C}_{\mathbb{T}}^{\text{reg}}$, $\mathcal{C}_{\mathbb{T}}^{\text{coh}}$, $\mathcal{C}_{\mathbb{T}}^{\text{fo}}$) of a cartesian (resp. regular, coherent, first-order) theory $\mathbb{T}$ by replacing the word 'geometric' with '$\mathbb{T}$-cartesian' (resp. 'regular', 'coherent', 'first-order') in Definition 1.4.1(a).

(c) We define the *classical syntactic category* $\mathcal{C}_{\mathbb{T}}^{\text{fo-cl}}$ of a first-order theory $\mathbb{T}$ by replacing, in Definition 1.4.1(a), the word 'geometric' with 'first-order' and the notion of (intuitionistic) provability with that of provability in classical first-order logic.

(d) Given a universal Horn theory $\mathbb{T}$, we define the *algebraic syntactic category* $\mathcal{C}_{\mathbb{T}}^{\text{alg}}$ of $\mathbb{T}$ as the category whose objects are the finite conjunctions of atomic formulae-in-context (up to α-equivalence) over the signature of $\mathbb{T}$ and whose arrows $\{\vec{x}.\phi\} \to \{\vec{y}.\psi\}$ (where the contexts $\vec{x} = (x_1, \ldots, x_n)$ and $\vec{y} = (y_1, \ldots, y_m)$ are supposed to be disjoint, without loss of generality) are sequences of terms $\vec{t} = (t_1(\vec{x}), \ldots, t_m(\vec{x}))$ such that the sequent $(\phi \vdash_{\vec{x}} \psi(t_1(\vec{x}), \ldots, t_m(\vec{x})))$ is provable in $\mathbb{T}$, modulo the equivalence relation which identifies two such sequences $\vec{t}$ and $\vec{t}'$ precisely when the sequent $(\phi \vdash_{\vec{x}} t_1(\vec{x}) = t_1'(\vec{x}) \wedge \cdots \wedge t_m(\vec{x}) = t_m'(\vec{x}))$ is provable in $\mathbb{T}$.

(e) We shall say that two geometric formulae-in-context $\{\vec{x}.\phi\}$ and $\{\vec{y}.\psi\}$, where $\vec{x}$ and $\vec{y}$ are disjoint, are $\mathbb{T}$-*equivalent* if they are isomorphic as objects of the syntactic category $\mathcal{C}_{\mathbb{T}}$, that is if there exists a geometric formula $\theta(\vec{x}, \vec{y})$ which is $\mathbb{T}$-provably functional from $\{\vec{x}.\phi\}$ to $\{\vec{y}.\psi\}$ and which moreover satisfies the property that the sequents $(\theta \wedge \theta[\vec{x}'/\vec{x}] \vdash_{\vec{x},\vec{x}',\vec{y}} \vec{x} = \vec{x}')$ and $(\psi \vdash_{\vec{y}} (\exists \vec{x})\theta)$ are provable in $\mathbb{T}$.

(f) We shall say that two geometric formulae $\{\vec{x}.\phi\}$ and $\{\vec{x}.\psi\}$ in the same context are $\mathbb{T}$-*provably equivalent*, and we write $\phi \dashv\vdash_{\mathbb{T}} \psi$, if both the sequents $(\phi \vdash_{\vec{x}} \psi)$ and $(\psi \vdash_{\vec{x}} \phi)$ are provable in $\mathbb{T}$.

Lemma 1.4.2 (Lemma D1.4.4(iv) [55]). *Any subobject of $\{\vec{x}.\phi\}$ in $\mathcal{C}_{\mathbb{T}}$ is isomorphic to one of the form*

$$\{\vec{x}'.\psi[\vec{x}'/\vec{x}]\} \xrightarrow{[\psi \wedge \vec{x}'=\vec{x}]} \{\vec{x}.\phi\},$$

where $\psi(\vec{x})$ is a geometric formula in the context $\vec{x}$ such that the sequent $(\psi \vdash_{\vec{x}} \phi)$ is provable in $\mathbb{T}$. We will denote this subobject simply by $[\psi]$.

Moreover, for two such subobjects $[\psi]$ and $[\chi]$ we have that $[\psi] \leq [\chi]$ in $\mathrm{Sub}_{\mathcal{C}_\mathbb{T}}(\{\vec{x}.\phi\})$ if and only if the sequent $(\psi \vdash_{\vec{x}} \chi)$ is provable in $\mathbb{T}$.

Theorem 1.4.3

(i) *For any cartesian theory $\mathbb{T}$, $\mathcal{C}_\mathbb{T}^{\mathrm{cart}}$ is a cartesian category.*
(ii) *For any regular theory $\mathbb{T}$, $\mathcal{C}_\mathbb{T}^{\mathrm{reg}}$ is a regular category.*
(iii) *For any coherent theory $\mathbb{T}$, $\mathcal{C}_\mathbb{T}^{\mathrm{coh}}$ is a coherent category.*
(iv) *For any first-order theory $\mathbb{T}$, $\mathcal{C}_\mathbb{T}^{\mathrm{fo}}$ is a Heyting category.*
(v) *For any geometric theory $\mathbb{T}$, $\mathcal{C}_\mathbb{T}$ is a geometric category.*
(vi) *For any first-order theory $\mathbb{T}$, $\mathcal{C}_\mathbb{T}^{\mathrm{fo\text{-}cl}}$ is a Boolean coherent category.*

Sketch of proof The structure of subobject lattices in syntactic categories directly reflects the structure of the formulae in the relevant fragment of logic, via the identification provided by Lemma 1.4.2; for instance, unions of subobjects correspond to disjunctions, intersections to conjunctions, top element of a lattice $\mathrm{Sub}(\{\vec{x}.\phi\})$ to the associated formula $\{\vec{x}.\phi\}$ (regarded as the identical subobject $[\phi]$) and bottom element of a lattice $\mathrm{Sub}(\{\vec{x}.\phi\})$, if it exists, to the formula $\{\vec{x}.\bot\}$. Pullbacks correspond in a natural way to substitutions, and cover-mono factorizations of morphisms to existential quantifications (the cover-mono factorization of a morphism $[\theta(\vec{x},\vec{y})] : \{\vec{x}.\phi\} \to \{\vec{y}.\psi\}$ is given by the canonical arrows $\{\vec{x}.\phi\} \twoheadrightarrow \{\vec{y}.(\exists \vec{x})\theta(\vec{x},\vec{y})\} \rightarrowtail \{\vec{y}.\psi\}$). For more details, we refer the reader to the proof of Lemma D1.4.10 [55]. □

Definition 1.4.4 Let $\mathbb{T}$ be a geometric theory over a signature Σ. The *universal model* of $\mathbb{T}$ in $\mathcal{C}_\mathbb{T}$ is the structure $M_\mathbb{T}$ which assigns

- to a sort A the object $\{x^A.\top\}$ where x^A is a variable of sort A;
- to a function symbol $f : A_1 \cdots A_n \to B$ the morphism

$$\{x_1^{A_1}, \ldots, x_n^{A_n}.\top\} \xrightarrow{[f(x_1^{A_1},\ldots,x_n^{A_n})=y^B]} \{y^B.\top\};$$

- to a relation symbol $R \rightarrowtail A_1 \cdots A_n$ the subobject

$$\{x_1^{A_1}, \ldots, x_n^{A_n}.R(x_1^{A_1}, \ldots, x_n^{A_n})\} \xrightarrow{[R(x_1^{A_1},\ldots,x_n^{A_n})]} \{x_1^{A_1}, \ldots, x_n^{A_n}.\top\}.$$

The attribute 'universal' for the structure $M_\mathbb{T}$ is justified by Theorem 1.4.10 below.

Theorem 1.4.5 *Let $\mathbb{T}$ be a geometric theory over a signature Σ. Then*

(i) *For any geometric formula-in-context $\{\vec{x}.\phi\}$ over Σ, its interpretation $[[\vec{x}.\phi]]_{M_\mathbb{T}}$ in $M_\mathbb{T}$ is the subobject $[\phi] : \{\vec{x}.\phi\} \rightarrowtail \{\vec{x}.\top\}$.*
(ii) *A geometric sequent $(\phi \vdash_{\vec{x}} \psi)$ over Σ is satisfied in $M_\mathbb{T}$ if and only if it is provable in $\mathbb{T}$.*

Sketch of proof The first part of the theorem can be easily proved by induction on the structure of ϕ, while the second follows from the first by using Lemma 1.4.2. □

Remark 1.4.6 Definition 1.4.4 and Theorem 1.4.5 admit obvious variants for the other fragments of first-order logic, namely cartesian, regular, coherent, intuitionistic and classical first-order logic.

1.4.2 Syntactic sites

The idea behind the notion of syntactic site is that the property of a functor to preserve the logical structure on a given syntactic category can be interpreted as a form of continuity with respect to a Grothendieck topology naturally defined on it. Before introducing such topologies, a few remarks are in order.

Recall that every arrow in a regular category can be factored uniquely as a cover followed by a monomorphism, and that covers are precisely the arrows g such that $\mathrm{Im}(g) = 1_{\mathrm{cod}(g)}$. In a coherent (resp. geometric) category, a finite (resp. small) *covering family* is a family of arrows such that the union of their images is the maximal subobject.

Definition 1.4.7 (a) For a regular category $\mathcal{C}$, the *regular topology* $J_{\mathcal{C}}^{\mathrm{reg}}$ is the Grothendieck topology on $\mathcal{C}$ whose covering sieves are those which contain a cover.

(b) For a coherent category $\mathcal{C}$, the *coherent topology* $J_{\mathcal{C}}^{\mathrm{coh}}$ is the Grothendieck topology on $\mathcal{C}$ whose covering sieves are those which contain finite covering families.

(c) For a geometric category $\mathcal{C}$, the *geometric topology* $J_{\mathcal{C}}^{\mathrm{geom}}$ is the Grothendieck topology on $\mathcal{C}$ whose covering sieves are those which contain small covering families.

Proposition 1.4.8 *Let $\mathcal{C}$ and $\mathcal{D}$ be regular (resp. coherent, geometric) categories. Then a cartesian functor is regular (resp. coherent, geometric) if and only if it sends $J_{\mathcal{C}}^{\mathrm{reg}}$-covering (resp. $J_{\mathcal{C}}^{\mathrm{coh}}$-covering, $J_{\mathcal{C}}^{\mathrm{geom}}$-covering) sieves to covering families.*

Remark 1.4.9 The Grothendieck topologies $J_{\mathcal{C}}^{\mathrm{reg}}, J_{\mathcal{C}}^{\mathrm{coh}}$ and $J_{\mathcal{C}}^{\mathrm{geom}}$ are all subcanonical, whence for any regular (resp. coherent, geometric) category we have a Yoneda embedding $y^{\mathrm{reg}} : \mathcal{C} \hookrightarrow \mathbf{Sh}(\mathcal{C}, J_{\mathcal{C}}^{\mathrm{reg}})$ (resp. $y^{\mathrm{coh}} : \mathcal{C} \hookrightarrow \mathbf{Sh}(\mathcal{C}, J_{\mathcal{C}}^{\mathrm{coh}})$, $y^{\mathrm{geom}} : \mathcal{C} \hookrightarrow \mathbf{Sh}(\mathcal{C}, J_{\mathcal{C}}^{\mathrm{geom}}))$.

We shall denote by $J_{\mathbb{T}}^{\mathrm{reg}}$ (resp. by $J_{\mathbb{T}}^{\mathrm{coh}}$, by $J_{\mathbb{T}}$) the regular (resp. coherent, geometric) topology on the regular (resp. coherent, geometric) category $\mathcal{C}_{\mathbb{T}}^{\mathrm{reg}}$ (resp. $\mathcal{C}_{\mathbb{T}}^{\mathrm{coh}}$, $\mathcal{C}_{\mathbb{T}}$) of a regular (resp. coherent, geometric) theory $\mathbb{T}$ and refer to them as the *syntactic topologies* on the relevant syntactic categories.

We shall denote by $\mathbf{Cart}(\mathcal{C}_{\mathbb{T}}^{\mathrm{cart}}, \mathcal{D})$ (resp. $\mathbf{Reg}(\mathcal{C}_{\mathbb{T}}^{\mathrm{reg}}, \mathcal{D})$, $\mathbf{Coh}(\mathcal{C}_{\mathbb{T}}^{\mathrm{coh}}, \mathcal{D})$, $\mathbf{Geom}(\mathcal{C}_{\mathbb{T}}, \mathcal{D})$, $\mathbf{Heyt}(\mathcal{C}_{\mathbb{T}}^{\mathrm{fo}}, \mathcal{D})$, $\mathbf{Bool}(\mathcal{C}_{\mathbb{T}}^{\mathrm{fo\text{-}cl}}, \mathcal{D})$) the category of cartesian (resp. regular, coherent, geometric, Heyting, coherent) functors from $\mathcal{C}_{\mathbb{T}}^{\mathrm{cart}}$ (resp. $\mathcal{C}_{\mathbb{T}}^{\mathrm{reg}}$, $\mathcal{C}_{\mathbb{T}}^{\mathrm{coh}}$, $\mathcal{C}_{\mathbb{T}}$, $\mathcal{C}_{\mathbb{T}}^{\mathrm{fo}}$, $\mathcal{C}_{\mathbb{T}}^{\mathrm{fo\text{-}cl}}$) to a cartesian (resp. regular, coherent, geometric, Heyting, Boolean coherent) category $\mathcal{D}$.

1.4.3 Models as functors

The importance of syntactic categories lies in the fact that they allow us to associate with a theory (in the sense of axiomatic presentation), which is a 'linguistic', unstructured kind of entity, a well-structured mathematical object whose 'geometry' embodies the syntactic aspects of the theory. A most notable fact is that the models of the theory can be recovered as functors defined on its syntactic category which respect the 'logical' structure on it. More specifically, we have the following

Theorem 1.4.10

(i) *For any cartesian theory $\mathbb{T}$ and cartesian category $\mathcal{D}$, we have an equivalence of categories* $\mathbf{Cart}(\mathcal{C}_{\mathbb{T}}^{\mathrm{cart}}, \mathcal{D}) \simeq \mathbb{T}\text{-mod}(\mathcal{D})$ *natural in $\mathcal{D}$.*

(ii) *For any regular theory $\mathbb{T}$ and regular category $\mathcal{D}$, we have an equivalence of categories* $\mathbf{Reg}(\mathcal{C}_{\mathbb{T}}^{\mathrm{reg}}, \mathcal{D}) \simeq \mathbb{T}\text{-mod}(\mathcal{D})$ *natural in $\mathcal{D}$.*

(iii) *For any coherent theory $\mathbb{T}$ and coherent category $\mathcal{D}$, we have an equivalence of categories* $\mathbf{Coh}(\mathcal{C}_{\mathbb{T}}^{\mathrm{coh}}, \mathcal{D}) \simeq \mathbb{T}\text{-mod}(\mathcal{D})$ *natural in $\mathcal{D}$.*

(iv) *For any geometric theory $\mathbb{T}$ and geometric category $\mathcal{D}$, we have an equivalence of categories* $\mathbf{Geom}(\mathcal{C}_{\mathbb{T}}, \mathcal{D}) \simeq \mathbb{T}\text{-mod}(\mathcal{D})$ *natural in $\mathcal{D}$.*

(v) *For any finitary first-order theory $\mathbb{T}$ and Heyting category $\mathcal{D}$, we have an equivalence of categories* $\mathbf{Heyt}(\mathcal{C}_{\mathbb{T}}^{\mathrm{fo\text{-}cl}}, \mathcal{D}) \simeq \mathbb{T}\text{-mod}_e(\mathcal{D})$ *natural in $\mathcal{D}$.*

(vi) *For any finitary first-order theory $\mathbb{T}$ and Boolean category $\mathcal{D}$, we have an equivalence of categories* $\mathbf{Bool}(\mathcal{C}_{\mathbb{T}}^{\mathrm{fo\text{-}cl}}, \mathcal{D}) \simeq \mathbb{T}\text{-mod}_e(\mathcal{D})$ *natural in $\mathcal{D}$.*

Sketch of proof One half of the equivalence sends a model $M \in \mathbb{T}\text{-mod}(\mathcal{D})$ to the functor $F_M : \mathcal{C}_{\mathbb{T}} \to \mathcal{D}$ assigning to a formula $\{\vec{x}.\phi\}$ (the domain of) its interpretation $[[\vec{x}.\phi]]_M$ in M.

The other half of the equivalence sends a functor $F : \mathcal{C}_{\mathbb{T}} \to \mathcal{D}$ to the image $F(M_{\mathbb{T}})$ of the universal model $M_{\mathbb{T}}$ of $\mathbb{T}$ (cf. Definition 1.4.4) under F.

For more details, see the proof of Theorem D1.4.7 [55]. □

Remark 1.4.11 By Theorem 1.4.3, the properties of Theorem 1.4.10 characterize syntactic categories up to equivalence, as representing objects for the (2-)functors $\mathcal{D} \to \mathbb{T}\text{-mod}(\mathcal{D})$ (resp. $\mathcal{D} \to \mathbb{T}\text{-mod}_e(\mathcal{D})$ in the case of intuitionistic and classical first-order syntactic categories).

The concept of syntactic category of a first-order theory also allows us to formalize the idea of a 'dictionary' between two theories providing a means of translating formulae in the language of the former into formulae in the language of the latter in such a way as to induce functors (resp. an equivalence) between their categories of models (inside categories possessing the required categorical structure). This is realized by the notions of interpretation (resp. bi-interpretation) of one theory into another.

Definition 1.4.12 Within a given fragment of first-order logic, an *interpretation* (resp. a *bi-interpretation*) of a theory $\mathbb{T}$ into a theory $\mathbb{S}$ is a functor (resp.

an equivalence) between their respective syntactic categories which respects the logical structure on them.

Notice that any interpretation $I : \mathcal{C}_\mathbb{T} \to \mathcal{C}_\mathbb{S}$ induces a functor $s_I : \mathbb{S}\text{-mod}(\mathcal{D}) \to \mathbb{T}\text{-mod}(\mathcal{D})$, for any 'logical' category $\mathcal{D}$ of the appropriate kind. However, it is not true that every functor $\mathbb{S}\text{-mod}(\mathcal{D}) \to \mathbb{T}\text{-mod}(\mathcal{D})$ is induced by an interpretation.

1.4.4 Categories with 'logical structure' as syntactic categories

Theorem 1.4.13 *Any small cartesian (resp. regular, coherent, geometric) category is, up to categorical equivalence, the cartesian (resp. regular, coherent, geometric) syntactic category of some cartesian (resp. regular, coherent, geometric) theory.*

Proof Let $\mathbb{T}^\mathcal{C}$ be the theory over the internal language $\Sigma^{\mathrm{f}}_\mathcal{C}$ of $\mathcal{C}$ (cf. section 1.3.1.1) consisting of the following sequents:

$$(\top \vdash_x (\ulcorner f \urcorner(x) = x))$$

for any identity arrow f in $\mathcal{C}$;

$$(\top \vdash_x (\ulcorner f \urcorner(x) = \ulcorner h \urcorner(\ulcorner g \urcorner(x))))$$

for any triple of arrows f, g, h of $\mathcal{C}$ such that f is equal to the composite $h \circ g$;

$$(\top \vdash_{[]} (\exists x)\top) \text{ and } (\top \vdash_{x,x'} (x = x'))$$

where x and x' are of sort $\ulcorner 1 \urcorner$, 1 being the terminal object of $\mathcal{C}$;

$$(\top \vdash_x (\ulcorner h \urcorner(\ulcorner f \urcorner(x)) = \ulcorner k \urcorner(\ulcorner g \urcorner(x)))),$$
$$(((\ulcorner f \urcorner(x) = \ulcorner f \urcorner(x')) \wedge (\ulcorner g \urcorner(x) = \ulcorner g \urcorner(x'))) \vdash_{x,x'} (x = x')), \text{ and}$$
$$((\ulcorner h \urcorner(y) = \ulcorner k \urcorner(z)) \vdash_{y,z} (\exists x)((\ulcorner f \urcorner(x) = y) \wedge (\ulcorner g \urcorner(x) = z)))$$

for any pullback square

$$\begin{array}{ccc} a & \xrightarrow{f} & b \\ {\scriptstyle g}\downarrow & & \downarrow{\scriptstyle h} \\ c & \xrightarrow{k} & d \end{array}$$

in $\mathcal{C}$.

It is immediate to see that for any cartesian category $\mathcal{D}$, $\mathbb{T}^\mathcal{C}$-models are 'the same thing' as cartesian functors $\mathcal{C} \to \mathcal{D}$. So we have an equivalence of categories $\mathbb{T}^\mathcal{C}\text{-mod}(\mathcal{D}) \simeq \mathbf{Cart}(\mathcal{C}, \mathcal{D})$ natural in $\mathcal{D} \in \mathbf{Cart}$. Since we also have an equivalence $\mathbf{Cart}(\mathcal{C}^{\mathrm{cart}}_{\mathbb{T}^\mathcal{C}}, \mathcal{D}) \simeq \mathbb{T}^\mathcal{C}\text{-mod}(\mathcal{D})$ natural in $\mathcal{D} \in \mathbf{Cart}$ (by Theorem 1.4.10), by composing the two we find an equivalence $\mathbf{Cart}(\mathcal{C}, \mathcal{D}) \simeq \mathbf{Cart}(\mathcal{C}^{\mathrm{cart}}_{\mathbb{T}^\mathcal{C}}, \mathcal{D})$ natural in $\mathcal{D} \in \mathbf{Cart}$ and hence, by the (2-dimensional) Yoneda lemma, an equivalence of categories $\mathcal{C}^{\mathrm{cart}}_{\mathbb{T}^\mathcal{C}} \simeq \mathcal{C}$, one half of which sends a formula $\phi(\vec{x})$ to (the domain of) its interpretation $[[\vec{x} \, . \, \phi]]$ in the tautological $\Sigma^{\mathrm{f}}_\mathcal{C}$-structure in $\mathcal{C}$.

Given a Grothendieck topology J on a cartesian category $\mathcal{C}$, let us denote by $\mathbb{T}^{\mathcal{C}}_J$ the theory obtained from $\mathbb{T}^{\mathcal{C}}$ by adding all the sequents of the form

$$\left(\top \vdash_x \bigvee_{i \in I} (\exists y_i)(\ulcorner f_i \urcorner(y_i) = x)\right)$$

for a J-covering family $\{f_i : B_i \to A \mid i \in I\}$. By Proposition 1.4.8, if $\mathcal{C}$ is a regular (resp. coherent, geometric) category then for any regular (resp. coherent, geometric) category $\mathcal{D}$ the regular (resp. coherent, geometric) functors $\mathcal{C} \to \mathcal{D}$ are exactly the cartesian functors on $\mathcal{C}$ which are J-continuous, where J is the regular (resp. coherent, geometric) topology on $\mathcal{C}$. So we conclude as above that if $\mathcal{C}$ is a regular (resp. coherent, geometric) category then there is an equivalence of categories $\mathcal{C}^{\text{reg}}_{\mathbb{T}^{\mathcal{C}}_J} \simeq \mathcal{C}$ (resp. $\mathcal{C}^{\text{coh}}_{\mathbb{T}^{\mathcal{C}}_J} \simeq \mathcal{C}$, $\mathcal{C}^{\text{geom}}_{\mathbb{T}^{\mathcal{C}}_J} \simeq \mathcal{C}$) one half of which sends a formula $\phi(\vec{x})$ to (the domain of) its interpretation $[[\vec{x}.\phi]]$ in the canonical $\Sigma^{\text{f}}_{\mathcal{C}}$-structure in $\mathcal{C}$. □

Remark 1.4.14 The fact that every cartesian (resp. regular, coherent, geometric) category $\mathcal{C}$ is equivalent to the syntactic category of a theory $\mathbb{T}$ allows one to interpret categorical constructions on $\mathcal{C}$ as logical operations in $\mathbb{T}$.

1.4.5 Soundness and completeness

Theorem 1.4.15 (Soundness). *Let $\mathbb{T}$ be a cartesian (resp. regular, coherent, geometric, first-order) theory over a signature Σ, and let M be a model of $\mathbb{T}$ in a cartesian (resp. regular, coherent, geometric, Heyting) category $\mathcal{C}$. If σ is a sequent (in the appropriate fragment of first-order logic over Σ) which is provable in $\mathbb{T}$, then σ is satisfied in M.*

Proof The thesis immediately follows from Theorems 1.4.5 and 1.4.10 in view of Remark 1.4.6. □

Theorem 1.4.16 (Strong completeness). *Let $\mathbb{T}$ be a cartesian (resp. regular, coherent, geometric, first-order) theory. If a cartesian (resp. regular, coherent, geometric, Heyting,) sequent σ is satisfied in all models of $\mathbb{T}$ in cartesian (resp. regular, coherent, geometric, Heyting) categories, then it is provable in $\mathbb{T}$.*

Proof Thanks to the construction of syntactic categories, the property of completeness with respect to all the models in categories of the appropriate kind becomes a tautology (cf. Theorem 1.4.5). □

Let us now discuss classical completeness, that is, completeness with respect to the class of set-based models.

Definition 1.4.17 Within a given fragment of geometric logic, a theory $\mathbb{T}$ is said to have *enough set-based models* if for any geometric sequent σ over its signature which is valid in all the set-based models of $\mathbb{T}$, σ is provable in $\mathbb{T}$.

The following theorem is equivalent, via the notion of Morleyization of a first-order theory discussed in section 2.1.1, to Gödel's classical completeness theorem for classical first-order logic.

Theorem 1.4.18 (Classical completeness for coherent logic, cf. Corollary D1.5.10(ii) [55]). *Assuming the axiom of choice, every coherent theory has enough set-based models.*

Remark 1.4.19 Having extended the notion of model from **Set** to an arbitrary 'logical' category, it is no longer necessary, as in classical finitary first-order logic, to appeal to non-constructive principles such as the axiom of choice to ensure the existence of 'enough' models of the theory (that is, of a class of models such that validity in all of them amounts to provability in the theory). Indeed, the universal model $M_\mathbb{T}$ in the syntactic category of $\mathbb{T}$ exists constructively and enjoys a strong form of completeness (cf. Theorem 1.4.5): it is literally made up of formulae, and represents a natural first-order analogue of the Lindenbaum-Tarski algebra of a propositional theory (cf. Remark 2.1.20 below).

2
CLASSIFYING TOPOSES AND THE 'BRIDGE' TECHNIQUE

This chapter consists of two parts. In the first part we review the fundamental notion of classifying topos of a geometric theory and discuss the appropriate kinds of interpretations between theories which induce morphisms between the associated classifying toposes; the theoretical presentation is accompanied by a few concrete examples of classifying toposes of theories naturally arising in mathematics. We then establish a characterization theorem for universal models of geometric theories inside classifying toposes. In the second part we explain the general unifying technique 'toposes as bridges' originally introduced in [14]. This technique, which allows one to extract 'concrete' information from the existence of different representations for the classifying topos of a geometric theory, will be systematically exploited in the course of the book to establish theoretical results as well as applications.

2.1 Geometric logic and classifying toposes

2.1.1 Geometric theories

In this book we shall be mostly concerned with geometric theories, because of their connection with the theory of Grothendieck toposes.

Geometric theories are linked to Grothendieck toposes via the notion of classifying topos. Indeed, as we shall see in section 2.1.2, every geometric theory has a classifying topos and conversely any Grothendieck topos can be seen as the classifying topos of some geometric theory.

Recall from section 1.2 that a *geometric theory* over a first-order signature Σ is a theory whose axioms are sequents of the form $(\phi \vdash_{\vec{x}} \psi)$, where $\phi(\vec{x})$ and $\psi(\vec{x})$ are *geometric formulae* over Σ in the context $\vec{x}$ i.e. formulae with a finite number of free variables in $\vec{x}$ built up from atomic formulae over Σ by only using finitary conjunctions, infinitary disjunctions and existential quantifications.

Whilst the notion of geometric theory might seem quite restrictive at first sight, it turns out that most of the (first-order) theories naturally arising in mathematics have a geometric axiomatization (over their signature). Also, the possibility of employing infinitary disjunctions (of an arbitrary cardinality) in the construction of geometric formulae makes geometric logic particularly expressive and suitable for axiomatizing theories which do not belong to the realm of classical first-order logic (think for example of the notion of algebraic extension of a given field, or of that of ℓ-group with strong unit).

Rather remarkably, if a finitary first-order theory $\mathbb{T}$ is not geometric, one can canonically construct a coherent theory $\mathbb{T}'$ over a larger signature, called the *Morleyization* of $\mathbb{T}$, whose models in the category of sets (and, more generally, in any Boolean coherent category) can be identified with those of $\mathbb{T}$:

Proposition 2.1.1 (Lemma D1.5.13 [55], cf. also [2]). *Let $\mathbb{T}$ be a first-order theory over a signature Σ. Then there is a signature Σ' containing Σ and a coherent theory $\mathbb{T}'$ over Σ', called the* Morleyization *of $\mathbb{T}$, such that we have*

$$\mathbb{T}\text{-mod}_e(\mathcal{C}) \simeq \mathbb{T}'\text{-mod}(\mathcal{C})$$

for any Boolean coherent category $\mathcal{C}$.

Sketch of proof The signature Σ' of $\mathbb{T}'$ has, in addition to all the sorts, function symbols and relation symbols of the signature Σ of $\mathbb{T}$, two relation symbols $C_\phi \rightarrowtail A_1 \cdots A_n$ and $D_\phi \rightarrowtail A_1 \cdots A_n$ for each first-order formula ϕ over Σ (where $A_1 \cdots A_n$ is the string of sorts corresponding to the canonical context of ϕ), while the axioms of $\mathbb{T}'$ are given by the sequents of the form $(C_\phi \vdash_{\vec{x}} C_\psi)$ for any axiom $(\phi \vdash_{\vec{x}} \psi)$ of $\mathbb{T}$, plus a set of coherent sequents involving the new relation symbols C_ϕ and D_ϕ which ensure that in any model M of $\mathbb{T}'$ in a Boolean coherent category $\mathcal{C}$, the interpretation of C_ϕ coincides with the interpretation of ϕ and the interpretation of D_ϕ coincides with the complement of the interpretation of ϕ (cf. pp. 859-860 of [55] for more details). □

Proposition 2.1.2 *Let $\mathbb{T}$ be a finitary first-order theory over a signature Σ and $\mathbb{T}'$ its Morleyization. Then*

(i) *For any finitary first-order sequent $\sigma \equiv (\phi \vdash_{\vec{x}} \psi)$ over Σ, σ is provable in $\mathbb{T}$ using classical first-order logic if and only if the sequent $(C_\phi \vdash_{\vec{x}} C_\psi)$ is provable in $\mathbb{T}'$ using coherent logic.*

(ii) *The classical first-order syntactic category $\mathcal{C}_\mathbb{T}^{\text{fo-cl}}$ of $\mathbb{T}$ is isomorphic to the coherent syntactic category of $\mathbb{T}'$ and to the classical first-order syntactic category of $\mathbb{T}'$.*

Proof

(i) This follows immediately from the axioms defining $\mathbb{T}'$.

(ii) It is readily seen that every finitary first-order (resp. coherent) formula over the signature of $\mathbb{T}$ is classically provably equivalent (resp. provably equivalent in coherent logic) to a coherent formula over the signature of $\mathbb{T}'$. It follows that the coherent syntactic category of $\mathbb{T}'$ is Boolean and that every morphism between models of $\mathbb{T}'$ in Boolean coherent categories is an elementary morphism. Hence, by Proposition 2.1.1, both the coherent syntactic category of $\mathbb{T}'$ and the classical first-order syntactic category of $\mathbb{T}'$ satisfy the universal property of the category $\mathcal{C}_\mathbb{T}^{\text{fo-cl}}$ with respect to models of $\mathbb{T}$ in Boolean coherent categories whence, by universality, these categories are both equivalent to $\mathcal{C}_\mathbb{T}^{\text{fo-cl}}$ (cf. Remark 1.4.11), as required.

□

Remark 2.1.3 It follows at once from the proof of part (ii) of the proposition that a first-order theory is complete in the sense of classical model theory (i.e. every first-order sentence over its signature is either provably false or provably true, but not both) if and only if its Morleyization is complete in the sense of geometric logic (i.e. every geometric sentence over its signature is either provably false or provably true, but not both).

The notion of Morleyization is important because it enables us to study any kind of first-order theory by using the methods of topos theory. In fact, we can expect many important properties of first-order theories to be naturally expressible as properties of their Morleyizations, and these latter properties to be in turn expressible in terms of invariant properties of their classifying toposes (cf. for instance Remarks 2.1.3 and 4.2.22). Still, one can 'turn' a finitary first-order theory $\mathbb{T}$ into a geometric one in alternative ways, i.e. by simply adding some sorts to the signature of $\mathbb{T}$ and axioms over the extended signature so as to ensure that each of the first-order formulae which appear in the axioms of $\mathbb{T}$ becomes equivalent to a geometric formula in the new theory and the set-based models of the latter can be identified with those of $\mathbb{T}$. The Morleyization construction just represents a canonical, generally 'non-economical' way of doing this which works uniformly for any finitary first-order theory.

2.1.2   *The notion of classifying topos*

We have seen in section 1.3.1 (cf. Proposition 1.3.15) that every Grothendieck topos is a geometric category. Thus we can consider models of geometric theories in any Grothendieck topos. Inverse image functors of geometric morphisms of toposes preserve finite limits (by definition) and arbitrary colimits (having a right adjoint); so they are geometric functors and hence they preserve the interpretation of (arbitrary) geometric formulae (cf. the proof of Lemma 1.3.23). In general such functors are *not* Heyting functors, which explains why the following definition only makes sense for geometric theories.

Definition 2.1.4 Let $\mathbb{T}$ be a geometric theory over a given signature. A *classifying topos* of $\mathbb{T}$ is a Grothendieck topos $\mathbf{Set}[\mathbb{T}]$ (also denoted by $\mathcal{E}_{\mathbb{T}}$) such that for any Grothendieck topos $\mathcal{E}$ we have an equivalence of categories

$$\mathbf{Geom}(\mathcal{E}, \mathbf{Set}[\mathbb{T}]) \simeq \mathbb{T}\text{-mod}(\mathcal{E})$$

natural in $\mathcal{E}$.

Naturality means that for any geometric morphism $f : \mathcal{E} \to \mathcal{F}$, we have a commutative square (up to natural isomorphism)

$$\begin{array}{ccc} \mathbf{Geom}(\mathcal{F}, \mathbf{Set}[\mathbb{T}]) & \xrightarrow{\simeq} & \mathbb{T}\text{-mod}(\mathcal{F}) \\ \downarrow{\scriptstyle -\circ f} & & \downarrow{\scriptstyle \mathbb{T}\text{-mod}(f^*)} \\ \mathbf{Geom}(\mathcal{E}, \mathbf{Set}[\mathbb{T}]) & \xrightarrow{\simeq} & \mathbb{T}\text{-mod}(\mathcal{E}) \end{array}$$

in the (meta-)2-category $\mathfrak{CAT}$ of categories.

In other words, there is a model U of $\mathbb{T}$ in $\mathcal{E}_{\mathbb{T}}$, called 'the' *universal model* of $\mathbb{T}$, characterized by the universal property that any model M in a Grothendieck topos can be obtained, up to isomorphism, as the pullback $f^*(U)$ of U along the inverse image f^* of a unique (up to isomorphism) geometric morphism f from $\mathcal{E}$ to $\mathbf{Set}[\mathbb{T}]$.

Remark 2.1.5 The classifying topos of a geometric theory $\mathbb{T}$ can be seen as a *representing object* for the (pseudo-)functor

$$\mathbb{T}\text{-mod} : \mathfrak{BTop}^{\mathrm{op}} \to \mathfrak{CAT}$$

which assigns

- to a topos $\mathcal{E}$ the category $\mathbb{T}\text{-mod}(\mathcal{E})$ of models of $\mathbb{T}$ in $\mathcal{E}$ and
- to a geometric morphism $f : \mathcal{E} \to \mathcal{F}$ the functor $\mathbb{T}\text{-mod}(f^*) : \mathbb{T}\text{-mod}(\mathcal{F}) \to \mathbb{T}\text{-mod}(\mathcal{E})$ sending a model $M \in \mathbb{T}\text{-mod}(\mathcal{F})$ to its image $f^*(M)$ under f^*.

In particular, classifying toposes are unique up to categorical equivalence.

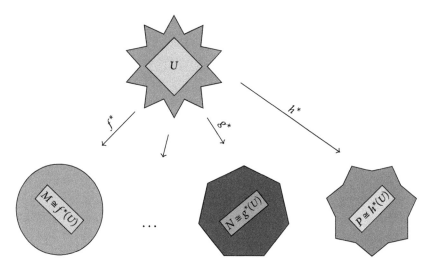

Fig. 2.1: Classifying topos

In Figure 2.1, the big shapes represent different toposes while the inner lighter shapes represent models of a given theory inside them; in particular, the dark grey star represents the classifying topos of a given theory and the light grey diamond represents 'the' universal model of the theory inside it. As the picture illustrates, all the models of the theory, including all the classical set-based models, are sorts of 'shadows' of the universal model lying in the classifying topos. This indicates that the symmetries of the theory are best understood by adopting the point of view of its classifying topos, since all the other models are images of this inner 'core' under 'deformations' realized by structure-preserving functors.

Contrary to what happens in classical logic, where one is forced to appeal to non-constructive principles such as the axiom of choice in order to ensure the existence of 'enough' set-based models for faithfully representing a finitary first-order theory, here we dispose of a constructively defined universe in which the syntax and semantics of a geometric theory meet yielding a strong form of completeness (cf. Theorem 1.4.5) and definability (cf. Theorem 6.1.3). Moreover, classifying toposes not only exist for finitary geometric theories, but also for arbitrary infinitary ones which do not necessarily satisfy a classical completeness theorem (one can exhibit non-contradictory infinitary geometric theories without any set-based models).

Definition 2.1.6 A *universal model* of a geometric theory $\mathbb{T}$ is a model $U_{\mathbb{T}}$ of $\mathbb{T}$ in a Grothendieck topos $\mathcal{G}$ such that for any $\mathbb{T}$-model M in a Grothendieck topos $\mathcal{F}$ there exists a unique (up to isomorphism) geometric morphism $f_M : \mathcal{F} \to \mathcal{G}$ such that $f_M^*(U_{\mathbb{T}}) \cong M$.

Remarks 2.1.7 (a) By the (2-dimensional) Yoneda lemma, if a Grothendieck topos $\mathcal{G}$ contains a universal model of a geometric theory $\mathbb{T}$ then $\mathcal{G}$ satisfies the universal property of the classifying topos of $\mathbb{T}$. Conversely, if a Grothendieck topos $\mathcal{E}$ classifies a geometric theory $\mathbb{T}$ then $\mathcal{E}$ contains a universal model of $\mathbb{T}$.

(b) If M and N are universal models of a geometric theory $\mathbb{T}$ lying respectively in toposes $\mathcal{F}$ and $\mathcal{G}$ then there exists a unique up to isomorphism (geometric) equivalence between $\mathcal{F}$ and $\mathcal{G}$ such that its inverse image functors send M and N to each other (up to isomorphism).

The classifying topos of a geometric theory can be canonically built as the category of sheaves on its geometric syntactic category with respect to the geometric topology on it. For smaller fragments of geometric logic, such as for example cartesian (resp. regular, coherent) logic, there exist variants of this syntactic construction, consisting in replacing the geometric syntactic site of the theory with its cartesian (resp. regular, coherent) syntactic site. More specifically, the following result holds (cf. also section D3.1 of [55]).

Theorem 2.1.8

(i) *For any universal Horn theory $\mathbb{T}$, the topos $[(\mathcal{C}_{\mathbb{T}}^{\mathrm{alg}})^{\mathrm{op}}, \mathbf{Set}]$ classifies $\mathbb{T}$.*
(ii) *For any cartesian theory $\mathbb{T}$, the topos $[(\mathcal{C}_{\mathbb{T}}^{\mathrm{cart}})^{\mathrm{op}}, \mathbf{Set}]$ classifies $\mathbb{T}$.*
(iii) *For any regular theory $\mathbb{T}$, the topos $\mathbf{Sh}(\mathcal{C}_{\mathbb{T}}^{\mathrm{reg}}, J_{\mathbb{T}}^{\mathrm{reg}})$ classifies $\mathbb{T}$.*
(iv) *For any coherent theory $\mathbb{T}$, the topos $\mathbf{Sh}(\mathcal{C}_{\mathbb{T}}^{\mathrm{coh}}, J_{\mathbb{T}}^{\mathrm{coh}})$ classifies $\mathbb{T}$.*
(v) *For any geometric theory $\mathbb{T}$, the topos $\mathbf{Sh}(\mathcal{C}_{\mathbb{T}}, J_{\mathbb{T}})$ classifies $\mathbb{T}$.*

Proof The proof of (i) requires a slightly different argument from that of the other points and can be found in [7] (cf. Theorem 1 therein).

By Diaconescu's equivalence (cf. Theorem 1.1.34) and Proposition 1.1.32, for any cartesian (resp. regular, coherent, geometric) theory $\mathbb{T}$ the geometric morphisms $\mathcal{E} \to [(\mathcal{C}_{\mathbb{T}}^{\mathrm{cart}})^{\mathrm{op}}, \mathbf{Set}]$ (resp. $\mathcal{E} \to \mathbf{Sh}(\mathcal{C}_{\mathbb{T}}^{\mathrm{reg}}, J_{\mathbb{T}}^{\mathrm{reg}})$, $\mathcal{E} \to \mathbf{Sh}(\mathcal{C}_{\mathbb{T}}^{\mathrm{coh}}, J_{\mathbb{T}}^{\mathrm{coh}})$,

$\mathcal{E} \to \mathbf{Sh}(\mathcal{C}_\mathbb{T}, J_\mathbb{T}))$ correspond, naturally in $\mathcal{E}$, to the cartesian functors $\mathcal{C}_\mathbb{T}^{\text{cart}} \to \mathcal{E}$ (resp. to the cartesian $J_\mathbb{T}^{\text{reg}}$-continuous functors $\mathcal{C}_\mathbb{T}^{\text{reg}} \to \mathcal{E}$, to the cartesian $J_\mathbb{T}^{\text{coh}}$-continuous functors $\mathcal{C}_\mathbb{T}^{\text{coh}} \to \mathcal{E}$, to the cartesian $J_\mathbb{T}$-continuous functors $\mathcal{C}_\mathbb{T} \to \mathcal{E}$). So (ii), (iii), (iv) and (v) follow from Theorem 1.4.10 and Proposition 1.4.8. □

Remark 2.1.9 For each of the points of Theorem 2.1.8, the image of the universal model $M_\mathbb{T}$ of $\mathbb{T}$ in its syntactic category (as in Definition 1.4.4) under the Yoneda embedding from it to the classifying topos of $\mathbb{T}$ yields a universal model $U_\mathbb{T}$ of $\mathbb{T}$ (recall from Remark 1.4.9 that all the 'logical' topologies are subcanonical). It follows in particular from Theorem 1.4.5(ii) that $U_\mathbb{T}$ is a conservative model of $\mathbb{T}$.

As a corollary one obtains the following fundamental result due to A. Joyal, G. Reyes and M. Makkai:

Theorem 2.1.10 *Every geometric theory has a classifying topos.* □

In the converse direction, we have the following well-known result:

Theorem 2.1.11 *Every Grothendieck topos is the classifying topos of some geometric theory.*

Proof Let $\mathbf{Sh}(\mathcal{C}, J)$ be a Grothendieck topos, where $(\mathcal{C}, J)$ is a small site. Diaconescu's equivalence provides, for any Grothendieck topos $\mathcal{E}$, an equivalence, natural in $\mathcal{E}$, between the category $\mathbf{Geom}(\mathcal{E}, \mathbf{Sh}(\mathcal{C}, J))$ of geometric morphisms from $\mathcal{E}$ to $\mathbf{Sh}(\mathcal{C}, J)$ and the category $\mathbf{Flat}_J(\mathcal{C}, \mathcal{E})$ of J-continuous flat functors from $\mathcal{C}$ to $\mathcal{E}$. Now, we can construct a geometric theory $\mathbb{T}_J^\mathcal{C}$ such that its models in any Grothendieck topos $\mathcal{E}$ can be identified precisely with the J-continuous flat functors from $\mathcal{C}$ to $\mathcal{E}$ (and the homomorphisms of $\mathbb{T}_J^\mathcal{C}$-models with the natural transformations between the corresponding flat functors); clearly, $\mathbb{T}_J^\mathcal{C}$ will be classified by the topos $\mathbf{Sh}(\mathcal{C}, J)$. We shall call $\mathbb{T}_J^\mathcal{C}$ the *theory of J-continuous flat functors on $\mathcal{C}$*. It is instructive to write down an axiomatization for $\mathbb{T}_J^\mathcal{C}$.

The signature of $\mathbb{T}_J^\mathcal{C}$ has one sort $\ulcorner a \urcorner$ for each object a of $\mathcal{C}$ and one function symbol $\ulcorner f \urcorner : \ulcorner a \urcorner \to \ulcorner b \urcorner$ for each arrow $f : a \to b$ in $\mathcal{C}$. The axioms of $\mathbb{T}_J^\mathcal{C}$ are the following (to indicate that a variable x has sort $\ulcorner a \urcorner$ we write x^a):

$$(\top \vdash_x \ulcorner f \urcorner(x) = x)$$

for any identity arrow f in $\mathcal{C}$;

$$(\top \vdash_x \ulcorner f \urcorner(x) = \ulcorner h \urcorner(\ulcorner g \urcorner(x)))$$

for any triple of arrows f, g, h of $\mathcal{C}$ such that f is equal to the composite $h \circ g$;

$$\left(\top \vdash_{[]} \bigvee_{a \in Ob(\mathcal{C})} (\exists x^a)\top\right)$$

(where the disjunction ranges over all the objects of $\mathcal{C}$);

$$\left(\top \vdash_{x^a, y^b} \bigvee_{a \xleftarrow{f} c \xrightarrow{g} b} (\exists z^c)(\ulcorner f \urcorner(z^c) = x^a \wedge \ulcorner g \urcorner(z^c) = y^b)\right)$$

for any objects a, b of $\mathcal{C}$ (where the disjunction ranges over all the cones $a \xleftarrow{f} c \xrightarrow{g} b$ on the discrete diagram on the two objects a and b);

$$\left(\ulcorner f \urcorner(x^a) = \ulcorner g \urcorner(x^a) \vdash_{x^a} \bigvee_{\substack{c \xrightarrow{h} a \\ f \circ h = g \circ h}} (\exists z^c)(\ulcorner h \urcorner(z^c) = x^a)\right)$$

for any pair of arrows $f, g : a \to b$ in $\mathcal{C}$ with common domain and codomain (where the disjunction ranges over all the arrows h which equalize f and g);

$$\left(\top \vdash_{x^a} \bigvee_{i \in I} (\exists y_i^{b_i})(\ulcorner f_i \urcorner(y_i^{b_i}) = x^a)\right)$$

for each J-covering family $\{f_i : b_i \to a \mid i \in I\}$.

Notice that the first two groups of axioms express functoriality, the third, fourth and fifth the filtering property (cf. Proposition 1.1.32), and the sixth the property of J-continuity. □

Two geometric theories are said to be *Morita-equivalent* if they have equivalent classifying toposes (cf. Definition 2.2.1 below). From the above discussion it follows that Grothendieck toposes can be thought of as canonical representatives for Morita equivalence classes of geometric theories.

It is important to note that the method of constructing classifying toposes via syntactic sites is by no means the only one for 'calculating' the classifying topos of a geometric theory. Alternative techniques, of more 'semantic' or 'geometric' nature, have been developed. For instance, as we shall see in Chapter 3, every representation of a geometric theory as an extension of another geometric theory over the same signature leads to a representation of its classifying topos as a subtopos of the classifying topos of the latter theory; when applied to extensions $\mathbb{S}$ of theories $\mathbb{T}$ classified by a presheaf topos, this leads to a 'semantic' representation for the classifying topos of $\mathbb{S}$ as a topos of sheaves on the opposite of the category of finitely presentable $\mathbb{T}$-models (cf. Chapter 8). More generally, as we shall argue in section 2.2, it is reasonable to expect 'different ways of looking at a certain theory' to materialize into different representations of its classifying topos.

It should also be mentioned that the notions of geometric theory and classifying topos can be 'relativized' to an arbitrary base topos (cf. for instance sections 6.5 of [50] and B4.2 of [55]). Still, a theory of internal geometric theories and relative classifying toposes has not yet been formally developed.

Via the syntactic construction of classifying toposes of coherent theories, the classical completeness theorem for coherent theories (cf. Theorem 1.4.18) translates into the following theorem about coherent toposes. Recall that a Grothendieck topos is said to be *coherent* if it admits a small site of definition of the form $(\mathcal{C}, J)$, where $\mathcal{C}$ is a cartesian category and J is a finite-type topology on $\mathcal{C}$ (cf. section 8.1.4 for the notion of topology of finite type).

Theorem 2.1.12 (Deligne's theorem – Exposé VI, Proposition 9.0 [3]). *Assuming the axiom of choice, every coherent topos has enough points.* □

2.1.3 Interpretations and geometric morphisms

The assignment from a collection of geometric theories to a collection of Grothendieck toposes sending a theory to its classifying topos can be made functorial, as follows. A natural notion of morphism between theories is given by the notion of interpretation of one theory into another. As we shall see below, there are actually many variants of this notion, all of which induce geometric morphisms between the respective classifying toposes.

Definition 2.1.13 Let $\mathbb{T}_1$ and $\mathbb{T}_2$ be two cartesian (resp. regular, coherent, geometric) theories.

(a) A *cartesian* (resp. *regular, coherent, geometric*) *interpretation* of $\mathbb{T}_1$ in $\mathbb{T}_2$ is a cartesian (resp. regular, coherent, geometric) functor from $\mathcal{C}_{\mathbb{T}_1}^{\text{cart}}$ (resp. $\mathcal{C}_{\mathbb{T}_1}^{\text{reg}}$, $\mathcal{C}_{\mathbb{T}_1}^{\text{coh}}$, $\mathcal{C}_{\mathbb{T}_1}$) to $\mathcal{C}_{\mathbb{T}_2}^{\text{cart}}$ (resp. $\mathcal{C}_{\mathbb{T}_2}^{\text{reg}}$, $\mathcal{C}_{\mathbb{T}_2}^{\text{coh}}$, $\mathcal{C}_{\mathbb{T}_2}$).
(b) The cartesian (resp. regular, coherent, geometric) theories $\mathbb{T}_1$ and $\mathbb{T}_2$ are said to be *cartesianly bi-interpretable* (resp. *regularly bi-interpretable, coherently bi-interpretable, geometrically bi-interpretable*) if their cartesian (resp. regular, coherent, geometric) syntactic categories are equivalent.

If $\mathbb{T}_1$ and $\mathbb{T}_2$ are cartesian theories then any cartesian functor $\mathcal{C}_{\mathbb{T}_1}^{\text{cart}} \to \mathcal{C}_{\mathbb{T}_2}^{\text{cart}}$ induces a geometric morphism

$$[(\mathcal{C}_{\mathbb{T}_2}^{\text{cart}})^{\text{op}}, \mathbf{Set}] \to [(\mathcal{C}_{\mathbb{T}_1}^{\text{cart}})^{\text{op}}, \mathbf{Set}]$$

(cf. Example 1.1.15(c)).

If $\mathbb{T}_1$ and $\mathbb{T}_2$ are regular (resp. coherent, geometric) theories then, since by Proposition 1.4.8 the regular (resp. coherent, geometric) functors $\mathcal{C}_{\mathbb{T}_1}^{\text{reg}} \to \mathcal{C}_{\mathbb{T}_2}^{\text{reg}}$ (resp. $\mathcal{C}_{\mathbb{T}_1}^{\text{coh}} \to \mathcal{C}_{\mathbb{T}_2}^{\text{coh}}$, $\mathcal{C}_{\mathbb{T}_1} \to \mathcal{C}_{\mathbb{T}_2}$) are exactly the cartesian functors which send $J_{\mathbb{T}_1}^{\text{reg}}$-covering sieves (resp. $J_{\mathbb{T}_1}^{\text{coh}}$-covering sieves, $J_{\mathbb{T}_1}$-covering sieves) on $\mathcal{C}_{\mathbb{T}_1}^{\text{reg}}$ (resp. $\mathcal{C}_{\mathbb{T}_1}^{\text{coh}}$, $\mathcal{C}_{\mathbb{T}_1}$) to $J_{\mathbb{T}_2}^{\text{reg}}$-covering sieves (resp. $J_{\mathbb{T}_2}^{\text{coh}}$-covering sieves, $J_{\mathbb{T}_2}$-covering sieves) on $\mathcal{C}_{\mathbb{T}_2}^{\text{reg}}$ (resp. $\mathcal{C}_{\mathbb{T}_2}^{\text{coh}}$, $\mathcal{C}_{\mathbb{T}_2}$), any regular (resp. coherent, geometric) interpretation of $\mathbb{T}_1$ in $\mathbb{T}_2$ induces a geometric morphism $\mathbf{Sh}(\mathcal{C}_{\mathbb{T}_2}^{\text{reg}}, J_{\mathbb{T}_2}^{\text{reg}}) \to \mathbf{Sh}(\mathcal{C}_{\mathbb{T}_1}^{\text{reg}}, J_{\mathbb{T}_1}^{\text{reg}})$ (resp. $\mathbf{Sh}(\mathcal{C}_{\mathbb{T}_2}^{\text{coh}}, J_{\mathbb{T}_2}^{\text{coh}}) \to \mathbf{Sh}(\mathcal{C}_{\mathbb{T}_1}^{\text{coh}}, J_{\mathbb{T}_1}^{\text{coh}})$, $\mathbf{Sh}(\mathcal{C}_{\mathbb{T}_2}, J_{\mathbb{T}_2}) \to \mathbf{Sh}(\mathcal{C}_{\mathbb{T}_1}, J_{\mathbb{T}_1})$) (again by Example 1.1.15(c)).

Let us now discuss the relationship between syntactic categories and classifying toposes. We shall see that, whilst it is always possible to recover the cartesian syntactic category of a cartesian theory from its classifying topos (up to equivalence), it is only possible in general to recover the effectivization of the regular syntactic category (resp. the effective positivization of the coherent syntactic category) of a regular (resp. coherent) theory from its classifying topos.

The topos-theoretic invariants which are useful in this respect are the following:

Definition 2.1.14 (a) An object A of a Grothendieck topos is said to be *irreducible* if every epimorphic family in the topos with codomain A contains a split epimorphism.
(b) An object A of a Grothendieck topos is said to be *compact* if every epimorphic family with codomain A contains a finite epimorphic subfamily.
(c) An object A of a Grothendieck topos is said to be *supercompact* if every epimorphic family with codomain A contains an epimorphism.
(d) An object A of a Grothendieck topos $\mathcal{E}$ is said to be *coherent* if it is compact and for any arrow $u : B \to A$ in $\mathcal{E}$ such that B is compact, the domain of the kernel pair of u is compact.
(e) An object A of a Grothendieck topos $\mathcal{E}$ is said to be *supercoherent* if it is supercompact and for any arrow $u : B \to A$ in $\mathcal{E}$ such that B is supercompact, the domain of the kernel pair of u is supercompact.

For a proof of the following proposition we refer the reader to section D3.3 of [55] for points (ii) and (iii) and to section 4 of [20] for point (i); for the notions involved in the proposition, see section 1.3.2.

Proposition 2.1.15
(i) The cartesian syntactic category $\mathcal{C}_{\mathbb{T}}^{\text{cart}}$ of a cartesian theory $\mathbb{T}$ can be recovered, up to equivalence, from its classifying topos $[(\mathcal{C}_{\mathbb{T}}^{\text{cart}})^{\text{op}}, \mathbf{Set}]$ as the full subcategory on the irreducible objects.
(ii) The effectivization $\mathbf{Eff}(\mathcal{C}_{\mathbb{T}}^{\text{reg}})$ of the regular syntactic category $\mathcal{C}_{\mathbb{T}}^{\text{reg}}$ of a regular theory $\mathbb{T}$ can be recovered, up to equivalence, from its classifying topos $\mathbf{Sh}(\mathcal{C}_{\mathbb{T}}^{\text{reg}}, J_{\mathbb{T}}^{\text{reg}})$ as the full subcategory on the supercoherent objects.
(iii) The effective positivization $\mathbf{Eff}(\mathbf{Pos}(\mathcal{C}_{\mathbb{T}}^{\text{coh}}))$ of the coherent syntactic category $\mathcal{C}_{\mathbb{T}}^{\text{coh}}$ of a coherent theory $\mathbb{T}$ can be recovered, up to equivalence, from its classifying topos $\mathbf{Sh}(\mathcal{C}_{\mathbb{T}}^{\text{coh}}, J_{\mathbb{T}}^{\text{coh}})$ as the full subcategory on the coherent objects.

Notice that every object of $\mathbf{Eff}(\mathcal{C}_{\mathbb{T}}^{\text{reg}})$ has a syntactic description as a 'formal quotient' of an object of $\mathcal{C}_{\mathbb{T}}^{\text{reg}}$ by an equivalence relation in $\mathcal{C}_{\mathbb{T}}^{\text{reg}}$, and similarly any object of $\mathbf{Eff}(\mathbf{Pos}(\mathcal{C}_{\mathbb{T}}^{\text{coh}}))$ is a 'formal quotient' in $\mathbf{Pos}(\mathcal{C}_{\mathbb{T}}^{\text{coh}})$ of a 'formal finite coproduct' of objects of $\mathcal{C}_{\mathbb{T}}^{\text{coh}}$ by an equivalence relation in $\mathbf{Pos}(\mathcal{C}_{\mathbb{T}}^{\text{coh}})$.

This motivates the following definition:

Definition 2.1.16 (a) A *generalized regular interpretation* of a regular theory $\mathbb{T}_1$ in a regular theory $\mathbb{T}_2$ is a regular functor $\mathbf{Eff}(\mathcal{C}_{\mathbb{T}_1}^{\text{reg}}) \to \mathbf{Eff}(\mathcal{C}_{\mathbb{T}_2}^{\text{reg}})$.
(b) A *generalized coherent interpretation* of a coherent theory $\mathbb{T}_1$ in a coherent theory $\mathbb{T}_2$ is a coherent functor $\mathbf{Eff}(\mathbf{Pos}(\mathcal{C}_{\mathbb{T}_1}^{\text{coh}})) \to \mathbf{Eff}(\mathbf{Pos}(\mathcal{C}_{\mathbb{T}_2}^{\text{coh}}))$.

As is the case for their classical counterparts, generalized interpretations also induce geometric morphisms between the classifying toposes of the two theories.

For geometric theories, as we saw in section 1.3.2, the analogue of the effective positivization construction is the ∞-pretopos completion: applied to the syntactic category of a geometric theory, this construction yields precisely the classifying topos of the theory (by Giraud's theorem and Theorem 1.4.10).

The following proposition represents the 'functorialization' of Proposition 2.1.15. Before stating it, we need to introduce some definitions.

A geometric morphism $f : \mathcal{E} \to \mathcal{F}$ is said to be *coherent* if the inverse image functor $f^* : \mathcal{F} \to \mathcal{E}$ of f sends coherent objects of $\mathcal{F}$ to coherent objects of $\mathcal{E}$. Similarly, a geometric morphism $f : \mathcal{E} \to \mathcal{F}$ is said to be *regular* if $f^* : \mathcal{F} \to \mathcal{E}$ sends supercoherent objects of $\mathcal{F}$ to supercoherent objects of $\mathcal{E}$.

Proposition 2.1.17

(i) *Every essential geometric morphism $\mathcal{E}_{\mathbb{T}_2} \to \mathcal{E}_{\mathbb{T}_1}$ between the classifying toposes of two cartesian theories $\mathbb{T}_1$ and $\mathbb{T}_2$ is induced by a (unique up to isomorphism) cartesian interpretation of $\mathbb{T}_1$ in $\mathbb{T}_2$.*

(ii) *Every regular geometric morphism $\mathcal{E}_{\mathbb{T}_2} \to \mathcal{E}_{\mathbb{T}_1}$ between the classifying toposes of two regular theories $\mathbb{T}_1$ and $\mathbb{T}_2$ is induced by a (unique up to isomorphism) generalized regular interpretation of $\mathbb{T}_1$ in $\mathbb{T}_2$.*

(iii) *Every coherent geometric morphism $\mathcal{E}_{\mathbb{T}_2} \to \mathcal{E}_{\mathbb{T}_1}$ between the classifying toposes of two coherent theories $\mathbb{T}_1$ and $\mathbb{T}_2$ is induced by a (unique up to isomorphism) generalized coherent interpretation of $\mathbb{T}_1$ in $\mathbb{T}_2$.*

Note that, since the classifying topos $\mathbf{Sh}(\mathcal{C}_\mathbb{T}, J_\mathbb{T})$ of a geometric theory $\mathbb{T}$ is the ∞-pretopos completion of $\mathcal{C}_\mathbb{T}$, we can regard any geometric morphism $\mathbf{Sh}(\mathcal{C}_{\mathbb{T}_2}, J_{\mathbb{T}_2}) \to \mathbf{Sh}(\mathcal{C}_{\mathbb{T}_1}, J_{\mathbb{T}_1})$ between the classifying toposes of two geometric theories $\mathbb{T}_1$ and $\mathbb{T}_2$ as a generalized geometric interpretation of $\mathbb{T}_1$ in $\mathbb{T}_2$.

We can give an alternative description of the notions of interpretation defined above in terms of internal models, as follows. If $\mathbb{T}$ is a cartesian (resp. regular, coherent, geometric) theory then, by Theorem 1.4.10, the category of cartesian (resp. regular, coherent, geometric) functors from $\mathcal{C}_\mathbb{T}^{\text{cart}}$ (resp. $\mathcal{C}_\mathbb{T}^{\text{reg}}$, $\mathcal{C}_\mathbb{T}^{\text{coh}}$, $\mathcal{C}_\mathbb{T}$) to any cartesian (resp. regular, coherent, geometric) category $\mathcal{D}$ is equivalent to the category of models of $\mathbb{T}$ in $\mathcal{D}$, where the equivalence sends each model $M \in \mathbb{T}\text{-mod}(\mathcal{D})$ to the functor $F_M : \mathcal{C}_\mathbb{T} \to \mathcal{D}$ assigning to each object $\{\vec{x}.\phi\}$ of $\mathcal{C}_\mathbb{T}$ its interpretation $[[\vec{x}.\phi]]_M$ in M. Thus, any geometric interpretation of a geometric theory $\mathbb{T}_1$ in a geometric theory $\mathbb{T}_2$ corresponds to an internal $\mathbb{T}_1$-model in the syntactic category $\mathcal{C}_{\mathbb{T}_2}$ of $\mathbb{T}_2$, and similarly for cartesian, regular and coherent theories (and their generalized interpretations). Note that if $\mathbb{T}_2$ has enough set-based models (for example, if $\mathbb{T}_2$ is coherent assuming the axiom of choice) then the condition that a certain Σ_1-structure U, where Σ_1 is the signature of $\mathbb{T}_1$, in (a completion of) the syntactic category of $\mathbb{T}_2$ be a $\mathbb{T}_1$-model is equivalent to the requirement that for any $\mathbb{T}_2$-model M in $\mathbf{Set}$ the functor F_M send U to a $\mathbb{T}_1$-model.

Remarks 2.1.18 (a) Since a given theory may belong to more than one fragment of geometric logic, it is natural to wonder whether the notions of bi-interpretability of theories given above for different fragments of geometric logic are compatible with each other. We shall answer this question in the affirmative in section 2.2.4.

(b) Since the classification of models of geometric theories already takes place at the level of geometric categories (cf. Theorem 1.4.10), one may naturally wonder why we should focus our attention on more sophisticated structures such as classifying toposes. In fact, there are several reasons for this:

First, a Grothendieck topos is, in terms of internal categorical structure, much richer than an arbitrary geometric category and hence constitutes a mathematical environment which is naturally more amenable to computations and which enjoys higher degrees of symmetry. In fact, it is a standard process in mathematics to complete mathematical entities with respect to natural operations through the formal addition of 'imaginaries' and work in the extended, more 'symmetric' environment to solve problems posed in the original context (think for instance of the real line and the complex plane).

Second, a very important aspect of Grothendieck toposes, on which the theory of topos-theoretic 'bridges' described in section 2.2 is based, is the fact that they admit a very well-behaved, although highly non-trivial, representation theory. The abstract relationship between an object and its different 'presentations' is geometrically embodied in the context of Grothendieck toposes in the form of relationships between a given topos and its different sites of definition (or, more generally, its different representations); the possibility of working at two levels, that of toposes and that of their presentations, makes the theory extremely fruitful and technically flexible.

Third, the 2-category $\mathfrak{BTop}$ of Grothendieck toposes is itself quite rich in terms of (2-)categorical structure (cf. for instance section B4 of [55]).

2.1.4 Classifying toposes for propositional theories

Recall that a *propositional theory* is a geometric theory over a signature which has no sorts. The signature of a propositional theory thus merely consists of a set of 0-ary relation symbols, which define the atomic propositions; all the other formulae (i.e. sentences) are built up from them by using finitary conjunctions and infinitary disjunctions.

Propositional theories can be used for axiomatizing subsets of a given set satisfying particular properties, e.g. filters on a frame or prime ideals of a commutative ring.

Proposition 2.1.19 (cf. Remark D3.1.14 [55]). *Localic toposes are precisely the classifying toposes of propositional theories.*

Specifically, we can define, for any locale L, a propositional theory $\mathbb{P}_L$ classified by it, as follows. The signature of $\mathbb{P}_L$ consists of one atomic proposition F_a (to be thought of as the assertion that a is in the filter) for each $a \in L$; the axioms of $\mathbb{P}_L$ are

$$(\top \vdash F_1),$$

all the sequents of the form

$$(F_a \wedge F_b \vdash F_{a \wedge b})$$

for any $a, b \in L$, and all the sequents of the form

$$\left(F_a \vdash \bigvee_{i \in I} F_{a_i}\right)$$

whenever $\bigvee_{i \in I} a_i = a$ in L.

Notice that the set-based models of the theory $\mathbb{P}_L$ are the completely prime filters on L, that is the subsets $F \subseteq L$ such that $1 \in F$, $a \wedge b \in F$ whenever $a, b \in F$ and for any family of elements $\{a_i \in L \mid i \in I\}$ such that $\bigvee_{i \in I} a_i \in F$, there exists $i \in I$ such that $a_i \in F$.

Remark 2.1.20 Recall that the *Lindenbaum-Tarski algebra* $\mathcal{L}_\mathbb{T}$ of a propositional theory $\mathbb{T}$ is the algebra consisting of the provable-equivalence classes of sentences over its signature. In the case of a geometric propositional theory $\mathbb{T}$, $\mathcal{L}_\mathbb{T}$ is clearly a frame, equivalent to the geometric syntactic category of $\mathbb{T}$. The classifying topos of $\mathbb{T}$ is given by the category $\mathbf{Sh}(\mathcal{L}_\mathbb{T}, J_{\mathcal{L}_\mathbb{T}})$ of sheaves on this locale with respect to the canonical topology $J_{\mathcal{L}_\mathbb{T}}$ on it. This indicates that the notion of classifying topos of a geometric theory represents a natural first-order generalization of that of Lindenbaum-Tarski of a propositional theory.

We shall encounter propositional theories again in section 2.1.6.

2.1.5 Classifying toposes for cartesian theories

We can describe the classifying topos of a cartesian theory directly in terms of its finitely presented models.

Recall that, given a cartesian theory $\mathbb{T}$ over a signature Σ, a $\mathbb{T}$-model M in **Set** is said to be *finitely presented* by a cartesian formula-in-context $\phi(\vec{x})$ over Σ if there exists a string of elements $(\xi_1, \ldots, \xi_n) \in MA_1 \times \cdots \times MA_n$ (where $A_1 \cdots A_n$ is the string of sorts associated with $\vec{x}$), called the *generators* of M, such that for any $\mathbb{T}$-model N in **Set** and any string of elements $(b_1, \ldots, b_n) \in NA_1 \times \cdots \times NA_n$ such that $(b_1, \ldots, b_n) \in [[\vec{x} . \phi]]_N$, there exists a unique arrow $f : M \to N$ in $\mathbb{T}$-mod(**Set**) such that $(f_{A_1} \times \cdots \times f_{A_n})((\xi_1, \ldots, \xi_n)) = (b_1, \ldots, b_n)$.

We denote by f.p.$\mathbb{T}$-mod(**Set**) the full subcategory of $\mathbb{T}$-mod(**Set**) on the finitely presented models.

Theorem 2.1.21 (Corollary D3.1.2 [55]). *For any cartesian theory $\mathbb{T}$, we have an equivalence of categories*

$$\text{f.p.}\mathbb{T}\text{-mod}(\mathbf{Set}) \simeq (\mathcal{C}_\mathbb{T}^{\text{cart}})^{\text{op}}.$$

In particular, $\mathbb{T}$ is classified by the topos [f.p.$\mathbb{T}$-mod(**Set**), **Set**] *(cf. Theorem 2.1.8(ii)).*

Remark 2.1.22 If $\mathbb{T}$ is a universal Horn theory (in particular, a finitary algebraic theory) then the category $\mathcal{C}_{\mathbb{T}}^{\text{alg}}$ is dual to the category of finitely presented $\mathbb{T}$-algebras, and the presheaf topos $[(\mathcal{C}_{\mathbb{T}}^{\text{alg}})^{\text{op}}, \mathbf{Set}]$ satisfies the universal property of the classifying topos for $\mathbb{T}$ (cf. Theorem 1 of [7]).

Examples 2.1.23
(a) The (algebraic) theory of Boolean algebras is classified by the topos $[\mathbf{Bool}_{\text{fin}}, \mathbf{Set}]$ where $\mathbf{Bool}_{\text{fin}}$ is the category of finite Boolean algebras and Boolean algebra homomorphisms between them.
(b) The (algebraic) theory of commutative rings with unit is classified by the topos $[\mathbf{Rng}_{\text{f.g.}}, \mathbf{Set}]$, where $\mathbf{Rng}_{\text{f.g.}}$ is the category of finitely generated rings and ring homomorphisms between them.

2.1.6 Further examples

2.1.6.1 The Zariski topos

Let Σ be the one-sorted signature for the theory $\mathbb{T}$ of commutative rings with unit, i.e. the signature consisting of two binary function symbols $+$ and $\cdot$, one unary function symbol $-$ and two constants 0 and 1.

The *coherent theory of local rings* is obtained from $\mathbb{T}$ by adding the sequents

$$(0 = 1 \vdash_{[]} \bot)$$

and

$$((\exists z)((x+y) \cdot z = 1) \vdash_{x,y} ((\exists z)(x \cdot z = 1) \vee (\exists z)(y \cdot z = 1))).$$

Definition 2.1.24 The *Zariski topos* is the topos $\mathbf{Sh}(\mathbf{Rng}_{\text{f.g.}}^{\text{op}}, J)$ of sheaves on the opposite of the category $\mathbf{Rng}_{\text{f.g.}}$ of finitely generated rings with respect to the topology J on $\mathbf{Rng}_{\text{f.g.}}^{\text{op}}$ defined as follows: for any cosieve S in $\mathbf{Rng}_{\text{f.g.}}$ on an object A, $S \in J(A)$ if and only if S contains a finite family $\{\xi_i : A \to A[s_i^{-1}] \mid 1 \leq i \leq n\}$ of canonical inclusions $\xi_i : A \to A[s_i^{-1}]$ in $\mathbf{Rng}_{\text{f.g.}}$ where $\{s_1, \ldots, s_n\}$ is a set of elements of A which is not contained in any proper ideal of A.

Proposition 2.1.25 (Theorem VII.6.3 [63]). *The (coherent) theory of local rings is classified by the Zariski topos.*

More generally, one can prove that the big Zariski topos of an affine scheme $\text{Spec}(A)$ classifies the theory of local A-algebras. On the other hand, the small Zariski topos $\text{Spec}(A)$ classifies the theory of localizations of the ring A, equivalently the propositional theory $\mathbb{P}_A$ of *prime filters* on A defined as follows. The signature of $\mathbb{P}_A$ consists of a propositional symbol P_a for each element $a \in A$, and the axioms of $\mathbb{P}_A$ are the following:

$$(\top \vdash P_{1_A});$$
$$(P_{0_A} \vdash \bot);$$
$$(P_{a \cdot b} \dashv\vdash P_a \wedge P_b)$$

for any a, b in A;

$$(P_{a+b} \vdash P_a \vee P_b)$$

for any $a, b \in A$.

The models of $\mathbb{P}_A$ in **Set** are precisely the prime filters on A, that is the subsets S of A such that the complement $A \setminus S$ is a prime ideal. In fact, the Zariski topology is homeomorphic to the subterminal topology (in the sense of Definition 1.1.23) on the set of points of the classifying topos of $\mathbb{P}_A$. If, instead of taking the theory $\mathbb{P}_A$ of prime filters, we had considered the propositional theory of prime ideals (axiomatized over the same signature in the obvious way), we would have obtained a classifying topos inequivalent to the small Zariski topos of A, in spite of the fact that the two theories have the same models in **Set**. In fact, in order to prove that their categories of set-based models are equivalent, one has to use the Law of Excluded Middle, a principle which is not sound in general toposes, even in very simple ones such as the Sierpiński topos (i.e. the topos of sheaves on the Sierpiński space).

2.1.6.2 The classifying topos for integral domains

The *theory of integral domains* is the theory obtained from the theory of commutative rings with unit by adding the axioms

$$(0 = 1 \vdash_{[]} \bot);$$

$$(x \cdot y = 0 \vdash_{x,y} x = 0 \vee y = 0).$$

Proposition 2.1.26 *The theory of integral domains is classified by the topos* $\mathbf{Sh}(\mathbf{Rng}_{\text{f.g.}}^{\text{op}}, J)$ *of sheaves on the opposite of the category* $\mathbf{Rng}_{\text{f.g.}}$ *of finitely generated rings with respect to the topology J on* $\mathbf{Rng}_{\text{f.g.}}^{\text{op}}$ *defined as follows: for any cosieve S in* $\mathbf{Rng}_{\text{f.g.}}$ *on an object A, $S \in J(A)$ if and only if either*

- *A is the zero ring and S is the empty sieve on it, or*
- *S contains a non-empty finite family $\{\pi_{a_i} : A \to A/(a_i) \mid 1 \leq i \leq n\}$ of canonical projections $\pi_{a_i} : A \to A/(a_i)$ in $\mathbf{Rng}_{\text{f.g.}}$ where $\{a_1, \ldots, a_n\}$ is a set of elements of A such that $a_1 \cdot \ldots \cdot a_n = 0$.*

For a proof of this proposition, see section 8.1.5 below.

2.1.6.3 The topos of simplicial sets

Definition 2.1.27 The theory $\mathbb{I}$ of *intervals* is written over a one-sorted signature having a binary relation symbol $\leq$ and two constants b and t, and has as axioms the following sequents:

$$(\top \vdash_x x \leq x);$$
$$(x \leq y \wedge y \leq z \vdash_{x,y,z} x \leq z);$$
$$(x \leq y \wedge y \leq x \vdash_{x,y} x = y);$$
$$(\top \vdash_x x \leq t \wedge b \leq x);$$
$$(b = t \vdash \bot);$$
$$(\top \vdash_{x,y} x \leq y \vee y \leq x).$$

Proposition 2.1.28 *The theory $\mathbb{I}$ is classified by the topos $[\Delta^{\mathrm{op}}, \mathbf{Set}]$ of simplicial sets.*

A detailed proof of this proposition may be found in section VIII.8 of [63].

2.1.7 A characterization theorem for universal models in classifying toposes

It is natural to wonder whether, given a geometric theory $\mathbb{T}$, one can give conditions on a pair $(\mathcal{E}, M)$ consisting of a Grothendieck topos $\mathcal{E}$ and a model M of $\mathbb{T}$ in $\mathcal{E}$ for $\mathcal{E}$ to be a classifying topos of $\mathbb{T}$ and M a universal $\mathbb{T}$-model in it. The following result provides a positive answer to this question.

Theorem 2.1.29 *Let $\mathbb{T}$ be a geometric theory, $\mathcal{E}$ a Grothendieck topos and M a model of $\mathbb{T}$ in $\mathcal{E}$. Then $\mathcal{E}$ is a classifying topos of $\mathbb{T}$ and M a universal model of $\mathbb{T}$ if and only if the following conditions are satisfied:*

(i) *The family $\mathcal{F}$ of objects which can be built from the interpretations in M of the sorts, function symbols and relation symbols over the signature of $\mathbb{T}$ by using geometric logic constructions (i.e. the objects given by the domains of the interpretations in M of geometric formulae over the signature of $\mathbb{T}$) is separating for $\mathcal{E}$.*

(ii) *The model M is conservative for $\mathbb{T}$, that is for any geometric sequent σ over the signature of $\mathbb{T}$, σ is valid in M if and only if it is provable in $\mathbb{T}$.*

(iii) *Any arrow k in $\mathcal{E}$ between objects A and B in the family $\mathcal{F}$ of point (i) is definable; that is, if A (resp. B) is equal to the interpretation of a geometric formula $\phi(\vec{x})$ (resp. $\psi(\vec{y})$) over the signature of $\mathbb{T}$, there exists a $\mathbb{T}$-provably functional formula θ from $\phi(\vec{x})$ to $\psi(\vec{y})$ such that the interpretation of θ in M is equal to the graph of k.*

Proof The fact that conditions (i), (ii) and (iii) are necessary for $\mathcal{E}$ to be a classifying topos of $\mathbb{T}$ and M a universal $\mathbb{T}$-model in it follows at once from Theorem 2.1.8 and Remark 2.1.9; so it remains to prove that they are also sufficient.

By the universal property of the geometric syntactic category $\mathcal{C}_\mathbb{T}$ of $\mathbb{T}$, the $\mathbb{T}$-model M corresponds to a geometric functor $F_M : \mathcal{C}_\mathbb{T} \to \mathcal{E}$ assigning to each object of $\mathcal{C}_\mathbb{T}$ (the domain of) its interpretation in M and acting accordingly on arrows. Condition (ii) in the statement of the theorem is equivalent to the assertion that the functor F_M be faithful, while condition (iii) is equivalent to saying that F_M is full. Therefore under conditions (ii) and (iii) $\mathcal{C}_\mathbb{T}$ embeds, up to equivalence, as a full subcategory of $\mathcal{E}$. Now, condition (i) ensures that $\mathcal{C}_\mathbb{T}$ is dense in $\mathcal{E}$, whence the Comparison Lemma (Theorem 1.1.8) yields an equivalence $\mathcal{E} \simeq \mathbf{Sh}(\mathcal{C}_\mathbb{T}, J)$, where J is the Grothendieck topology on $\mathcal{C}_\mathbb{T}$ induced by the canonical topology on $\mathcal{E}$, that is the geometric topology on $\mathcal{C}_\mathbb{T}$. By the syntactic construction of classifying toposes and universal models, we can thus conclude that $\mathcal{E}$ is 'the' classifying topos of $\mathbb{T}$ and M is a universal model for $\mathbb{T}$. □

As immediate corollaries of Theorem 2.1.29, we recover the following known results:

- Let $\mathcal{C}$ be a separating set of objects for a Grothendieck topos $\mathcal{E}$, and $\Sigma_\mathcal{C}^\text{f}$ the signature consisting of one sort $\ulcorner c \urcorner$ for each object c of $\mathcal{C}$ and one function symbol $\ulcorner f \urcorner$ for each arrow f in $\mathcal{E}$ between objects in $\mathcal{C}$ (cf. section 1.3.1.1). Then there exists a geometric theory $\mathbb{T}$ over the signature $\Sigma_\mathcal{C}^\text{f}$ classified by $\mathcal{E}$, whose universal model is given by the 'tautological' $\Sigma_\mathcal{C}^\text{f}$-structure in $\mathcal{E}$ (cf. p. 837 of [55]).
- Let B be a pre-bound for $\mathcal{E}$ over **Set** (that is, an object such that the subobjects of its finite powers form a separating set of objects for $\mathcal{E}$); then there exists a one-sorted geometric theory $\mathbb{T}$ classified by $\mathcal{E}$ and a universal model of $\mathbb{T}$ in it whose underlying object is B (cf. Theorem D3.2.5 [55]).

The first result can be obtained from Theorem 2.1.29 by observing that the theory $\mathbb{T}$ of the tautological $\Sigma_\mathcal{C}^\text{f}$-structure satisfies all the hypotheses of the theorem; that the first two conditions are satisfied is obvious, while the fact that the third holds can be proved as follows. Since every object of the syntactic category of $\mathbb{T}$ is a subobject of a finite product of objects of the form $\{x^{\ulcorner c \urcorner} . \top\}$ (for $c \in \mathcal{C}$), it is enough to prove that every arrow $k : R \to c$ in $\mathcal{E}$ having as domain the interpretation $R = [[\vec{x} . \phi]]_M$ of a geometric formula $\phi(\vec{x})$ over the signature of $\mathbb{T}$ and as codomain an object c in $\mathcal{C}$ is definable. Suppose that the sorts of the variables in $\vec{x} = (x_1, \ldots, x_n)$ are respectively $\ulcorner c'_1 \urcorner, \ldots, \ulcorner c'_n \urcorner$ and let $r : R \to c'_1 \times \cdots \times c'_n$ denote the corresponding subobject. Since $\mathcal{C}$ is a separating set of objects for $\mathcal{E}$, the family of arrows $\{ f_i : c_i \to R \mid i \in I \}$ from objects of $\mathcal{C}$ to R is epimorphic, whence the geometric formula

$$\bigvee_{i \in I} (\exists z_i^{\ulcorner c_i \urcorner})(x_1 = \ulcorner g_1 \urcorner (z_i) \wedge \cdots \wedge x_n = \ulcorner g_n \urcorner (z_i) \wedge x^{\ulcorner c \urcorner} = \ulcorner k \circ f_i \urcorner (z_i)),$$

where $r \circ f_i = <g_1, \ldots, g_n>$, is $\mathbb{T}$-provably functional from $\{\vec{x} . \phi\}$ to $\{x^{\ulcorner c \urcorner} . \top\}$ and its interpretation coincides with k. By Diaconescu's equivalence, the theory $\mathbb{T}$ can be explicitly characterized as the theory of flat $J_\mathcal{E}^\mathcal{C}$-continuous functors on $\mathcal{C}$, where $J_\mathcal{E}^\mathcal{C}$ is the Grothendieck topology on $\mathcal{C}$ induced by the canonical topology on $\mathcal{E}$.

The second result can be deduced from Theorem 2.1.29 by taking $\mathbb{T}$ to be the theory of the tautological structure over the one-sorted signature Σ_B consisting of one n-ary relation symbol $\ulcorner R \urcorner$ for each subobject $R \rightarrowtail B^n$ in $\mathcal{E}$. The fact that $\mathbb{T}$ satisfies the first two conditions of the theorem is obvious, while the validity of the third condition follows from the fact that the graphs of morphisms $R \rightarrowtail B^n \to B$ in $\mathcal{E}$ are interpretations of $(n+1)$-ary relation symbols over Σ_B.

2.2 Toposes as 'bridges'

In this section we describe the unifying methodology 'toposes as bridges' introduced in [16]. This general technique, which has already found applications in different fields of mathematics (see [28]), will be systematically applied throughout the book, either implicitly or explicitly, in a variety of different contexts, so it is essential for the reader to acquire familiarity with it.

2.2.1 The 'bridge-building' technique

The 'bridge-building' technique allows one to construct topos-theoretic 'bridges' connecting distinct mathematical theories with each other.

Specifically, if $\mathbb{T}$ and $\mathbb{T}'$ are two Morita-equivalent theories (that is, geometric theories classified by the same topos) then their common classifying topos can be used as a 'bridge' for transferring information between them:

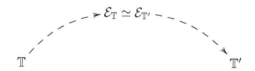

The transfer of information between $\mathbb{T}$ and $\mathbb{T}'$ takes place by expressing topos-theoretic *invariants* (that is, properties of (Grothendieck) toposes or constructions on them which are stable under categorical equivalence) defined on their common classifying topos directly in terms of the theories $\mathbb{T}$ and $\mathbb{T}'$. This is usually done by associating with each of the two theories a site of definition for its classifying topos (for example, its geometric syntactic site) and then considering invariants on the common classifying topos from the points of view of the two sites of definition. More precisely, suppose that $(\mathcal{C}, J)$ and $(\mathcal{D}, K)$ are two sites of definition for the same topos, and that I is a topos-theoretic invariant. Then one can seek *site characterizations* for I, that is, in the case I is a property (the case of I being a 'construction' admits an analogous treatment), logical equivalences of the kind 'the topos $\mathcal{E}$ satisfies I if and only if $(\mathcal{C}, J)$ satisfies a property $P_{(\mathcal{C},J)}$ (written in the language of the site $(\mathcal{C}, J)$)' and, similarly for $(\mathcal{D}, K)$, logical equivalences of the kind 'the topos $\mathbf{Sh}(\mathcal{D}, K)$ satisfies I if and only if $(\mathcal{D}, K)$ satisfies a property $Q_{(\mathcal{D},K)}$':

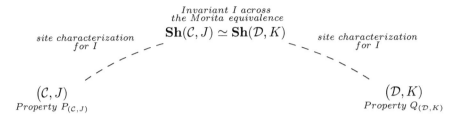

Clearly, such characterizations immediately lead to a logical equivalence between the properties $P_{(\mathcal{C},J)}$ and $Q_{(\mathcal{D},K)}$, which can thus be seen as different manifestations of a unique property, namely I, in the context of the two different sites $(\mathcal{C}, J)$ and $(\mathcal{D}, K)$.

In fact, one does not necessarily need 'if and only if' site characterizations in order to build 'bridges': in order to establish an implication between a property $P_{(\mathcal{C},J)}$ of a site $(\mathcal{C}, J)$ and a property $Q_{(\mathcal{D},K)}$ of another site of definition $(\mathcal{D}, K)$ of the same topos, it suffices to find an invariant I such that, on the one hand, $P_{(\mathcal{C},J)}$ implies I on $\mathbf{Sh}(\mathcal{C}, J)$ and, on the other hand, I on $\mathbf{Sh}(\mathcal{D}, K)$ implies $Q_{(\mathcal{D},K)}$.

The 'bridge' technique allows us to interpret and study many correspondences, dualities and equivalences arising in different fields of mathematics by means of the investigation of the way topos-theoretic invariants express themselves in terms of sites. In other words, the representation theory of Grothendieck toposes becomes a sort of 'meta-theory of mathematical duality' which makes it possible to effectively compare distinct mathematical theories with each other and transfer knowledge between them. In the following sections we discuss in more detail the subject of *Morita equivalences*, which play the role of 'decks' of our 'bridges', and of *site characterizations* for topos-theoretic invariants, which constitute their 'arches'.

Incidentally, it should be noted that this method could be generalized to the case of 'bridges' whose decks are provided by some kind of relationship between toposes which is not necessarily an equivalence, in the presence of properties or constructions of toposes which are invariant with respect to such a relation. Nonetheless, the advantage of focusing on Morita equivalences is twofold; on one hand, due to the fact that every property expressed in categorical language is automatically invariant with respect to categorical equivalence, we dispose of an unlimited number of invariants readily available for consideration, while on the other hand Morita equivalences realize a *unification* of 'concrete' properties of different theories by means of their interpretation as different *manifestations* of a unique property lying at the topos-theoretic level.

2.2.2 Decks of 'bridges': *Morita equivalences*

Let us first recall the following classical definition (cf. sections D1.4 and D3.1 of [55]):

Definition 2.2.1 Two geometric theories $\mathbb{T}$ and $\mathbb{T}'$ are said to be *Morita-equivalent* if they have equivalent classifying toposes, i.e. if they have equivalent categories of models in every Grothendieck topos $\mathcal{E}$, naturally in $\mathcal{E}$, that is for each Grothendieck topos $\mathcal{E}$ there is an equivalence of categories

$$\tau_{\mathcal{E}} : \mathbb{T}\text{-mod}(\mathcal{E}) \to \mathbb{T}'\text{-mod}(\mathcal{E})$$

such that for any geometric morphism $f : \mathcal{F} \to \mathcal{E}$ the following diagram commutes (up to isomorphism):

$$\begin{array}{ccc} \mathbb{T}\text{-mod}(\mathcal{E}) & \xrightarrow{\tau_{\mathcal{E}}} & \mathbb{T}'\text{-mod}(\mathcal{E}) \\ f^* \downarrow & & \downarrow f^* \\ \mathbb{T}\text{-mod}(\mathcal{F}) & \xrightarrow{\tau_{\mathcal{F}}} & \mathbb{T}'\text{-mod}(\mathcal{F}). \end{array}$$

Note that 'to be Morita-equivalent to each other' defines an equivalence relation on the collection of all geometric theories.

Given the level of technical sophistication of this definition, it is reasonable to wonder whether Morita equivalences naturally arise in mathematics and, if that is the case, whether there are systematic ways of 'generating' them. The following remarks are meant to show that the answer to both questions is positive.

- If two geometric theories $\mathbb{T}$ and $\mathbb{T}'$ have equivalent categories of models in the category **Set** then, provided that the given categorical equivalence is established by only using constructive logic (that is, by avoiding in particular the law of excluded middle and the axiom of choice) and geometric constructions (that is, by only using set-theoretic constructions which involve finite limits and arbitrary colimits, equivalently which admit a syntactic formulation involving only equalities, finite conjunctions, (possibly) infinitary disjunctions and existential quantifications), one can reasonably expect the original equivalence to 'lift' to a Morita equivalence between $\mathbb{T}$ and $\mathbb{T}'$. Indeed, as we have seen in section 1.3.5, a Grothendieck topos behaves logically as a 'generalized universe of sets' in which one can perform most of the classical set-theoretic arguments and constructions, with the significant exception of those requiring non-constructive principles. One could thus expect to be able to generalize the original equivalence between the categories of set-based models of the two theories to the case of models in arbitrary Grothendieck toposes; the fact that the constructions involved in the definition of the equivalence are geometric would then ensure that the above-mentioned naturality condition for Morita equivalences be satisfied (since geometric constructions are preserved by inverse image functors of geometric morphisms). As examples of 'liftings' of naturally arising categorical equivalences to Morita equivalences we mention the Morita equivalence between MV-algebras and abelian ℓ-groups with strong unit (cf. [30]) and that between abelian ℓ-groups and perfect MV-algebras (cf. [31]).

- Two cartesian (in particular, finitary algebraic) theories $\mathbb{T}$ and $\mathbb{T}'$ have equivalent categories of models in **Set** if and only if they are Morita-equivalent (or, equivalently, if and only if their cartesian syntactic categories are equivalent). Indeed, $\mathbb{T}$-mod(**Set**) $\simeq \mathbb{T}'$-mod(**Set**) if and only if f.p.$\mathbb{T}$-mod(**Set**) $\simeq$ f.p.$\mathbb{T}'$-mod(**Set**), if and only if $\mathcal{C}_{\mathbb{T}}^{\text{cart}} \simeq \mathcal{C}_{\mathbb{T}'}^{\text{cart}}$, if and only if $\mathcal{E}_{\mathbb{T}} \simeq \mathcal{E}_{\mathbb{T}'}$ (cf. Theorem 2.1.21).

- If two geometric (resp. regular, coherent) theories have equivalent geometric (resp. regular, coherent) syntactic categories (i.e. they are bi-interpretable in the sense of Definition 2.1.13) then they are Morita-equivalent. This follows at once from Theorem 2.1.8 by observing that the 'logical' topologies in the syntactic sites are defined intrinsically in terms of the categorical structure present on the relevant syntactic categories. Anyway, as can be naturally expected, the most interesting Morita equivalences do not arise from bi-interpretations.

  In particular, if two finitary first-order theories are bi-interpretable (in the sense of classical model theory) then their Morleyizations (in the sense of Proposition 2.1.1) are Morita-equivalent. In fact, from Proposition 2.1.2(ii) it follows at once that the syntactic Boolean pretoposes of two classical first-order theories are equivalent if and only if their Morleyizations are Morita-equivalent.

- Two associative rings with unit are Morita-equivalent (in the classical, ring-theoretic, sense) if and only if the algebraic theories axiomatizing the (left) modules over them are Morita-equivalent (in the topos-theoretic sense). Indeed, by the second remark above, these theories are Morita-equivalent if and only if their categories of set-based models are equivalent, that is if and only if the categories of (left) modules over the two rings are equivalent. Specifically, for each associative ring with unit R, the theory axiomatizing its (left) R-modules can be defined as the theory obtained from the algebraic theory of abelian groups by adding one unary function symbol for each element of the ring and writing down the obvious equational axioms which express the conditions in the definition of an R-module.

- Other notions of Morita equivalence for various kinds of algebraic or geometric structures considered in the literature can be naturally reformulated as equivalences between different representations of the same topos, and hence as Morita equivalences between different geometric theories. For instance:
    - Two topological groups are Morita-equivalent (in the sense of [65]) if and only if their toposes of continuous actions are equivalent. Natural analogues of this notion for topological and localic groupoids have been studied by several authors and a summary of the main results is contained in section C5.3 of [55].
    - Two small categories are Morita-equivalent (in the sense of [43]) if and only if the corresponding presheaf toposes are equivalent, that

is, if and only if their Cauchy completions (also called Karoubian completions) are equivalent (cf. [10]).
- Two inverse semigroups are Morita-equivalent (in the sense of [70] or, equivalently, of [44]) if and only if their classifying toposes (as defined in [45]) are equivalent (cf. [44]).

- Categorical dualities or equivalences between 'concrete' categories can often be seen as arising from the process of 'functorializing' Morita equivalences which express structural relationships between each pair of objects corresponding to each other under the given duality or equivalence (cf. for instance [16], [26] and [22]). In fact, the theory of geometric morphisms of toposes provides various natural ways of 'functorializing' bunches of Morita equivalences.

- Different small sites of definition for a given topos can be interpreted logically as Morita equivalences between different theories (cf. Theorem 2.1.11); in fact, the converse also holds, in the sense that any Morita equivalence between two geometric theories gives rise canonically to two different sites of definition of their common classifying topos. The representation theory of Grothendieck toposes in terms of sites and, more generally, any technique that allows one to obtain different sites of definition or representations for a given topos (for example, Theorem 1.1.8) thus constitutes a tool for generating Morita equivalences.

- The usual notions of *spectra* for mathematical structures can be naturally interpreted in terms of classifying toposes, and the resulting sheaf representations as arising from Morita equivalences between 'algebraic' and 'topological' representations of such toposes. More specifically, Cole's general theory of spectra [34] (cf. also section 6.5 of [50] for a succinct overview of this theory) is based on the construction of classifying toposes of suitable theories. Coste introduced in [38] alternative representations of such classifying toposes, identifying in particular simple sets of conditions under which they can be represented as toposes of sheaves on a topological space; he then derived from the equivalence between two of these representations, one of essentially algebraic nature and the other of topological nature, a criterion for the canonical homomorphism from the given algebraic structure to the structure of global sections of the associated structure sheaf to be an isomorphism.

- The notion of Morita equivalences materializes in many situations the intuitive feeling of 'looking at the same thing in different ways', meaning, for instance, describing the same structure(s) in different languages or constructing a given object in different ways. Concrete examples of this general remark can be found for instance in [16] and [22], where the different constructions of the Zariski spectrum of a ring, of the Gelfand spectrum of a C^*-algebra and of the Stone-Čech compactification of a topological space are interpreted as Morita equivalences between different theories.

- Different ways of looking at a given mathematical theory can often be formalized as Morita equivalences. Indeed, a different way of describing the structures axiomatized by a certain theory often gives rise to a theory written in a different language whose models (in any Grothendieck topos) can be identified, in a natural way, with those of the former theory and which is therefore Morita-equivalent to it.

- A geometric theory *alone* generates an infinite number of Morita equivalences via its 'internal dynamics'. In fact, any way of looking at a geometric theory as an extension of a geometric theory written over its signature provides a different representation of its classifying topos as a subtopos of the classifying topos of the latter theory (cf. the duality theorem of Chapter 3).

- As we remarked in section 1.1.2, different separating sets of objects for a given topos give rise to different sites of definition for it; indeed, for any separating set of objects $\mathcal{C}$ of a Grothendieck topos we have an equivalence $\mathcal{E} \simeq \mathbf{Sh}(\mathcal{C}, J_{\mathcal{E}}|_{\mathcal{C}})$, where $J_{\mathcal{E}}|_{\mathcal{C}}$ is the Grothendieck topology on $\mathcal{C}$ induced by the canonical topology $J_{\mathcal{E}}$ on $\mathcal{E}$. In particular, for any topological space X and any basis $\mathcal{B}$ for it, we have an equivalence $\mathbf{Sh}(X) \simeq \mathbf{Sh}(\mathcal{B}, J_{\mathbf{Sh}(X)}|_{\mathcal{B}})$ (cf. [16] and [22] for examples of dualities arising from Morita equivalences of this form where the induced topologies $J_{\mathbf{Sh}(X)}|_{\mathcal{B}}$ can be characterized intrinsically by means of topos-theoretic invariants).

2.2.3 Arches of 'bridges': *site characterizations*

As we remarked above, the 'arches' of topos-theoretic 'bridges' should be provided by site characterizations for topos-theoretic invariants, that is results connecting invariant properties of (resp. constructions on) toposes and properties of (resp. constructions on) their sites of definition (written in their respective languages).

It thus becomes crucial to investigate the behaviour of topos-theoretic invariants with respect to sites. As a matter of fact, this behaviour is often very natural, in the sense that topos-theoretic invariants generally admit natural site characterizations. For instance, 'if and only if' characterizations for a wide class of geometric invariants of toposes, notably including the property of a topos being atomic (resp. locally connected, localic, equivalent to a presheaf topos, compact, two-valued), were obtained in [21]. On the other hand, it was shown in [25] that a wide class of logically inspired invariants of toposes, obtained by interpreting first-order formulae written in the language of Heyting algebras in the internal Heyting algebra given by the subobject classifier (cf. Theorem 1.3.29), admit elementary 'if and only if' site characterizations. Moreover, we shall see in Chapters 3 and 4 that several notable invariants of subtoposes admit natural site chacterizations as well as explicit logical descriptions in terms of the theories classified by them. Topos-theoretic invariants relevant in algebraic geometry and homotopy theory, such as cohomology and homotopy groups of toposes, also admit, at least in many important cases, natural characterizations in terms of sites.

It should be noted that, whilst it is often possible to obtain, by using topos-theoretic methods, site characterizations for topos-theoretic invariants holding for large classes of sites, such criteria can be highly non-trivial as far as their mathematical depth is concerned (since the representation theory of toposes is by all means a non-trivial subject) and hence, when combined with specific Morita equivalences to form 'bridges', they can lead to deep results on the relevant theories, especially when the given Morita equivalence is a non-trivial one. These insights can actually be quite surprising when observed from a concrete point of view (that is from the point of view of the two theories related by the Morita equivalence), since a given topos-theoretic invariant may manifest itself in very different ways in the context of different sites. For instance, as shown in [13], for any (small) site $(\mathcal{C}, J)$ the topos $\mathbf{Sh}(\mathcal{C}, J)$ satisfies the invariant property of being De Morgan if and only if for any object c of the category $\mathcal{C}$ and any J-closed sieve R on c, the sieve

$$\{ f : d \to c \mid (f^*(R) = R_d) \text{ or } (\text{for any } g : e \to d, g^*(f^*(R)) = R_e \Rightarrow g \in R_d)\}$$

is J-covering, where $R_c := \{ f : d \to c \mid \emptyset \in J(d)\}$ (for any $c \in \mathcal{C}$).

This property specializes, on a presheaf topos $[\mathcal{C}^{\mathrm{op}}, \mathbf{Set}]$, to the property of $\mathcal{C}$ to satisfy the right Ore condition, and on a topos $\mathbf{Sh}(X)$ of sheaves on a topological space X, to the property of X to be extremely disconnected (i.e. the property that the closure of any open set is open) – cf. also [52]. We shall see in the book many other examples enlightening the natural behaviour of topos-theoretic invariants with respect to sites.

The 'centrality' of topos-theoretic invariants in mathematics is well illustrated by the fact that, in spite of their apparent remoteness from the more 'concrete' objects of study in mathematics, once translated at the level of sites or theories, they often specialize to construction of natural mathematical or logical interest. Besides cohomology and homotopy groups of toposes, whose 'concrete' instantiations in the context of topological spaces and schemes have been of central importance in topology and algebraic geometry throughout the last decades, a great number of other invariants, including many which might seem too abstract to be connected to any problem of natural mathematical interest, can be profitably used to shed light on classical theories. For example, a logically inspired construction such as the *DeMorganization* of a topos (introduced in [13] as the largest dense subtopos of a given topos satisfying De Morgan's law) was shown in [29] to yield, when applied to a specific topos such as the classifying topos of the coherent theory of fields, the classifying topos of a very natural mathematical theory, namely the theory of fields of finite characteristic which are algebraic over their prime field. The author's papers contain several other examples, some of which will be presented in the later chapters of the book.

It should be noted that the arches of our 'bridges' need not necessarily be 'symmetric', that is arising from an instantiation in the context of two given sites of a unique site characterization holding for both of them. As an example, take the property of a topos to be coherent: this invariant admits an 'elementary' site characterization holding for all trivial sites (i.e. sites whose underlying Grothendieck topology is trivial), and the following implication holds: if a theory is coherent then its classifying topos is coherent. These criteria can for instance be combined to obtain a 'bridge' yielding a result on coherent theories classified by a presheaf topos.

The level of mathematical 'depth' of the results obtained by applying the 'bridge' technique can vary enormously, depending on the complexity of the site characterizations and of the given Morita equivalence. Still, as we shall see at various points of the book, even very simple invariants applied to easily established Morita equivalences can yield surprising insights which would be hardly attainable, or not even imaginable, otherwise.

Lastly, it is worth noting that one can apply the 'bridge' technique, described in section 2.2.1 for sites, also in the case of alternative mathematical objects used for representing Grothendieck toposes (such as topological or localic groupoids, or quantales), replacing site characterizations with appropriate characterizations of the given invariant in terms of such objects.

2.2.4 Some simple examples

In this section we shall derive, as applications of the 'bridge' technique, some logical results which settle questions left unanswered in the previous sections of the chapter (cf. Remarks 1.2.14 and 2.1.18(a)).

Recall that for any fragment of geometric logic there is a corresponding notion of provability for theories in that fragment; for instance, we have a notion of provability of regular sequents in regular logic and a notion of provability of coherent sequents in coherent logic. A natural question is whether these notions of provability are compatible with each other, that is, if the notion of provability in a given fragment of logic specializes to the notion of provability in a smaller fragment when applied to sequents lying in that fragment (note that a theory in a given fragment can always be considered as a theory in any larger fragment). Similar questions can be posed concerning the notion of bi-interpretability and of completeness of a theory in a given fragment of logic with respect to its set-based models.

There is a natural topos-theoretic way of dealing with all these questions, which exploits the fact that for a theory in a given fragment of geometric logic one has multiple syntactic representations of its classifying topos, each corresponding to a particular larger fragment of logic in which the theory can be considered (cf. Theorem 2.1.8). For example, given a coherent theory $\mathbb{T}$ over a signature Σ, we have two syntactic representations of its classifying topos: as the category of sheaves $\mathbf{Sh}(\mathcal{C}_{\mathbb{T}}^{\text{coh}}, J_{\mathbb{T}}^{\text{coh}})$ on the coherent syntactic site $(\mathcal{C}_{\mathbb{T}}^{\text{coh}}, J_{\mathbb{T}}^{\text{coh}})$ of $\mathbb{T}$ and as the category of sheaves $\mathbf{Sh}(\mathcal{C}_{\mathbb{T}}, J_{\mathbb{T}})$ on the geometric syntactic site $(\mathcal{C}_{\mathbb{T}}, J_{\mathbb{T}})$ of $\mathbb{T}$. Transferring appropriate invariants across such different representations will yield the desired results:

Toposes as 'bridges' 77

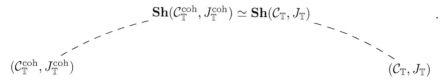

We shall address these questions one by one, highlighting in each case the relevant topos-theoretic invariants and site characterizations.

Below, by a *fragment L* of geometric logic we mean either cartesian, regular, coherent or geometric logic. Clearly, each of this fragments is contained in the successive in the list. By a sequent in cartesian (resp. regular, coherent, geometric) logic we mean a cartesian (resp. regular, coherent, geometric) sequent.

Theorem 2.2.2 *Let $\mathbb{T}$ be a theory in a fragment L of geometric logic and σ a sequent in L over its signature. Then σ is provable in $\mathbb{T}$, regarded as a theory in L, if and only if it is provable in $\mathbb{T}$, regarded as a theory in any fragment of geometric logic containing L.*

Proof We shall give the proof of the theorem in the case when $\mathbb{T}$ is a coherent theory, regarded in both coherent logic and geometric logic. The other cases can be proved in a completely analogous way.

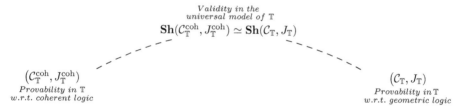

From the fact that the Yoneda embedding $\mathcal{C}_{\mathbb{T}}^{\text{coh}} \to \mathbf{Sh}(\mathcal{C}_{\mathbb{T}}^{\text{coh}}, J_{\mathbb{T}}^{\text{coh}})$ is conservative, it follows that a coherent sequent over Σ is valid in the universal model of $\mathbb{T}$ lying in the topos $\mathbf{Sh}(\mathcal{C}_{\mathbb{T}}^{\text{coh}}, J_{\mathbb{T}}^{\text{coh}})$ if and only if it is valid in the universal model $M_{\mathbb{T}}^{\text{coh}}$ of $\mathbb{T}$ lying in its coherent syntactic category $\mathcal{C}_{\mathbb{T}}^{\text{coh}}$, i.e. if and only if it is provable in $\mathbb{T}$ by using coherent logic (cf. Theorem 1.4.5 and Remark 2.1.9). Similarly, one obtains that a geometric sequent over Σ is valid in the universal model of $\mathbb{T}$ lying in the topos $\mathbf{Sh}(\mathcal{C}_{\mathbb{T}}, J_{\mathbb{T}})$ if and only if it is provable in $\mathbb{T}$ by using geometric logic. But the property of validity of a given sequent in the universal model of a geometric theory is a topos-theoretic invariant, whence a coherent sequent is provable in $\mathbb{T}$ by using coherent logic if and only if it is provable in $\mathbb{T}$ by using geometric logic, as required. □

Let us now turn to the topic of 'classical completeness' of theories in fragments of geometric logic. In section 1.4.5 we introduced the property, for a theory considered within a given fragment of geometric logic, of having enough set-based models. It is natural to wonder whether this property depends on the fragment of logic in which the theory is considered. We shall now see, again by applying the 'bridge technique', that this is not the case.

Theorem 2.2.3 *Let $\mathbb{T}$ be a theory in a given fragment L of geometric logic. Then $\mathbb{T}$ has enough set-based models, regarded as a theory in L, if and only if it has enough set-based models, regarded as a theory in any larger fragment of geometric logic containing L.*

Proof As we did for the previous result, we shall give the proof of the theorem in the case of a coherent theory $\mathbb{T}$, regarded both in coherent logic and in geometric logic, the other cases being completely analogous.

The result naturally arises from the transfer of an invariant, namely the property of a topos of having enough points, across the two different representations $\mathbf{Sh}(\mathcal{C}_{\mathbb{T}}^{\mathrm{coh}}, J_{\mathbb{T}}^{\mathrm{coh}}) \simeq \mathbf{Sh}(\mathcal{C}_{\mathbb{T}}, J_{\mathbb{T}})$ of the classifying topos of $\mathbb{T}$. Indeed, in terms of the site $(\mathcal{C}_{\mathbb{T}}^{\mathrm{coh}}, J_{\mathbb{T}}^{\mathrm{coh}})$ the invariant rephrases as the condition that every coherent sequent over Σ which is satisfied in all the set-based models of $\mathbb{T}$ should be provable in $\mathbb{T}$ by using coherent logic, while in terms of the site $(\mathcal{C}_{\mathbb{T}}, J_{\mathbb{T}})$ it rephrases as the condition that every geometric sequent over Σ which is satisfied in all the set-based models of $\mathbb{T}$ should be provable in $\mathbb{T}$ by using geometric logic:

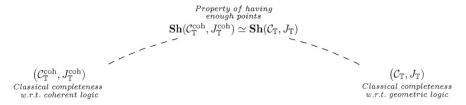

The above-mentioned site characterizations for the property of the classifying topos of having enough points can be proved as follows. By definition, $\mathbf{Sh}(\mathcal{C}_{\mathbb{T}}^{\mathrm{coh}}, J_{\mathbb{T}}^{\mathrm{coh}})$ has enough points if and only if the inverse image functors f^* of the geometric morphisms $f : \mathbf{Set} \to \mathbf{Sh}(\mathcal{C}_{\mathbb{T}}^{\mathrm{coh}}, J_{\mathbb{T}}^{\mathrm{coh}})$ are jointly conservative. Now, since the geometric morphism $f_M : \mathbf{Set} \to \mathbf{Sh}(\mathcal{C}_{\mathbb{T}}^{\mathrm{coh}}, J_{\mathbb{T}}^{\mathrm{coh}})$ corresponding to a $\mathbb{T}$-model M in $\mathbf{Set}$ satisfies $f^*(y^{\mathrm{coh}}(M_{\mathbb{T}}^{\mathrm{coh}})) = M$ (where $M_{\mathbb{T}}^{\mathrm{coh}}$ is the universal model of $\mathbb{T}$ lying in the coherent syntactic category $\mathcal{C}_{\mathbb{T}}^{\mathrm{coh}}$ of $\mathbb{T}$ – as in Definition 1.4.4 and Remark 1.4.6 – and $y^{\mathrm{coh}} : \mathcal{C}_{\mathbb{T}}^{\mathrm{coh}} \to \mathbf{Sh}(\mathcal{C}_{\mathbb{T}}^{\mathrm{coh}}, J_{\mathbb{T}}^{\mathrm{coh}})$ is the Yoneda embedding) we have that, by Theorem 1.4.5 and the fact that y^{coh} is conservative, if the topos $\mathbf{Sh}(\mathcal{C}_{\mathbb{T}}^{\mathrm{coh}}, J_{\mathbb{T}}^{\mathrm{coh}})$ has enough points then any coherent sequent over Σ which is satisfied in every $\mathbb{T}$-model M in $\mathbf{Set}$ is satisfied in $M_{\mathbb{T}}^{\mathrm{coh}}$ i.e. provable in $\mathbb{T}$ within coherent logic. Conversely, the fact that $\mathcal{C}_{\mathbb{T}}^{\mathrm{coh}}$ is a separating set of objects for $\mathbf{Sh}(\mathcal{C}_{\mathbb{T}}^{\mathrm{coh}}, J_{\mathbb{T}}^{\mathrm{coh}})$ easily implies that if $\mathbb{T}$ satisfies classical completeness with respect to coherent logic then $\mathbf{Sh}(\mathcal{C}_{\mathbb{T}}^{\mathrm{coh}}, J_{\mathbb{T}}^{\mathrm{coh}})$ has enough points. A similar characterization can be obtained for the geometric syntactic site of $\mathbb{T}$ (cf. Proposition 4.2.20 below). □

Finally, let us discuss the independence of the notion of bi-interpretability of theories (in the sense of Definition 1.4.12) from the fragment of logic in which they are considered.

Theorem 2.2.4 *Let $\mathbb{T}$ and $\mathbb{S}$ be two theories in a fragment L of geometric logic. Then $\mathbb{T}$ and $\mathbb{S}$ are bi-interpretable within the fragment L if and only if they are bi-interpretable within any fragment of geometric logic containing L.*

Proof We shall first give the proof of the theorem in the case of two regular theories $\mathbb{T}$ and $\mathbb{S}$, considered both in regular logic and in coherent logic, and then discuss how to modify the proof in the other cases.

The 'only if' direction follows immediately from the universal properties of regular and coherent syntactic categories. Indeed, by Theorem 1.4.10 any equivalence $\mathcal{C}_{\mathbb{T}}^{\text{reg}} \simeq \mathcal{C}_{\mathbb{S}}^{\text{reg}}$ induces an equivalence between the categories of models of $\mathbb{T}$ and $\mathbb{S}$ in any regular, and hence in particular in any coherent, category $\mathcal{D}$, naturally in $\mathcal{D}$. The 2-dimensional Yoneda lemma (cf. section 3.1.1) thus yields an equivalence $\mathcal{C}_{\mathbb{T}}^{\text{coh}} \simeq \mathcal{C}_{\mathbb{S}}^{\text{coh}}$, as desired.

The 'if' direction can be proved by using the 'bridge' technique, as follows.

In order to deduce that $\mathcal{C}_{\mathbb{T}}^{\text{reg}} \simeq \mathcal{C}_{\mathbb{S}}^{\text{reg}}$ from $\mathcal{C}_{\mathbb{T}}^{\text{coh}} \simeq \mathcal{C}_{\mathbb{S}}^{\text{coh}}$, it will be sufficient to characterize $\mathcal{C}_{\mathbb{T}}^{\text{reg}}$ as a full subcategory of $\mathcal{C}_{\mathbb{T}}^{\text{coh}}$ in an invariant way. Since both these categories fully embed in the classifying topos of $\mathbb{T}$ (as the topologies $J_{\mathbb{T}}^{\text{reg}}$ and $J_{\mathbb{T}}^{\text{coh}}$ are subcanonical) it would be enough to characterize in invariant terms the objects of the classifying topos of $\mathbb{T}$ which come from the regular syntactic site of $\mathbb{T}$ among those which come from the coherent syntactic site of $\mathbb{T}$. We notice that all the objects of $\mathbf{Sh}(\mathcal{C}_{\mathbb{T}}^{\text{reg}}, J_{\mathbb{T}}^{\text{reg}})$ coming from $\mathcal{C}_{\mathbb{T}}^{\text{reg}}$ are supercompact (cf. Definition 2.1.14). Indeed, the property of the site $(\mathcal{C}_{\mathbb{T}}^{\text{reg}}, J_{\mathbb{T}}^{\text{reg}})$ to be regular implies that the image under the Yoneda embedding $y^{\text{reg}} : \mathcal{C}_{\mathbb{T}}^{\text{reg}} \to \mathbf{Sh}(\mathcal{C}_{\mathbb{T}}^{\text{reg}}, J_{\mathbb{T}}^{\text{reg}})$ of any object of $\mathcal{C}_{\mathbb{T}}^{\text{reg}}$ is a supercoherent and *a fortiori* supercompact object of $\mathbf{Sh}(\mathcal{C}_{\mathbb{T}}^{\text{reg}}, J_{\mathbb{T}}^{\text{reg}})$ (cf. Proposition 10.8.1(ii)).

On the other hand, one can easily characterize in terms of the coherent syntactic site of $\mathbb{T}$ the property of an object $\{\vec{x} \,.\, \phi\}$ of $\mathcal{C}_{\mathbb{T}}^{\text{coh}}$ to be sent by the Yoneda embedding $y^{\text{coh}} : \mathcal{C}_{\mathbb{T}}^{\text{coh}} \to \mathbf{Sh}(\mathcal{C}_{\mathbb{T}}^{\text{coh}}, J_{\mathbb{T}}^{\text{coh}})$ to a supercompact object of $\mathbf{Sh}(\mathcal{C}_{\mathbb{T}}^{\text{coh}}, J_{\mathbb{T}}^{\text{coh}})$: $y^{\text{coh}}(\{\vec{x} \,.\, \phi\})$ is supercompact in $\mathbf{Sh}(\mathcal{C}_{\mathbb{T}}^{\text{coh}}, J_{\mathbb{T}}^{\text{coh}})$ if and only if for any finite family $\{\psi_i(\vec{x}) \mid i \in I\}$ of coherent formulae in the same context, $(\phi \vdash_{\vec{x}} \bigvee_{i \in I} \psi_i)$ provable in $\mathbb{T}$ implies $(\phi \vdash_{\vec{x}} \psi_i)$ provable in $\mathbb{T}$ for some $i \in I$. Let us refer to this property of objects $\{\vec{x} \,.\, \phi\}$ of $\mathcal{C}_{\mathbb{T}}^{\text{coh}}$ as to 'finite regularity'. So, we have that an object of $\mathcal{C}_{\mathbb{T}}^{\text{coh}}$ is finitely regular if and only if every finite covering of it by subobjects in $\mathcal{C}_{\mathbb{T}}^{\text{coh}}$ is trivial.

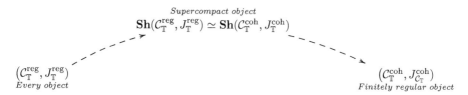

Now, it can easily be proved by induction on the structure of coherent formulae that every coherent formula is provably equivalent in coherent logic to a finite disjunction of regular formulae in the same context. This clearly implies that every finitely regular coherent formula $\{\vec{x}.\phi\}$ is isomorphic in $\mathcal{C}_{\mathbb{T}}^{\mathrm{coh}}$ to a regular formula. We can thus conclude that the full subcategory of $\mathcal{C}_{\mathbb{T}}^{\mathrm{coh}}$ on the finitely regular objects is equivalent to the category $\mathcal{C}_{\mathbb{T}}^{\mathrm{reg}}$, and similarly for the theory $\mathbb{S}$. It follows that the equivalence $\mathcal{C}_{\mathbb{T}}^{\mathrm{coh}} \simeq \mathcal{C}_{\mathbb{S}}^{\mathrm{coh}}$ restricts to an equivalence $\mathcal{C}_{\mathbb{T}}^{\mathrm{reg}} \simeq \mathcal{C}_{\mathbb{S}}^{\mathrm{reg}}$, as desired.

In the case of a cartesian theory considered in a larger fragment of geometric logic, the notion to use in the place of that of supercompact object is that of irreducible object (the only difference in the proof being that, in order to show that every formula which is sent by the Yoneda embedding to an irreducible object of the classifying topos is isomorphic in the relevant syntactic category to a cartesian formula, one uses Proposition 2.1.15).

In the case of coherent theories, the notion to use is instead that of compact object. □

2.2.5 A theory of 'structural translations'

Let us conclude this chapter with a few methodological remarks.

The view underlying the 'bridge' technique consists in regarding a topos as an object which, together with all its different representations, embodies a great amount of relationships existing between the different theories classified by it. Any topos-theoretic invariant behaves in this context like a 'pair of glasses' which allows us to discern information which is 'hidden' in a given Morita equivalence. Toposes can thus act as 'universal translators' across different mathematical theories which have the same 'semantic content'.

From a technical point of view, the main reason for the effectiveness of the 'bridge' technique is twofold: on the one hand, as we have argued in section 2.2.3, topos-theoretic invariants usually manifest themselves in significantly different ways in the context of different sites; on the other hand, due to the very well-behaved nature of the representation theory of Grothendieck toposes, site characterizations are often derivable by means of rather canonical 'calculations'.

Unlike the traditional, 'dictionary-oriented' method of translation based on a 'renaming', according to a given 'dictionary', of the primitive constituents of the information as expressed in a given language, the 'invariant-oriented' translations realized by topos-theoretic 'bridges' consist of 'structural unravelings' of invariants across different representations of the relevant toposes. Interestingly, for the transfer of 'global' properties of toposes, it is only the *existence* of a Morita equivalence that matters, rather than its explicit description, since, by its very definition, a topos-theoretic invariant is stable under *any* categorical equivalence. If one wants to establish more 'specific' results, one can use invariant properties of *objects* of toposes rather than properties of the whole toposes, in which case the knowledge of at least the 'local' behaviour of the Morita equivalence at such objects is needed, but for investigating most of the 'global' properties of theories such information is not at all necessary.

We have already hinted above to the fact that there is a strong element of *automatism* implicit in the 'bridge' technique. In fact, in order to obtain insights on the Morita equivalence under consideration, in many cases one can just readily apply to it general characterizations connecting properties of sites and topos-theoretic invariants. Still, the results generated in this way are in general non-trivial; in some cases they can be rather 'weird' according to the usual mathematical standards (although they might still be quite deep) but, with a careful choice of Morita equivalences and invariants, one can easily get interesting and natural mathematical results. In fact, a lot of information that is not visible with the usual 'glasses' is revealed by the application of this machinery.

The range of applicability of the 'bridge' technique is very broad within mathematics, by the very generality of the notion of topos (and of that of geometric theory). Through this method, results are generated transversally to the various mathematical fields in a 'uniform' way which is determined by the form of the toposes involved and the invariants considered on them. Notice that this way of doing mathematics is inherently 'upside down': instead of starting with simple ingredients and combining them to build more complicated structures, one assumes as primitive ingredients rich and sophisticated (meta-)mathematical entities, namely Morita equivalences and topos-theoretic invariants, and proceeds to extract from them 'concrete' information relevant for classical mathematics.

3
A DUALITY THEOREM

In this chapter, after establishing a number of preliminary results on Grothendieck topologies, we state and prove a duality theorem between the subtoposes of the classifying topos of a geometric theory and the *quotients* of the theory, i.e. the geometric theory extensions over its signature. We then discuss the proof-theoretic significance of the theorem, obtaining an alternative proof of it based on proof-theoretic ideas. We conclude the chapter with a deduction theorem for geometric logic analogous to the well-known one for classical first-order logic, which we establish by using the above-mentioned proof-theoretic interpretation of the duality theorem.

3.1 Preliminary results

3.1.1 A 2-dimensional Yoneda lemma

In this section we present a 2-dimensional version of the Yoneda lemma which will be useful in the sequel.

Recall that there are a number of 2-categories which naturally play a role in topos theory; among them, there are the 2-category $\mathfrak{Cat}$ of small categories, functors and natural transformations between them and the 2-category $\mathfrak{BTop}$ of Grothendieck toposes, geometric morphisms and geometric transformations between them. Also, we have all the 2-categories arising from notable fragments of geometric logic, namely the 2-category $\mathfrak{Cart}$ of cartesian categories, cartesian functors and natural transformations between them, the 2-category $\mathfrak{Reg}$ of regular categories, regular functors and natural transformations between them, the 2-category $\mathfrak{Coh}$ of coherent categories, coherent functors and natural transformations between them, and the 2-category $\mathfrak{Geom}$ of geometric categories, geometric functors and natural transformations between them.

Given a strict 2-category $\mathcal{R}$ and two 0-cells a and b in $\mathcal{R}$, we say that a and b are *equivalent* if there exist 1-cells $f : a \to b$ and $g : b \to a$ and invertible 2-cells $\alpha : f \circ g \Rightarrow 1_b$ and $\beta : g \circ f \Rightarrow 1_a$. Given a 2-category $\mathcal{R}$, we have an obvious 2-functor $Y : \mathcal{R} \to [\mathcal{R}^{\text{op}}, \mathfrak{Cat}]$ (where $[\mathcal{R}^{\text{op}}, \mathfrak{Cat}]$ is the 2-category of 2-functors $\mathcal{R}^{\text{op}} \to \mathfrak{Cat}$), which sends a 0-cell a to the (obviously defined) 2-functor $Y(a) := \mathcal{R}(-, a) : \mathcal{R}^{\text{op}} \to \mathfrak{Cat}$. Notice that the above notion of equivalence specializes in $\mathfrak{Cat}$ to the well-known notion of equivalence between small categories.

The following result is well-known, but we present a proof for the reader's convenience.

Theories, Sites, Toposes. Olivia Caramello.
© Olivia Caramello 2018. Published 2018 by Oxford University Press.

Proposition 3.1.1 *With the above notation, for any $a, b \in \mathcal{R}$, the functors $Y(a)$ and $Y(b)$ are equivalent (as 0-cells in the 2-category $[\mathcal{R}^{\mathrm{op}}, \mathfrak{Cat}]$) if and only if a and b are equivalent (as 0-cells in $\mathcal{R}$).*

Proof It is easy to see that two 2-functors $F, G : \mathcal{R}^{\mathrm{op}} \to \mathfrak{Cat}$ are equivalent if and only if, for each $c \in \mathcal{R}$, the categories $F(c)$ and $G(c)$ are equivalent via functors $K(c) : F(c) \to G(c)$ and $L(c) : G(c) \to F(c)$, naturally in $c \in \mathcal{R}$, i.e. for any 1-cell $f : c \to d$ in $\mathcal{R}$ the obvious naturality squares for both K and L commute up to an invertible natural transformation.

Now suppose that for $a, b \in \mathcal{R}$ we have that $Y(a)$ and $Y(b)$ are equivalent via transformations $K : Y(a) \Rightarrow Y(b)$ and $L : Y(b) \Rightarrow Y(a)$ such that $K \circ L \cong Y(b)$ and $L \circ K \cong Y(a)$. Then we have $K(a) : \mathcal{R}(a,a) \to \mathcal{R}(b,a)$ and $L(b) : \mathcal{R}(b,b) \to \mathcal{R}(a,b)$; let us put $f := K(a)(1_a) : a \to b$ and $g := L(b)(1_b) : b \to a$. We want to prove that $g \circ f \simeq 1_a$ and $f \circ g \simeq 1_b$. Consider the naturality square for K corresponding to the arrow $g : b \to a$:

$$\begin{array}{ccc} \mathcal{R}(a,a) & \xrightarrow{K(a)} & \mathcal{R}(a,b) \\ {\scriptstyle -\circ g}\downarrow & & \downarrow{\scriptstyle -\circ g} \\ \mathcal{R}(b,a) & \xrightarrow{K(b)} & \mathcal{R}(b,b). \end{array}$$

By our hypothesis, this square commutes up to an invertible natural transformation, so $f \circ g = K(a)(1_a) \circ g \cong K(b)(g) \cong K(b)(L(b)(1_b)) \cong 1_b$. Dually, or more explicitly by replacing K with L and f with g in the above argument, one obtains an isomorphism $g \circ f \cong 1_a$. So the 1-cells f and g give an equivalence between a and b, as required.

The 'if' direction of the proposition is trivial. □

3.1.2 *An alternative view of Grothendieck topologies*

In a category $\mathcal{C}$ we call a collection of arrows in $\mathcal{C}$ with common codomain a *presieve*; given a presieve P on $c \in \mathcal{C}$, we define the sieve $\overline{P}$ generated by P as the collection of arrows in $\mathcal{C}$ with codomain c which factor through an arrow in P.

Given a collection $\mathcal{U}$ of presieves on $\mathcal{C}$, we define the Grothendieck topology *generated by $\mathcal{U}$* as the smallest Grothendieck topology J on $\mathcal{C}$ such that all the sieves generated by the presieves in $\mathcal{U}$ are J-covering.

Given two Grothendieck topologies J and J' on a category $\mathcal{C}$ such that $J' \supseteq J$, we say that J' *is generated over J by a collection $\mathcal{U}$ of sieves in $\mathcal{C}$* if J' is generated by the collection of sieves on $\mathcal{C}$ which are either J-covering or belonging to $\mathcal{U}$.

Given a site $(\mathcal{C}, J)$ and a sieve S on an object c of $\mathcal{C}$, we denote by $\overline{S}^J$ the J-*closure* $\{f : \mathrm{dom}(f) \to c \mid f^*(S) \in J(\mathrm{dom}(f))\}$ of S; notice that $\overline{S}^J$ is the smallest J-closed sieve on c containing S. Recall that, if $\mathcal{C}$ is small, via the identification of sieves on an object c with subobjects in $[\mathcal{C}^{\mathrm{op}}, \mathbf{Set}]$ of $\mathcal{C}(-, c)$, $\overline{S}^J$

corresponds to the closure of $S \rightarrowtail \mathcal{C}(-,c)$ with respect to the universal closure operation on $[\mathcal{C}^{\mathrm{op}}, \mathbf{Set}]$ associated with the (local operator corresponding to the) Grothendieck topology J on $\mathcal{C}$ as in Theorem 1.3.36.

Remark 3.1.2 Given a functor $F : \mathcal{C} \rightarrow \mathcal{E}$, where $\mathcal{E}$ is a Grothendieck topos, and a presieve P in $\mathcal{C}$, F sends P to an epimorphic family if and only if it sends $\overline{P}$ to an epimorphic family.

We note that the classical definition of a Grothendieck topology (cf. Definition 1.1.1(c)) can also be put in the following alternative form (cf. the notion of pretopology in [3], vol. 269 II.1.3):

Definition 3.1.3 A Grothendieck topology on a category $\mathcal{C}$ is an assignment J sending any object c of $\mathcal{C}$ to a collection $J(c)$ of sieves on c in such a way that

(a) the maximal sieve M_c belongs to $J(c)$;
(b) for each pair of sieves S and T on c such that $T \in J(c)$ and $S \supseteq T$, $S \in J(c)$;
(c) if $R \in J(c)$ then for any arrow $g : d \rightarrow c$ there exists a sieve $S \in J(d)$ such that for each arrow f in S, $g \circ f \in R$;
(d) if the sieve S generated by a presieve $\{f_i : c_i \rightarrow c \mid i \in I\}$ belongs to $J(c)$ and for each $i \in I$ we have a presieve $\{g_{ij} : d_{ij} \rightarrow c_i \mid j \in I_i\}$ such that the sieve T_i generated by it belongs to $J(c_i)$, then the sieve R generated by the family of composites $\{f_i \circ g_{ij} : d_{ij} \rightarrow c \mid i \in I, j \in I_i\}$ belongs to $J(c)$.

In point (d) of this definition, the sieve R will be called the *composite* of the sieve S with the sieves T_i for $i \in I$ and denoted by $S * \{T_i \mid i \in I\}$.

Let us prove the equivalence of the two definitions.

First, let us assume the classical definition and derive the second. To prove property (b), let us assume that $S \supseteq T$ with $T \in J(c)$; then for every arrow f in T we have $f^*(S) \supseteq f^*(T) = M_c \in J(c)$ so by the transitivity axiom $S \in J(c)$, as required. Property (c) immediately follows from the stability axiom. Property (d) follows from the transitivity axiom by observing that for all arrows f in S, $f^*(R)$ is J-covering. Indeed, if $f \in S$ then $f = f_i \circ h$ for some $i \in I$ and arrow h; so $f^*(R) = h^*(f_i^*(R)) \supseteq h^*(T_i) \in J(\mathrm{dom}(h))$ and hence $f^*(R) \in J(\mathrm{dom}(f))$ by property (b) and the stability axiom.

Conversely, let us assume the second definition and derive the classical one. The stability axiom easily follows from (b) and (c); indeed, if $R \in J(c)$ and $g : d \rightarrow c$ is an arrow with codomain c, then $h^*(R)$ contains the sieve S given by property (c) and hence it is J-covering by property (b). To prove the transitivity axiom it suffices to observe that, given a sieve R on c and a sieve $S \in J(c)$ such that $h^*(R)$ is J-covering for all arrows h in S, R contains the composite of the sieve S with the sieves of the form $h^*(R)$ for h in S.

Note that, in Definition 3.1.3, one can equivalently require in property (d) that the presieves $\{f_i : c_i \rightarrow c \mid i \in I\}$ and $\{g_{ij} : d_{ij} \rightarrow c_i \mid j \in I_i\}$ be sieves; indeed, it is clear from the above arguments that both versions of the condition are equivalent, assuming properties (a), (b) and (c), to the transitivity axiom.

In the case when the category $\mathcal{C}$ has pullbacks, property (c) (equivalently, the stability axiom) may be replaced by the following condition: if the sieve generated by $\{f_i : c_i \to c \mid i \in I\}$ belongs to $J(c)$ then, for any arrow $g : d \to c$, the sieve generated by the family obtained by pulling back the arrows f_i along g belongs to $J(d)$.

Remark 3.1.4 The operation of composition of sieves in $\mathcal{C}$ as in Definition 3.1.3(d) behaves naturally with respect to the operator $\overline{(-)}^J$ of J-closure of sieves for a Grothendieck topology J on $\mathcal{C}$; that is, with the above notation, we have

$$\overline{S * \{T_i \mid i \in I\}}^J = \overline{S * \{\overline{T_i}^J \mid i \in I\}}^J.$$

To verify this equality, it clearly suffices to prove that $S * \{\overline{T_i}^J \mid i \in I\} \subseteq \overline{S * \{T_i \mid i \in I\}}^J$; but this easily follows from property (b) in Definition 3.1.3.

3.1.3 *Generators for Grothendieck topologies*

Recall from section 1.4.2 that, if $\mathcal{C}$ is a regular category, we may define the regular topology $J_{\mathcal{C}}^{\text{reg}}$ as the Grothendieck topology on $\mathcal{C}$ having as sieves exactly those which contain a cover, while if $\mathcal{C}$ is a geometric category, we may define the geometric topology $J_{\mathcal{C}}^{\text{geom}}$ as the Grothendieck topology on $\mathcal{C}$ having as sieves exactly those which contain a small covering family.

The following result about these topologies holds. Below, by a *principal sieve* we mean a sieve which is generated by a single arrow; we denote by (f) the principal sieve generated by an arrow f.

Proposition 3.1.5 *Let $\mathcal{C}$ be a category and J a Grothendieck topology on it.*

(i) *If $\mathcal{C}$ is regular and $J \supseteq J_{\mathcal{C}}^{\text{reg}}$ then J is generated over $J_{\mathcal{C}}^{\text{reg}}$ by a collection of sieves generated by a family of monomorphisms.*

(ii) *If $\mathcal{C}$ is geometric and $J \supseteq J_{\mathcal{C}}^{\text{geom}}$ then J is generated over $J_{\mathcal{C}}^{\text{geom}}$ by a collection of principal sieves generated by single monomorphisms.*

Proof

(i) Given an object $c \in \mathcal{C}$ and a sieve R on c in $\mathcal{C}$, let us denote, for each arrow r in R, by $\text{dom}(r) \overset{r''}{\twoheadrightarrow} x \overset{r'}{\rightarrowtail} c$ its cover-mono factorization in $\mathcal{C}$ and by R' the sieve in $\mathcal{C}$ generated by the arrows r' (for r in R). Clearly, it is enough to prove that $R \in J(c)$ if and only if $R' \in J(c)$. The 'only if' part follows from property (b) in Definition 3.1.3 since $R' \supseteq R$, while the 'if' part follows from property (d) in Definition 3.1.3 by noticing that, since $J \supseteq J_{\mathcal{C}}^{\text{reg}}$, all the principal sieves generated by the arrows r'' (for $r \in R$) are J-covering.

(ii) Given an object $c \in \mathcal{C}$ and a sieve R on $c \in \mathcal{C}$, let r be the subobject of c given by the union in $\text{Sub}_{\mathcal{C}}(c)$ of all the images in $\mathcal{C}$ of the morphisms in R (this union exists because, $\mathcal{C}$ being well-powered, there is, up to isomorphism, only a *set* of monomorphisms with a given codomain).

Clearly, it is enough to prove that $R \in J(c)$ if and only if $(r) \in J(c)$. The 'only if' part follows from property (b) in Definition 3.1.3 since $(r) \supseteq R$, while the 'if' part follows from property (d) in Definition 3.1.3 by observing that, since $J \supseteq J_{\mathcal{C}}^{\mathrm{geom}}$, the sieve generated by the inclusions into r of the images of the morphisms in R is J-covering.

□

Given a sieve R on a regular category $\mathcal{C}$, it is natural to consider the sieve R^{reg} generated by all the images of morphisms in R; similarly, if $\mathcal{C}$ is a geometric category, it is natural to consider the sieve R^{geom} generated by the union (in the appropriate subobject lattice) of all the images of morphisms in R. The following result provides a link between these constructions and the notions of regular and geometric topology:

Proposition 3.1.6 *Let R be a sieve on a category $\mathcal{C}$.*

(i) *If $\mathcal{C}$ is a regular category then $R^{\mathrm{reg}} = \overline{R}^{J_{\mathcal{C}}^{\mathrm{reg}}}$.*

(ii) *If $\mathcal{C}$ is a geometric category then $R^{\mathrm{geom}} = \overline{R}^{J_{\mathcal{C}}^{\mathrm{geom}}}$.*

Proof

(i) Let us begin by proving that the inclusion $R^{\mathrm{reg}} \subseteq \overline{R}^{J_{\mathcal{C}}^{\mathrm{reg}}}$ holds. Clearly, it suffices to show that for any f in R, $\mathrm{Im}(f) \in \overline{R}^{J_{\mathcal{C}}^{\mathrm{reg}}}$. Now, if $a \overset{f'}{\twoheadrightarrow} a' \overset{\mathrm{Im}(f)}{\rightarrowtail} b$ is the cover-mono factorization of f then $(f') \subseteq \mathrm{Im}(f)^{*}(R)$; but f' is a cover, so $(f') \in J_{\mathcal{C}}^{\mathrm{reg}}(a')$ and hence $\mathrm{Im}(f)^{*}(R)$ is $J_{\mathcal{C}}^{\mathrm{reg}}$-covering by property (b) in Definition 3.1.3. It remains to prove the converse inclusion. If $f \in \overline{R}^{J_{\mathcal{C}}^{\mathrm{reg}}}$ then $f^{*}(R)$ contains a cover, call it h. Since composition of covers is a cover then f factors through $\mathrm{Im}(f \circ h)$ and hence $f \in R^{\mathrm{reg}}$, as required.

(ii) Let R be a sieve generated by a presieve $\{r_i \mid i \in I\}$ on an object $c \in \mathcal{C}$ (for our purposes we can suppose I to be a set without loss of generality, every geometric category being well-powered). Let us denote by $r : d \rightarrowtail c$ the union in $\mathrm{Sub}_{\mathcal{C}}(c)$ of the subobjects $\mathrm{Im}(r_i)$ as i varies in I and by h_i the (unique) factorization of r_i through r (for each $i \in I$). To prove the inclusion $R^{\mathrm{geom}} \subseteq \overline{R}^{J_{\mathcal{C}}^{\mathrm{geom}}}$, it is enough to show that $r \in \overline{R}^{J_{\mathcal{C}}^{\mathrm{geom}}}$. Now, $r = \bigcup_{i \in I} \mathrm{Im}(r_i)$ so $1_d = r^{*}(\bigcup_{i \in I} \mathrm{Im}(r_i)) = \bigcup_{i \in I} r^{*}(\mathrm{Im}(r \circ h_i)) = \bigcup_{i \in I} \mathrm{Im}(r^{*}(r \circ h_i)) = \bigcup_{i \in I} \mathrm{Im}(h_i)$, where the second and third equalities follow from the fact that in any geometric category cover-mono factorizations and small unions of subobjects are stable under pullback and the last equality follows from the fact that r is monic. So $\{h_i \mid i \in I\}$ is a small covering family contained in $r^{*}(R)$ and hence $r^{*}(R)$ is $J_{\mathcal{C}}^{\mathrm{geom}}$-covering, as required. Conversely, let us suppose that, given $f : d \to c$, $f^{*}(R)$ contains a small covering family $\{h_j \mid j \in J\}$. We want to prove that f factors through r. Since r is monic, this condition is clearly equivalent to requiring that $f^*(r) = 1_d$. Now, $f^*(r) = f^*(\bigcup_{i \in I}\mathrm{Im}(r_i)) = \bigcup_{i \in I}\mathrm{Im}(f^*(r_i))$. For each $j \in J$ there exists $i \in I$ such that $f \circ h_j \in (r_i)$ and hence h_j factors through $f^*(r_i)$; this in turn implies that $\mathrm{Im}(h_j)$ factors through $\mathrm{Im}(f^*(r_i))$, whence $\bigcup_{i \in I}\mathrm{Im}(f^*(r_i)) \supseteq \bigcup_{j \in J}\mathrm{Im}(h_j) = 1_d$. Therefore $f^*(r) = 1_d$, as required. □

Remark 3.1.7 It follows as an immediate consequence of Proposition 3.1.6 that if $\mathcal{C}$ is a regular (resp. geometric) category then, for any sieve R on c and any arrow $f : d \to c$, $f^*(R^{\mathrm{reg}}) = f^*(R)^{\mathrm{reg}}$ (resp. $f^*(R^{\mathrm{geom}}) = f^*(R)^{\mathrm{geom}}$); indeed, universal closure operations always commute with pullbacks.

3.2 Quotients and subtoposes

3.2.1 The duality theorem

In this section, we establish a duality theorem providing a bijection between the subtoposes of the classifying topos of a given geometric theory $\mathbb{T}$ and the 'quotients' of $\mathbb{T}$ (considered up to a syntactic notion of equivalence).

Let us start with an easy remark: every subtopos (in the Lawvere-Tierney sense) of a Grothendieck topos is a Grothendieck topos. This can be seen in (at least) two different ways, as follows.

Recall that a subtopos of a topos $\mathcal{E}$ is a geometric inclusion of the form $\mathbf{sh}_j(\mathcal{E}) \hookrightarrow \mathcal{E}$ for a local operator j on $\mathcal{E}$, equivalently an equivalence class of geometric inclusions to the topos $\mathcal{E}$. By Theorem 1.3.36, the subtoposes of a presheaf topos $[\mathcal{C}^{\mathrm{op}}, \mathbf{Set}]$ are in bijection with the Grothendieck topologies J on the category $\mathcal{C}$, i.e. every geometric inclusion to $[\mathcal{C}^{\mathrm{op}}, \mathbf{Set}]$ is, up to equivalence, of the canonical form $\mathbf{Sh}(\mathcal{C}, J) \hookrightarrow [\mathcal{C}^{\mathrm{op}}, \mathbf{Set}]$ for a unique Grothendieck topology J on $\mathcal{C}$; moreover, such a geometric inclusion $\mathbf{Sh}(\mathcal{C}, J) \hookrightarrow [\mathcal{C}^{\mathrm{op}}, \mathbf{Set}]$ factors through another canonical geometric inclusion $\mathbf{Sh}(\mathcal{C}, J') \hookrightarrow [\mathcal{C}^{\mathrm{op}}, \mathbf{Set}]$ if and only if $J' \subseteq J$ (i.e. every J'-covering sieve is a J-covering sieve). Now, the geometric inclusions to a Grothendieck topos $\mathbf{Sh}(\mathcal{C}, J)$ can be clearly identified with the geometric inclusions to $[\mathcal{C}^{\mathrm{op}}, \mathbf{Set}]$ which factor through $\mathbf{Sh}(\mathcal{C}, J) \hookrightarrow [\mathcal{C}^{\mathrm{op}}, \mathbf{Set}]$ and hence the subtoposes of $\mathbf{Sh}(\mathcal{C}, J)$ correspond precisely to the Grothendieck topologies J' on $\mathcal{C}$ such that $J' \supseteq J$. This provides us with the first proof of our remark. Alternatively, we can argue as follows. By Theorem C2.2.8 [55], an elementary topos $\mathcal{E}$ is a Grothendieck topos if and only if there exists a bounded geometric morphism $\mathcal{E} \to \mathbf{Set}$ (in the sense of Definition B3.1.7 [55]). Now, a geometric inclusion is always a localic morphism (cf. Example A4.6.2(a) [55]), and hence a bounded morphism (cf. Example B3.1.8 [50]); but a composite of bounded morphism is a bounded morphism (by Lemma B3.1.10(i) [55]), whence our claim follows immediately.

This remark is fundamental for our purposes for the following reason. For each elementary topos $\mathcal{E}$, the collection of subtoposes of $\mathcal{E}$ has the structure of a coHeyting algebra (cf. Example A4.5.13(f) [55]), and there are many important concepts in topos theory that apply in this context (cf. section A4 of [50]); it is therefore natural to investigate their meaning in the setting of Grothendieck toposes. In fact, thanks to the duality theorem established below, we will also be able to interpret all these concepts in the context of geometric theories (cf. Chapter 4).

Before we can state our duality theorem, which describes how the relationship between Grothendieck toposes and geometric theories given by the theory of classifying toposes 'restricts' to the context of subtoposes of a given Grothendieck topos, we need to introduce some definitions. Concerning terminology, we use the term *theory* in the sense of axiomatic presentation (as in Definition 1.2.5); accordingly, we consider two theories over a given signature equal when they have exactly the same axioms.

Definition 3.2.1 Let $\mathbb{T}$ be a geometric theory over a signature Σ and σ, σ' two geometric sequents over Σ. Then σ and σ' are said to be $\mathbb{T}$-*equivalent* if σ is provable in $\mathbb{T} \cup \{\sigma'\}$ and σ' is provable in $\mathbb{T} \cup \{\sigma\}$.

Definition 3.2.2 Let $\mathbb{T}$ be a geometric theory over a signature Σ. A *quotient* of $\mathbb{T}$ is a geometric theory $\mathbb{T}'$ over Σ such that every axiom of $\mathbb{T}$ is provable in $\mathbb{T}'$.

Remark 3.2.3 The notion of provability in geometric logic to which we refer here is the one defined in section 1.2.3; that system is essentially constructive, but, by Theorem 1.2.13, we may add the Law of Excluded Middle to it (thus making it classical) without affecting the corresponding notion of provability.

Definition 3.2.4 Let $\mathbb{T}$ and $\mathbb{T}'$ be geometric theories over a signature Σ. We say that $\mathbb{T}$ and $\mathbb{T}'$ are *syntactically equivalent*, and we write $\mathbb{T} \equiv_s \mathbb{T}'$, if for every geometric sequent σ over Σ, σ is provable in $\mathbb{T}$ if and only if σ is provable in $\mathbb{T}'$.

We are now ready to state the duality theorem between quotients and subtoposes.

Theorem 3.2.5 *Let $\mathbb{T}$ be a geometric theory over a signature Σ. Then the assignment sending a quotient of $\mathbb{T}$ to its classifying topos yields a bijection between the $\equiv_s$-equivalence classes of quotients of $\mathbb{T}$ and the subtoposes of the classifying topos $\mathbf{Set}[\mathbb{T}]$ of $\mathbb{T}$.*

Proof Recall from section 2.1.2 that the classifying topos $\mathbf{Set}[\mathbb{T}]$ of $\mathbb{T}$ can be represented as $\mathbf{Sh}(\mathcal{C}_{\mathbb{T}}, J_{\mathbb{T}})$, where $\mathcal{C}_{\mathbb{T}}$ is the geometric syntactic category of $\mathbb{T}$ and $J_{\mathbb{T}}$ is the canonical topology on it (i.e. the Grothendieck topology on $\mathcal{C}_{\mathbb{T}}$ having as covering sieves exactly those which contain small covering families) and that we have an equivalence of categories $\mathbb{T}\text{-mod}(\mathcal{E}) \simeq \mathbf{Flat}_{J_{\mathbb{T}}}(\mathcal{C}_{\mathbb{T}}, \mathcal{E})$ (natural in $\mathcal{E} \in \mathfrak{BTop}$) which sends each model $M \in \mathbb{T}\text{-mod}(\mathcal{E})$ to the functor $F_M : \mathcal{C}_{\mathbb{T}} \to \mathcal{E}$ assigning to a formula $\{\vec{x} \,.\, \phi\}$ (the domain of) its interpretation $[[\vec{x} \,.\, \phi]]_M$ in M.

Notice that $\mathcal{C}_{\mathbb{T}}$ is an essentially small category, i.e. it is equivalent to a small category (cf. the proof of Lemma D1.4.10(iv) [55]); hence all the results valid for small sites naturally extend to sites with underlying category $\mathcal{C}_{\mathbb{T}}$.

Let us recall the construction of pullbacks in $\mathcal{C}_\mathbb{T}$. Given two morphisms

$$\{\vec{x}.\phi\} \xrightarrow{[\theta]} \{\vec{y}.\psi\}$$

and

$$\{\vec{x'}.\phi'\} \xrightarrow{[\theta']} \{\vec{y}.\psi\}$$

in $\mathcal{C}_\mathbb{T}$ with common codomain, we have the following pullback square in $\mathcal{C}_\mathbb{T}$:

$$\{\vec{x},\vec{x'}.(\exists\vec{y})(\theta[\vec{x}/\vec{x}] \wedge \theta'[\vec{x'}/\vec{x'}])\} \xrightarrow{[(\exists\vec{y})(\theta \wedge \theta' \wedge \vec{x'}=\vec{x'})]} \{\vec{x'}.\phi'\}$$
$$\downarrow{\scriptstyle[(\exists\vec{y})(\theta\wedge\theta'\wedge\vec{x}=\vec{x})]} \qquad\qquad\qquad\qquad \downarrow{\scriptstyle[\theta']}$$
$$\{\vec{x}.\phi\} \xrightarrow{\qquad\qquad[\theta]\qquad\qquad} \{\vec{y}.\psi\}.$$

We note that the sequent $(\phi' \vdash_{\vec{x'}} (\exists \vec{x},\vec{x'})((\exists\vec{y})(\theta \wedge \theta' \wedge \vec{x'} = \vec{x'})))$ is provable in $\mathbb{T}$ from the sequent $(\psi \vdash_{\vec{y}} (\exists\vec{x})\theta)$. Indeed, it is clearly equivalent in geometric logic to the sequent $(\phi' \vdash_{\vec{x'}} (\exists\vec{y})((\exists\vec{x})\theta \wedge \theta'))$, and the sequents $(\phi' \vdash_{\vec{x'}} (\exists\vec{y})\theta')$ and $(\theta' \vdash_{\vec{x'},\vec{y}} \psi)$ are provable in $\mathbb{T}$.

Next, we observe that, given a $\mathbb{T}$-model M in a Grothendieck topos $\mathcal{E}$, $F_M : \mathcal{C}_\mathbb{T} \to \mathcal{E}$ sends a small family $\{\theta_i \mid i \in I\}$ of morphisms

$$\{\vec{x_i}.\phi_i\} \xrightarrow{[\theta_i]} \{\vec{y}.\psi\}$$

in $\mathcal{C}_\mathbb{T}$ with common codomain to an epimorphic family in $\mathcal{E}$ if and only if $[[\vec{y}.\psi]]_M = [[\vec{y}. \bigvee_{i \in I}(\exists\vec{x_i})\theta_i]]_M$, equivalently if and only if the sequent $(\psi \vdash_{\vec{y}} \bigvee_{i \in I}(\exists\vec{x_i})\theta_i)$ is satisfied in M.

These remarks show, by the soundness theorem for geometric logic (cf. Theorem 1.4.15), that for any small presieve R in $\mathcal{C}_\mathbb{T}$, the $J_\mathbb{T}$-continuous flat functors on $\mathcal{C}_\mathbb{T}$ sending R to an epimorphic family also send all the pullbacks of R along arrows in $\mathcal{C}_\mathbb{T}$ to epimorphic families. It thus follows from Remark 3.1.2 and Lemma 3 [7] that the $J_\mathbb{T}$-continuous flat functors on $\mathcal{C}_\mathbb{T}$ which send each of the small presieves in a given collection $\mathcal{F}$ to an epimorphic family coincide with the $J_\mathbb{T}$-continuous flat functors on $\mathcal{C}_\mathbb{T}$ which are $J_\mathcal{F}$-continuous, where $J_\mathcal{F}$ is the Grothendieck topology on $\mathcal{C}_\mathbb{T}$ generated over $J_\mathbb{T}$ by the sieves generated by the presieves in $\mathcal{F}$.

Given a quotient $\mathbb{T}'$ of $\mathbb{T}$, we may construct its classifying topos as follows. Let $\mathbb{T}'$ be obtained from $\mathbb{T}$ by adding a number of geometric sequents over Σ; of course, up to syntactic equivalence, there are many possible ways of presenting $\mathbb{T}'$ in such a form (for example one may take as axioms all the axioms of $\mathbb{T}'$ or, more economically, all the axioms of $\mathbb{T}'$ which are not provable in $\mathbb{T}$), but our construction will be independent from any particular presentation. For each of these axioms ($\phi \vdash_{\vec{x}} \psi$), consider the corresponding monomorphism

$$\{\vec{x'} . \phi \wedge \psi\} \xrightarrow{[(\phi \wedge \psi \wedge \vec{x'}=\vec{x})]} \{\vec{x} . \phi\}$$

in the geometric syntactic category $\mathcal{C}_{\mathbb{T}}$ of $\mathbb{T}$.

It is clear that, given a $\mathbb{T}$-model M in a Grothendieck topos $\mathcal{E}$, the functor $F_M : \mathcal{C}_{\mathbb{T}} \to \mathcal{E}$ sends the morphism

$$\{\vec{x'} . \phi \wedge \psi\} \xrightarrow{[(\phi \wedge \psi \wedge \vec{x'}=\vec{x})]} \{\vec{x} . \phi\}$$

to an epimorphism if and only if $[[\vec{x} . \phi]]_M \leq [[\vec{x} . \psi]]_M$ i.e. if and only if the sequent ($\phi \vdash_{\vec{x}} \psi$) holds in M.

So the $J_{\mathbb{T}}$-continuous flat functors on $\mathcal{C}_{\mathbb{T}}$ which send each of the monomorphisms corresponding to the axioms of $\mathbb{T}'$ to an epimorphism correspond to the models of $\mathbb{T}'$. Therefore, if $J_{\mathbb{T}'}^{\mathbb{T}}$ is the smallest Grothendieck topology on $\mathcal{C}_{\mathbb{T}}$ for which all the $J_{\mathbb{T}}$-covering sieves and the sieves containing a morphism corresponding to an axiom of $\mathbb{T}'$ are covering then, by Diaconescu's equivalence, the topos $\mathbf{Sh}(\mathcal{C}_{\mathbb{T}}, J_{\mathbb{T}'}^{\mathbb{T}})$ classifies the theory $\mathbb{T}'$, and the canonical geometric inclusion $\mathbf{Sh}(\mathcal{C}_{\mathbb{T}}, J_{\mathbb{T}'}^{\mathbb{T}}) \hookrightarrow \mathbf{Sh}(\mathcal{C}_{\mathbb{T}}, J_{\mathbb{T}})$ corresponding to the inclusion $J_{\mathbb{T}} \subseteq J_{\mathbb{T}'}^{\mathbb{T}}$ realizes $\mathbf{Sh}(\mathcal{C}_{\mathbb{T}}, J_{\mathbb{T}'}^{\mathbb{T}})$ as a subtopos of $\mathbf{Set}[\mathbb{T}]$.

Now, to have a well-defined assignment from the $\equiv_s$-equivalence classes of quotients of $\mathbb{T}$ to the subtoposes of $\mathbf{Set}[\mathbb{T}]$, it remains to verify that the topology $J_{\mathbb{T}'}^{\mathbb{T}}$ does not depend on the particular choice of axioms for $\mathbb{T}'$, i.e. that it is the same for all the quotients in a given $\equiv_s$-equivalence class.

Let $\mathbb{T}_1$ and $\mathbb{T}_2$ be quotients of $\mathbb{T}$ such that $\mathbb{T}_1 \equiv_s \mathbb{T}_2$; we want to prove that $J_{\mathbb{T}_1}^{\mathbb{T}} = J_{\mathbb{T}_2}^{\mathbb{T}}$. We will prove the existence of a geometric equivalence $\tau : \mathbf{Sh}(\mathcal{C}_{\mathbb{T}}, J_{\mathbb{T}_1}^{\mathbb{T}}) \to \mathbf{Sh}(\mathcal{C}_{\mathbb{T}}, J_{\mathbb{T}_2}^{\mathbb{T}})$ such that the diagram in $\mathfrak{BTop}$

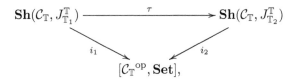

where the geometric inclusions

$$\mathbf{Sh}(\mathcal{C}_{\mathbb{T}}, J_{\mathbb{T}_1}^{\mathbb{T}}) \to [(\mathcal{C}_{\mathbb{T}})^{\mathrm{op}}, \mathbf{Set}]$$

and

$$\mathbf{Sh}(\mathcal{C}_{\mathbb{T}}, J_{\mathbb{T}_2}^{\mathbb{T}}) \to [(\mathcal{C}_{\mathbb{T}})^{\mathrm{op}}, \mathbf{Set}]$$

are the canonical ones, commutes up to isomorphism.

From the identification of equivalence classes of geometric inclusions to a given topos with local operators on that topos (cf. section 1.3.4) the equality of the two topologies $J_{\mathbb{T}_1}^{\mathbb{T}}$ and $J_{\mathbb{T}_2}^{\mathbb{T}}$ will then follow. By the 2-dimensional Yoneda lemma recalled in the section 3.1.1, it is equivalent to show the existence of an equivalence of categories

$$l_{\mathcal{E}} : \mathbf{Geom}(\mathcal{E}, \mathbf{Sh}(\mathcal{C}_{\mathbb{T}}, J_{\mathbb{T}_1}^{\mathbb{T}})) \to \mathbf{Geom}(\mathcal{E}, \mathbf{Sh}(\mathcal{C}_{\mathbb{T}}, J_{\mathbb{T}_2}^{\mathbb{T}}))$$

natural in $\mathcal{E} \in \mathfrak{BTop}$ such that $(i_1 \circ -) \circ l_{\mathcal{E}} \cong (i_2 \circ -)$ for each $\mathcal{E} \in \mathfrak{BTop}$. Since $\mathbb{T}_1 \equiv_s \mathbb{T}_2$, $\mathbb{T}_1$ and $\mathbb{T}_2$ have the same models (in every Grothendieck topos) and hence we may obtain such an equivalence by composing

$$\mathbf{Geom}(\mathcal{E}, \mathbf{Sh}(\mathcal{C}_{\mathbb{T}}, J_{\mathbb{T}_1}^{\mathbb{T}})) \simeq \mathbf{Flat}_{J_{\mathbb{T}_1}}(\mathcal{C}_{\mathbb{T}}, \mathcal{E}) \simeq \mathbb{T}_1\text{-mod}(\mathcal{E}) = \mathbb{T}_2\text{-mod}(\mathcal{E})$$

with

$$\mathbb{T}_2\text{-mod}(\mathcal{E}) \simeq \mathbf{Flat}_{J_{\mathbb{T}_2}}(\mathcal{C}_{\mathbb{T}}, \mathcal{E}) \simeq \mathbf{Geom}(\mathcal{E}, \mathbf{Sh}(\mathcal{C}_{\mathbb{T}}, J_{\mathbb{T}_2}^{\mathbb{T}})),$$

where the first and last equivalences are instances of Diaconescu's equivalence.

Conversely, let $\mathcal{E} \hookrightarrow \mathbf{Set}[\mathbb{T}]$ be a subtopos of $\mathbf{Set}[\mathbb{T}]$; up to equivalence, this subtopos has the form $\mathbf{Sh}(\mathcal{C}_{\mathbb{T}}, J) \hookrightarrow \mathbf{Sh}(\mathcal{C}_{\mathbb{T}}, J_{\mathbb{T}})$ for a unique Grothendieck topology J such that $J \supseteq J_{\mathbb{T}}$. Let us prove that there exists a quotient $\mathbb{T}^J$ of $\mathbb{T}$ classified by it. Let $\mathbb{T}^J$ be the theory consisting of all (the axioms of $\mathbb{T}$ and) the geometric sequents over Σ of the form $(\psi \vdash_{\vec{y}} (\exists \vec{x})\theta)$, where $[\theta]$ is any monomorphism

$$\{\vec{x}.\phi\} \xrightarrow{[\theta]} \{\vec{y}.\psi\}$$

in $\mathcal{C}_{\mathbb{T}}$ generating a J-covering sieve.

Since, for any $\mathbb{T}$-model M in a Grothendieck topos $\mathcal{E}$, the functor F_M sends $[\theta]$ to an epimorphism if and only if the sequent $(\psi \vdash_{\vec{y}} (\exists \vec{x})\theta)$ holds in M, it follows from Remark 3.1.2, Proposition 3.1.5 and Lemma 3 [7] that the equivalence $\mathbb{T}\text{-mod}(\mathcal{E}) \simeq \mathbf{Flat}_{J_{\mathbb{T}}}(\mathcal{C}_{\mathbb{T}}, \mathcal{E})$ restricts to an equivalence $\mathbb{T}^J\text{-mod}(\mathcal{E}) \simeq \mathbf{Flat}_J(\mathcal{C}_{\mathbb{T}}, \mathcal{E})$ (naturally in $\mathcal{E} \in \mathfrak{BTop}$) and hence that $\mathcal{E} = \mathbf{Sh}(\mathcal{C}_{\mathbb{T}}, J)$ classifies the theory $\mathbb{T}^J$.

To conclude the proof of the theorem it remains to show that the two assignments $\mathbb{T}' \to J_{\mathbb{T}'}^{\mathbb{T}}$ and $J \to \mathbb{T}^J$ are bijections inverse to each other between the $\equiv_s$-equivalence classes of quotients of $\mathbb{T}$ and the subtoposes of the classifying topos $\mathbf{Set}[\mathbb{T}]$ of $\mathbb{T}$.

To prove that for any quotient $\mathbb{T}'$ of $\mathbb{T}$ we have $\mathbb{T}' \equiv_s \mathbb{T}^{J_{\mathbb{T}'}^{\mathbb{T}}}$, we argue as follows. First, we observe that for any $\mathbb{T}$-model M in a Grothendieck topos $\mathcal{E}$, M is a $\mathbb{T}'$-model if and only if it is a $\mathbb{T}^{J_{\mathbb{T}'}^{\mathbb{T}}}$-model; indeed, by definition of $J_{\mathbb{T}'}^{\mathbb{T}}$ and of $\mathbb{T}^{J_{\mathbb{T}'}^{\mathbb{T}}}$, both the $\mathbb{T}'$-models and the $\mathbb{T}^{J_{\mathbb{T}'}^{\mathbb{T}}}$-models in $\mathcal{E}$ correspond to the functors in $\mathbf{Flat}_{J_{\mathbb{T}'}^{\mathbb{T}}}(\mathcal{C}_{\mathbb{T}'}, \mathcal{E})$ via the equivalence $\mathbf{Flat}_{J_{\mathbb{T}}}(\mathcal{C}_{\mathbb{T}}, \mathcal{E}) \simeq \mathbb{T}\text{-mod}(\mathcal{E})$.

Now, consider the image $U_{\mathbb{T}'}^{\mathbb{T}}$ of $a_{J_{\mathbb{T}'}^{\mathbb{T}}} \circ y^{\mathbb{T}}$ in $\mathbb{T}'\text{-mod}(\mathcal{E})$ through the equivalence $\mathbf{Flat}_{J_{\mathbb{T}'}^{\mathbb{T}}}(\mathcal{C}_{\mathbb{T}}, \mathcal{E}) \simeq \mathbb{T}'\text{-mod}(\mathcal{E})$, where $y^{\mathbb{T}} : \mathcal{C}_{\mathbb{T}} \to [(\mathcal{C}_{\mathbb{T}})^{\mathrm{op}}, \mathbf{Set}]$ is the Yoneda embedding and $a_{J_{\mathbb{T}'}^{\mathbb{T}}} : [(\mathcal{C}_{\mathbb{T}})^{\mathrm{op}}, \mathbf{Set}] \to \mathbf{Sh}(\mathcal{C}_{\mathbb{T}}, J_{\mathbb{T}'}^{\mathbb{T}})$ is the associated sheaf functor. By Diaconescu's equivalence and the naturality in $\mathcal{E} \in \mathfrak{BTop}$ of the equivalences

$$\mathbb{T}'\text{-mod}(\mathcal{E}) = \mathbb{T}^{J_{\mathbb{T}'}^{\mathbb{T}}}\text{-mod}(\mathcal{E}) \simeq \mathbf{Flat}_{J_{\mathbb{T}'}^{\mathbb{T}}}(\mathcal{C}_{\mathbb{T}}, \mathcal{E}) \simeq \mathbf{Geom}(\mathcal{E}, \mathbf{Sh}(\mathcal{C}_{\mathbb{T}}, J_{\mathbb{T}'}^{\mathbb{T}})),$$

the Σ-structure $U_{\mathbb{T}'}^{\mathbb{T}}$ is a universal model for both $\mathbb{T}'$ and $\mathbb{T}^{J_{\mathbb{T}'}^{\mathbb{T}}}$; in particular, it is conservative both as a $\mathbb{T}'$-model and as a $\mathbb{T}^{J_{\mathbb{T}'}^{\mathbb{T}}}$-model (cf. Remark 2.1.9). From this it clearly follows that $\mathbb{T}' \equiv_s \mathbb{T}^{J_{\mathbb{T}'}^{\mathbb{T}}}$, as required.

On the other hand, the fact that $J = J_{\mathbb{T}^J}^{\mathbb{T}}$ directly follows from the definition of the assignment $\mathbb{T}' \to J_{\mathbb{T}'}^{\mathbb{T}}$. □

We shall refer to the topology $J_{\mathbb{T}'}^{\mathbb{T}}$ as the *associated $\mathbb{T}$-topology of $\mathbb{T}'$* and to the ($\equiv_s$-equivalence class of the) quotient $\mathbb{T}^J$ as the *associated $\mathbb{T}$-quotient of J*.

For each Grothendieck topos $\mathcal{E}$, we denote by

$$\tau^{\mathcal{E}} : \mathbb{T}\text{-mod}(\mathcal{E}) \simeq \mathbf{Geom}(\mathcal{E}, \mathbf{Sh}(\mathcal{C}_{\mathbb{T}}, J_{\mathbb{T}}))$$

the composite of the equivalence

$$\mathbb{T}\text{-mod}(\mathcal{E}) \simeq \mathbf{Flat}_{J_{\mathbb{T}}}(\mathcal{C}_{\mathbb{T}}, \mathcal{E})$$

considered in the proof of Theorem 3.2.5 with Diaconescu's equivalence

$$\mathbf{Flat}_{J_{\mathbb{T}}}(\mathcal{C}_{\mathbb{T}}, \mathcal{E}) \simeq \mathbf{Geom}(\mathcal{E}, \mathbf{Sh}(\mathcal{C}_{\mathbb{T}}, J_{\mathbb{T}})).$$

Given a quotient $\mathbb{T}'$ of a theory $\mathbb{T}$, we denote by $i_{\mathbb{T}'}^{\mathcal{E}} : \mathbb{T}'\text{-mod}(\mathcal{E}) \hookrightarrow \mathbb{T}\text{-mod}(\mathcal{E})$ the canonical inclusion functor.

Remark 3.2.6 With the above notation, given a quotient $\mathbb{T}'$ of $\mathbb{T}$ with associated $\mathbb{T}$-topology J on $\mathcal{C}_{\mathbb{T}}$ then, denoting by i_J the canonical geometric inclusion $\mathbf{Sh}(\mathcal{C}_{\mathbb{T}}, J) \hookrightarrow \mathbf{Sh}(\mathcal{C}_{\mathbb{T}}, J_{\mathbb{T}})$, the diagram in $\mathfrak{CAT}$

$$\begin{array}{ccc} \mathbb{T}'\text{-mod}(\mathcal{E}) & \xrightarrow{\simeq} & \mathbf{Geom}(\mathcal{E}, \mathbf{Sh}(\mathcal{C}_{\mathbb{T}}, J)) \\ \downarrow{i_{\mathbb{T}'}^{\mathcal{E}}} & & \downarrow{i_J \circ -} \\ \mathbb{T}\text{-mod}(\mathcal{E}) & \xrightarrow[\tau^{\mathcal{E}}]{\simeq} & \mathbf{Geom}(\mathcal{E}, \mathbf{Sh}(\mathcal{C}_{\mathbb{T}}, J_{\mathbb{T}})) \end{array}$$

commutes (up to isomorphism) naturally in $\mathcal{E} \in \mathfrak{BTop}$.

It should be noted that our method of constructing the associated $\mathbb{T}$-topology of a given quotient of $\mathbb{T}$ has points in common with the 'forcing' method expressed by Proposition D3.1.10 [55]; important precedents for this result are the paper [72] on 'forcing topologies' by M. Tierney and the work of M. and M.-F. Coste [39] on the classifying toposes of coherent theories (included in the Appendix of [64]). In fact, our arguments show that, more generally, it is always possible to construct the classifying topos of a quotient $\mathbb{T}'$ of a given cartesian (resp. regular, coherent, geometric) theory $\mathbb{T}$ as a category of sheaves on the cartesian (resp. regular, coherent, geometric) syntactic category of $\mathbb{T}$ starting from a way of expressing $\mathbb{T}'$ as a theory obtained from $\mathbb{T}$ by adding axioms of the form $(\psi \vdash_{\vec{y}} \bigvee_{i \in I} (\exists \vec{x}_i)\theta_i)$ where ψ and the θ_i are cartesian (resp. regular, coherent, geometric) formulae over the signature of $\mathbb{T}$ such that the θ_i are

T-provably functional formualae from a cartesian (resp. regular, coherent, geometric) formula-in-context to $\psi(\vec{y})$; see also Theorem 8.1.10, which arises from an implementation of the same idea in the case $\mathbb{T}$ is a theory of presheaf type, and which is an actual generalization of Proposition D3.1.10 [55].

Given a Grothendieck topos $\mathcal{E}$ and a signature Σ, it is natural to wonder under which conditions there exists a geometric theory $\mathbb{T}$ over Σ classified by $\mathcal{E}$. Theorem 3.2.5 gives an answer to this question: there exists such a theory if and only if there exists a geometric inclusion from $\mathcal{E}$ to the classifying topos of the empty theory over Σ.

3.2.2 The proof-theoretic interpretation

In this section we provide an alternative, syntactic, proof of Theorem 3.2.5. This will be based on a logical interpretation of the notion of Grothendieck topology. Specifically, given a collection $\mathcal{A}$ of sieves on a given category $\mathcal{C}$, the notion of Grothendieck topology on $\mathcal{C}$ naturally gives rise to a proof system $\mathcal{T}_{\mathcal{C}}^{\mathcal{A}}$, as follows: the axioms of $\mathcal{T}_{\mathcal{C}}^{\mathcal{A}}$ are the sieves in $\mathcal{A}$ plus all the maximal sieves, while the inference rules of $\mathcal{T}_{\mathcal{C}}^{\mathcal{A}}$ are the proof-theoretic versions of the well-known axioms for Grothendieck topologies, i.e.

Stability rule:
$$\frac{R}{f^*(R)}$$

where R is any sieve on an object c of $\mathcal{C}$ and f is any arrow in $\mathcal{C}$ with codomain c.

Transitivity rule:
$$\frac{Z \quad \{f^*(R) \mid f \in Z\}}{R}$$

where R and Z are sieves in $\mathcal{C}$ on a given object of $\mathcal{C}$.

Notice that the 'closed theories' of this proof system are precisely the Grothendieck topologies on $\mathcal{C}$ which contain the sieves in $\mathcal{A}$ as covering sieves and that the closure of a 'theory' in $\mathcal{T}_{\mathcal{C}}^{\mathcal{A}}$, i.e. of a collection $\mathcal{U}$ of sieves in $\mathcal{C}$, is exactly the Grothendieck topology on $\mathcal{C}$ generated by $\mathcal{A}$ and $\mathcal{U}$.

The following theorem will give a 'proof-theoretic equivalence' between the deduction system of geometric logic over a given geometric theory $\mathbb{T}$ (as described in section 1.2.2) and the system $\mathcal{T}_{\mathcal{C}_\mathbb{T}}^{J_\mathbb{T}}$.

Given a geometric theory $\mathbb{T}$ over a signature Σ, let $\mathcal{S}$ be the collection of geometric sequents over Σ, $\tilde{\mathcal{S}}$ the quotient of $\mathcal{S}$ by the relation of $\mathbb{T}$-equivalence and $S(\mathcal{C}_\mathbb{T})$ the collection of sieves in the geometric syntactic category $\mathcal{C}_\mathbb{T}$ of $\mathbb{T}$.

Motivated by the proof of the duality theorem in section 3.2.1, we define two correspondences $\mathcal{F} : \mathcal{S} \to S(\mathcal{C}_\mathbb{T})$ and $\mathcal{G} : S(\mathcal{C}_\mathbb{T}) \to \tilde{\mathcal{S}}$, as follows.

Given a geometric sequent $\sigma \equiv (\phi \vdash_{\vec{x}} \psi)$ over Σ, we set $\mathcal{F}(\sigma)$ equal to the principal sieve in $\mathcal{C}_\mathbb{T}$ generated by the monomorphism

$$\{\vec{x'} . \phi \wedge \psi\} \xrightarrow{[(\phi \wedge \psi \wedge \vec{x'} = \vec{x})]} \{\vec{x} . \phi\}.$$

Conversely, given a sieve R in $\mathcal{C}_{\mathbb{T}}$, we set $\mathcal{G}(R)$ equal to the $\mathbb{T}$-equivalence class of any geometric sequent $(\psi \vdash_{\vec{y}} (\exists \vec{x})\theta)$ such that $[\theta]$ is a monomorphism

$$\{\vec{x}.\phi\} \xrightarrow{[\theta]} \{\vec{y}.\psi\}$$

in $\mathcal{C}_{\mathbb{T}}$ generating the principal sieve $\overline{R}^{J_{\mathbb{T}}}$ (cf. Proposition 3.1.6).

Applying the covariant powerset image functor $\mathcal{P}$ to $\mathcal{F}$ and $\mathcal{G}$, we obtain maps of posets

$$\mathcal{P}(\mathcal{F}) : \mathcal{P}(\mathcal{S}) \to \mathcal{P}(S(\mathcal{C}_{\mathbb{T}}))$$

and

$$\mathcal{P}(\mathcal{G}) : \mathcal{P}(S(\mathcal{C}_{\mathbb{T}})) \to \mathcal{P}(\tilde{\mathcal{S}}).$$

From now on we will write $\mathcal{F}(U)$ for $\mathcal{P}(\mathcal{F})(U)$ and $\mathcal{G}(V)$ for $\mathcal{P}(\mathcal{G})(V)$.

We have closure operators

$$\overline{(-)}^{\mathcal{T}} : \mathcal{P}(\mathcal{S}) \to \mathcal{P}(\mathcal{S})$$

and

$$\overline{(-)}^{\mathcal{T}} : \mathcal{P}(S(\mathcal{C}_{\mathbb{T}})) \to \mathcal{P}(S(\mathcal{C}_{\mathbb{T}}))$$

defined as follows: for a collection U of geometric sequents over Σ, $\overline{U}^{\mathcal{T}}$ is the collection of geometric sequents σ which are provable in $\mathbb{T} \cup U$ using geometric logic, while, for a collection V of sieves in $\mathcal{C}_{\mathbb{T}}$, $\overline{V}^{\mathcal{T}}$ is the Grothendieck topology in $\mathcal{C}_{\mathbb{T}}$ generated by $J_{\mathbb{T}}$ and V (i.e. the smallest Grothendieck topology J on $\mathcal{C}_{\mathbb{T}}$ such that all the $J_{\mathbb{T}}$-covering sieves and the sieves in V are J-covering). Note that the relation of $\mathbb{T}$-equivalence on $\mathcal{S}$ is compatible with the closure operator $\overline{(-)}^{\mathcal{T}}$, that is, we have a factorization $\overline{(-)}^{\mathcal{T}}_{\tilde{\mathcal{S}}} : \mathcal{P}(\tilde{\mathcal{S}}) \to \mathcal{P}(\mathcal{S})$ of $\overline{(-)}^{\mathcal{T}} : \mathcal{P}(\mathcal{S}) \to \mathcal{P}(\mathcal{S})$ through the image $\mathcal{P}(\mathcal{S}) \to \mathcal{P}(\tilde{\mathcal{S}})$ under $\mathcal{P}$ of the natural projection map $\mathcal{S} \to \tilde{\mathcal{S}}$.

The closed points with respect to these closure operators are respectively the syntactic-equivalence classes of quotients of $\mathbb{T}$ and the Grothendieck topologies J on $\mathcal{C}_{\mathbb{T}}$ such that $J \supseteq J_{\mathbb{T}}$.

Let us define $F : \mathcal{P}(\mathcal{S}) \to \mathcal{P}(S(\mathcal{C}_{\mathbb{T}}))$ as the composite $\overline{(-)}^{\mathcal{T}} \circ \mathcal{P}(\mathcal{F})$ and $G : \mathcal{P}(S(\mathcal{C}_{\mathbb{T}})) \to \mathcal{P}(\mathcal{S})$ as the composite $\overline{(-)}^{\mathcal{T}}_{\tilde{\mathcal{S}}} \circ \mathcal{P}(\mathcal{G})$.

Given a collection U of geometric sequents over Σ, we define $\mathbb{T}^U$ to be the collection of all the geometric sequents σ over Σ such that $\mathcal{F}(\sigma)$ belongs to $\overline{\mathcal{F}(U)}^{\mathcal{T}}$. Similarly, given a collection V of sieves on $\mathcal{C}_{\mathbb{T}}$, we define J_V to be the collection of sieves R in $\mathcal{C}$ such that any sequent in $\mathcal{G}(R)$ is provable in $\mathbb{T} \cup \mathcal{G}(V)$ (by using geometric logic).

The following result shows that the maps $\mathcal{P}(\mathcal{F})$ and $\mathcal{P}(\mathcal{G})$ are compatible with respect to these closure operators and that F and G are inverse to each other on the subsets of closed points, that is between the collection of syntactic-equivalence classes of quotients of $\mathbb{T}$ and the collection of Grothendieck topologies on $\mathcal{C}_\mathbb{T}$ which contain $J_\mathbb{T}$. In fact, given a quotient $\mathbb{T}'$ of $\mathbb{T}$, $F(\mathbb{T}') = J^\mathbb{T}_{\mathbb{T}'}$, while, for a Grothendieck topology $J \supseteq J_\mathbb{T}$, $G(J) = \mathbb{T}^J$ (using the notations of section 3.2.1). This approach thus provides a different, entirely syntactic, way to arrive at the duality of Theorem 3.2.5.

Theorem 3.2.7 *With the above notation, we have that*

(i) *For any* $U \in \mathcal{P}(\mathcal{S})$, $\mathcal{F}(\overline{U}^\mathbb{T}) \subseteq \overline{\mathcal{F}(U)}^\mathcal{T}$.

(ii) *For any* $V \in \mathcal{P}(S(\mathcal{C}_\mathbb{T}))$, $\mathcal{G}(\overline{V}^\mathcal{T}) \subseteq \overline{\mathcal{G}(V)}^\mathbb{T}$.

(iii) *For any* $U \in \mathcal{P}(\mathcal{S})$, $G(F(U)) = \overline{U}^\mathbb{T} = \mathbb{T}^U$.

(iv) *For any* $V \in \mathcal{P}(S(\mathcal{C}_\mathbb{T}))$, $F(G(V)) = \overline{V}^\mathcal{T} = J_V$.

In other words, the maps $\mathcal{P}(F)$ *and* $\mathcal{P}(G)$ *define a proof-theoretic equivalence between the classical deduction system for geometric logic over* $\mathbb{T}$ *and the proof system* $\mathbb{T}^{J_\mathbb{T}}_{\mathcal{C}_\mathbb{T}}$.

Proof

(i) We have to prove that, given $U := \{\sigma_i \mid i \in I\} \in \mathcal{P}(\mathcal{S})$, if a geometric sequent σ is provable in $\mathbb{T} \cup U$ using geometric logic, then $\mathcal{F}(\{\sigma\})$ belongs to $\overline{\mathcal{F}(U)}^\mathcal{T}$. Let us show this by induction on the complexity of a proof of $\sigma \equiv (\phi \vdash_{\vec{x}} \psi)$ in $\mathbb{T} \cup U$.

If $\sigma \in U$ then the thesis is clear.

If σ belongs to $\mathbb{T}$ or, more generally, is provable in $\mathbb{T}$, then the morphism

$$\{\vec{x}'.\phi \wedge \psi\} \xrightarrow{[\phi \wedge \psi \wedge \vec{x}'=\vec{x}]} \{\vec{x}.\phi\}$$

in $\mathcal{C}_\mathbb{T}$ is isomorphic to the identity morphism on $\{\vec{x}.\phi\}$ and hence belongs to $\overline{\mathcal{F}(U)}^\mathcal{T}$ by the maximality axiom for Grothendieck topologies. Notice in particular that if σ is an axiom of geometric logic then $\mathcal{F}(\sigma)$ belongs to $\overline{\mathcal{F}(U)}^\mathcal{T}$.

Let us now verify that all the inference rules for geometric logic are 'sound' with respect to the operation $\mathcal{F}$, i.e. that if each of the premises σ of an inference rule satisfies '$\mathcal{F}(\sigma)$ belongs to $\overline{\mathcal{F}(U)}^\mathcal{T}$' then its conclusion σ' also satisfies '$\mathcal{F}(\sigma')$ belongs to $\overline{\mathcal{F}(U)}^\mathcal{T}$'.

Substitution rule:

$$\frac{(\phi \vdash_{\vec{x}} \psi)}{(\phi[\vec{s}/\vec{x}] \vdash_{\vec{y}} \psi[\vec{s}/\vec{x}])}$$

where $\vec{y}$ is any string of variables including all the variables occurring in a string of terms $\vec{s}$.

We have to prove that if the principal sieve in $\mathcal{C}_\mathbb{T}$ generated by the morphism

$$\{\vec{x}'.\phi \wedge \psi\} \xrightarrow{[\phi \wedge \psi \wedge \vec{x}'=\vec{x}]} \{\vec{x}.\phi\}$$

is $\overline{\mathcal{F}(U)}^{\mathbb{T}}$-covering then the principal sieve generated by the morphism

$$\{\vec{y}'\,.\,\phi[\vec{s}/\vec{x}]\wedge\psi[\vec{s}/\vec{x}]\}\xrightarrow{[(\phi[\vec{s}/\vec{x}]\wedge\psi[\vec{s}/\vec{x}]\wedge\vec{y}'=\vec{y})]}\{\vec{y}\,.\,\phi[\vec{s}/\vec{x}]\}$$

is also $\overline{\mathcal{F}(U)}^{\mathbb{T}}$-covering.

For any geometric formula $\phi(\vec{x})$ and a term $s(\vec{y})$ over Σ, the diagram

$$\begin{array}{ccc}
\{\vec{y}'\,.\,\phi[\vec{s}/\vec{x}]\} & \xrightarrow{[(s(\vec{y}')=\vec{x})\wedge\phi]} & \{\vec{x}'\,.\,\phi[\vec{x}'/\vec{x}]\} \\
{\scriptstyle [\phi[\vec{s}/\vec{x}]\wedge\vec{y}'=\vec{y}]}\downarrow & & \downarrow{\scriptstyle [\phi\wedge\vec{x}'=\vec{x}]} \\
\{\vec{y}\,.\,\top\} & \xrightarrow{[s(\vec{y})=\vec{x}]} & \{\vec{x}\,.\,\top\}
\end{array}$$

is a pullback in $\mathcal{C}_{\mathbb{T}}$. To prove this, we preliminarily observe that if $\chi\equiv(\exists\vec{y})\xi$ is a geometric formula in a context $\vec{x}$ such that the sequent

$$(\xi\wedge\xi[\vec{z}/\vec{y}]\vdash_{\vec{x},\vec{y},\vec{z}}\vec{y}=\vec{z})$$

is provable in $\mathbb{T}$ then the objects $\{\vec{x}\,.\,\chi\}$ and $\{\vec{x},\vec{y}\,.\,\xi\}$ are isomorphic in $\mathcal{C}_{\mathbb{T}}$. Indeed, it is an easy consequence of Lemma 1.4.2 that the arrow

$$\{\vec{x},\vec{y}\,.\,\xi\}\xrightarrow{[\xi\wedge(\vec{x}'=\vec{x})]}\{\vec{x}'\,.\,\chi[\vec{x}'/\vec{x}]\}$$

is an isomorphism.

Now, it immediately follows from the substitution rule (and the equality axioms) that the bisequent $((\exists\vec{x})((s(\vec{y})=\vec{x})\wedge\phi(\vec{x}))\dashv\vdash_{\vec{y}}\phi[\vec{s}/\vec{x}])$ is provable in geometric logic.

In view of the construction of pullbacks in $\mathcal{C}_{\mathbb{T}}$ given in section 3.2.1, these two remarks together imply that the above square is a pullback in $\mathcal{C}_{\mathbb{T}}$, as required.

From this fact we immediately deduce that the morphism

$$\{\vec{x}'\,.\,\phi[\vec{s}/\vec{x}]\wedge\psi[\vec{s}/\vec{x}]\}\xrightarrow{[\phi[\vec{s}/\vec{x}]\wedge\psi[\vec{s}/\vec{x}]\wedge\vec{x}'=\vec{x}]}\{\vec{x}\,.\,\phi[\vec{s}/\vec{x}]\}$$

is (isomorphic to) the pullback in $\mathcal{C}_{\mathbb{T}}$ along $[(s(\vec{y})=\vec{x})\wedge\phi]:\{\vec{y}\,.\,\phi[\vec{s}/\vec{x}]\}\to\{\vec{x}\,.\,\phi\}$ of the morphism

$$\{\vec{x}'\,.\,\phi\wedge\psi\}\xrightarrow{[\phi\wedge\psi\wedge\vec{x}'=\vec{x}]}\{\vec{x}\,.\,\phi\}.$$

We note that, for a Grothendieck topology J on a category $\mathcal{C}$, if the diagram is a pullback square in $\mathcal{C}$ then $(f) \in J(c)$ implies $(f') \in J(d)$. Indeed, by the universal property of the pullback, we have $(f') = h^*((f))$. Applying this to our pullback square yields our thesis.

This concludes our proof that the substitution rule is 'sound' for the operation $\mathcal{F}$.

Cut rule:
$$\frac{(\phi \vdash_{\vec{x}} \psi)(\psi \vdash_{\vec{x}} \chi)}{(\phi \vdash_{\vec{x}} \chi)}$$

We have to prove that if the principal sieves in $\mathcal{C}_\mathbb{T}$ respectively generated by the morphisms

$$\{\vec{x'}.\phi \wedge \psi\} \xrightarrow{[\phi \wedge \psi \wedge \vec{x'}=\vec{x}]} \{\vec{x}.\phi\}$$

and

$$\{\vec{x'}.\psi \wedge \chi\} \xrightarrow{[\psi \wedge \chi \wedge \vec{x'}=\vec{x}]} \{\vec{x}.\psi\}$$

are $\overline{\mathcal{F}(U)}^\mathcal{T}$-covering then the principal sieve generated by the morphism

$$\{\vec{x'}.\phi \wedge \chi\} \xrightarrow{[\phi \wedge \chi \wedge \vec{x'}=\vec{x}]} \{\vec{x}.\phi\}$$

is also $\overline{\mathcal{F}(U)}^\mathcal{T}$-covering.

The diagrams

$$\begin{array}{ccc}
\{\vec{x'''}.\phi \wedge \psi \wedge \chi\} & \xrightarrow{[\phi \wedge \psi \wedge \chi \wedge \vec{x'''}=\vec{x''}]} & \{\vec{x''}.\phi \wedge \chi\} \\
{\scriptstyle [\phi \wedge \psi \wedge \chi \wedge \vec{x'''}=\vec{x'}]} \downarrow & & \downarrow {\scriptstyle [\phi \wedge \chi \wedge \vec{x''}=\vec{x}]} \\
\{\vec{x'}.\phi \wedge \psi\} & \xrightarrow{[\phi \wedge \psi \wedge \vec{x'}=\vec{x}]} & \{\vec{x}.\phi\}
\end{array}$$

and

$$\begin{array}{ccc}
\{\vec{x'''}.\phi \wedge \psi \wedge \chi\} & \xrightarrow{[\phi \wedge \psi \wedge \chi \wedge \vec{x'''}=\vec{x''}]} & \{\vec{x''}.\psi \wedge \chi\} \\
{\scriptstyle [\phi \wedge \psi \wedge \chi \wedge \vec{x'''}=\vec{x'}]} \downarrow & & \downarrow {\scriptstyle [\psi \wedge \chi \wedge \vec{x''}=\vec{x}]} \\
\{\vec{x'}.\phi \wedge \psi\} & \xrightarrow{[\phi \wedge \psi \wedge \vec{x'}=\vec{x}]} & \{\vec{x}.\psi\}
\end{array}$$

are clearly pullback squares in $\mathcal{C}_\mathbb{T}$.

By the stability axiom for Grothendieck topologies, the principal sieve generated by the morphism

$$\{\vec{x'''}.\phi \wedge \psi \wedge \chi\} \xrightarrow{[\phi \wedge \psi \wedge \chi \wedge \vec{x'''}=\vec{x'}]} \{\vec{x'}.\phi \wedge \psi\}$$

is $\overline{\mathcal{F}(U)}^T$-covering, since it is the pullback of the ($\overline{\mathcal{F}(U)}^T$-covering) principal sieve generated by the morphism

$$\{\vec{x'}.\psi \wedge \chi\} \xrightarrow{[\psi \wedge \chi \wedge \vec{x'}=\vec{x}]} \{\vec{x}.\psi\}$$

along the arrow

$$\{\vec{x'}.\phi \wedge \psi\} \xrightarrow{[\phi \wedge \psi \wedge \vec{x'}=\vec{x}]} \{\vec{x}.\psi\}.$$

Since the principal sieve generated by the morphism

$$\{\vec{x'}.\phi \wedge \psi\} \xrightarrow{[\phi \wedge \psi \wedge \vec{x'}=\vec{x}]} \{\vec{x}.\phi\}$$

is $\overline{\mathcal{F}(U)}^T$-covering, we can conclude, by the transitivity axiom for Grothendieck topologies and the fact that the first of the two squares above is a pullback, that the principal sieve generated by the morphism

$$\{\vec{x'}.\phi \wedge \chi\} \xrightarrow{[\phi \wedge \chi \wedge \vec{x'}=\vec{x}]} \{\vec{x}.\phi\}$$

is $\overline{\mathcal{F}(U)}^T$-covering, as required.
Rule for finite conjunction:

$$\frac{(\phi \vdash_{\vec{x}} \psi)(\phi \vdash_{\vec{x}} \chi)}{(\phi \vdash_{\vec{x}} (\psi \wedge \chi))}$$

We have to prove that if the principal sieves in $\mathcal{C}_\mathbb{T}$ respectively generated by the morphisms

$$\{\vec{x'}.\phi \wedge \psi\} \xrightarrow{[\phi \wedge \psi \wedge \vec{x'}=\vec{x}]} \{\vec{x}.\phi\}$$

and

$$\{\vec{x'}.\phi \wedge \chi\} \xrightarrow{[\phi \wedge \chi \wedge \vec{x'}=\vec{x}]} \{\vec{x}.\phi\}$$

are $\overline{\mathcal{F}(U)}^T$-covering then the principal sieve generated by the morphism

$$\{\vec{x'}.\phi \wedge (\psi \wedge \chi)\} \xrightarrow{[\phi \wedge (\psi \wedge \chi) \wedge \vec{x'}=\vec{x}]} \{\vec{x}.\phi\}$$

is also $\overline{\mathcal{F}(U)}^T$-covering.

We have observed that the diagram

$$\begin{array}{ccc}
\{\vec{x'''}.\phi \wedge \psi \wedge \chi\} & \xrightarrow{[\phi\wedge\psi\wedge\chi\wedge\vec{x'''}=\vec{x''}]} & \{\vec{x''}.\phi \wedge \chi\} \\
{\scriptstyle [\phi\wedge\psi\wedge\chi\wedge\vec{x'''}=\vec{x'}]}\Big\downarrow & & \Big\downarrow{\scriptstyle [\phi\wedge\chi\wedge\vec{x''}=\vec{x}]} \\
\{\vec{x'}.\phi \wedge \psi\} & \xrightarrow{[\phi\wedge\psi\wedge\vec{x'}=\vec{x}]} & \{\vec{x}.\phi\}
\end{array}$$

is a pullback square in $\mathcal{C}_\mathbb{T}$. The stability axiom for Grothendieck topologies then ensures that the principal sieve generated by the morphism

$$\{\vec{x'''}.\phi \wedge \psi \wedge \chi\} \xrightarrow{[\phi\wedge\psi\wedge\chi\wedge\vec{x'''}=\vec{x'}]} \{\vec{x'}.\phi \wedge \psi\}$$

is $\overline{\mathcal{F}(U)}^T$-covering, as it is the pullback of the ($\overline{\mathcal{F}(U)}^T$-covering) principal sieve generated by the morphism

$$\{\vec{x''}.\phi \wedge \chi\} \xrightarrow{[\phi\wedge\chi\wedge\vec{x''}=\vec{x}]} \{\vec{x}.\phi\}$$

along the arrow

$$\{\vec{x'}.\phi \wedge \psi\} \xrightarrow{[\phi\wedge\psi\wedge\vec{x'}=\vec{x}]} \{\vec{x}.\phi\}.$$

But the principal sieve generated by the morphism

$$\{\vec{x'}.\phi \wedge \psi\} \xrightarrow{[\phi\wedge\psi\wedge\vec{x'}=\vec{x}]} \{\vec{x}.\phi\}$$

is $\overline{\mathcal{F}(U)}^T$-covering and hence, since the morphism

$$\{\vec{x'}.\phi \wedge (\psi \wedge \chi)\} \xrightarrow{[\phi\wedge(\psi\wedge\chi)\wedge\vec{x'}=\vec{x}]} \{\vec{x}.\phi\}$$

is equal to the composite of

$$\{\vec{x'}.\phi \wedge \psi\} \xrightarrow{[\phi\wedge\psi\wedge\vec{x'}=\vec{x}]} \{\vec{x}.\phi\}$$

and

$$\{\vec{x'''}.\phi \wedge \psi \wedge \chi\} \xrightarrow{[\phi\wedge\psi\wedge\chi\wedge\vec{x'''}=\vec{x'}]} \{\vec{x'}.\phi \wedge \psi\},$$

it follows from property (d) in Definition 3.1.3 that the principal sieve generated by the morphism

$$\{\vec{x'}.\phi \wedge (\psi \wedge \chi)\} \xrightarrow{[\phi\wedge(\psi\wedge\chi)\wedge\vec{x'}=\vec{x}]} \{\vec{x}.\phi\}$$

is $\overline{\mathcal{F}(U)}^T$-covering, as required.

Rule for infinitary disjunction:

$$\frac{\{(\phi_i \vdash_{\vec{x}} \chi) \mid i \in I\}}{\left(\bigvee_{i \in I} \phi_i \vdash_{\vec{x}} \chi\right)}$$

We have to prove that if each of the principal sieves in $\mathcal{C}_\mathbb{T}$ generated by the morphisms

$$\{\vec{x'}.\phi_i \wedge \chi\} \xrightarrow{[\phi_i \wedge \chi \wedge \vec{x'}=\vec{x}]} \{\vec{x}.\phi_i\}$$

(as i varies in I) is $\overline{\mathcal{F}(U)}^T$-covering then the principal sieve generated by the morphism

$$\{\vec{x'}.\left(\bigvee_{i \in I}\phi_i\right) \wedge \chi\} \xrightarrow{\left[\left(\bigvee_{i \in I}\phi_i\right) \wedge \chi \wedge \vec{x'}=\vec{x}\right]} \{\vec{x}.\bigvee_{i \in I}\phi_i\}$$

is also $\overline{\mathcal{F}(U)}^T$-covering.

The sieve on $\{\vec{x}.\bigvee_{i \in I}\phi_i\}$ generated by the morphisms

$$j_i := \{\vec{x'}.\phi_i\} \xrightarrow{[\phi_i \wedge \vec{x'}=\vec{x}]} \{\vec{x}.\bigvee_{i \in I}\phi_i\}$$

(as i varies in I) is $\overline{\mathcal{F}(U)}^T$-covering by definition of $J_\mathbb{T}$, since $\overline{\mathcal{F}(U)}^T \supseteq J_\mathbb{T}$.

Now, for each $i \in I$, the diagram

$$\begin{array}{ccc} \{\vec{x'''}.\phi_i \wedge \chi\} & \xrightarrow{[\phi_i \wedge \chi \wedge \vec{x'''}=\vec{x''}]} & \{\vec{x''}.\left(\bigvee_{i \in I}\phi_i\right) \wedge \chi\} \\ {\scriptstyle [\phi_i \wedge \chi \wedge \vec{x'''}=\vec{x'}]}\downarrow & & \downarrow {\scriptstyle \left[\left(\bigvee_{i \in I}\phi_i\right) \wedge \chi \wedge \vec{x''}=\vec{x}\right]} \\ \{\vec{x'}.\phi_i\} & \xrightarrow{j_i} & \{\vec{x}.\bigvee_{i \in I}\phi_i\} \end{array}$$

is a pullback square in $\mathcal{C}_\mathbb{T}$. Our thesis thus follows from the transitivity axiom for Grothendieck topologies.

Rules for existential quantification:

$$\frac{(\phi \vdash_{\vec{x},\vec{y}} \psi)}{((\exists \vec{y})\phi \vdash_{\vec{x}} \psi)}$$

where $\vec{y}$ is not free in ψ.

We have to prove that the principal sieve in $\mathcal{C}_\mathbb{T}$ generated by the morphism

$$\{\vec{x'}, \vec{y'} . \phi \wedge \psi\} \xrightarrow{[\phi \wedge \psi \wedge \vec{x'} = \vec{x} \wedge \vec{y'} = \vec{y}]} \{\vec{x}, \vec{y} . \phi\}$$

is $\overline{\mathcal{F}(U)}^T$-covering if and only if the principal sieve generated by the morphism

$$\{\vec{x'} . ((\exists \vec{y})\phi) \wedge \psi\} \xrightarrow{[((\exists \vec{y})\phi) \wedge \vec{x'} = \vec{x}]} \{\vec{x'} . (\exists \vec{y})\phi\}$$

is $\overline{\mathcal{F}(U)}^T$-covering.
The diagram

$$\{\vec{x'''}, \vec{y'''} . (\phi \wedge \psi)[\vec{x'''}/\vec{x}, \vec{y'''}/\vec{y}]\} \xrightarrow{[\phi \wedge \psi \wedge \vec{x'''} = \vec{x'} \wedge \vec{y'''} = \vec{y'}]} \{\vec{x''}, \vec{y''} . \phi[\vec{x''}/\vec{x}, \vec{y''}/\vec{y}]\}$$

$$\downarrow {\scriptstyle [\phi \wedge \psi \wedge \vec{x'''} = \vec{x'}]} \qquad\qquad\qquad\qquad\qquad \downarrow {\scriptstyle [\phi \wedge \vec{x''} = \vec{x}]}$$

$$\{\vec{x'} . ((\exists \vec{y})\phi \wedge \psi)[\vec{x'}/\vec{x}]\} \xrightarrow{[((\exists \vec{y})\phi) \wedge \psi \wedge \vec{x'} = \vec{x}]} \{\vec{x} . (\exists \vec{y})\phi\}$$

is a pullback square in $\mathcal{C}_\mathbb{T}$ (this follows at once from the construction of pullbacks in $\mathcal{C}_\mathbb{T}$ reviewed in section 3.2.1 by invoking the rules for existential quantification, as in the proof for the substitution rule).

The 'if' part of our claim clearly follows from the stability axiom for Grothendieck topologies, so it remains to prove the 'only if' part. To this end, we notice that the morphism

$$\{\vec{x''}, \vec{y''} . \phi[\vec{x''}/\vec{x}, \vec{y''}/\vec{y}]\} \xrightarrow{[\phi \wedge \vec{x''} = \vec{x}]} \{\vec{x} . (\exists \vec{y})\phi\}$$

is a cover in $\mathcal{C}_\mathbb{T}$; this implies that the sieve generated by it is $\overline{\mathcal{F}(U)}^T$-covering by definition of $J_\mathbb{T}$, since $\overline{\mathcal{F}(U)}^T \supseteq J_\mathbb{T}$. Hence, since the above square is cartesian, the principal sieve generated by the morphism

$$\{\vec{x'} . ((\exists \vec{y})\phi) \wedge \psi\} \xrightarrow{[((\exists \vec{y})\phi) \wedge \psi \wedge \vec{x'} = \vec{x}]} \{\vec{x'} . (\exists \vec{y})\phi\}$$

is $\overline{\mathcal{F}(U)}^T$-covering by properties (b) and (d) in Definition 3.1.3.
This completes the proof of part (i) of the theorem.

(ii) We have to prove that, given $V \in \mathcal{P}(S(\mathcal{C}_\mathbb{T}))$, if a sieve R is $\overline{V}^T$-covering then any sequent in $\mathcal{G}(R)$ is provable in $\mathbb{T} \cup \mathcal{G}(V)$ using geometric logic, that is $J_V \supseteq \overline{V}^{\mathcal{T}}$. In fact, we will show that J_V is a Grothendieck topology containing $J_\mathbb{T}$ and all the sieves in V as its covering sieves; this will immediately imply our thesis.

By definition of J_V, the sieves in V are J_V-covering, and if R is a $J_\mathbb{T}$-covering sieve then, by definition of $J_\mathbb{T}$, any sequent in $\mathcal{G}(R)$ is provable in $\mathbb{T}$, whence R is J_V-covering. To prove that J_V is a Grothendieck topology, we use Definition 3.1.3. Property (a) is obvious and property (b) easily follows from the cut rule in geometric logic. Property (c) is an immediate consequence of the observation following the first cartesian square in the proof of Theorem 3.2.5. It remains to prove property (d). Since $\mathcal{G}(R) = \mathcal{G}(\overline{R}^{J_\mathbb{T}})$ for any sieve R in $\mathcal{C}_\mathbb{T}$ (by Proposition 3.1.6 and Remark 3.1.7), it suffices to prove that for any sieve S in $\mathcal{C}_\mathbb{T}$ generated by a monomorphism $m : d \to c$ and any sieve T in $\mathcal{C}_\mathbb{T}$ on d, if both S and T are J_V-covering then $S * \{T\}$ is J_V-covering. Now, in view of the equality $\mathcal{G}(R) = \mathcal{G}(\overline{R}^{J_\mathbb{T}})$, our thesis easily follows from the cut rule in geometric logic, by using Proposition 3.1.6, Remark 3.1.7 and Remark 3.1.4.

(iii) Let us first show that $\overline{\mathcal{G}(\mathcal{F}(U))}^\mathbb{T} = \overline{U}^\mathbb{T}$. Note that $\mathcal{G}(\mathcal{F}(U))$ is the collection of sequents of the form $\mathcal{G}(\mathcal{F}(\sigma))$ as σ varies in U. If $\sigma \equiv (\phi \vdash_{\vec{x}} \psi)$ then $\mathcal{G}(\mathcal{F}(\sigma))$ is the $\mathbb{T}$-equivalence class of the sequent $(\phi \vdash_{\vec{x}} \phi \wedge \psi)$; but this sequent is clearly $\mathbb{T}$-equivalent to σ, whence $\overline{\mathcal{G}(\mathcal{F}(U))}^\mathbb{T} = \overline{U}^\mathbb{T}$, as required.

We have that

$$G(F(U)) = \overline{\mathcal{G}(F(U))}^\mathbb{T} = \overline{\mathcal{G}(\overline{\mathcal{F}(U)}^\mathcal{T})}^\mathbb{T} = \overline{\mathcal{G}(\overline{\mathcal{F}(U)}^\mathbb{T})}^\mathbb{T} = \overline{\mathcal{G}(\mathcal{F}(U))}^\mathbb{T} = \overline{U}^\mathbb{T},$$

where the third equality follows from part (i) of the theorem. To complete the proof of part (iii) of the theorem, it thus remains to show that $\overline{U}^\mathbb{T} = \mathbb{T}^U$. The inclusion $\overline{U}^\mathbb{T} \subseteq \mathbb{T}^U$ follows from part (i) of the theorem, while the converse inclusion follows as a consequence of the equality $\overline{\mathcal{G}(\mathcal{F}(U))}^\mathbb{T} = \overline{U}^\mathbb{T}$ and of part (i) of the theorem: if $\sigma \in \mathbb{T}^U$ then $\sigma \in \overline{\{\sigma\}}^\mathbb{T} = \overline{\mathcal{G}(\mathcal{F}(\{\sigma\}))}^\mathbb{T} \subseteq \overline{\mathcal{G}(\mathcal{F}(U))}^\mathbb{T} = \overline{U}^\mathbb{T}$.

(iv) Let us first prove that $\overline{\mathcal{F}(\mathcal{G}(V))}^\mathcal{T} = \overline{V}^\mathcal{T}$. We observe that $\mathcal{F}(\mathcal{G}(V))$ is the collection of sieves of the form $\mathcal{F}(\mathcal{G}(R))$ as R varies in V. Moreover, we have that $\mathcal{F}(\mathcal{G}(R)) = \overline{R}^{J_\mathbb{T}}$. Our claim thus follows from Proposition 3.1.5. Now, by using the fact that $\overline{\mathcal{F}(\mathcal{G}(V))}^\mathcal{T} = \overline{V}^\mathcal{T}$, one can prove the required equalities as in the proof of part (iii) of the theorem, with part (ii) playing here the role of part (i) there.

$\square$

Remark 3.2.8 Let $\mathbb{T}$ be a geometric theory over a signature Σ. Given a quotient $\mathbb{T}'$ of $\mathbb{T}$, let $J_{\mathbb{T}'}^{\mathbb{T}}$ be the associated $\mathbb{T}$-topology of $\mathbb{T}'$. Then the equalities $\overline{U}^{\mathcal{T}} = \mathbb{T}^U$ and $\overline{V}^{\mathcal{T}} = J_V$ of Theorem 3.2.7 yield the following equivalences:

(i) For any sieve $R \in S(\mathcal{C}_{\mathbb{T}})$, $R \in J_{\mathbb{T}'}^{\mathbb{T}}$ if and only if any sequent in $\mathcal{G}(R)$ is provable in $\mathbb{T}'$.

(ii) For any geometric sequent σ over Σ, σ is provable in $\mathbb{T}'$ if and only if $\mathcal{F}(\sigma)$ is $J_{\mathbb{T}'}^{\mathbb{T}}$-covering.

In particular, we obtain the following characterization of the syntactic topology $J_{\mathbb{T}}$ on $\mathcal{C}_{\mathbb{T}}$: a sieve R is $J_{\mathbb{T}}$-covering if and only if some (equivalently, every) sequent in $\mathcal{G}(R)$ is provable in $\mathbb{T}$.

The interest of Theorem 3.2.7 lies in the fact that the proof system $\mathcal{T}_{\mathcal{C}_{\mathbb{T}}}^{J_{\mathbb{T}}}$ is, for many purposes, computationally better behaved than the classical deduction system for geometric logic over $\mathbb{T}$; indeed, it is often easier to determine whether a given sieve belongs to the Grothendieck topology generated by a certain family of sieves than to test whether a given geometric sequent is provable in a certain geometric theory.

Let us briefly consider how much of Theorem 3.2.7 survives for smaller fragments of geometric logic, e.g. cartesian, regular, or coherent logic. If $\mathbb{T}$ is a cartesian (resp. regular, coherent) theory over Σ, one can define exactly as above an assignment $\mathcal{F}$ from the collection $\mathcal{S}^{\text{cart}}$ (resp. $\mathcal{S}^{\text{reg}}$, $\mathcal{S}^{\text{coh}}$) of cartesian (resp. regular, coherent) sequents over Σ to the collection $S(\mathcal{C}_{\mathbb{T}}^{\text{cart}})$ (resp. $S(\mathcal{C}_{\mathbb{T}}^{\text{reg}})$, $S(\mathcal{C}_{\mathbb{T}}^{\text{coh}})$) of sieves in the cartesian (resp. regular, coherent) syntactic category of the theory $\mathbb{T}$. Accordingly, the closure operator on $\mathcal{P}(\mathcal{S}^{\text{cart}})$ (resp. $\mathcal{P}(\mathcal{S}^{\text{reg}})$, $\mathcal{P}(\mathcal{S}^{\text{coh}})$) sends a collection U of sequents in $\mathcal{S}^{\text{cart}}$ (resp. $\mathcal{S}^{\text{reg}}$, $\mathcal{S}^{\text{coh}}$) to the collection of cartesian (resp. regular, coherent) sequents over Σ which are derivable in $U \cup \mathbb{T}$ by using cartesian (resp. regular, coherent) logic, and it is immediate to see that the proof of part (i) of the theorem continues to hold. On the contrary, no assignment $\mathcal{G}$ with values in the class of cartesian (resp. regular, coherent) sequents over Σ can be defined, since one should restrict to sieves generated by a monomorphism (resp. a single arrow, a finite number of arrows); however, if we consider $\mathcal{G}$ to take values in the class of geometric sequents over Σ as in the geometric case then part (ii) of the theorem continues to hold and for any presieve V in the cartesian (resp. regular, coherent) category $\mathcal{C}_{\mathbb{T}}^{\text{cart}}$ (resp. $\mathcal{C}_{\mathbb{T}}^{\text{reg}}$, $\mathcal{C}_{\mathbb{T}}^{\text{coh}}$), the theory $\mathcal{G}(\overline{V}^{\mathcal{T}})$ is classified by the topos $\mathbf{Sh}(\mathcal{C}_{\mathbb{T}}^{\text{cart}}, \overline{V}^{\mathcal{T}})$ (resp. $\mathbf{Sh}(\mathcal{C}_{\mathbb{T}}^{\text{reg}}, \overline{V}^{\mathcal{T}})$, $\mathbf{Sh}(\mathcal{C}_{\mathbb{T}}^{\text{coh}}, \overline{V}^{\mathcal{T}})$) – cf. the proof of Theorem 3.2.5.

Remarks 3.2.9 (a) Given a geometric quotient $\mathbb{T}'$ of $\mathbb{T}$, it is natural to look for axiomatizations of $\mathbb{T}'$ over $\mathbb{T}$ which are 'as simple as possible'; this translates, via Theorem 3.2.5, into the problem of finding an 'as simple as possible' set of generators for the associated Grothendieck topology $J_{\mathbb{T}'}^{\mathbb{T}}$ over $J_{\mathbb{T}}$; in fact, if a collection V of presieves in $\mathcal{C}_{\mathbb{T}}$ generates a Grothendieck topology J, then, by Theorem 3.2.7(ii), $\mathbb{T}^J$ is axiomatized over $\mathbb{T}$ by the collection of sequents in $\mathcal{G}(V)$. Note that, conversely, if a collection U of geometric sequents axiomatizes a quotient $\mathbb{T}'$ then, by Theorem 3.2.7(i), the collection of presieves $\mathcal{F}(U)$ generates over $J_{\mathbb{T}}$ the Grothendieck topology $J_{\mathbb{T}'}^{\mathbb{T}}$. For instance, one may ask if $\mathbb{T}'$ can be axiomatized over $\mathbb{T}$ by geometric sequents of the form $(\top \vdash_{\vec{x}} \phi)$; this corresponds to requiring that $J_{\mathbb{T}'}^{\mathbb{T}}$ should be generated over $J_{\mathbb{T}}$ by a collection of principal sieves generated by subobjects of objects of the form $\{\vec{x}.\top\}$; two notable classes of theories with this property are given by the Booleanizations and DeMorganizations of geometric theories (cf. Remark 3.2.9(b) and section 4.2.3 below).

(b) One can often get interesting insights on theories by investigating the associated Grothendieck topologies. As an illustration of this, let us show that the Booleanization $\mathbb{T}'$ of a geometric theory $\mathbb{T}$, i.e. the quotient of $\mathbb{T}$ corresponding via the duality of Theorem 3.2.5 to the subtopos $\mathbf{sh}_{\neg\neg}(\mathbf{Set}[\mathbb{T}]) \hookrightarrow \mathbf{Set}[\mathbb{T}]$ (see section 1.3.4 above), is axiomatized over $\mathbb{T}$ by geometric sequents of the form $(\top \vdash_{\vec{x}} \phi)$ as claimed in (a). Given a Heyting category $\mathcal{C}$, let $\tilde{\mathcal{C}}$ be its full subcategory on the non-zero objects. Since the only objects which are $J_{\mathbb{T}}$-covered by the empty sieve are the zero ones, $\tilde{\mathcal{C}}_{\mathbb{T}}$ is $J_{\mathbb{T}}$-dense in $\mathcal{C}_{\mathbb{T}}$ and the induced Grothendieck topology $J_{\mathbb{T}}|_{\tilde{\mathcal{C}}_{\mathbb{T}}}$ is contained in the double negation topology on $\tilde{\mathcal{C}}_{\mathbb{T}}$; hence $J_{\mathbb{T}'}^{\mathbb{T}}$ is generated over $J_{\mathbb{T}}$ by the sieves generated in $\mathcal{C}_{\mathbb{T}}$ by the principal stably non-empty sieves in $\tilde{\mathcal{C}}_{\mathbb{T}}$ (cf. Proposition 4.2.11 below). Now, given a Heyting category $\mathcal{C}$ and monomorphisms $f : d \rightarrowtail c'$ and $g : c' \rightarrowtail c$ in $\tilde{\mathcal{C}}$, it is immediate to see that if (f), regarded as a sieve in $\tilde{\mathcal{C}}$, is stably non-empty then $f \cong g^*((g \circ f) \cup \neg(g \circ f))$ where $\cup$ and $\neg$ are respectively the union and pseudocomplementation in the Heyting algebra $\mathrm{Sub}_{\mathcal{C}}(c)$, whence (f) is the pullback of a stably non-empty sieve in $\tilde{\mathcal{C}}$ on c. Therefore, since every object in $\mathcal{C}_{\mathbb{T}}$ admits a monomorphism to an object of the form $\{\vec{x}.\top\}$, the topology $J_{\mathbb{T}'}^{\mathbb{T}}$ is generated over $J_{\mathbb{T}}$ by a collection of principal sieves generated by subobjects of objects of the form $\{\vec{x}.\top\}$, as required.

3.3 A deduction theorem for geometric logic

The following result represents an analogue for geometric logic of the deduction theorem in classical first-order logic; we shall derive it as an application of our proof-theoretic interpretation of Theorem 3.2.5.

Theorem 3.3.1 *Let $\mathbb{T}$ be a geometric theory over a signature Σ and ϕ, ψ geometric sentences over Σ such that the sequent $(\top \vdash_{[]} \psi)$ is provable in the theory $\mathbb{T} \cup \{(\top \vdash_{[]} \phi)\}$. Then the sequent $(\phi \vdash_{[]} \psi)$ is provable in the theory $\mathbb{T}$.*

Proof By Theorem 3.2.5 and Lemma 1.4.2 [50], we can rephrase our thesis as follows: if the principal sieve in $\mathcal{C}_{\mathbb{T}}$ generated by the arrow $\{[].\psi\} \xrightarrow{[\psi]} \{[].\top\}$ belongs to the Grothendieck topology on $\mathcal{C}_{\mathbb{T}}$ generated over $J_{\mathbb{T}}$ by the principal sieve generated by the arrow $\{[].\phi\} \xrightarrow{[\phi]} \{[].\top\}$, then $[\phi] \leq [\psi]$ in $\mathrm{Sub}_{\mathcal{C}_{\mathbb{T}}}(\{[].\top\})$.

Recalling Theorem 1.4.13 and the fact that the syntactic topology $J_\mathbb{T}$ is the geometric topology on the category $\mathcal{C}_\mathbb{T}$, we can further rewrite our thesis as follows: if $\mathcal{C}$ is a geometric category and $J_\mathcal{C}^{geom}$ is the geometric topology on it then, given subobjects $m : a \rightarrowtail 1$ and $n : b \rightarrowtail 1$ of the terminal object 1 of $\mathcal{C}$ such that (n) belongs to the Grothendieck topology generated over $J_\mathcal{C}^{geom}$ by the sieve (m), $m \leq n$ in $\text{Sub}_\mathcal{C}(1)$.

Let us apply the formula for the Grothendieck topology $(D^r)^l$ generated by a family of sieves D that is stable under pullback (cf. section 4.1.1 below) to the collection D of sieves which are either $J_\mathcal{C}^{geom}$-covering or of the form $f^*((m))$ for an arrow f with codomain 1; starting from the assumption that $(n) \in (D^r)^l(b)$, we want to deduce that $m \leq n$ in $\text{Sub}_\mathcal{C}(1)$.

We note that $m \leq n$ if and only if $m \leq (m \Rightarrow n)$, if and only if $m^*(m \Rightarrow n) \cong 1_a$ (where $\Rightarrow$ denotes the Heyting implication in $\text{Sub}_\mathcal{C}(1)$ – note that this operation is well-defined by Proposition 1.3.14).

Now, by the results of section 4.1.1, we have that for any $c \in \mathcal{C}$

$$(D^r)^l(c) = \{S \text{ sieve on } c \mid \text{for any arrow } d \xrightarrow{f} c \text{ and sieve } Z \text{ on } d,$$
$$[Z \in D^r(d) \text{ and } f^*(S) \subseteq Z] \text{ implies } Z = M_d\}.$$

Taking Z to be $m^*((m \Rightarrow n))$, S to be (n) and f to be m in this formula, we see that, since $n \leq (m \Rightarrow n)$, in order to prove that $m^*(m \Rightarrow n) \cong 1_a$ it suffices to show that $m^*((m \Rightarrow n)) \in D^r(a)$. In fact, we shall prove that $(m \Rightarrow n) \in D^r(1)$, which implies that $m^*((m \Rightarrow n)) \in D^r(a)$ since D^r is stable under pullback.

Again by the results of section 4.1.1, we have that for any $c \in \mathcal{C}$

$$D^r(c) = \{T \text{ sieve on } c \mid \text{for any arrow } d \xrightarrow{f} c \text{ and sieve } S \text{ on } d,$$
$$[S \in D(d) \text{ and } S \subseteq f^*(T)] \text{ implies } f \in T\}.$$

To show that $(m \Rightarrow n) \in D^r(1)$, we are thus reduced to prove that for any arrow $f : d \to 1$ with codomain 1 and any sieve S on d such that $S \in D(d)$, $S \subseteq f^*((m \Rightarrow n))$ implies $1_d \in f^*((m \Rightarrow n))$. Now, if $S \in D(d)$ then there are two possible cases: either S is $J_\mathcal{C}^{geom}$-covering or (since 1 is a terminal object) S is equal to $f^*((m))$. In the first case, we have that if $S \subseteq f^*((m \Rightarrow n))$ then $f^*((m \Rightarrow n))$ is $J_\mathcal{C}^{geom}$-covering and hence, it being generated by a monomorphism, maximal, as required. In the second case, we have that if $S \subseteq f^*((m \Rightarrow n))$ then $f^*(m) \leq f^*(m \Rightarrow n)$; but $f^*(m \Rightarrow n) = f^*(m) \Rightarrow f^*(n)$ (cf. Lemma A1.4.13 [55]) whence $f^*(m) \leq f^*(m \Rightarrow n)$ implies $f^*(m) \leq f^*(n)$ i.e. $1_d \in f^*((m \Rightarrow n))$. □

4
LATTICES OF THEORIES

In this chapter we use the duality theorem between quotients and subtoposes established in Chapter 3 to interpret many concepts of elementary topos theory which apply to the lattice of subtoposes of a given topos in the context of geometric theories.

Notions that we analyse in detail include the coHeyting algebra structure on the lattice of subtoposes of a given topos, open, closed, quasi-closed subtoposes, the Booleanization and DeMorganization of a topos, the dense-closed factorization of a geometric inclusion, skeletal inclusions, the surjection-inclusion factorization of a geometric morphism, atoms in the lattice of subtoposes of a given topos, and subtoposes with enough points.

We also obtain an explicit description of the Heyting operation between Grothendieck topologies on a given category and of the Grothendieck topology generated by a given collection of sieves, as well as a number of results about the problem of 'relativizing' a local operator with respect to a given subtopos.

4.1 The lattice operations on Grothendieck topologies and quotients

In this section we study the structure of the lattice of subtoposes of a given Grothendieck topos. It is well known that this lattice, endowed with the obvious order relation given by the inclusion of subtoposes, is a coHeyting algebra (see, for instance, section A4.5 of [55]). Our aim is to describe this structure in terms of Grothendieck topologies and then of theories in view of Theorem 3.2.5.

Given a Heyting algebra H and an element $a \in H$, the set $\uparrow(a)$ of elements $h \in H$ such that $h \geq a$ is closed under the operations of conjunction, disjunction and Heyting implication and is therefore a Heyting algebra with respect to these operations. Indeed, the assertion about the conjunction and disjunction is obvious, while the fact that $b \Rightarrow c$ is in $\uparrow(a)$ if b and c are follows from the inequality $c \leq (b \Rightarrow c)$.

This remark allows us to restrict our attention to the case of subtoposes of a presheaf topos in order to describe the operations of union, intersection and coHeyting implication on a pair of subtoposes of a given Grothendieck topos; indeed, the union (resp. intersection, coHeyting implication) of two subtoposes of $\mathbf{Sh}(\mathcal{C}, J)$ is the same as their union (resp. intersection, coHeyting implication) in the coHeyting algebra of subtoposes of $[\mathcal{C}^{\text{op}}, \mathbf{Set}]$, since the order relation in the former lattice is the restriction of the order relation in the latter.

Theories, Sites, Toposes. Olivia Caramello.
© Olivia Caramello 2018. Published 2018 by Oxford University Press.

4.1.1 The lattice operations on Grothendieck topologies

Let $\mathcal{E} = [\mathcal{C}^{\mathrm{op}}, \mathbf{Set}]$ be a presheaf topos, with subobject classifier Ω. Recall that $\Omega : \mathcal{C}^{\mathrm{op}} \to \mathbf{Set}$ is defined by

$$\Omega(c) = \{R \mid R \text{ is a sieve on } c\} \text{ (for any object } c \in \mathcal{C});$$
$$\Omega(f) = f^*(-) \text{ (for any arrow } f \text{ in } \mathcal{C}),$$

where $f^*(-)$ denotes the operation of pullback of sieves in $\mathcal{C}$ along f.

We know from Theorem 1.3.36 that, given a small category $\mathcal{C}$, the Grothendieck topologies J on $\mathcal{C}$ correspond exactly to local operators on the topos $[\mathcal{C}^{\mathrm{op}}, \mathbf{Set}]$; this correspondence, to which we refer as $(*)$, sends a local operator $j : \Omega \to \Omega$ to the subobject $J \rightarrowtail \Omega$ which it classifies, that is to the Grothendieck topology J on $\mathcal{C}$ defined by setting $S \in J(c)$ if and only if $j(c)(S) = M_c$, and conversely a subobject $J \rightarrowtail \Omega$ to the arrow $j : \Omega \to \Omega$ classifying it.

Given a subobject $A \rightarrowtail \Omega$, its classifying arrow $\chi_A : \Omega \to \Omega$ is given by the formula

$$\chi_A(c)(S) = \{f : d \to c \mid f^*(S) \in A(d)\}$$

(cf. the description of the classifying arrow of a subobject after Theorem 1.1.12).

It will be convenient for our purposes to give an explicit description of the internal Heyting operations $\wedge, \vee, \Rightarrow : \Omega \to \Omega$ on our topos $\mathcal{E}$. To this end, we recall from Lemma A1.6.3 [55] the abstract definitions of these operations (cf. also the proof of Theorem 1.3.29).

The internal conjunction map $\wedge : \Omega \times \Omega \to \Omega$ is the classifying arrow of the subobject $<\top, \top> : 1 \rightarrowtail \Omega \times \Omega$, whence

$$\wedge(c)(S, T) = S \cap T$$

for any object $c \in \mathcal{C}$ and sieves S and T on c.

The internal disjunction map $\vee : \Omega \times \Omega \to \Omega$ is the classifying arrow of the union of the subobjects $\pi_1^*(\top)$ and $\pi_2^*(\top)$ where π_1 and π_2 are the two product projections $\Omega \times \Omega \to \Omega$, whence

$$\vee(c)(S, T) = \{f : d \to c \mid f^*(S) \cup f^*(T) = M_d\}$$

for any object $c \in \mathcal{C}$ and sieves S and T on c.

The internal implication map $\Rightarrow : \Omega \times \Omega \to \Omega$ is the classifying arrow of the equalizer $\Omega_1 \rightarrowtail \Omega \times \Omega$ of $\wedge$ and π_1, whence

$$\Rightarrow(c)(S, T) = \{f : d \to c \mid f^*(S) \subseteq f^*(T)\}$$

for any object $c \in \mathcal{C}$ and sieves S and T on c.

It is immediate to see that the order relation between local operators on $\mathcal{E}$ given by the opposite of the natural order between subtoposes transfers via $(*)$ to the following order between Grothendieck topologies on $\mathcal{C}$: $J \leq J'$ if and only if $J(c) \subseteq J'(c)$ for every $c \in \mathcal{C}$, that is if and only if every J-covering sieve is

J'-covering. Hence, the relation $\leq$ defines a Heyting algebra structure on the collection of Grothendieck topologies on the category $\mathcal{C}$; in particular, for any Grothendieck topologies J and J' on $\mathcal{C}$, there exist a meet $J \wedge J'$, a join $J \vee J'$ and a Heyting implication $J \Rightarrow J'$. The bottom element of this lattice is the Grothendieck topology $\bot$ on $\mathcal{C}$ given by $\bot(c) = \{M_c\}$ for every $c \in \mathcal{C}$, while the top element is the topology $\top$ defined by $\top(c) = \{S \mid S \text{ sieve on } c\}$ for every $c \in \mathcal{C}$.

It is immediate to see that the meet $J \wedge J'$ of two Grothendieck topologies J and J' on $\mathcal{C}$ is given by the following formula: for any object c of $\mathcal{C}$ and any sieve S on c, $S \in (J \wedge J')(c)$ if and only if $S \in J(c)$ and $S \in J'(c)$; indeed, the class of Grothendieck topologies is clearly closed under intersection. The join $J \vee J'$ is the smallest Grothendieck topology K such that $J \leq K$ and $J' \leq K$, so it is the Grothendieck topology generated by the collection of sieves which are either J-covering or J'-covering. In order to get a more explicit description of this operation, as well as of that of Heyting implication, we apply A. Joyal's theory as described in section A4.5 of [55]; this will lead in particular to an explicit description of the Grothendieck topology generated by a family of sieves which is stable under pullback.

First, let us explicitly describe, in terms of the category $\mathcal{C}$, the Galois connection from $\mathrm{Sub}_{\mathcal{E}}(\Omega)$ to itself given by the mappings $D \to D^r$ and $D \to D^l$ described at p. 213 of [55].

Given a subobject $D \rightarrowtail \Omega$, the subobjects $D^r \rightarrowtail \Omega$ and $D^l \rightarrowtail \Omega$ are respectively defined to be

$$\forall_{\pi_2}((\pi_1^*(D) \Rightarrow \Theta) \rightarrowtail \Omega$$

and

$$\forall_{\pi_1}((\pi_2^*(D) \Rightarrow \Theta) \rightarrowtail \Omega,$$

where π_1 and π_2 are the two product projections $\Omega \times \Omega \to \Omega$, π_1^* and π_2^* are the pullback functors $\mathrm{Sub}(\Omega) \to \mathrm{Sub}(\Omega \times \Omega)$ respectively along π_1 and π_2, $\Theta \rightarrowtail \Omega \times \Omega$ is the equalizer of $\pi_2, \Rightarrow: \Omega \times \Omega \to \Omega$, and $\forall_{\pi_1}, \forall_{\pi_2}$ are the right adjoints respectively to π_1^* and π_2^*.

From the above formulas, we get the following expression for Θ:

$$\Theta(c) = \{(S,T) \mid S \text{ and } T \text{ are sieves on } c \text{ such that for all } f: d \to c, \\ f^*(S) \subseteq f^*(T) \text{ implies } f \in T\},$$

for any object $c \in \mathcal{C}$.

Now, by using formula (7) at p. 146 of [63], we obtain

$$(\pi_1^*(D) \Rightarrow \Theta)(c) = \{(S,T) \mid S \text{ and } T \text{ are sieves on } c \text{ such that for all } f: d \to c, \\ [f^*(S) \in D(d) \text{ and } f^*(S) \subseteq f^*(T)] \text{ implies } f \in T\}.$$

By using formula (15) at p. 148 of [63], we get the following description of $\forall_{\pi_2}(A)$ for a subobject A of $\Omega \times \Omega$:

110   *Lattices of theories*

$$\forall_{\pi_2}(A)(c) = \{R \text{ sieve on } c \mid \text{ for all } f : d \to c, \, \Omega \times f^*(R) \subseteq A\},$$

for any object $c \in \mathcal{C}$. If we apply this formula to the subobject $\pi_1^*(D) \Rightarrow \Theta$ calculated above we obtain

$$D^r(c) = \{T \text{ sieve on } c \mid \text{ for any arrows } e \xrightarrow{h} d \xrightarrow{g} c \text{ and sieve } S \text{ on } d,$$
$$[h^*(S) \in D(e) \text{ and } h^*(S) \subseteq h^*(g^*(T))] \text{ implies } h \in g^*(T)\}.$$

Similarly, we derive the following expression for D^l:

$$D^l(c) = \{S \text{ sieve on } c \mid \text{ for any arrows } e \xrightarrow{h} d \xrightarrow{g} c \text{ and sieve } T \text{ on } d,$$
$$[h^*(T) \in D(e) \text{ and } h^*(g^*(S)) \subseteq h^*(T)] \text{ implies } h \in T\}.$$

Notice that the above formulas can alternatively be put in the following form:

$$D^r(c) = \{T \text{ sieve on } c \mid \text{ for any arrow } d \xrightarrow{f} c \text{ and sieve } S \text{ on } d,$$
$$[S \in D(d) \text{ and } S \subseteq f^*(T)] \text{ implies } f \in T\};$$

$$D^l(c) = \{S \text{ sieve on } c \mid \text{ for any arrow } d \xrightarrow{f} c \text{ and sieve } Z \text{ on } d,$$
$$[Z \in D(d) \text{ and } f^*(S) \subseteq Z] \text{ implies } Z = M_d\}.$$

For instance, the equivalence of the former expression for D^l with this latter formulation can be verified by taking $g = f$, $h = 1_d$ and $T = Z$ in one direction and $f = g \circ h$ and $Z = h^*(T)$ in the other direction.

From these expressions one immediately obtains the following formula:

$$(D^r)^l(c) = \{S \text{ sieve on } c \mid \text{ for any arrow } d \xrightarrow{f} c \text{ and sieve } T \text{ on } d,$$
$$[(\text{for any arrow } e \xrightarrow{g} d \text{ and sieve } Z \text{ on } e$$
$$(Z \in D(e) \text{ and } Z \subseteq g^*(T)) \text{ implies } g \in T) \text{ and } (f^*(S) \subseteq T)]$$
$$\text{implies } T = M_d\}.$$

We recall from the proof of Corollary A4.5.13(i) [55] that the classifying arrow of $(D^r)^l$ is the smallest local operator on $\mathcal{E}$ for which all the monomorphisms in $\mathcal{E}$ whose classifying arrow factors through $D \rightarrowtail \Omega$ are dense. Let us now show that, via the identification (∗), this local operator corresponds precisely to the Grothendieck topology generated by D (i.e. the smallest Grothendieck topology J on $\mathcal{C}$ such that all the sieves in D, regarded here as a collection of sieves in $\mathcal{C}$, are J-covering). To this end, it suffices to recall that, given a local operator j on a topos $\mathcal{E}$, the c_j-dense monomorphisms are exactly the monomorphisms whose classifying arrow factors through the subobject classified by j; notice that if $\mathcal{E} = [\mathcal{C}^{\text{op}}, \textbf{Set}]$ and j corresponds to a Grothendieck topology J on $\mathcal{C}$, this subobject is exactly J (regarded as a subobject of $\Omega_{[\mathcal{C}^{\text{op}}, \textbf{Set}]}$). Now, all the sieves in D are J-covering if and only if $D \leq J$ as subobjects of Ω, so our claim immediately follows. So our formula for $(D^r)^l$ gives an explicit description of the Grothendieck topology generated by D, as recorded by the following result:

Proposition 4.1.1 *Let $\mathcal{C}$ be a small category and D a family of sieves in $\mathcal{C}$ which is stable under pullback. Then the Grothendieck topology G_D generated by D is equal to $(D^r)^l$ and given by*

$$G_D(c) = \{S \text{ sieve on } c \mid \text{for any arrow } d \xrightarrow{f} c \text{ and sieve } T \text{ on } d,$$
$$[(\text{for any arrow } e \xrightarrow{g} d \text{ and sieve } Z \text{ on } e$$
$$(Z \in D(e) \text{ and } Z \subseteq g^*(T)) \text{ implies } g \in T) \text{ and } (f^*(S) \subseteq T)]$$
$$\text{implies } T = M_d\}$$

for any object $c \in \mathcal{C}$. □

Remark 4.1.2 The Grothendieck topology on a small category $\mathcal{C}$ generated by a given family F of sieves in $\mathcal{C}$ is given by $G_{F_{\text{p.b.}}}$, where $F_{\text{p.b.}}$ is the collection of sieves in $\mathcal{C}$ which are pullbacks of sieves in F.

Starting from Corollary A4.5.13(i) [55], one can prove by similar means that our formula for D^l gives an explicit description of the largest Grothendieck topology J on $\mathcal{C}$ such that all the sieves in D are J-closed (one recalls that the classifying arrow of D^l is the largest local operator on $\mathcal{E}$ for which the monomorphisms in $\mathcal{E}$ whose classifying arrow factors through $D \rightarrowtail \Omega$ are closed and replaces, in the above discussion, the subobject J classifying c_j-dense monomorphisms with the subobject Ω_J classifying c_j-closed monomorphisms, namely the equalizer of the arrows $j, 1_\Omega : \Omega \to \Omega$).

As an application of the above formulas for D^l, let us derive an explicit formula for the Heyting operation on the collection of Grothendieck topologies on a given small category.

Example 4.5.14(f) [55] provides a description of the Heyting operation on the collection of local operators on a topos: given local operators j_1 and j_2 on a topos $\mathcal{E}$, $j_1 \Rightarrow j_2 = (J_1 \cap \Omega_{j_2})^l$ (where J_1 is the subobject of Ω corresponding to j_1 and Ω_{j_2} is the image of j_2). If $\mathcal{E} = [\mathcal{C}^{\text{op}}, \mathbf{Set}]$ and j_1, j_2 correspond to Grothendieck topologies J_1, J_2 on $\mathcal{C}$ via (∗) then our (second) formula for D^l gives the following expression for $J_1 \Rightarrow J_2$:

$$(J_1 \Rightarrow J_2)(c) = \{S \text{ sieve on } c \mid \text{for any arrow } d \xrightarrow{f} c \text{ and sieve } Z \text{ on } d,$$
$$[Z \text{ is } J_1\text{-covering and } J_2\text{-closed and } f^*(S) \subseteq Z]$$
$$\text{implies } Z = M_d\}.$$

In particular, the pseudocomplement $\neg J$ of a Grothendieck topology J on $\mathcal{C}$ is given by the following formula:

$$\neg J(c) = \{S \text{ sieve on } c \mid \text{for any arrow } d \xrightarrow{f} c \text{ and sieve } Z \text{ on } d,$$
$$[Z \text{ is } J\text{-covering and } f^*(S) \subseteq Z] \text{ implies } Z = M_d\}.$$

Let us now prove directly that, given a category $\mathcal{C}$ and a collection D of sieves in $\mathcal{C}$ which is closed under pullback, the above formula for D^l always defines a Grothendieck topology on $\mathcal{C}$ and $(D^r)^l$ is the Grothendieck topology on $\mathcal{C}$

generated by D. This will ensure that our results also hold for a general, not necessarily small, category $\mathcal{C}$.

It is clear that the above formula for D^l yields a collection of sieves satisfying the maximality and stability axioms for Grothendieck topologies, so it remains to verify that it satisfies the transitivity axiom. Let R and S be sieves on $c \in \mathcal{C}$ such that $S \in D^l(c)$ and for each $s : a \to c$ in S, $s^*(R) \in D^l(a)$; we want to prove that $R \in D^l(c)$, i.e. that for any arrow $f : d \to c$ and sieve Z on d, $(Z \in D(d)$ and $f^*(R) \subseteq Z)$ implies $Z = M_d$. Now, for any $h \in f^*(S)$, $h^*(f^*(R)) \subseteq h^*(Z)$ and hence $h \in Z$ since $(f \circ h)^*(R) \in D^l(\mathrm{dom}(h))$. So $f^*(S) \subseteq Z$, whence $Z = M_d$ since $S \in D^l(c)$.

Let us now show that $(D^r)^l$ is the Grothendieck topology on $\mathcal{C}$ generated by D; since we already know that $(D^r)^l$ is a Grothendieck topology, this amounts to verifying that $D \leq (D^r)^l$ and that for any Grothendieck topology K on $\mathcal{C}$ which contains D, $(D^r)^l \leq K$. The fact that $D \leq (D^r)^l$ follows immediately from the above formula for $(D^r)^l$ (to see that any $S \in D(c)$ belongs to $(D^r)^l(c)$ take $g = 1_d$ and $Z = f^*(S)$ in it). To show that any Grothendieck topology K on $\mathcal{C}$ containing D is contained in $(D^r)^l$, suppose that S is a sieve in $(D^r)^l(c)$; then S is K-covering if and only if $\overline{S}^K = M_c$. Now, if we take $f = 1_c$ and $T = \overline{S}^K$ in the formula for $(D^r)^l$ we obtain that $f^*(S) \subseteq T$ and that for any arrow $e \overset{g}{\to} d$ and sieve Z on e, $[Z \in D(e)$ and $Z \subseteq g^*(T)]$ implies that $g^*(T)$ is K-covering and hence maximal (it being K-closed); the formula thus implies that T is maximal, as required.

We can also verify directly that the above formula for $J_1 \Rightarrow J_2$ satisfies the property of the Heyting implication between J_1 and J_2, i.e. that for any Grothendieck topology K on $\mathcal{C}$, $K \wedge J_1 \leq J_2$ if and only if $K \leq (J_1 \Rightarrow J_2)$. Indeed, $(J_1 \Rightarrow J_2) \wedge J_1 \leq J_2$ since for every $((J_1 \Rightarrow J_2) \wedge J_1)$-covering sieve S, $S \subseteq \overline{S}^{J_2}$ and hence $\overline{S}^{J_2}$ is maximal, i.e. S is J_2-covering; in the other direction, if $K \wedge J_1 \leq J_2$ then, for any K-covering sieve S, $[Z$ is J_1-covering and J_2-closed, and $f^*(S) \subseteq Z]$ implies that Z is $K \wedge J_1$-covering and hence J_2-covering and J_2-closed, i.e. maximal. Incidentally, this argument yields the following alternative expression for $J_1 \Rightarrow J_2$:

$$(J_1 \Rightarrow J_2)(c) = \{S \text{ sieve on } c \mid \text{for any arrow } d \overset{f}{\to} c \text{ and sieve } Z \text{ on } d,$$
$$[Z \text{ is } J_1\text{-covering and } f^*(S) \subseteq Z] \text{ implies } Z \text{ is } J_2\text{-covering}\}$$

for any $c \in \mathcal{C}$.

4.1.2 The lattice operations on theories

By using Theorem 3.2.5 we can transfer the lattice operations on the collection of Grothendieck topologies on the geometric syntactic category $\mathcal{C}_\mathbb{T}$ of a geometric theory $\mathbb{T}$ which contain the syntactic topology $J_\mathbb{T}$ into the context of quotients of $\mathbb{T}$.

Let us denote by $\mathfrak{Th}_\Sigma^\mathbb{T}$ the collection of syntactic-equivalence classes of geometric theories over Σ which are quotients of $\mathbb{T}$. By definition of the duality of

Theorem 3.2.5, the order on $\mathfrak{Th}_\Sigma^{\mathbb{T}}$ corresponding to the order $\leq$ between Grothendieck topologies on $\mathcal{C}_\mathbb{T}$ is the following: $\mathbb{T}' \leq \mathbb{T}''$ if and only if all the geometric sequents provable in $\mathbb{T}'$ (equivalently, all the geometric sequents in a given collection of axioms for $\mathbb{T}'$) are provable in $\mathbb{T}''$. So Theorem 3.2.5 gives the following result:

Theorem 4.1.3 *Let $\mathbb{T}$ be a geometric theory over a signature Σ. Then the collection $\mathfrak{Th}_\Sigma^{\mathbb{T}}$ of (syntactic-equivalence classes of) geometric theories over Σ which are quotients of $\mathbb{T}$, endowed with the order defined by '$\mathbb{T}' \leq \mathbb{T}''$ if and only if all the axioms of $\mathbb{T}'$ are provable in $\mathbb{T}''$', is a Heyting algebra.* □

By taking $\mathbb{T}$ to be the empty (geometric) theory $\mathbb{O}_\Sigma$ over Σ, we obtain in particular that the collection $\mathfrak{Th}_\Sigma^{\mathbb{O}_\Sigma}$ of all the (syntactic-equivalence classes of) geometric theories over Σ is a Heyting algebra.

Recalling the definition of the order in $\mathfrak{Th}_\Sigma^{\mathbb{T}}$, we obtain the following descriptions of the lattice operations in $\mathfrak{Th}_\Sigma^{\mathbb{T}}$:

(i) The bottom element is $\mathbb{T}$.

(ii) The top element is the *contradictory theory*, i.e. the theory axiomatized by $(\top \vdash_{[]} \bot)$ (in which all the geometric sequents over Σ are provable).

(iii) The wedge $\mathbb{T}' \wedge \mathbb{T}''$ is the largest geometric theory over Σ which is contained in both $\mathbb{T}'$ and $\mathbb{T}''$, i.e. the collection of geometric sequents σ over Σ such that σ is provable in both $\mathbb{T}'$ and $\mathbb{T}''$.

(iv) The join $\mathbb{T}' \vee \mathbb{T}''$ is the (syntactic-equivalence class of) the smallest geometric theory over Σ which contains both $\mathbb{T}'$ and $\mathbb{T}''$, i.e. the theory whose axioms are given by the union of the axioms of $\mathbb{T}'$ with those of $\mathbb{T}''$.

(v) The Heyting implication $\mathbb{T}' \Rightarrow \mathbb{T}''$ is the (syntactic-equivalence class of the) largest geometric theory $\mathbb{S}$ over Σ such that $\mathbb{S} \wedge \mathbb{T}' \leq \mathbb{T}''$, i.e. such that every geometric sequent σ which is provable in both $\mathbb{S}$ and $\mathbb{T}'$ is provable in $\mathbb{T}''$; in particular, the pseudocomplement $\neg \mathbb{T}'$ is the (syntactic-equivalence class of the) largest geometric theory over Σ such that every geometric sequent σ which is provable in both $\mathbb{S}$ and $\mathbb{T}'$ is provable in $\mathbb{T}$.

We note that these operations are quite natural from a logical viewpoint; however it is by no means obvious from the perspective of geometric logic that there should exist a Heyting operation on the lattice of syntactic-equivalence classes of geometric theories over a given signature, while this follows as a formal consequence of Theorem 3.2.5. Another consequence of the theorem is the fact that the lattices $\mathfrak{Th}_\Sigma^{\mathbb{T}}$ are complete; indeed, any intersection of Grothendieck topologies is a Grothendieck topology.

Figure 4.1 represents the isomorphic lattice structures present on the collection of geometric quotients of theories $\mathbb{T}$, $\mathbb{S}$, $\mathbb{R}$, etc. classified by a given topos $\mathcal{E}$.

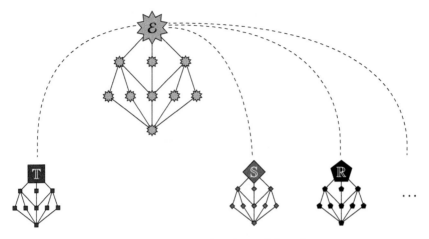

Fig. 4.1: Lattices of theories

Let us discuss, from the point of view of geometric logic, the fact that our lattice $\mathfrak{Th}_\Sigma^{\mathbb{T}}$ is distributive. This is a formal consequence of the fact that it is a Heyting algebra, which is ensured by Theorem 3.2.5, but it is instructive to establish this result directly by means of purely logical arguments. Our condition means that for any (syntactic-equivalence classes of) geometric theories $\mathbb{T}'$ and $\{\mathbb{T}_k \mid k \in K\}$ in $\mathfrak{Th}_\Sigma^{\mathbb{T}}$, $\mathbb{T}' \wedge (\bigvee_{k \in K} \mathbb{T}_k) = \bigvee_{k \in K}(\mathbb{T}' \wedge \mathbb{T}_k)$. Since the inequality $\geq$ is trivially satisfied, we are reduced to showing that for any geometric sequent σ over Σ, if σ is provable in $\mathbb{T}'$ and is derivable from axioms of the $\mathbb{T}_k$, then σ is derivable from axioms of the $\mathbb{T}' \wedge \mathbb{T}_k$. To this end, the following lemma is useful.

Lemma 4.1.4 *Let Σ be a first-order signature. If a geometric sequent $\sigma \equiv (\phi \vdash_{\vec{x}} \psi)$ over Σ is provable in the theory $\mathbb{S} = \{\tau \equiv (\phi_\tau \vdash \psi_\tau) \mid \tau \in \mathbb{S}\}$ using geometric logic then σ is provable in the theory $\mathbb{S}_\sigma = \{(\phi_\tau \wedge \phi \vdash \psi_\tau \vee \psi) \mid \tau \in \mathbb{S}\}$ using geometric logic.*

Proof Given a geometric sequent $\tau \equiv (\chi \vdash_{\vec{x}} \xi)$ over Σ, for a string of variables $\vec{x}'$ of the same kind as $\vec{x}$ we denote by $\mathcal{W}_{\vec{x}'}(\tau)$ the sequent $(\chi \wedge \phi[\vec{x}'/\vec{x}] \vdash \xi \vee \psi[\vec{x}'/\vec{x}])$. It is easy to check that for any instance of an inference rule of geometric logic, if we choose a string $\vec{x}'$ of variables that do not appear in any of the sequents involved in it then the image via $\mathcal{W}_{\vec{x}'}$ of the conclusion of the rule is derivable in geometric logic from the images via $\mathcal{W}_{\vec{x}'}$ of the premises of the rule. This clearly implies our thesis. □

Now, our claim can be deduced as follows. If we have a derivation of a sequent $\sigma \equiv (\phi \vdash_{\vec{x}} \psi)$ over Σ which is provable in $\mathbb{T}'$ from sequents $\tau \equiv (\phi_\tau \vdash \psi_\tau)$ each of which is an axiom of some $\mathbb{T}_k$ then by Lemma 4.1.4 we have a derivation of σ from the sequents $(\phi \wedge \phi_\tau \vdash \psi \vee \psi_\tau)$, each of which is, on the one hand, provable in $\mathbb{T}'$ since it is derivable from σ, and, on the other hand, provable in $\mathbb{T}_k$ whenever τ lies in $\mathbb{T}_k$ since it is derivable from τ.

This is an illustration of the usefulness of Theorem 3.2.5 in providing insights into geometric logic; we shall discuss other applications of it below.

4.1.3 The Heyting implication in $\mathfrak{Th}_\Sigma^\mathbb{T}$

The purpose of this section is to give an explicit description of the Heyting operation on quotients of a given geometric theory $\mathbb{T}$. We will achieve this by interpreting the formula for the Heyting implication of Grothendieck topologies obtained in section 4.1.1 at the level of theories via the duality of Theorem 3.2.5.

Given two local operators j and j' on a topos $\mathcal{E}$, we write $j \leq j'$ to mean $\mathbf{sh}_{c_{j'}}(\mathcal{E}) \subseteq \mathbf{sh}_{c_j}(\mathcal{E})$ where c_j and c'_j are the universal closure operations corresponding to j and j' (as in Theorem 1.3.32).

The following lemma will be useful for our purposes.

Lemma 4.1.5 *Let $\mathcal{E}$ be an elementary topos and j, j' two local operators on $\mathcal{E}$ with associated universal closure operations c_j and $c_{j'}$. Then $j \leq j'$ if and only if, for every subobject m in $\mathcal{E}$, $c_j(m) \leq c_{j'}(m)$ (equivalently, $c_{j'}(m) = c_{j'}(c_j(m))$).*

Proof Let L_j and $L_{j'}$ be the cartesian reflectors on $\mathcal{E}$ associated respectively with the local operators j and j'. Recall from section 1.3.4 that, given a monomorphism $m : A' \rightarrowtail A$ in $\mathcal{E}$, $c_j(m)$ is defined by the pullback square

$$\begin{array}{ccc} c_j(A') & \longrightarrow & L_j A' \\ {\scriptstyle c_j(m)} \downarrow & & \downarrow {\scriptstyle L_j m} \\ A & \xrightarrow{\eta_A} & L_j A, \end{array}$$

where η is the unit of the reflection induced by L_j. If $j \leq j'$ then $L_{j'}(L_j(m)) \cong L_{j'}(m)$ since $L_{j'}$ factors through L_j and both L_j and $L_{j'}$ are cartesian reflectors. If we apply the pullback-preserving functor $L_{j'}$ to the above pullback square we thus obtain that $L_{j'}(c_j(m)) \cong L_{j'}(L_j(m)) \cong L_{j'}(m)$; from this it immediately follows by definition of $c_{j'}$ in terms of $L_{j'}$ that $c_{j'}(m) = c_{j'}(c_j(m))$. In particular, $c_j(m) \leq c_{j'}(m)$.

The converse direction immediately follows from the fact that j is the classifying arrow of $c_j(\top)$ for each local operator j (cf. the proof of Theorem 1.3.32). □

Remark 4.1.6 It follows at once from Lemma 4.1.5 that if $j \leq j'$ then, for any subobject m, if m is $c_{j'}$-closed then m is c_j-closed.

We shall also need the following results.

Proposition 4.1.7 *Let $\mathcal{C}$ be a regular category, J a Grothendieck topology on $\mathcal{C}$ such that $J \supseteq J_\mathcal{C}^{\mathrm{reg}}$ and $r : d \to c$ a cover in $\mathcal{C}$. Then*

(i) For any sieve R on c, $R \in J(c)$ if and only if $r^*(R) \in J(d)$.
(ii) For any sieve R on c generated by a monomorphism, R is J-closed if and only if $r^*(R)$ is J-closed.
(iii) For any sieve R on c, R is J-closed if and only if, for any monomorphism $f : b \to c$, $f^*(R) \in J(b)$ implies $f \in R$.
(iv) For any sieves R and T on c such that T is generated by a monomorphism, $r^*(R) \subseteq r^*(T)$ if and only if $R \subseteq T$.

Proof

(i) This immediately follows from the stability and transitivity axioms for Grothendieck topologies.
(ii) The 'only if' part is obvious; let us prove the 'if' part. Given an arrow $f : b \to c$ such that $f^*(R) \in J(b)$, we want to prove that $f \in R$. Consider the pullback

$$\begin{array}{ccc} a & \xrightarrow{h} & d \\ {\scriptstyle g}\downarrow & & \downarrow{\scriptstyle r} \\ b & \xrightarrow{f} & c \end{array}$$

in $\mathcal{C}$. From the commutativity of this square and the stability axiom for Grothendieck topologies it follows that $h^*(r^*(R)) \in J(a)$ and hence that $h \in r^*(R)$ i.e. $r \circ h \in R$. But $f \circ g = r \circ h$, so $f \circ g \in R$. Since g is a cover, R is generated by a monomorphism by our hypothesis and covers are orthogonal to monomorphisms (cf. Lemma A1.3.2 [55]), we conclude that $f \in R$, as required.
(iii) The 'only if' part is obvious, so it remains to prove that if for any monomorphism $f : b \rightarrowtail c$, $f^*(R) \in J(b)$ implies $f \in R$, then R is J-closed. Let $g : e \to c$ be an arrow such that $g^*(R) \in J(e)$; we want to prove that $g \in R$. Denoting by $e \xrightarrow{g''} u \xrightarrow{g'} c$ the cover-mono factorization of g, we have by part (i) of the proposition that $g'^*(R) \in J(u)$; so $g' \in R$ by our hypothesis and hence $g \in R$, as required.
(iv) The 'if' part is obvious, so it remains to prove that if $r^*(R) \subseteq r^*(T)$ then $R \subseteq T$. Given $f \in R$, consider the pullback

$$\begin{array}{ccc} a & \xrightarrow{h} & d \\ {\scriptstyle g}\downarrow & & \downarrow{\scriptstyle r} \\ b & \xrightarrow{f} & c \end{array}$$

in $\mathcal{C}$. Now, h belongs to $r^*(R)$ and hence to $r^*(T)$, so $f \circ g = r \circ h \in T$. But g is a cover and T is generated by a monomorphism so, since covers are orthogonal to monomorphisms (cf. Lemma A1.3.2 [55]), $f \in T$, as required.

□

Proposition 4.1.8 *Let $\mathbb{T}$ be a geometric theory over a signature Σ, $\mathbb{T}'$ a quotient of $\mathbb{T}$ and $\{\{\vec{x_i}.\phi_i\} \stackrel{[\theta_i]}{\rightarrow} \{\vec{y}.\psi\} \mid i \in I\}$ a set of generators for a sieve S in the syntactic category $\mathcal{C}_\mathbb{T}$ of $\mathbb{T}$. Then*

(i) *S is $J_{\mathbb{T}'}^{\mathbb{T}}$-covering if and only if the sequent $(\psi \vdash_{\vec{y}} \bigvee_{i \in I} (\exists \vec{x_i}) \theta_i)$ is provable in $\mathbb{T}'$.*

(ii) *S is $J_{\mathbb{T}'}^{\mathbb{T}}$-closed if and only if it is generated by a single monomorphism and for any geometric formula $\psi'(\vec{y})$ over Σ such that $(\psi' \vdash_{\vec{y}} \psi)$ is provable in $\mathbb{T}$, the sequent $(\psi' \vdash_{\vec{y}} \bigvee_{i \in I} (\exists \vec{x_i}) \theta_i)$ is provable in $\mathbb{T}'$ (if and) only if it is provable in $\mathbb{T}$.*

Proof

(i) This is precisely equivalence (i) in Remark 3.2.8.

(ii) This follows from Remark 4.1.6, Proposition 3.1.6(ii), Proposition 4.1.7 (iii) and part (i) of this proposition, by recalling the identification of subobjects of $\{\vec{y}.\psi\}$ in $\mathcal{C}_\mathbb{T}$ with $\mathbb{T}$-provable equivalence classes of geometric formulae $\psi'(\vec{y})$ over Σ such that $(\psi' \vdash_{\vec{y}} \psi)$ is provable in $\mathbb{T}$ (cf. Lemma 1.4.2).

□

Now, let us suppose that $\mathcal{C}$ is the syntactic category $\mathcal{C}_\mathbb{T}$ of a geometric theory $\mathbb{T}$ and that J_1 and J_2 are respectively the associated $\mathbb{T}$-topologies $J_{\mathbb{T}_1}^{\mathbb{T}}$ and $J_{\mathbb{T}_2}^{\mathbb{T}}$ of two quotients $\mathbb{T}_1$ and $\mathbb{T}_2$ of $\mathbb{T}$ (in the sense of section 3.2.1).
Recall from section 4.1.1 that

$(J_1 \Rightarrow J_2)(c) = \{S$ sieve on $c \mid$ for every arrow $d \stackrel{f}{\rightarrow} c$ and sieve Z on d, $[Z$ is J_1-covering and $f^*(S) \subseteq Z]$ implies Z is J_2-covering$\}$

for any $c \in \mathcal{C}$. In this formula, the quantification over all the sieves Z can be restricted to all the sieves Z of the form (t) for a monic arrow t. To show this, we notice that, by Proposition 3.1.6, $\overline{Z}^{J_\mathbb{T}}$ is generated by a monic arrow, and Z is J_1-covering (resp. J_2-covering) if and only if $\overline{Z}^{J_\mathbb{T}}$ is. Indeed, if Z is J_1-covering (resp. J_2-covering) then clearly $\overline{Z}^{J_\mathbb{T}}$ is J_1-covering (resp. J_2-covering) while the converse implication follows from the transitivity axiom for Grothendieck topologies in light of the fact that $J_\mathbb{T} \subseteq J_1$ (resp. $J_\mathbb{T} \subseteq J_2$).
We thus obtain the following simplified formula:

$(J_1 \Rightarrow J_2)(c) = \{S$ sieve on $c \mid$ for any arrow $d \stackrel{f}{\rightarrow} c$ and sieve $T = (t)$ on d with t monic, $[T$ is J_1-covering and $f^*(S) \subseteq T]$ implies T is J_2-covering$\}$

for any $c \in \mathcal{C}$.

Let us apply this formula to the syntactic category of a geometric theory $\mathbb{T}$. By Proposition 3.1.5(ii), $J_1 \Rightarrow J_2$ is generated over $J_\mathbb{T}$ by sieves generated by a single monic arrow. This remark enables us to arrive at a simplified axiomatization of the Heyting implication $\mathbb{T}_1 \Rightarrow \mathbb{T}_2$, as follows.

Recall that, by Lemma 1.4.2, any subobject in $\mathcal{C}_\mathbb{T}$ is isomorphic to one of the form

$$\{\vec{y}'.\psi[\vec{y}'/\vec{y}]\} \xrightarrow{[\psi' \wedge \vec{y}'=\vec{y}]} \{\vec{y}.\psi\},$$

where $\psi(\vec{y})$ is a geometric formula such that the sequent $(\psi' \vdash_{\vec{y}} \psi)$ is provable in $\mathbb{T}$.

On the other hand, for any arrows

$$\{\vec{x}.\phi\} \xrightarrow{[\theta]} \{\vec{y}.\psi\}$$

and

$$\{\vec{y}'.\psi[\vec{y}'/\vec{y}]\} \xrightarrow{[\psi' \wedge \vec{y}'=\vec{y}]} \{\vec{y}.\psi\}$$

in $\mathcal{C}_\mathbb{T}$ with common codomain, we have the following pullback square in $\mathcal{C}_\mathbb{T}$:

$$\begin{array}{ccc} \{\vec{x}.(\exists \vec{y})(\theta[\underline{x}/\vec{x}] \wedge \psi')\} & \xrightarrow{[(\exists \vec{y})(\theta \wedge \psi' \wedge \underline{\vec{y}}=\vec{y}')]} & \{\vec{y}'.\psi[\vec{y}'/\vec{y}]\} \\ {\scriptstyle [(\exists \vec{y})(\theta \wedge \psi' \wedge \underline{\vec{x}}=\vec{x})]} \downarrow & & \downarrow {\scriptstyle [\psi' \wedge \vec{y}'=\vec{y}]} \\ \{\vec{x}.\phi\} & \xrightarrow{[\theta]} & \{\vec{y}.\psi\}. \end{array}$$

By Lemma 1.4.2, the sieve generated by the arrow

$$[(\exists \vec{y})(\theta \wedge \psi' \wedge \underline{\vec{x}} = \vec{x})] : \{\vec{x}.(\exists \vec{y})(\theta[\underline{x}/\vec{x}] \wedge \psi')\} \to \{\vec{x}.\phi\}$$

is contained in the sieve generated by an arrow

$$\{\vec{x}'.\chi[\vec{x}'/\vec{x}]\} \xrightarrow{[\chi \wedge \vec{x}'=\vec{x}]} \{\vec{x}.\phi\}$$

if and only if the sequent $((\exists \vec{y})(\theta(\vec{x},\vec{y}) \wedge \psi'(\vec{y})) \vdash_{\vec{x}} \chi)$ is provable in $\mathbb{T}$.

In light of Proposition 4.1.8(i), we thus obtain the following result:

Theorem 4.1.9 *Let $\mathbb{T}$ be a geometric theory over a signature Σ and $\mathbb{T}_1, \mathbb{T}_2$ two quotients of $\mathbb{T}$. Then $\mathbb{T}_1 \Rightarrow \mathbb{T}_2$ is the theory obtained from $\mathbb{T}$ by adding all the geometric sequents $(\psi \vdash_{\vec{y}} \psi')$ over Σ with the property that $(\psi' \vdash_{\vec{y}} \psi)$ is provable in $\mathbb{T}$ and for any $\mathbb{T}$-provably functional geometric formula $\theta(\vec{x}, \vec{y})$ from a geometric formula-in-context $\{\vec{x}.\phi\}$ to $\{\vec{y}.\psi\}$ and any geometric formula χ in the context $\vec{x}$ such that $(\chi \vdash_{\vec{x}} \phi)$ is provable in $\mathbb{T}$, the conjunction of the facts*

(i) *$(\phi \vdash_{\vec{x}} \chi)$ is provable in $\mathbb{T}_1$*
(ii) *$((\exists \vec{y})(\theta(\vec{x},\vec{y}) \wedge \psi'(\vec{y})) \vdash_{\vec{x}} \chi)$ is provable in $\mathbb{T}$*

implies that $(\phi \vdash_{\vec{x}} \chi)$ is provable in $\mathbb{T}_2$. □

In particular, we obtain that the pseudocomplement of a quotient $\mathbb{T}'$ in $\mathfrak{Th}_\Sigma^\mathbb{T}$ is the theory $\neg \mathbb{T}' = (\mathbb{T}' \Rightarrow \mathbb{T})$ obtained from $\mathbb{T}$ by adding all the geometric sequents $(\psi \vdash_{\vec{y}} \psi')$ over Σ with the property that $(\psi' \vdash_{\vec{y}} \psi)$ is provable in $\mathbb{T}$ and for any $\mathbb{T}$-provably functional geometric formula $\theta(\vec{x}, \vec{y})$ from a geometric formula-in-context $\{\vec{x}.\phi\}$ to $\{\vec{y}.\psi\}$ and any geometric formula χ in the context $\vec{x}$ such that $(\chi \vdash_{\vec{x}} \phi)$ is provable in $\mathbb{T}$, the conjunction of the facts

(i) $(\phi \vdash_{\vec{x}} \chi)$ is provable in $\mathbb{T}'$
(ii) $((\exists \vec{y})(\theta(\vec{x}, \vec{y}) \wedge \psi'(\vec{y})) \vdash_{\vec{x}} \chi)$ is provable in $\mathbb{T}$

implies that $(\phi \vdash_{\vec{x}} \chi)$ is provable in $\mathbb{T}$.

4.2 Transfer of notions from topos theory to logic

In this section we describe in detail the logical counterparts, via the duality of Theorem 3.2.5, of a number of important topos-theoretic notions which apply to the lattice of subtoposes of a given topos.

The following preliminary section, which concerns the problem of relativizing a local operator with respect to another one, will be instrumental in connection with calculations concerning open and quasi-closed local operators on a topos.

4.2.1 *Relativization of local operators*

Recall from Theorem 1.3.32 that for any topos $\mathcal{E}$ there is a bijection between universal closure operations on $\mathcal{E}$ and local operators on $\mathcal{E}$: this bijection sends a local operator $j : \mathcal{E} \to \mathcal{E}$ to the universal closure operation c_j (also denoted c_L, where L is the corresponding reflector on $\mathcal{E}$) defined, for each monomorphism $m : A' \rightarrowtail A$ in $\mathcal{E}$, by the pullback square

$$\begin{array}{ccc} c_L(A') & \longrightarrow & LA' \\ \downarrow & & \downarrow Lm \\ A & \xrightarrow{\eta_A^L} & LA, \end{array}$$

where L is the cartesian reflector on $\mathcal{E}$ corresponding to j and η_A^L is the unit of the reflection, and a universal closure operation c on $\mathcal{E}$ to the local operator $j_c : \Omega \to \Omega$ given by the classifying arrow of the subobject $c(1 \xrightarrow{\top} \top)$. Recall also that, given a local operator j on $\mathcal{E}$, the domain Ω_j of the equalizer $e_j : \Omega_j \rightarrowtail \Omega$ of the arrows $1_\Omega, j : \Omega \to \Omega$ is the subobject classifier of the topos $\mathbf{sh}_j(\mathcal{E})$ and the classifying arrow $\chi_m : A \to \Omega$ of a monomorphism m in $\mathcal{E}$ factors through e_j if and only if m is c_j-closed.

Given geometric inclusions $\mathcal{F}' \underset{L'}{\overset{i'}{\rightleftarrows}} \mathcal{F}$ and $\mathcal{F} \underset{L}{\overset{i}{\rightleftarrows}} \mathcal{E}$, let us denote by $j_{L'}$ and j_L the corresponding local operators respectively on $\mathcal{F}$ and on $\mathcal{E}$. Denoting by Ω the subobject classifier of $\mathcal{E}$, we define $e_L : \Omega_L \rightarrowtail \Omega$ to be the equalizer of $1_\Omega, j_L : \Omega \to \Omega$, $e_{L'} : (\Omega_L)_{L'} \rightarrowtail \Omega_L$ to be the equalizer of $1_\Omega, j_{L'} : \Omega_{L'} \to \Omega_{L'}$ and $e_{L' \circ L} : \Omega_{L' \circ L} \rightarrowtail \Omega$ to be the equalizer of $1_\Omega, j_{L' \circ L} : \Omega \to \Omega$.

Lemma 4.2.1 *With the above notation, the composite arrow*

$$(\Omega_L)_{L'} \xrightarrow{e_{L'}} \Omega_L \xrightarrow{e_L} \Omega$$

and the arrow

$$\Omega_{L' \circ L} \xrightarrow{e_{L' \circ L}} \Omega$$

are isomorphic (as objects of $\mathcal{E}/\Omega$).

Proof Let us first prove that, given a subobject $m : A' \rightarrowtail A$ in $\mathcal{E}$ with classifying arrow $\chi_m : A \to \Omega$,

(1) χ_m factors through $e_L \circ e_{L'}$ if and only if ($c_L(m) = m$ and $c_{L'}(Lm) = Lm$);
(2) χ_m factors through $e_{L' \circ L}$ if and only if $c_{L' \circ L}(m) = m$;
(3) ($c_L(m) = m$ and $c_{L'}(Lm) = Lm$) if and only if $c_{L' \circ L}(m) = m$.

(1) χ_m factors through $e_L \circ e_{L'}$ if and only if χ_m factors through e_L and the factorization χ_m^L of χ_m through e_L factors through $e_{L'}$; by definition of e_L, the first condition means precisely that $c_L(m) = m$, while the second, in view of the adjunction $\mathrm{Hom}_{\mathbf{sh}_{j_L}(\mathcal{E})}(LA, \Omega_L) \cong \mathrm{Hom}_{\mathcal{E}}(A, \Omega_L)$, is equivalent to the requirement that the subobject in $\mathbf{sh}_{j_L}(\mathcal{E})$ classified by the factorization $\overline{\chi_m^L} : LA \to \Omega_L$ of χ_m^L through $\eta_A^L : A \to LA$ be $c_{L'}$-closed (by definition of $e_{L'}$).

Consider the diagram

$$\begin{array}{ccccc} A' & \xrightarrow{!} & 1 & \xrightarrow{!} & 1 \\ \downarrow m & & \downarrow \top_L & & \downarrow \top \\ A & \xrightarrow{\chi_m^L} & \Omega_L & \xrightarrow{e_L} & \Omega, \end{array}$$

where $\top_L$ is the factorization of $\top : 1 \to \Omega$ through e_L. The outer rectangle is the pullback square witnessing that χ_m classifies m, while the right-hand square is trivially a pullback (e_L being monic); the pullback lemma thus implies that the left-hand square is a pullback. Since L preserves pullbacks, the square

$$\begin{array}{ccc} LA' & \xrightarrow{!} & 1 \\ \downarrow Lm & & \downarrow \top_L \\ LA & \xrightarrow{\overline{\chi_m^L}} & \Omega_L \end{array}$$

is a pullback, i.e. $\overline{\chi_m^L}$ classifies the subobject Lm in $\mathbf{sh}_{j_L}(\mathcal{E})$. This concludes the proof of (1).

(2) This is immediate by definition of $\Omega_{L' \circ L}$.

(3) By definition of $c_{L'}$ and $c_{L'\circ L}$, we have a rectangle

$$\begin{array}{ccccc} c_{L'\circ L}(A) & \longrightarrow & c_{L'}(LA') & \longrightarrow & L'(LA') \\ \downarrow {\scriptstyle c_{L'\circ L}(m)} & & \downarrow {\scriptstyle c_{L'}(Lm)} & & \downarrow {\scriptstyle L'(Lm)} \\ A & \xrightarrow{\eta_A^L} & LA & \xrightarrow{\eta_{LA}^{L'}} & L'(LA) \end{array}$$

in which both squares are pullbacks; this follows as an immediate consequence of the pullback lemma, since $\eta_A^{L'\circ L} = \eta_{LA}^{L'} \circ \eta_A^L$. Notice in particular that if A is an L-sheaf then $c_{L'\circ L}(m) = c_{L'}(Lm)$.

Suppose that $c_L(m) = m$ and $c_{L'}(Lm) = Lm$. The fact that $c_{L'}(Lm) = Lm$ implies, by definition of $c_L(m)$ and the fact that the left-hand square above is a pullback, that $c_{L'\circ L}(m) = c_L(m)$; hence, $c_L(m) = m$ implies $c_{L'\circ L}(m) = m$, as required. Conversely, suppose that $c_{L'\circ L}(m) = m$. By applying the pullback-preserving functor L to the left-hand square above, we obtain that $Lm = c_{L'}(Lm)$; but then, by definition of $c_L(m)$, we have $c_{L'\circ L}(m) = c_L(m)$ and hence $c_L(m) = m$. This concludes the proof of (3).

Now, (1), (2) and (3) together imply that for any subobject m in $\mathcal{E}$, χ_m factors through $e_L \circ e_{L'}$ if and only if it factors through $e_{L'\circ L}$. Our thesis thus follows from the Yoneda lemma. $\square$

The following definition will be central for the results in this section:

Definition 4.2.2 Given a topos $\mathcal{E}$, two local operators j and k on $\mathcal{E}$ and a local operator $k' : \Omega_j \to \Omega_j$ in $\mathbf{sh}_j(\mathcal{E})$, we say that k *relativizes to* k' *at* j (or that k' is the *relativization of* k *at* j) if the diagram

$$\begin{array}{ccc} \Omega_j & \xrightarrow{k'} & \Omega_j \\ \downarrow {\scriptstyle e_j} & & \downarrow {\scriptstyle e_j} \\ \Omega & \xrightarrow{k} & \Omega \end{array}$$

in $\mathcal{E}$ commutes.

Notice that, in the above definition, since e_j is monic, there can be at most one relativization of k at j.

The fundamental properties of relativizations are expressed by the following result:

Theorem 4.2.3 *Let k' be the relativization of k at j as above. Then*

(i) $\mathbf{sh}_{k'}(\mathbf{sh}_j(\mathcal{E})) = \mathbf{sh}_{k\vee j}(\mathcal{E})$ *(where $k \vee j$ is the join of k and j in the lattice of local operators on $\mathcal{E}$).*

(ii) *For any subobject m in $\mathbf{sh}_j(\mathcal{E})$, $c_{k'}(m) = c_k(m)$.*

(iii) *If $k \geq j$ then, for any subobject m in $\mathcal{E}$, $c_{k'}(L_j m) = c_k(m)$.*

Proof

(i) Let s be the local operator on $\mathcal{E}$ corresponding to $\mathbf{sh}_{k'}(\mathbf{sh}_j(\mathcal{E}))$, regarded as a subtopos of $\mathcal{E}$ via the composite geometric inclusion

$$\mathbf{sh}_{k'}(\mathbf{sh}_j(\mathcal{E})) \hookrightarrow \mathbf{sh}_j(\mathcal{E}) \hookrightarrow \mathcal{E}.$$

We have to prove that $e_s : \Omega_s \rightarrowtail \Omega$ is isomorphic to $e_{k \vee j} : \Omega_{k \vee j} \rightarrowtail \Omega$. By the Yoneda lemma, it is equivalent to prove that for any subobject $m : A' \rightarrowtail A$, χ_m factors through e_s if and only if it factors through $e_{k \vee j}$. Now, by Lemma 4.2.1, χ_m factors through e_s if and only if m is c_j-closed and Lm is $c_{k'}$-closed, where L is the cartesian reflector corresponding to j, while, by Example A4.5.13 [55], χ_m factors through $e_{k \vee j}$ if and only if m is both c_j-closed and c_k-closed. So we have to prove that, given a c_j-closed subobject $m : A' \rightarrowtail A$, Lm is $c_{k'}$-closed if and only if m is c_k-closed. Consider the commutative diagram

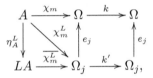

where the notation is that of the proof of Lemma 4.2.1.
From the proof of Lemma 4.2.1 we know that $\overline{\chi_m^L}$ is the characteristic map in $\mathbf{sh}_j(\mathcal{E})$ of the subobject Lm. By definition of Ω_k, χ_m factors through e_k (i.e. m is c_k-closed) if and only if $k \circ \chi_m = \chi_m$, while, by definition of $\Omega_{k'}^{\mathbf{sh}_j(\mathcal{E})}$, $\overline{\chi_m^L}$ factors through $\Omega_{k'}^{\mathbf{sh}_j(\mathcal{E})} \rightarrowtail \Omega_j$ (i.e. Lm is $c_{k'}$-closed) if and only if $k' \circ \overline{\chi_m^L} = \overline{\chi_m^L}$. Now, since e_j is monic and η_A^L is the unit of the reflection corresponding to j, $k' \circ \overline{\chi_m^L} = \overline{\chi_m^L}$ if and only if $e_j \circ k' \circ \overline{\chi_m^L} \circ \eta_A^L = e_j \circ \overline{\chi_m^L} \circ \eta_A^L$. But by the commutativity of the above diagram, this is precisely equivalent to $k \circ \chi_m = \chi_m$.

(ii) The condition $k \circ e_j = e_j \circ k' : \Omega_j \to \Omega$ is equivalent to the assertion that the subobjects classified by the maps $k \circ e_j$ and $e_j \circ k'$ are equal. Now, since k classifies $c_k(\top)$, $k \circ e_j$ classifies $e_j^*(c_k(\top)) = c_k(e_j^*(\top)) = c_k(\top_j)$, where $\top_j$ is the factorization of $\top$ through e_j, while $e_j \circ k'$ is easily seen to classify $c_{k'}(\top_j)$. So the condition amounts to requiring that $c_{k'}(\top_j) = c_k(\top_j)$. But every subobject in $\mathbf{sh}_j(\mathcal{E})$ is a pullback (both in $\mathbf{sh}_j(\mathcal{E})$ and in $\mathcal{E}$) of $\top_j$; hence for any subobject m in $\mathbf{sh}_j(\mathcal{E})$, $c_{k'}(m) = c_k(m)$, as required.

(iii) By (ii), it suffices to prove that if m is a subobject in $\mathcal{E}$ then $c_k(L_j m) = c_k(m)$; this immediately follows from the definition of $c_k(-)$ as the pullback of $L_k(-)$ along the unit of the reflector L_k and the fact that if $k \geq j$ then $L_k(m) \cong L_k(L_j(m))$.

□

Let us now consider some instances of relativizations.

Proposition 4.2.4 *With the notation of the proof of Lemma 4.2.1, $j_{L'} : \Omega_L \to \Omega_L$ is the relativization of $j_{L' \circ L}$ at j_L; that is, the square*

$$\begin{array}{ccc} \Omega_{j_L} & \xrightarrow{j_{L'}} & \Omega_{j_L} \\ {\scriptstyle e_{j_L}} \downarrow & & \downarrow {\scriptstyle e_{j_L}} \\ \Omega & \xrightarrow{j_{L' \circ L}} & \Omega \end{array}$$

commutes.

Proof Let us prove that the composites $e_{j_L} \circ j_{L'}$ and $j_{L' \circ L} \circ e_{j_L}$ classify the same subobject of $\mathbf{sh}_{j_L}(\mathcal{E})$, namely $c_{L'}(\top_L)$.

Consider the diagram

$$\begin{array}{ccccc} c_{L'}(1) & \xrightarrow{!} & 1 & \xrightarrow{!} & 1 \\ {\scriptstyle c_{L'}(\top_L)} \downarrow & & \downarrow {\scriptstyle \top_L} & & \downarrow {\scriptstyle \top} \\ \Omega_L & \xrightarrow{j_{L'}} & \Omega_L & \xrightarrow{e_{j_L}} & \Omega. \end{array}$$

Since both squares in it are pullbacks it follows from the pullback lemma that $e_{j_L} \circ j_{L'}$ classifies $c_{L'}(\top_L)$. On the other hand, if $j_{L' \circ L}$ classifies $c_{L' \circ L}(\top)$ then $j_{L' \circ L} \circ e_{j_L}$ classifies $e_{j_L}^*(c_{L' \circ L}(\top)) = c_{L' \circ L}(e_{j_L}^*(\top)) = c_{L' \circ L}(\top_L)$; but $c_{L' \circ L}(\top_L) = c_{L'}(\top_L)$ (cf. the proof of Lemma 4.2.1), which implies our thesis. □

Remarks 4.2.5 (a) All the relativizations arising as in Proposition 4.2.4 have the property that $k \geq j$. We shall see below instances of relativizations not enjoying this property. For the moment, we note that if k' is the relativization at j of two local operators k_1 and k_2 then $k_1 \vee j = k_2 \vee j$. This follows at once from Theorem 4.2.3 by recalling the identification between subcategories of sheaves on a topos and local operators on it.

(b) Given k and j local operators on a topos $\mathcal{E}$, there exists a relativization of k at j if and only if $j \circ k \circ e_j = k \circ e_j$ (equivalently, $c_k(\top_j)$ being classified by $k \circ e_j$, if and only if $c_k(\top_j)$ is c_j-closed); in particular, if $k \geq j$ then k relativizes at j. Conversely, given a local operator k' on $\mathbf{sh}_j(\mathcal{E})$, there always exists a local operator k on $\mathcal{E}$ such that k relativizes to k' at j. Take, for instance, k to be the local operator on $\mathcal{E}$ corresponding to the composite of the canonical geometric inclusions $\mathbf{sh}_{k'}(\mathbf{sh}_j(\mathcal{E})) \hookrightarrow \mathbf{sh}_j(\mathcal{E})$ and $\mathbf{sh}_j(\mathcal{E}) \hookrightarrow \mathcal{E}$; then, by Proposition 4.2.4 and Remark 4.2.5(a), k relativizes to k' at j.

Proposition 4.2.6 *Under the hypotheses of Theorem 4.2.3, if k relativizes to k' at j then $k \vee j$ relativizes to k' at j.*

Proof The condition $(k \vee j) \circ e_j = e_j \circ k'$ is equivalent to the assertion that the arrows $(k \vee j) \circ e_j$ and $e_j \circ k'$ classify the same subobject, i.e. that $c_{k \vee j}(\top_j) =$

$c_k(\top_j)$. Now, since $k \leq k \vee j$, $c_{k \vee j}(\top_j) \geq c_k(\top_j)$. To show that $c_{k \vee j}(\top_j) \leq c_k(\top_j)$ it is enough to prove, by the characterization of the closure of a subobject as the smallest closed subobject containing it, that $c_k(\top_j)$ is $c_{k \vee j}$-closed. We have observed in the proof of Theorem 4.2.3 that the $c_{k \vee j}$-closed subobjects are exactly those which are both c_j-closed and c_k-closed; our thesis thus follows at once from Remark 4.2.5(b).

Alternatively, our thesis follows as a consequence of Theorem 4.2.3(i) and Proposition 4.2.4. □

Let us now review the notions of open and quasi-closed local operator on a topos (cf. also section 4.2.2 below) and show that they behave naturally with respect to relativizations.

Recall from section A4.5 of [55] that the *open local operator* $o^{\mathcal{E}}(U)$ associated with a subterminal object U of a topos $\mathcal{E}$ is given by the composite arrow

$$\Omega \cong 1 \times \Omega \xrightarrow{u \times 1} \Omega \times \Omega \xrightarrow{\Rightarrow} \Omega,$$

where $u : 1 \to \Omega$ is the classifying arrow of the subobject $U \rightarrowtail 1$, while the *quasi-closed local operator* $qc^{\mathcal{E}}(U)$ associated with U is the composite arrow

$$\Omega \cong \Omega \times 1 \xrightarrow{1 \times <u,u>} \Omega \times \Omega \times \Omega \xrightarrow{\Rightarrow \times 1} \Omega \times \Omega \xrightarrow{\Rightarrow} \Omega.$$

Given a local operator j on a topos $\mathcal{E}$ and a subterminal object U of $\mathbf{sh}_j(\mathcal{E})$, since the inclusion $\mathbf{sh}_j(\mathcal{E}) \hookrightarrow \mathcal{E}$ preserves monomorphisms, we can regard U as a subterminal object of $\mathcal{E}$ as well. Then we have the following result:

Proposition 4.2.7 *Let $\mathcal{E}$ be a topos and j a local operator on $\mathcal{E}$. Given a subterminal object U of $\mathbf{sh}_j(\mathcal{E})$, the open (resp. quasi-closed) local operator $o^{\mathbf{sh}_j(\mathcal{E})}(U)$ (resp. $qc^{\mathbf{sh}_j(\mathcal{E})}(U)$) in $\mathbf{sh}_j(\mathcal{E})$ associated with U is the relativization at j of the open (resp. quasi-closed) local operator $o^{\mathcal{E}}(U)$ (resp. $qc^{\mathcal{E}}(U)$) in $\mathcal{E}$ associated with U (regarded as a subterminal object of $\mathcal{E}$).*

Proof From the description of the internal Heyting operations $\wedge_{\mathcal{E}}, \vee_{\mathcal{E}}, \Rightarrow_{\mathcal{E}}$: $\Omega \to \Omega$ on $\mathcal{E}$ given in the proof of Theorem 1.3.29, it follows that the diagrams

commute. Let us first show that the left-hand square commutes. The arrow $\wedge_{\mathcal{E}} : \Omega \times \Omega \to \Omega$ is the classifying arrow of $<\top, \top> : 1 \rightarrowtail \Omega \times \Omega$ and $\wedge_{\mathbf{sh}_j(\mathcal{E})} : \Omega_j \times \Omega_j \to \Omega$ is the classifying arrow of $<\top_j, \top_j> : 1 \rightarrowtail \Omega_j \times \Omega_j$, that is of the factorization of $<\top, \top>$ through $e_j \times e_j$. To prove that the composites

$e_j \circ \wedge_{\mathbf{sh}_j(\mathcal{E})}$ and $\wedge_{\mathcal{E}} \circ (e_j \times e_j)$ classify the same subobject of $\mathbf{sh}_j(\mathcal{E})$, namely $<\top_j, \top_j>: 1 \rightarrowtail \Omega_j \times \Omega_j$, consider the following diagram:

$$\begin{array}{ccccc} 1 & \xrightarrow{!} & 1 & \xrightarrow{!} & 1 \\ {\scriptstyle <\top_j,\top_j>}\downarrow & & {\scriptstyle \top_j}\downarrow & & \downarrow {\scriptstyle \top} \\ \Omega_j \times \Omega_j & \xrightarrow{\wedge_{\mathbf{sh}_j(\mathcal{E})}} & \Omega_j & \xrightarrow{e_j} & \Omega. \end{array}$$

Since both squares in it are pullbacks, it follows from the pullback lemma that $e_j \circ \wedge_{\mathbf{sh}_j(\mathcal{E})}$ classifies $<\top_j, \top_j>$. On the other hand, if $\wedge_{\mathcal{E}}$ classifies $<\top, \top>$ then $\wedge_{\mathcal{E}} \circ e_j$ classifies $e_j^*(<\top, \top>) = <\top_j, \top_j>$. This proves that the square for $\wedge$ commutes.

Let us now show that the square for $\Rightarrow$ commutes. We observe that $\Omega \times \Omega \xrightarrow{\Rightarrow_{\mathcal{E}}} \Omega$ is the classifying arrow of $r: E \rightarrowtail \Omega \times \Omega$ and $\Omega_j \times \Omega_j \xrightarrow{\Rightarrow_{\mathbf{sh}_j(\mathcal{E})}} \Omega_j$ is the classifying arrow of $r_j: E_j \rightarrowtail \Omega_j \times \Omega_j$ where r and r_j are respectively the equalizers of $\wedge_{\mathcal{E}}, \pi_1^{\mathcal{E}}: \Omega \times \Omega \to \Omega$ and of $\wedge_{\mathbf{sh}_j(\mathcal{E})}, \pi_1^{\mathbf{sh}_j(\mathcal{E})}: \Omega_j \times \Omega_j \to \Omega_j$.

It is easy to see, by invoking the commutativity of the square for $\wedge$, that the pullback of r along $e_j \times e_j$ is an equalizer for $\wedge_{\mathbf{sh}_j(\mathcal{E})}, \pi_1^{\mathbf{sh}_j(\mathcal{E})}: \Omega_j \times \Omega_j \to \Omega_j$ and hence that it is isomorphic to r_j; from this our claim follows immediately.

Now, by definition of open and quasi-closed local operators, the commutativity of the diagram for $\Rightarrow$ immediately implies our thesis, since if U is a subterminal object of $\mathbf{sh}_j(\mathcal{E})$ then the classifying arrow of $U \rightarrowtail 1$ in $\mathbf{sh}_j(\mathcal{E})$ is the factorization of its classifying arrow in $\mathcal{E}$ through $e_j: \Omega_j \rightarrowtail \Omega$. □

Definition 4.2.8 A local operator j on a topos $\mathcal{E}$ is said to be *dense* if $j \leq \neg\neg$.

A subtopos of a topos $\mathcal{E}$ is said to be *dense* if the corresponding local operator on $\mathcal{E}$ is dense.

As an application of Theorem 4.2.3 and Proposition 4.2.7, we recover the following well-known result:

Corollary 4.2.9 *Let $\mathcal{E}$ be topos and j be a dense local operator on $\mathcal{E}$. Then $\mathbf{sh}_{\neg\neg}(\mathbf{sh}_j(\mathcal{E})) = \mathbf{sh}_{\neg\neg}(\mathcal{E})$.*

Proof For any topos $\mathcal{E}$, $qc^{\mathcal{E}}(0_{\mathcal{E}}) = \neg\neg_{\mathcal{E}}$ (cf. p. 220 of [55]). Our thesis thus follows from Theorem 4.2.3 and Proposition 4.2.7 by invoking the fact (remarked at p. 219 of [55]) that for a dense local operator j on $\mathcal{E}$, the inclusion $\mathbf{sh}_j(\mathcal{E}) \hookrightarrow \mathcal{E}$ preserves the initial object. □

Given an elementary topos $\mathcal{E}$, let us denote by $\mathbf{Lop}(\mathcal{E})$ the collection of local operators on $\mathcal{E}$, endowed with the Heyting algebra structure given by the canonical order between them. For any local operator j on $\mathcal{E}$, there is a bijection between the collection of local operators k in $\mathcal{E}$ such that $k \geq j$ and the collection of local operators on $\mathbf{sh}_j(\mathcal{E})$. Indeed, if $k \geq j$ then the geometric inclusion $\mathbf{sh}_k(\mathcal{E}) \hookrightarrow \mathcal{E}$ factors (uniquely up to isomorphism) through $\mathbf{sh}_j(\mathcal{E}) \hookrightarrow \mathcal{E}$ and hence it corresponds to a unique local operator k_j on $\mathbf{sh}_j(\mathcal{E})$ such that $\mathbf{sh}_{k_j}(\mathbf{sh}_j(\mathcal{E})) = \mathbf{sh}_k(\mathcal{E})$, while conversely, given a local operator s on $\mathbf{sh}_j(\mathcal{E})$, the geometric inclusion given by the composite $\mathbf{sh}_s(\mathbf{sh}_j(\mathcal{E})) \hookrightarrow \mathbf{sh}_j(\mathcal{E}) \hookrightarrow \mathcal{E}$ corresponds to a unique local operator s^j on $\mathcal{E}$ such that $\mathbf{sh}_s(\mathbf{sh}_j(\mathcal{E})) = \mathbf{sh}_{s^j}(\mathcal{E})$. It is clear that these two correspondences are inverse to each other. Moreover, since the order between local operators on a topos corresponds to the reverse inclusion between the corresponding subcategories of sheaves, these bijections are also order-preserving (where the order between local operators k on $\mathcal{E}$ such that $k \geq j$ is the restriction of the order in $\mathbf{Lop}(\mathcal{E})$ and the order between local operators on $\mathbf{sh}_j(\mathcal{E})$ is the canonical one on $\mathbf{Lop}(\mathbf{sh}_j(\mathcal{E}))$). Recall that, given a Heyting algebra H and an element $a \in H$, $\uparrow(a) = \{b \in H \mid b \geq a\}$ is a Heyting algebra which is closed under the operations of conjunction, disjunction and Heyting implication in H and hence the map $a \vee (-) : H \to \uparrow(a)$ is a Heyting algebra homomorphism. So the bijections $(-)_j$ and $(-)^j$ are isomorphisms of Heyting algebras between the subalgebra $\uparrow(j)$ of $\mathbf{Lop}(\mathcal{E})$ and the algebra $\mathbf{Lop}(\mathbf{sh}_j(\mathcal{E}))$, whence the map $(j \vee (-))_j : \mathbf{Lop}(\mathcal{E}) \to \mathbf{Lop}(\mathbf{sh}_j(\mathcal{E}))$ is a Heyting algebra homomorphism.

4.2.2 Open, closed, quasi-closed subtoposes

Open, closed and quasi-closed subtoposes are particularly important classes of subtoposes, built from a given topos and a subterminal object of it. In this section we analyse the logical counterparts of these concepts in geometric logic, that is we describe the quotients of a given geometric theory corresponding via the duality of Theorem 3.2.5 to such subtoposes of its classifying topos.

4.2.2.1 Open subtoposes

Recall from section A4.5 of [55] that an open subtopos of a topos $\mathcal{E}$ is a geometric inclusion of the form $\mathcal{E}/U \hookrightarrow \mathcal{E}$ for a subterminal object U of $\mathcal{E}$. The associated universal closure operation sends a subobject $A' \rightarrowtail A$ to the implication $(A \times U) \Rightarrow A'$ in the Heyting algebra $\mathrm{Sub}_{\mathcal{E}}(A)$. If $L_U : \mathcal{E} \to \mathcal{E}$ is the corresponding cartesian reflector, we thus have that a monomorphism $A' \rightarrowtail A$ is L_U-dense if and only if $(A \times U) \leq A'$ in $\mathrm{Sub}_{\mathcal{E}}(A)$. So $A \times U$ is the smallest L_U-dense subobject of A, and hence L_U is the smallest local operator on $\mathcal{E}$ for which the monomorphism $U \rightarrowtail 1$ is dense (cf. the discussion preceding Lemma A4.5.10 [55]). From Proposition A4.3.11 [55] we thus deduce that a geometric morphism $f : \mathcal{F} \to \mathcal{E}$ factors through the inclusion $\mathcal{E}/U \hookrightarrow \mathcal{E}$ if and only if $f^*(U) = 1$.

Let $\mathcal{E}$ be the classifying topos $\mathbf{Set}[\mathbb{T}] \simeq \mathbf{Sh}(\mathcal{C}_\mathbb{T}, J_\mathbb{T})$ of a geometric theory $\mathbb{T}$ over a signature Σ. We shall now describe the quotient of $\mathbb{T}$ corresponding via Theorem 3.2.5 to an open subtopos $\mathcal{E}/U \hookrightarrow \mathcal{E}$ of $\mathcal{E}$. Recall that the geometric syntactic category $\mathcal{C}_\mathbb{T}$ of $\mathbb{T}$ embeds into its ∞-pretopos completion $\mathbf{Sh}(\mathcal{C}_\mathbb{T}, J_\mathbb{T})$ via the Yoneda embedding $y : \mathcal{C}_\mathbb{T} \hookrightarrow \mathbf{Sh}(\mathcal{C}_\mathbb{T}, J_\mathbb{T})$, and under this identification all the subobjects in $\mathbf{Set}[\mathbb{T}]$ of an object in $\mathcal{C}_\mathbb{T}$ lie again in $\mathcal{C}_\mathbb{T}$. Since the terminal object of $\mathcal{E}$ can be identified with $\{[] . \top\}$ and the subobjects of a given object $\{\vec{x} . \psi\}$ of $\mathcal{C}_\mathbb{T}$ can be identified with the geometric formulae $\phi(\vec{x})$ which $\mathbb{T}$-provably imply $\psi(\vec{x})$ (cf. Lemma 1.4.2), we conclude that the subterminal object U of 1 in $\mathcal{E}$ corresponds to a unique (up to $\mathbb{T}$-provable equivalence) geometric sentence (i.e.

geometric formula with no free variables) ϕ over Σ. Alternatively, $\{[] \, . \, \phi\}$ arises as the domain of the subobject of $\{[] \, . \, \top\}$ given by the union of all the images of the morphisms from objects in $\{c \in \mathcal{C}_\mathbb{T} \mid U(c) = 1_{\mathbf{Set}} = \{*\}\}$ to the terminal object of $\mathcal{C}_\mathbb{T}$ (cf. Proposition 3.1.5).

Recall that Diaconescu's equivalence

$$\mathbf{Geom}(\mathcal{F}, \mathbf{Sh}(\mathcal{C}_\mathbb{T}, J_\mathbb{T})) \simeq \mathbf{Flat}_{J_\mathbb{T}}(\mathcal{C}_\mathbb{T}, \mathcal{F})$$

sends a geometric morphism $f : \mathcal{F} \to \mathbf{Sh}(\mathcal{C}_\mathbb{T}, J_\mathbb{T})$ to the functor $f^* \circ y : \mathcal{C}_\mathbb{T} \to \mathcal{F}$ (where $y : \mathcal{C}_\mathbb{T} \to \mathbf{Sh}(\mathcal{C}_\mathbb{T}, J_\mathbb{T})$ is the Yoneda embedding), while the equivalence

$$\mathbb{T}\text{-mod}(\mathcal{F}) \simeq \mathbf{Flat}_{J_\mathbb{T}}(\mathcal{C}_\mathbb{T}, \mathcal{F})$$

sends each model $M \in \mathbb{T}\text{-mod}(\mathcal{F})$ to the functor $F_M : \mathcal{C}_\mathbb{T} \to \mathcal{F}$ assigning to a formula $\{\vec{x} \, . \, \phi\}$ its interpretation $[[\vec{x} \, . \, \phi]]_M$ in M. Thus, via the composite equivalence

$$\mathbf{Geom}(\mathcal{F}, \mathbf{Sh}(\mathcal{C}_\mathbb{T}, J_\mathbb{T})) \simeq \mathbb{T}\text{-mod}(\mathcal{F}),$$

the geometric morphisms $\mathcal{F} \to \mathcal{E} = \mathbf{Sh}(\mathcal{C}_\mathbb{T}, J_\mathbb{T})$ which factor through $\mathcal{E}/U \hookrightarrow \mathcal{E}$ correspond to the $\mathbb{T}$-models M such that $[[\phi]]_M = 1$, i.e. such that ϕ is satisfied in M. It follows that the quotient $\mathbb{T}_\phi$ of $\mathbb{T}$ obtained by adding to $\mathbb{T}$ the axiom $(\top \vdash_{[]} \phi)$ is classified by the topos $\mathcal{E}/U$ and corresponds to it via the duality of Theorem 3.2.5.

Given a topos $\mathbf{Sh}(\mathcal{C}, J)$ and a subterminal object U of it, the subtopos $\mathbf{Sh}(\mathcal{C}, J)/U \hookrightarrow \mathbf{Sh}(\mathcal{C}, J)$ corresponds to a unique Grothendieck topology J_U^{open} on $\mathcal{C}$ such that $J_U^{\text{open}} \supseteq J$; in particular, we have $\mathbf{Sh}(\mathcal{C}, J)/U \simeq \mathbf{Sh}(\mathcal{C}, J_U^{\text{open}})$. By Theorem 4.2.3 and Proposition 4.2.7, $J_U^{\text{open}} = J \vee J_{o(U)}$, where $J_{o(U)}$ is the Grothendieck topology on $\mathcal{C}$ corresponding to the open local operator $o(U)$ on $[\mathcal{C}^{\text{op}}, \mathbf{Set}]$ associated with U. Now, $o(U)$ is by definition given by the composite arrow

$$\Omega \cong 1 \times \Omega \xrightarrow{u \times 1} \Omega \times \Omega \xrightarrow{\Rightarrow} \Omega,$$

where Ω is the subobject classifier of $[\mathcal{C}^{\text{op}}, \mathbf{Set}]$ and $u : 1 \to \Omega$ is the classifying arrow of the subobject U. As a subterminal object of $[\mathcal{C}^{\text{op}}, \mathbf{Set}]$, U can be identified with the full subcategory $\mathcal{C}_U$ of $\mathcal{C}$ on the objects c such that $U(c) = \{*\}$. So $u(c)(*) = \{f : d \to c \mid d \in \mathcal{C}_U\}$ for any object $c \in \mathcal{C}$. Let us set, for any $c \in \mathcal{C}$, $Z(c) = u(c)(*)$. Then, by applying the general formula for $\Rightarrow$ recalled at the beginning of section 4.1.1, we obtain that $o(U)$ sends a sieve R on an object $c \in \mathcal{C}$ to $\{g : e \to c \mid g^*(Z(c)) \subseteq g^*(R)\}$. Hence $J_{o(U)}$ is given by

$$R \in J_{o(U)}(c) \text{ if and only if } R \supseteq Z(c)$$

for any $c \in \mathcal{C}$. In particular, by property (b) in Definition 3.1.3, $J_{o(U)}$ is generated by the sieves $Z(c)$, as c varies in $\mathcal{C}$. Notice in passing that for any arrow $f : d \to c$ in $\mathcal{C}$, $f^*(Z(c)) = Z(d)$.

Let us apply this discussion to the syntactic representation

$$\mathbf{Set}[\mathbb{T}] \simeq \mathbf{Sh}(\mathcal{C}_\mathbb{T}, J_\mathbb{T})$$

of the classifying topos $\mathbf{Set}[\mathbb{T}]$ of a geometric theory $\mathbb{T}$ over a signature Σ. Identifying the subterminal U of $\mathbf{Sh}(\mathcal{C}_\mathbb{T}, J_\mathbb{T})$ with (the $\mathbb{T}$-provable equivalence class of) a geometric sentence ϕ as above, we can describe the subcategory $\mathcal{C}_\phi$ corresponding to it as the full subcategory of $\mathcal{C}_\mathbb{T}$ on the objects $\{\vec{x}.\psi\}$ such that there exists (exactly) one morphism $\{\vec{x}.\psi\} \to \{[].\phi\}$ in $\mathcal{C}_\mathbb{T}$. By recalling the definition of morphisms in the syntactic category $\mathcal{C}_\mathbb{T}$, one then immediately obtains the following characterization for the objects of $\mathcal{C}_\phi$: $\{\vec{x}.\psi\} \in \mathcal{C}_\phi$ if and only if the sequent $(\psi \vdash_{\vec{x}} \phi)$ is provable in $\mathbb{T}$. By definition of $\mathcal{C}_\phi$, the sieve $Z(\{[].\top\})$ is generated by the morphism $\{[].\phi\} \rightarrowtail \{[].\top\}$. As the topology J_U^{open} is generated over $J_\mathbb{T}$ by the sieves $Z(c)$, and $Z(c)$ is the pullback of $Z(\{[].\top\})$ along the unique arrow $c \to \{[].\top\}$ for any $c \in \mathcal{C}_\mathbb{T}$, the quotient of $\mathbb{T}$ corresponding to the subtopos $\mathbf{Sh}(\mathcal{C}_\mathbb{T}, J_\mathbb{T})/U \hookrightarrow \mathbf{Sh}(\mathcal{C}_\mathbb{T}, J_\mathbb{T})$ is axiomatized over $\mathbb{T}$ by the sequent $(\top \vdash_{[]} \phi)$ (cf. Remark 3.2.9).

4.2.2.2 Closed subtoposes

Recall from section A4.5 of [55] that, given an elementary topos $\mathcal{E}$ and a subterminal object U of $\mathcal{E}$, the *closed local operator* $c^{\mathcal{E}}(U)$ associated with U is the composite arrow

$$\Omega \cong 1 \times \Omega \xrightarrow{u \times 1} \Omega \times \Omega \xrightarrow{\vee} \Omega,$$

where $u : 1 \to \Omega$ is the classifying arrow of the subobject U. Unlike open and quasi-closed local operators, a closed local operator on $\mathcal{E}$ associated with a subterminal object U lying in a subtopos $\mathbf{sh}_j(\mathcal{E})$ does not relativize in general to the closed local operator on $\mathbf{sh}_j(\mathcal{E})$ associated with U; however, if $\mathcal{E}$ is the topos $[\mathcal{C}^{\text{op}}, \mathbf{Set}]$ we can easily find a local operator on $\mathcal{E}$ which relativizes to $c^{\mathbf{Sh}(\mathcal{C},J)}(U)$, as follows. The local operator $c^{\mathbf{Sh}(\mathcal{C},J)}(U)$ sends any J-closed sieve R on $c \in \mathcal{C}$ to the (J-closed) sieve $\{f : d \to c \mid f^*(Z(c)) \cup f^*(R) \in J(d)\}$. This naturally leads us to consider the arrow $\Omega_{[\mathcal{C}^{\text{op}}, \mathbf{Set}]} \to \Omega_{[\mathcal{C}^{\text{op}}, \mathbf{Set}]}$ in $[\mathcal{C}^{\text{op}}, \mathbf{Set}]$ sending a sieve R on $c \in \mathcal{C}$ to the sieve $\{f : d \to c \mid f^*(Z(c)) \cup f^*(R) \in J(d)\}$. It is easily checked that this arrow is a local operator on $[\mathcal{C}^{\text{op}}, \mathbf{Set}]$ which relativizes to $c^{\mathbf{Sh}(\mathcal{C},J)}(U)$. Indeed, this arrow corresponds to the Grothendieck topology J_U^{closed} on $\mathcal{C}$ given by

$$R \in J_U^{\text{closed}}(c) \text{ if and only if } Z(c) \cup R \in J(c)$$

for any $c \in \mathcal{C}$. Since $J_U^{\text{closed}} \supseteq J$, J_U^{closed} is, by Theorem 4.2.3, the Grothendieck topology on $\mathcal{C}$ corresponding to the subtopos of $[\mathcal{C}^{\text{op}}, \mathbf{Set}]$ given by the composite of the closed subtopos of $\mathbf{Sh}(\mathcal{C}, J)$ associated with U with the canonical geometric inclusion $\mathbf{Sh}(\mathcal{C}, J) \hookrightarrow [\mathcal{C}^{\text{op}}, \mathbf{Set}]$.

Let us now give a description of the theory $\mathbb{T}_\phi^{\text{closed}}$ over Σ corresponding via Theorem 3.2.5 to the closed subtopos

$$\mathbf{sh}_{c^{\mathbf{Sh}(\mathcal{C}_\mathbb{T}, J_\mathbb{T})}(U)}(\mathbf{Sh}(\mathcal{C}_\mathbb{T}, J_\mathbb{T})) \hookrightarrow \mathbf{Sh}(\mathcal{C}_\mathbb{T}, J_\mathbb{T})$$

of the classifying topos of $\mathbb{T}$ associated with the subterminal object U of the topos $\mathbf{Sh}(\mathcal{C}_\mathbb{T}, J_\mathbb{T})$, identified as above with a geometric sentence ϕ over Σ. Since $Z(c)$ is the pullback of $Z(\{[] . \top\})$ along the unique arrow $c \to \{[] . \top\}$ for any $c \in \mathcal{C}_\mathbb{T}$, Theorem 3.2.5 and Proposition 3.2.9 allow us to identify $\mathbb{T}_\phi^{\text{closed}}$ with the quotient of $\mathbb{T}$ obtained by adding all the the geometric sequents

$$(\psi \vdash_{\vec{y}} \psi')$$

such that the sequents $(\psi' \vdash_{\vec{y}} \psi)$ and $(\psi \vdash_{\vec{y}} \psi' \lor (\phi \land \psi))$ are provable in $\mathbb{T}$. A minimal axiomatization for $\mathbb{T}_\phi^{\text{closed}}$ over $\mathbb{T}$ is given by the unique axiom $(\phi \vdash_{[]} \bot)$.

4.2.2.3 Quasi-closed subtoposes

If $\mathcal{E}$ is the topos $\mathbf{Sh}(\mathcal{C}, J)$ and U is a subterminal object of $\mathbf{Sh}(\mathcal{C}, J)$ then $qc^{\mathcal{E}}(U)$ corresponds to a unique Grothendieck topology J_U^{qc} on $\mathcal{C}$ such that $J_U^{\text{qc}} \supseteq J$. By Theorem 4.2.3 and Proposition 4.2.7, this topology is $J \lor J_{qc(U)}$, where $J_{qc(U)}$ is the Grothendieck topology on $\mathcal{C}$ corresponding to the quasi-closed local operator $qc^{[\mathcal{C}^{\text{op}}, \mathbf{Set}]}(U)$ on $[\mathcal{C}^{\text{op}}, \mathbf{Set}]$ associated with U (regarded here as a subterminal object of $[\mathcal{C}^{\text{op}}, \mathbf{Set}]$).

As above, let us identify U with the full subcategory $\mathcal{C}_U$ of $\mathcal{C}$ on the objects c such that $U(c) = \{*\}$ and set, for any $c \in \mathcal{C}$, $Z(c) = u(c) = \{f : d \to c \mid d \in \mathcal{C}_U\}$. In the case $\mathcal{E} = [\mathcal{C}^{\text{op}}, \mathbf{Set}]$ the local operator $qc^{\mathcal{E}}(U)$ is easily seen to send a sieve R on $c \in \mathcal{C}$ to the sieve $\Rightarrow (c)(\{f : d \to c \mid f^*(R) \subseteq f^*(Z(c))\}, Z(c))$. Hence $J_{qc(U)}$ is given by

$$R \in J_{qc(U)}(c) \text{ iff for any } f : d \to c, (f^*(R) \subseteq Z(d) \text{ implies } f \in Z(c))$$

for any $c \in \mathcal{C}$.

In order to specialize the above expression to the syntactic site of a geometric theory, we observe that, if $\mathcal{C}$ is a geometric category and J contains the geometric topology J^{geom} on $\mathcal{C}$ then the condition on the right-hand side of the equivalence is satisfied by $f : d \to c$ if and only if it is satisfied by the image $f' : d' \rightarrowtail c$ of f in $\mathcal{C}$. Indeed, since $\mathcal{C}_U$ is a J-ideal and every cover generates a J-covering sieve, $f \in Z(c)$ if and only if $f' \in Z(c)$. Let us now prove that for any $f : d \to c$, $f'^*(R) \subseteq Z(d')$ if and only if $f^*(R) \subseteq Z(d)$.

Since the $Z(c)$ are stable under pullback, $f'^*(R) \subseteq Z(d')$ implies $f^*(R) \subseteq Z(d)$. Conversely, let $r : d \twoheadrightarrow d'$ be the factorization of f through f'; given $g \in f'^*(R)$, consider the pullback

$$\begin{array}{ccc} e & \xrightarrow{g'} & d \\ {\scriptstyle r'}\downarrow & & \downarrow{\scriptstyle r} \\ e' & \xrightarrow{g} & d' \end{array}$$

in $\mathcal{C}$. Clearly, since R is a sieve, $g' \in f^*(R) \subseteq Z(d)$ whence $e \in \mathcal{C}_U$; but r' is a cover, which implies that $e' \in \mathcal{C}_U$ and hence that $g \in Z(d')$, as required.

This remark enables us to achieve a simplified description of the theory $\mathbb{T}_\phi^{qc}$ over Σ corresponding via Theorem 3.2.5 to the quasi-closed subtopos $qc^{\mathbf{Set}[\mathbb{T}]}(U)$ of the classifying topos $\mathbf{Set}[\mathbb{T}] \simeq \mathbf{Sh}(\mathcal{C}_\mathbb{T}, J_\mathbb{T})$ of $\mathbb{T}$, where ϕ is the geometric sentence over Σ corresponding to a subterminal object U of $\mathbf{Set}[\mathbb{T}]$. Indeed, by recalling the identification between $\mathbb{T}$-provable equivalence classes of geometric formulae $\psi'(\vec{y})$ such that $(\psi' \vdash_{\vec{y}} \psi)$ is provable in $\mathbb{T}$ and subobjects of $\{\vec{y}.\psi\}$ in $\mathcal{C}_\mathbb{T}$ given by Lemma 1.4.2, we get, by Theorem 3.2.5, Proposition 3.1.5(ii) and the syntactic characterization of $\mathcal{C}_\phi$ given above, the following axiomatization for $\mathbb{T}_\phi^{qc}$: $\mathbb{T}_\phi^{qc}$ is obtained from $\mathbb{T}$ by adding all the geometric sequents

$$(\psi \vdash_{\vec{y}} \psi')$$

such that $(\psi' \vdash_{\vec{y}} \psi)$ is provable in $\mathbb{T}$ and for any geometric formula $\chi(\vec{y})$ over Σ such that $(\chi \vdash_{\vec{y}} \psi)$ is provable in $\mathbb{T}$, if $(\chi \wedge \psi' \vdash_{\vec{y}} \phi)$ is provable in $\mathbb{T}$ then $(\chi \vdash_{\vec{y}} \phi)$ is provable in $\mathbb{T}$.

Remark 4.2.10 If ϕ is $\bot$ then, in view of Remark 3.2.9(b), $\mathbb{T}_\phi^{qc}$ specializes to the Booleanization of $\mathbb{T}$ described in section 4.2.3.

4.2.2.4 Open and closed quotients

Let $\mathbb{T}$ be a geometric theory over a signature Σ. Given an elementary topos $\mathcal{E}$, it is well known that the open and closed subtoposes associated with a given subterminal object are complementary to each other in $\mathbf{Lop}(\mathcal{E})$. From this we deduce, by Theorem 3.2.5, that the open and closed quotients $\mathbb{T}_\phi$ and $\mathbb{T}_\phi^{closed}$ of $\mathbb{T}$ corresponding to a given geometric sentence ϕ are complementary to each other in $\mathfrak{Th}_\Sigma^\mathbb{T}$; in fact, this can also be proved directly by means of purely logical arguments. We also know from the theory of elementary toposes that if U and V are complemented subterminals of a topos $\mathcal{E}$ then $o(U) = c(V)$; this implies, by Theorem 3.2.5, that if ϕ and ψ are two geometric sentences such that $(\top \vdash_{[]} \phi \vee \psi)$ and $(\phi \wedge \psi \vdash_{[]} \bot)$ then $\mathbb{T}_\phi = \mathbb{T}_\psi^{closed}$; this as well is not hard to prove by using logical arguments.

4.2.3 The Booleanization and DeMorganization of a geometric theory

Recall that the *Booleanization* of a topos $\mathcal{E}$ is the subtopos $\mathbf{sh}_{\neg\neg}(\mathcal{E}) \hookrightarrow \mathcal{E}$. This subtopos was characterized in [13] (cf. Theorem 2.4 therein) as the smallest (in fact, unique) Boolean dense subtopos of $\mathcal{E}$. The Booleanization of a topos $\mathcal{E}$ also coincides with the quasi-closed subtopos $\mathbf{sh}_{qc^\mathcal{E}(0_\mathcal{E})}(\mathcal{E}) \hookrightarrow \mathcal{E}$ (in the sense of section 4.2.2.3, cf. also Remark 4.2.10).

The notion of Booleanization is a topos-theoretic invariant admitting natural site characterizations; in particular, we have the following

Proposition 4.2.11

(i) *For any small category $\mathcal{C}$, the Booleanization of the topos $[\mathcal{C}^{op}, \mathbf{Set}]$ is isomorphic to the canonical geometric inclusion $\mathbf{Sh}(\mathcal{C}, D) \hookrightarrow [\mathcal{C}^{op}, \mathbf{Set}]$, where D is the dense topology on $\mathcal{C}$, that is the Grothendieck topology* on $\mathcal{C}$ whose covering sieves are exactly the stably non-empty ones (cf. Corollary VI.5 [63]); in particular, if $\mathcal{C}$ satisfies the right Ore condition then the Booleanization of $[\mathcal{C}^{\mathrm{op}}, \mathbf{Set}]$ coincides with the canonical geometric inclusion

$$\mathbf{Sh}(\mathcal{C}, J_{\mathrm{at}}) \hookrightarrow [\mathcal{C}^{\mathrm{op}}, \mathbf{Set}],$$

where J_{at} is the atomic topology on $\mathcal{C}$.

(ii) For any geometric theory $\mathbb{T}$, the Booleanization of its classifying topos can be identified with the subtopos corresponding via the duality of Theorem 3.2.5 to the quotient $\mathbb{T}_b$ of $\mathbb{T}$ obtained from $\mathbb{T}$ by adding all the sequents of the form

$$(\top \vdash_{\vec{y}} \psi)$$

for a geometric formula-in-context $\psi(\vec{y})$ over the signature of $\mathbb{T}$ which is $\mathbb{T}$-stably consistent, i.e. such that for any geometric formula $\chi(\vec{y})$ in the same context, if $(\chi \vdash_{\vec{y}} \bot)$ is not provable in $\mathbb{T}$ then $(\chi \wedge \psi \vdash_{\vec{y}} \bot)$ is not provable in $\mathbb{T}$ (cf. Theorem 5.7 [13] or Remark 4.2.10).

We call $\mathbb{T}_b$ the Booleanization of the theory $\mathbb{T}$.

A number of concrete examples of Booleanizations of theories arising in different fields of mathematics is provided in section 10.4 below.

Recall that a topos $\mathcal{E}$ satisfies De Morgan's law if and only if, in every subobject lattice in the topos, the identity $\neg m \vee \neg\neg m = 1$ holds. In [13] (cf. Theorem 2.3 therein) it is shown that every topos $\mathcal{E}$ admits a largest dense subtopos satisfying De Morgan's law; this topos is called the *DeMorganization* of $\mathcal{E}$.

In view of the duality of Theorem 3.2.5, it is natural to wonder whether the logical counterpart of this construction can be explicitly described in syntactic terms. A uniform axiomatization of the *DeMorganization* $\mathbb{T}_m$ of a geometric theory $\mathbb{T}$ (i.e. of the quotient of $\mathbb{T}$ corresponding to the DeMorganization of its classifying topos) was provided in [13] (cf. Theorem 5.6 therein) and reads as follows: $\mathbb{T}_m$ is obtained from $\mathbb{T}$ by adding all the geometric sequents of the form

$$(\top \vdash_{\vec{x}} \bigvee_{\psi(\vec{x}) \in I_{(\phi(\vec{x}), \phi'(\vec{x}))}} \psi)$$

for any two geometric formulae $\phi(\vec{x})$ and $\phi'(\vec{x})$ over the signature of $\mathbb{T}$ such that $(\phi \wedge \phi' \vdash_{\vec{x}} \bot)$ is provable in $\mathbb{T}$, where $I_{(\phi(\vec{x}), \phi'(\vec{x}))}$ is the collection of geometric formulae $\psi(\vec{x})$ over the signature of $\mathbb{T}$ such that either $(\psi \wedge \phi \vdash_{\vec{x}} \bot)$ or $(\psi \wedge \phi' \vdash_{\vec{x}} \bot)$ is provable in $\mathbb{T}$. On the other hand, in the case of specific theories of natural mathematical interest, it is often possible to establish more natural and 'economical' axiomatizations, by exploiting the specific features of the theory under consideration. For instance, Proposition 2.3 [29] identifies the DeMorganization of the coherent theory of fields as the geometric theory of fields of finite characteristic which are algebraic over their prime fields (and its Booleanization as the theory of fields of finite characteristic which are algebraic over their prime field and algebraically closed). This surprising result provides by itself a compelling illustration of the 'centrality' of topos-theoretic invariants in mathematics, including those which might appear too abstract to be of any 'concrete' relevance.

We should mention that there exist natural analogues of the Booleanization or DeMorganization construction for propositional logics other than Boolean and De Morgan logic, including all the usual superintuitionistic ones; these are introduced and characterized in terms of sites in [25].

4.2.4 The dense-closed factorization of a geometric inclusion

Let us continue our analysis of how well-known notions from elementary topos theory transfer to geometric logic through the duality of Theorem 3.2.5.

Recall from section A4.5 of [55] that the dense-closed factorization of a geometric inclusion $\mathbf{sh}_j(\mathcal{E}) \hookrightarrow \mathcal{E}$ is given by

$$\mathbf{sh}_j(\mathcal{E}) \hookrightarrow \mathbf{sh}_{c(ext(j))}(\mathcal{E}) \hookrightarrow \mathcal{E},$$

where $ext(j)$ is the c_j-closure of the subobject $0 \rightarrowtail 1$; the local operator $c(ext(j))$ is said to be the *closure* of j. In this section we shall interpret the meaning of this construction at the level of sites and then, via the duality of Theorem 3.2.5, in terms of theories.

Let $\mathbf{Sh}(\mathcal{C}, J)$ be a Grothendieck topos, $a_J : [\mathcal{C}^{op}, \mathbf{Set}] \to \mathbf{Sh}(\mathcal{C}, J)$ the associated sheaf functor and J' a Grothendieck topology on $\mathcal{C}$ which contains J. Let us calculate the dense-closed factorization of the canonical geometric inclusion $\mathbf{Sh}(\mathcal{C}, J') \hookrightarrow \mathbf{Sh}(\mathcal{C}, J)$ in terms of the corresponding local operator $\tau_{J'}^J$ on $\mathbf{Sh}(\mathcal{C}, J)$.

The monomorphism $0 \rightarrowtail 1$ in $\mathbf{Sh}(\mathcal{C}, J)$ is the image of the morphism $0 \rightarrowtail 1$ in $[\mathcal{C}^{op}, \mathbf{Set}]$ via the associated sheaf functor $a_J : [\mathcal{C}^{op}, \mathbf{Set}] \to \mathbf{Sh}(\mathcal{C}, J)$; from Proposition 4.2.4 and Theorem 4.2.3(iii) it thus follows that the closure of $0 \rightarrowtail 1$ in $\mathbf{Sh}(\mathcal{C}, J)$ with respect to the local operator corresponding to the geometric inclusion $\mathbf{Sh}(\mathcal{C}, J') \hookrightarrow \mathbf{Sh}(\mathcal{C}, J)$ is equal to the J'-closure of $0 \rightarrowtail 1$ in $[\mathcal{C}^{op}, \mathbf{Set}]$. Now, recall from [63] (formula (6) at p. 235) that, for any Grothendieck topology K on $\mathcal{C}$, the K-closure $c_K(A')$ of a subobject $A' \rightarrowtail E$ in $[\mathcal{C}^{op}, \mathbf{Set}]$ is given by

$$e \in c_K(A')(c) \text{ if and only if } \{f : d \to c \mid E(f)(e) \in A'(d)\} \in K(c).$$

Given a subterminal object U of $[\mathcal{C}^{op}, \mathbf{Set}]$, identified with the full subcategory $\mathcal{C}_U$ of $\mathcal{C}$ as in section 4.2.2.1 above, it is easy to verify that the K-closure $c_K(U \rightarrowtail 1)$ of $U \rightarrowtail 1$ in $[\mathcal{C}^{op}, \mathbf{Set}]$ identifies with the full subcategory $\mathcal{C}_U^K$ on the objects $c \in \mathcal{C}$ such that $\{f : d \to c \mid d \in \mathcal{C}_U\} \in K(c)$; in particular, if $U = 0$ then the objects of $\mathcal{C}_0^K$ are exactly the objects $c \in \mathcal{C}$ such that $\emptyset \in K(c)$.

By applying this to our topology J' we obtain that $ext(\tau_{J'}^J)$ identifies (as a subterminal object of $\mathbf{Sh}(\mathcal{C}, J)$) with the J-ideal $\mathcal{C}_0^{J'} = \{c \in \mathcal{C} \mid \emptyset \in J'(c)\}$. Recalling the description of closed local operators on Grothendieck toposes given in section 4.2.2.2, we deduce that the dense-closed factorization of the inclusion $\mathbf{Sh}(\mathcal{C}, J') \hookrightarrow \mathbf{Sh}(\mathcal{C}, J)$ is given by

$$\mathbf{Sh}(\mathcal{C}, J') \hookrightarrow \mathbf{Sh}(\mathcal{C}, J_{\mathcal{C}_0^{J'}}^{\text{closed}}) \hookrightarrow \mathbf{Sh}(\mathcal{C}, J),$$

where the topology $J_{\mathcal{C}_0^{J'}}^{\text{closed}}$ is defined by

$$R \in J_{\mathcal{C}_0^{J'}}^{\text{closed}}(c) \text{ if and only if } Z(c) \cup R \in J(c)$$

(where, for any $c \in \mathcal{C}$, $Z(c) = \{f : d \to c \mid \emptyset \in J'(d)\}$).

Finally, we study the effect of the dense-closed factorization on theories via the duality of Theorem 3.2.5.

Given a geometric theory $\mathbb{T}$ over a signature Σ and a quotient $\mathbb{T}'$ of $\mathbb{T}$, we can describe the quotient $\mathbb{T}_{\mathbb{T}'}^{\text{dc}}$ of $\mathbb{T}$ such that

$$\mathbf{Sh}(\mathcal{C}_\mathbb{T}, J_{\mathbb{T}'}^\mathbb{T}) \hookrightarrow \mathbf{Sh}(\mathcal{C}_\mathbb{T}, J_{\mathbb{T}_{\mathbb{T}'}^{\text{dc}}}^\mathbb{T}) \hookrightarrow \mathbf{Sh}(\mathcal{C}_\mathbb{T}, J_\mathbb{T})$$

is the dense-closed factorization of the canonical geometric inclusion

$$\mathbf{Sh}(\mathcal{C}_\mathbb{T}, J_{\mathbb{T}'}^\mathbb{T}) \hookrightarrow \mathbf{Sh}(\mathcal{C}_\mathbb{T}, J_\mathbb{T}).$$

By equivalence (i) in Remark 3.2.8 we have that $\emptyset \in J_{\mathbb{T}'}^\mathbb{T}(\{\vec{y}.\psi\})$ if and only if $(\psi \vdash_{\vec{y}} \bot)$ is provable in $\mathbb{T}'$. So the geometric sentence ϕ corresponding to the subterminal object given by $\mathcal{C}_0^{J_{\mathbb{T}'}^\mathbb{T}}$ is characterized by the property that $\{[].\phi\} \rightarrowtail \{[].\top\}$ is the union in $\mathcal{C}_\mathbb{T}$ of the images of all the arrows $\{\vec{y}.\psi\} \to \{[].\top\}$ such that $(\psi \vdash_{\vec{y}} \bot)$ is provable in $\mathbb{T}'$ (cf. section 4.2.2.1). In view of the results of section 4.2.2.2, we can thus conclude that $\mathbb{T}_{\mathbb{T}'}^{\text{dc}}$ is obtained from $\mathbb{T}$ by adding the sequent

$$(\phi \vdash_{[]} \bot)$$

or equivalently all the sequents of the form $(\psi \vdash_{\vec{y}} \bot)$ which are provable in $\mathbb{T}'$.

4.2.5 Skeletal inclusions

Recall from section D4.6 of [55] that a geometric morphism $f : \mathcal{F} \to \mathcal{E}$ is said to be *skeletal* if it restricts to a geometric morphism $\mathbf{sh}_{\neg\neg}(\mathcal{F}) \to \mathbf{sh}_{\neg\neg}(\mathcal{E})$. By Lemma D4.6.10 [55], a geometric inclusion $f : \mathcal{F} \to \mathcal{E}$ corresponding to a local operator j on $\mathcal{E}$ is skeletal if and only if $ext(j)$ (cf. section 4.2.4 for its definition) is a $\neg\neg$-closed subterminal object of $\mathcal{E}$.

Let us use the notation of section 4.2.4. Given the canonical geometric inclusion $\mathbf{Sh}(\mathcal{C}, J') \hookrightarrow \mathbf{Sh}(\mathcal{C}, J)$ corresponding to an inclusion $J \subseteq J'$, $ext(\tau_{J'}^J)$ identifies (as a subterminal object of $\mathbf{Sh}(\mathcal{C}, J)$) with the J-ideal $\mathcal{C}_0^{J'} = \{c \in \mathcal{C} \mid \emptyset \in J'(c)\}$. Now, consider the full subcategory $\tilde{\mathcal{C}}$ of $\mathcal{C}$ on the objects which are not J-covered by the empty sieve; $\tilde{\mathcal{C}}$ is J-dense in $\mathcal{C}$ and hence, by the Comparison Lemma (cf. Theorem 1.1.8), $\mathbf{Sh}(\mathcal{C}, J) \simeq \mathbf{Sh}(\tilde{\mathcal{C}}, J|_{\tilde{\mathcal{C}}})$, where $J|_{\tilde{\mathcal{C}}}$ is the induced Grothendieck topology on $\tilde{\mathcal{C}}$. Moreover, $J|_{\tilde{\mathcal{C}}}$ is dense, i.e. $J|_{\tilde{\mathcal{C}}} \leq \neg\neg_{[\tilde{\mathcal{C}}^{\text{op}}, \mathbf{Set}]}$. Therefore, by Corollary 4.2.9 and Theorem 4.2.3(ii), $ext(\tau_{J'}^J)$ is $\neg\neg_{\mathbf{Sh}(\tilde{\mathcal{C}}, J|_{\tilde{\mathcal{C}}})}$-closed (as a subterminal object of $\mathbf{Sh}(\tilde{\mathcal{C}}, J|_{\tilde{\mathcal{C}}})$) if and only if $ext(\tau_{J'}^J)$ is $\neg\neg_{[\tilde{\mathcal{C}}^{\text{op}}, \mathbf{Set}]}$-closed (as a subterminal object of $[\tilde{\mathcal{C}}^{\mathrm{op}}, \mathbf{Set}]$). But $\neg\neg_{[\tilde{\mathcal{C}}^{\mathrm{op}}, \mathbf{Set}]}$ is well known to correspond to the dense topology on $\tilde{\mathcal{C}}$ (cf. Example 1.1.4(b)); so, by formula (6) at p. 235 of [63], we obtain that $ext(\tau_{J'}^{J})$ is $\neg\neg_{[\tilde{\mathcal{C}}^{\mathrm{op}}, \mathbf{Set}]}$-closed if and only if, for any $c \in \tilde{\mathcal{C}}$, $c \in \mathcal{C}_0^{J'}$ if $\{f : d \to c \text{ in } \tilde{\mathcal{C}} \mid d \in \mathcal{C}_0^{J'}\}$ is stably non-empty in $\tilde{\mathcal{C}}$.

Hence the geometric inclusion $\mathbf{Sh}(\mathcal{C}, J') \hookrightarrow \mathbf{Sh}(\mathcal{C}, J)$ is skeletal if and only if, for any $c \in \tilde{\mathcal{C}}$, $\emptyset \in J'(c)$ if $Z(c) = \{f : d \to c \text{ in } \tilde{\mathcal{C}} \mid \emptyset \in J'(d)\}$ is stably non-empty in $\tilde{\mathcal{C}}$.

Let us interpret the notion of skeletal inclusion at the level of theories, via the duality of Theorem 3.2.5. Given a geometric theory $\mathbb{T}$ over a signature Σ, we want to characterize the quotients $\mathbb{T}'$ of $\mathbb{T}$ such that the canonical geometric inclusion $\mathbf{Sh}(\mathcal{C}_{\mathbb{T}}, J_{\mathbb{T}'}^{\mathbb{T}}) \hookrightarrow \mathbf{Sh}(\mathcal{C}_{\mathbb{T}}, J_{\mathbb{T}})$ is skeletal.

We have that $\emptyset \in J_{\mathbb{T}'}^{\mathbb{T}}(\{\vec{y}.\psi\})$ if and only if $(\psi \vdash_{\vec{y}} \bot)$ is provable in $\mathbb{T}'$. Given an object $\{\vec{y}.\psi\} \in \mathcal{C}_{\mathbb{T}}$, let us denote by $\{\vec{y}.\psi_{\mathbb{T}'}\} \rightarrowtail \{\vec{y}.\psi\}$ the subobject in $\mathcal{C}_{\mathbb{T}}$ given by the union in $\mathcal{C}_{\mathbb{T}}$ of all the subobjects $\{\vec{y}.\psi'\} \to \{\vec{y}.\psi\}$ such that $(\psi' \vdash_{\vec{y}} \bot)$ is provable in $\mathbb{T}'$. We thus obtain the following necessary and sufficient condition for $\mathbf{Sh}(\mathcal{C}_{\mathbb{T}}, J_{\mathbb{T}'}^{\mathbb{T}}) \hookrightarrow \mathbf{Sh}(\mathcal{C}_{\mathbb{T}}, J_{\mathbb{T}})$ to be skeletal (below by a $\mathbb{T}$-consistent geometric formula-in-context we mean a geometric formula $\phi(\vec{x})$ such that $(\phi \vdash_{\vec{x}} \bot)$ is not provable in $\mathbb{T}$): for any $\mathbb{T}$-consistent geometric formula $\psi(\vec{y})$ over Σ, if $(\chi \wedge \psi_{\mathbb{T}'})(\vec{y})$ is $\mathbb{T}$-consistent for any $\mathbb{T}$-consistent geometric formula $\chi(\vec{y})$ over Σ such that $(\chi \vdash_{\vec{y}} \psi)$ is provable in $\mathbb{T}$ then $(\psi \vdash_{\vec{y}} \bot)$ is provable in $\mathbb{T}'$.

4.2.6 The surjection-inclusion factorization

Recall that every geometric morphism can be factored, uniquely up to canonical equivalence, as a surjection followed by an inclusion (cf. for instance Theorem A4.2.10 [55]). In this section we shall discuss the meaning of this factorization in terms of theories via the duality of Theorem 3.2.5.

Definition 4.2.12 Let M be a Σ-structure in a Grothendieck topos $\mathcal{E}$. The *theory* $\mathrm{Th}(M)$ *of* M is the geometric theory over Σ consisting of all the geometric sequents σ over Σ which hold in M.

Theorem 4.2.13 *Let $\mathbb{T}$ be a geometric theory over a signature Σ and $f : \mathcal{F} \to \mathbf{Set}[\mathbb{T}]$ a geometric morphism corresponding to a $\mathbb{T}$-model M in $\mathcal{F}$ via the universal property of the classifying topos of $\mathbb{T}$. Then the topos $\mathcal{E}$ in the surjection-inclusion factorization $\mathcal{F} \twoheadrightarrow \mathcal{E} \hookrightarrow \mathbf{Set}[\mathbb{T}]$ of f classifies the quotient $\mathrm{Th}(M)$ of $\mathbb{T}$ via Theorem 3.2.5.*

Proof Let us denote by $\mathcal{F} \xrightarrow{f'} \mathcal{E} \xrightarrow{i} \mathbf{Set}[\mathbb{T}]$ the surjection-inclusion factorization of f. Since i is a geometric inclusion to the classifying topos of $\mathbb{T}$, i corresponds via the duality of Theorem 3.2.5 to a unique quotient $\mathbb{T}'$ of $\mathbb{T}$ (up to syntactic equivalence) whose classifying topos is $\mathcal{E}$. We want to prove that $\mathbb{T}' = \mathrm{Th}(M)$. From the proof of Theorem 3.2.5 we know that, if U is a universal model of $\mathbb{T}$ in $\mathbf{Set}[\mathbb{T}]$ then $i^*(U)$ is a universal model V for $\mathbb{T}'$. So f' corresponds to the $\mathbb{T}'$-model M via the universal property of the classifying topos of $\mathbb{T}'$, since $f'^*(V) =$

$f'^*(i^*(U)) \cong f^*(U) = M$. Now, since f' is a surjection, M is a conservative $\mathbb{T}'$-model (cf. Lemma 1.3.23 and Remark 2.1.9), from which it follows that $\mathbb{T}' = \mathrm{Th}(M)$. □

Remark 4.2.14 The theorem implies that if $\mathbb{T}$ is a geometric theory over a (possibly many-sorted) signature Σ and M is a $\mathbb{T}$-model in a Grothendieck topos then f_M is a surjection if and only if M is conservative. This result generalizes Corollary D3.2.6 [55], which was proved under the assumption that Σ be one-sorted.

4.2.7 Atoms

In this section we describe the atoms of the lattice of subtoposes of a given elementary topos, that is the non-trivial toposes having no proper subtoposes. Recall that a topos $\mathcal{E}$ is said to be *trivial* if it is equivalent to the category having just one object and the identity morphism on it, equivalently if it is *degenerate* i.e. $0_{\mathcal{E}} \cong 1_{\mathcal{E}}$.

Definition 4.2.15 A topos $\mathcal{E}$ is said to be *two-valued* if it has exactly two (non-isomorphic) subobjects of $1_{\mathcal{E}}$ (namely, the zero subobject and the identity one).

Proposition 4.2.16 *Let $\mathcal{E}$ be an elementary topos. Then the atoms of the lattice of subtoposes of $\mathcal{E}$ are exactly the two-valued Boolean subtoposes of $\mathcal{E}$.*

Proof Our thesis follows as an immediate consequence of the following two facts: first, every non-trivial topos contains a non-trivial Boolean subtopos, and second, a non-trivial Boolean topos does not contain any proper subtoposes if and only if it is two-valued. To prove the first assertion, we observe that if $\mathcal{E}$ is non-trivial then $\mathbf{sh}_{\neg\neg}(\mathcal{E})$ is again non-trivial; indeed, $1_{\mathcal{E}}$ clearly belongs to $\mathbf{sh}_{\neg\neg}(\mathcal{E})$ while $0_{\mathcal{E}}$ belongs to $\mathbf{sh}_{\neg\neg}(\mathcal{E})$ since $\neg\neg$ is a dense local operator on $\mathcal{E}$ (cf. p. 219 of [55]), so if $\mathbf{sh}_{\neg\neg}(\mathcal{E})$ is trivial then $0_{\mathcal{E}} \cong 1_{\mathcal{E}}$ i.e. $\mathcal{E}$ is trivial. The fact that $\mathbf{sh}_{\neg\neg}(\mathcal{E})$ is Boolean is well-known (see for example Lemma A4.5.22 [55]). This completes the proof of the first fact. It remains to prove the second assertion. Given two subterminal objects U and V of $\mathcal{E}$, the subtopos $\mathcal{E}/U \hookrightarrow \mathcal{E}$ is contained in the subtopos $\mathcal{E}/V \hookrightarrow \mathcal{E}$ if and only if $U \leq V$ in the lattice $\mathrm{Sub}_{\mathcal{E}}(1_{\mathcal{E}})$. Indeed, $\mathcal{E}/U \hookrightarrow \mathcal{E}$ factors through $\mathcal{E}/V \hookrightarrow \mathcal{E}$ if and only if the projection $U \times V \to U$ is isomorphic to the terminal object $1_U : U \to U$ in $\mathcal{E}/U$ and, since any object admits at most one morphism to a given subterminal object, this condition is equivalent to requiring that $U \leq V$ (equivalently, $U \leq U \times V$). Now, if $\mathcal{E}$ is Boolean then all the subtoposes of $\mathcal{E}$ are open (by Proposition A4.5.22 [55]), whence we have a lattice isomorphism between $\mathrm{Sub}_{\mathcal{E}}(1_{\mathcal{E}})$ and the lattice of subtoposes of $\mathcal{E}$; therefore a non-trivial Boolean topos does not contain any proper subtoposes if and only if it is two-valued. □

Remark 4.2.17 For any Grothendieck topos $\mathcal{E}$ with enough points, $\mathcal{E}$ is Boolean and two-valued if and only if it is atomic and connected. Indeed, we know from

Corollary D3.5.2 [55] that every Boolean Grothendieck topos with enough points is atomic; furthermore, any atomic topos is Boolean and it is two-valued if and only if it is connected (cf. the proof of Theorem 2.5 [23]).

Now we want to understand, via Theorem 3.2.5, the meaning of Proposition 4.2.16 in terms of theories.

Definition 4.2.18 (cf. section 4.1 of [13]) A geometric theory $\mathbb{T}$ is said to be *Boolean* if its classifying topos is a Boolean topos.

Remarks 4.2.19 (a) It is shown in [13] (Theorem 4.2 therein) that a geometric theory $\mathbb{T}$ over a signature Σ is Boolean if and only if for every geometric formula $\phi(\vec{x})$ over Σ there exists a geometric formula $\psi(\vec{x})$ over Σ in the same context such that $\phi \wedge \psi \dashv\vdash_{\mathbb{T}} \bot$ and $\phi \vee \psi \dashv\vdash_{\mathbb{T}} \top$ (where the notation is that of Definition 1.4.1(f)). From this criterion it easily follows that if $\mathbb{T}$ is a Boolean theory then every infinitary first-order formula over Σ is $\mathbb{T}$-provably equivalent using classical logic to a geometric formula in the same context; indeed, this can be proved by an inductive argument as in the proof of Theorem D3.4.6 [55] (in the case of an infinitary conjunction $\bigwedge_{i \in I} \phi_i$, we observe that this formula is equivalent in classical logic to the formula $\neg(\bigvee_{i \in I} \neg \phi_i)$, where the symbol $\neg$ denotes the first-order negation). Notice that, since every infinitary first-order formula is classically equivalent in $\mathbb{T}$ to a geometric formula, it follows from the axioms of infinitary first-order logic for implication and infinitary conjunction that the first-order implication between geometric formulae is classically provably equivalent in $\mathbb{T}$ to the Heyting implication between them in the relevant subobject lattice of $\mathcal{C}_{\mathbb{T}}$; moreover, the infinitary conjunction of a family of geometric formulae is classically provably equivalent in $\mathbb{T}$ to its infimum in that lattice.

(b) Recall the following fact about elementary toposes (cf. Proposition A4.5.22 [55]): an elementary topos is Boolean if and only if every subtopos of it is open. It is interesting to interpret the 'only if' part of this statement at the level of theories via the duality of Theorem 3.2.5. If $\mathbb{T}$ is a Boolean geometric theory over a signature Σ and $\mathbb{T}'$ is a quotient of $\mathbb{T}$, we want to show that there exists a geometric sentence ϕ over Σ such that $\mathbb{T}'$ is syntactically equivalent to $\mathbb{T}_\phi$. For any axiom $\sigma \equiv (\phi \vdash_{\vec{x}} \psi)$ of $\mathbb{T}'$, consider the geometric formula $U(\sigma)$ over Σ classically equivalent in $\mathbb{T}$ (as in Remark 4.2.19(a)) to the infinitary first-order sentence $\forall \vec{x}(\phi \Rightarrow \psi)$. Now, there is only a set of such formulae $U(\sigma)$ over Σ up to $\mathbb{T}$-provable equivalence, the geometric syntactic category $\mathcal{C}_{\mathbb{T}}$ being well-powered, so we can take ϕ to be a geometric sentence which is classically equivalent in $\mathbb{T}$ to their infinitary conjunction; it is immediate to see that ϕ has the required property.

Proposition 4.2.20 (Proposition 3.3 [18]). *Let $\mathbb{T}$ be a geometric theory over a signature Σ. Then $\mathbb{T}$ has enough set-based models (in the sense of Definition 1.4.17) if and only if its classifying topos $\mathbf{Set}[\mathbb{T}]$ has enough points.*

Proof By definition, **Set**[$\mathbb{T}$] has enough points if and only if the inverse image functors f^* of the geometric morphisms $f : \mathbf{Set} \to \mathbf{Set}[\mathbb{T}]$ are jointly conservative.

Since the geometric morphism $f_M : \mathbf{Set} \to \mathbf{Set}[\mathbb{T}]$ corresponding to a $\mathbb{T}$-model M in **Set** satisfies $f^*(U_\mathbb{T}) = M$ (where $U_\mathbb{T}$ is 'the' universal model of $\mathbb{T}$ lying in **Set**[$\mathbb{T}$]), if **Set**[$\mathbb{T}$] has enough points then any geometric sequent over Σ which is satisfied in every $\mathbb{T}$-model M in **Set** is satisfied in $U_\mathbb{T}$, equivalently it is provable in $\mathbb{T}$ (cf. Remark 2.1.9).

Conversely, if $\mathbb{T}$ has enough set-based models then, since the objects of the geometric syntactic category $\mathcal{C}_\mathbb{T}$ form a separating family of objects for the topos **Set**[$\mathbb{T}$] $\simeq \mathbf{Sh}(\mathcal{C}_\mathbb{T}, J_\mathbb{T})$, the inverse image functors of the points of **Set**[$\mathbb{T}$] jointly reflect isomorphisms of subobjects (equivalently, **Set**[$\mathbb{T}$] has enough points).

□

Definition 4.2.21 A geometric theory $\mathbb{T}$ over a signature Σ is said to be *complete* if every geometric sentence ϕ over Σ is $\mathbb{T}$-provably equivalent to $\top$ or $\bot$, but not both.

Remark 4.2.22 A geometric theory $\mathbb{T}$ over a signature Σ is complete if and only if its classifying topos is two-valued; indeed, as we observed in section 4.2.2.1, the subterminal objects of the classifying topos **Set**[$\mathbb{T}$] can be identified with the $\mathbb{T}$-provable equivalence classes of geometric sentences over Σ.

Atomic toposes were originally introduced in [4]. They can be characterized as the Grothendieck toposes whose objects can be written as coproducts of *atoms*, i.e. of non-zero objects which do not have any proper subobjects, or equivalently as the toposes which can be represented as categories of sheaves on a site of the form $(\mathcal{C}, J_{\mathrm{at}})$, where $\mathcal{C}$ is a category satisfying the right Ore condition and J_{at} is the atomic topology on it. Every atomic topos is Boolean and every Boolean topos with enough points is atomic (cf. Corollary C3.5.2 [55]).

Definition 4.2.23 A geometric theory $\mathbb{T}$ is said to be *atomic* if its classifying topos **Set**[$\mathbb{T}$] is an atomic topos.

Recall that a geometric theory $\mathbb{T}$ over a signature Σ is said to be contradictory if the sequent $(\top \vdash_{[]} \bot)$ is provable in $\mathbb{T}$.

Remark 4.2.24 A geometric theory is contradictory if and only if its classifying topos is trivial. Indeed, if $\mathbb{T}$ is contradictory then the trivial topos satisfies the universal property of the classifying topos of $\mathbb{T}$; conversely, if the classifying topos of $\mathbb{T}$ is trivial then $\bot$ holds in it and hence $(\top \vdash_{[]} \bot)$ is provable in $\mathbb{T}$.

The following proposition represents the translation of Proposition 4.2.16 in terms of theories via Theorem 3.2.5.

Proposition 4.2.25 *Let $\mathbb{T}$ be a geometric theory over a signature Σ. Then the maximal quotients of $\mathbb{T}$ (i.e. the non-contradictory quotients $\mathbb{T}'$ of $\mathbb{T}$ such that for every geometric sequent σ over Σ either σ is provable in $\mathbb{T}$ or the theory $\mathbb{T} \cup \{\sigma\}$ is contradictory) are exactly the Boolean and complete ones.* □

Remarks 4.2.26 (a) The 'if' direction in Proposition 4.2.25 can be easily proved without appealing to Theorem 3.2.5, as follows. If $\mathbb{T}$ is Boolean then a geometric sequent $(\phi \vdash_{\vec{x}} \psi)$ over Σ is provable in $\mathbb{T}$ if and only if the infinitary first-order sentence $\forall \vec{x}(\phi \Rightarrow \psi)$ is. But by Remark 4.2.19 this formula is $\mathbb{T}$-provably equivalent using classical logic to a geometric sentence, and this sentence is $\mathbb{T}$-provably equivalent to $\top$ or $\bot$ since $\mathbb{T}$ is complete.

(b) The theories whose classifying toposes admit the Galois-type representations described in section 10.5 satisfy the equivalent conditions of the proposition (when considered as quotients of the empty theory over their signature); indeed, their classifying toposes are atomic and two-valued. More generally, the maximal quotients of an atomic theory $\mathbb{T}$ are precisely the theories of the form $\mathbb{T} \cup \{(\top \vdash_{[]} \phi)\}$, where ϕ is any of the geometric formulae arising in the decomposition of the terminal object $\{[] . \top\}$ of $\mathbf{Set}[\mathbb{T}]$ as a coproduct of atoms (cf. also Remark 10.4.6).

4.2.8 Subtoposes with enough points

We have seen that the classifying topos $\mathbf{Set}[\mathbb{T}]$ of a geometric theory $\mathbb{T}$ has enough points if and only if $\mathbb{T}$ has enough set-based models (cf. Proposition 4.2.20).

Given a point p of a topos $\mathcal{E}$, let us denote by $\mathcal{E}_p \hookrightarrow \mathcal{E}$ the inclusion part of the surjection-inclusion factorization of p. By Theorem 4.2.13, if $\mathcal{E} = \mathbf{Set}[\mathbb{T}]$ then $\mathcal{E}_p$ classifies $\mathrm{Th}(M)$ where M is the $\mathbb{T}$-model corresponding to p.

Given a Grothendieck topos $\mathcal{E}$, let us define the *subtopos* $\mathcal{E}^{\mathrm{points}}$ *of points* of $\mathcal{E}$ to be the union of all the subtoposes $\mathcal{E}_p$ of $\mathcal{E}$ as p varies among the points of $\mathcal{E}$ (such a union exists because, dually, any intersection of Grothendieck topologies is a Grothendieck topology).

From Theorem 3.2.5 and the description of the (infinitary) meet in $\mathfrak{Th}_\Sigma^\mathbb{T}$, it follows that the topos $\mathbf{Set}[\mathbb{T}]^{\mathrm{points}}$ classifies the intersection of all the theories $\mathrm{Th}(M)$ as M varies among the $\mathbb{T}$-models M in $\mathbf{Set}$; in particular $\mathbf{Set}[\mathbb{T}]$ coincides with $\mathbf{Set}[\mathbb{T}]^{\mathrm{points}}$ if and only if it has enough points. Notice that, obviously, any intersection in $\mathfrak{Th}_\Sigma^\mathbb{T}$ of theories of the form $\mathrm{Th}(M)$ (for a $\mathbb{T}$-model M in $\mathbf{Set}$) has enough set-based models; so all the toposes of the form $\mathbf{Set}[\mathbb{T}]^{\mathrm{points}}$ have enough points. Therefore, given a geometric theory $\mathbb{T}$, the quotients of $\mathbb{T}$ having enough set-based models are exactly the intersections in $\mathfrak{Th}_\Sigma^\mathbb{T}$ of theories of the form $\mathrm{Th}(M)$ (where M is a $\mathbb{T}$-model in $\mathbf{Set}$). In light of the fact that every Grothendieck topos is (equivalent to) the classifying topos of a geometric theory, this translates into the following equivalent topos-theoretic statement: the subtoposes of a Grothendieck topos $\mathcal{E}$ which have enough points are exactly the unions of subtoposes of the form $\mathcal{E}_p$ where p is a point of $\mathcal{E}$.

Finally, we note that, given an atom $\mathcal{F}$ in the lattice of subtoposes of a Grothendieck topos $\mathcal{E}$, i.e. a Boolean and two-valued subtopos $\mathcal{F}$ of $\mathcal{E}$ (cf. section 4.2.7 above), if $\mathcal{F}$ has enough points then $\mathcal{F}$ is of the form $\mathcal{E}_p$ for a point p of $\mathcal{E}$. Indeed, a topos with enough points has a point if and only if it is non-trivial.

5

FLAT FUNCTORS AND CLASSIFYING TOPOSES

In this chapter we develop a general theory of extensions of flat functors along geometric morphisms of toposes. This theory will be applied in the next chapter to characterize the class of theories whose classifying topos is equivalent to a presheaf topos. In fact, the necessary and sufficient conditions for a theory $\mathbb{T}$ to be of presheaf type (i.e. classified by a presheaf topos) provided by the characterization theorem of Chapter 6 arise precisely from the requirement that for any Grothendieck topos $\mathcal{E}$ the operation of extension of flat functors with values in $\mathcal{E}$ from the opposite of the category of finitely presentable models of $\mathbb{T}$ to the geometric syntactic category of $\mathbb{T}$ should yield an equivalence of categories onto the category of $\mathbb{T}$-models in $\mathcal{E}$, naturally in $\mathcal{E}$.

The plan of the chapter is as follows.

In the first section, we investigate $\mathcal{E}$-indexed colimits of internal diagrams in a Grothendieck topos $\mathcal{E}$, analysing in particular their behaviour with respect to final $\mathcal{E}$-indexed functors (cf. section 5.1.5) and establishing explicit characterizations for an $\mathcal{E}$-indexed cocone to be colimiting (cf. section 5.1.6). Also, we exploit the abstract interpretation of colimits as generalized tensor products to derive commutativity results which allow us to interpret certain set-indexed colimits arising in the context of the above-mentioned characterization theorem as special kinds of filtered indexed colimits (cf. section 5.1.4).

In the second section, we investigate the properties of the operation on flat functors induced by a geometric morphism via Diaconescu's equivalence. We focus in particular on geometric morphisms between presheaf toposes induced by embeddings between small categories and on geometric morphisms to the classifying topos of a geometric theory induced by a small category of set-based models of the theory. We also establish, in section 5.2.4, a general 'hom-tensor' adjunction between categories of $\mathcal{E}$-valued functors (for $\mathcal{E}$ a Grothedieck topos) $[\mathcal{C}, \mathcal{E}]$ and $[\mathcal{D}, \mathcal{E}]$ induced by a functor $P : \mathcal{C} \to [\mathcal{D}^{\mathrm{op}}, \mathbf{Set}]$; this generalizes the well-known adjunction induced by left Kan extensions along a given functor.

In the third section, we describe a way of representing flat functors into arbitrary Grothendieck toposes by using a suitable internalized version of the Yoneda lemma.

5.1 Preliminary results on indexed colimits in toposes

5.1.1 Background on indexed categories

In this section we recall some standard notions and facts from the theory of indexed categories that will be useful in the sequel; we refer the reader to [55] (especially sections B1.2, B2.3 and B3 therein) and to [67] for the relevant prerequisites.

Given an internal category $\mathbb{C}$ in a topos $\mathcal{E}$, we denote by $\mathbb{C}_0$ (resp. $\mathbb{C}_1$) its object of objects (resp. of arrows), and by $d_0^{\mathbb{C}}, d_1^{\mathbb{C}} : \mathbb{C}_1 \to \mathbb{C}_0$ the domain and codomain arrows.

Given a small category $\mathcal{C}$ and a Grothendieck topos $\mathcal{E}$, we can always internalize $\mathcal{C}$ into $\mathcal{E}$ by means of the inverse image functor $\gamma_{\mathcal{E}}^*$ of the unique geometric morphism $\gamma_{\mathcal{E}} : \mathcal{E} \to \mathbf{Set}$; the resulting internal category in $\mathcal{E}$ will be called the *internalization* of $\mathcal{C}$ in $\mathcal{E}$ and denoted by the symbol $\mathbb{C}$.

Given a category $\mathcal{E}$, an *$\mathcal{E}$-indexed category* is by definition a pseudofunctor $\mathcal{E}^{\mathrm{op}} \to \mathfrak{CAT}$, where $\mathfrak{CAT}$ is the meta-2-category of categories. Indexed categories will be denoted by underlined letters, possibly with a subscript indicating the indexing category; the fibre at an object E of an $\mathcal{E}$-indexed category $\underline{\mathcal{A}}$ will be denoted by $\mathcal{A}_E$, and the functor $\mathcal{A}_E \to \mathcal{A}_{E'}$ corresponding to an arrow $e : E' \to E$ in $\mathcal{E}$ will be denoted by $\mathcal{A}_e$.

An $\mathcal{E}$-indexed subcategory $\underline{\mathcal{B}}_{\mathcal{E}}$ of an $\mathcal{E}$-indexed category $\underline{\mathcal{A}}_{\mathcal{E}}$ consists, for each object $E \in \mathcal{E}$, of a subcategory $\mathcal{B}_E$ of the category $\mathcal{A}_E$ such that for any arrow $\alpha : E' \to E$ in $\mathcal{E}$ the functor $\mathcal{A}_\alpha : \mathcal{A}_E \to \mathcal{A}_{E'}$ restricts to the subcategories $\mathcal{B}_E$ and $\mathcal{B}_{E'}$. This clearly defines an $\mathcal{E}$-indexed category $\underline{\mathcal{B}}_{\mathcal{E}}$ with an $\mathcal{E}$-indexed inclusion functor $\underline{\mathcal{B}}_{\mathcal{E}} \hookrightarrow \underline{\mathcal{A}}_{\mathcal{E}}$.

An $\mathcal{E}$-indexed functor $F : \underline{\mathcal{B}}_{\mathcal{E}} \to \underline{\mathcal{A}}_{\mathcal{E}}$ is said to be *full* if for every object E of $\mathcal{E}$ the functor $F_E : \mathcal{A}_E \to \mathcal{B}_E$ is full. An $\mathcal{E}$-indexed subcategory $\underline{\mathcal{B}}_{\mathcal{E}}$ of an $\mathcal{E}$-indexed category $\underline{\mathcal{A}}_{\mathcal{E}}$ is said to be a *full $\mathcal{E}$-indexed subcategory* of $\underline{\mathcal{A}}_{\mathcal{E}}$ if the associated $\mathcal{E}$-indexed inclusion functor is full.

Every Grothendieck topos $\mathcal{E}$ gives rise to an $\mathcal{E}$-indexed category $\underline{\mathcal{E}}_{\mathcal{E}}$ obtained by indexing $\mathcal{E}$ over itself, i.e. by setting $\mathcal{E}_E = \mathcal{E}/E$ for any $E \in \mathcal{E}$ and $\mathcal{E}_e = e^* : \mathcal{E}/E \to \mathcal{E}/E'$ for any arrow $e : E' \to E$ in $\mathcal{E}$.

Given an internal category $\mathbb{C} = (d_0^{\mathbb{C}}, d_1^{\mathbb{C}} : \mathbb{C}_1 \to \mathbb{C}_0)$ in a topos $\mathcal{E}$, a *diagram of shape $\mathbb{C}$ in $\mathcal{E}$* is a pair $(f : F \to \mathbb{C}_0, \phi : \mathbb{C}_1 \times_{\mathbb{C}_0} F \to F)$ of arrows in $\mathcal{E}$ satisfying appropriate conditions which generalize those defining the notion of set-based diagram (cf. Definition 2.14 [50]), where the pullback $\mathbb{C}_1 \times_{\mathbb{C}_0} F$ is taken relatively to the arrow $d_0^{\mathbb{C}} : \mathbb{C}_1 \to \mathbb{C}_0$ and the arrow $f : F \to \mathbb{C}_0$:

$$\begin{array}{ccc} \mathbb{C}_1 \times_{\mathbb{C}_0} F & \xrightarrow{\pi_F} & F \\ \pi_1 \downarrow & & \downarrow f \\ \mathbb{C}_1 & \xrightarrow{d_0^{\mathbb{C}}} & \mathbb{C}_0. \end{array}$$

For any Grothendieck topos $\mathcal{E}$ and any internal category $\mathbb{C}$ in $\mathcal{E}$, we have an $\mathcal{E}$-indexed category $[\mathbb{C}, \mathcal{E}]$ whose underlying category is the category $[\mathbb{C}, \mathcal{E}]$ of diagrams of shape $\mathbb{C}$ in $\mathcal{E}$ and morphisms between them.

Any internal category $\mathbb{C} = (d_0^{\mathbb{C}}, d_1^{\mathbb{C}} : \mathbb{C}_1 \to \mathbb{C}_0)$ in $\mathcal{E}$ naturally gives rise to an $\mathcal{E}$-indexed category $\underline{\mathbb{C}}_{\mathcal{E}}$, called the $\mathcal{E}$-*externalization* of $\mathbb{C}$, whose fibre $\mathbb{C}_E$ at any object E of $\mathcal{E}$ is the category whose objects are the arrows $E \to \mathbb{C}_0$ in $\mathcal{E}$ and whose arrows $x \to x'$ are the arrows $a : E \to \mathbb{C}_1$ in $\mathcal{E}$ such that $d_0^{\mathbb{C}} \circ a = x$ and $d_1^{\mathbb{C}} \circ a = x'$. An $\mathcal{E}$-indexed category is said to be *small* if it is of the form $\underline{\mathbb{C}}_{\mathcal{E}}$ for some internal category $\mathbb{C}$ in $\mathcal{E}$.

The category $[\mathbb{C}, \mathcal{E}]$ is equivalent (naturally in $\mathcal{E}$) to the category $[\underline{\mathbb{C}}_{\mathcal{E}}, \underline{\mathcal{E}}_{\mathcal{E}}]_{\mathcal{E}}$ of $\mathcal{E}$-indexed functors $\underline{\mathbb{C}}_{\mathcal{E}} \to \underline{\mathcal{E}}_{\mathcal{E}}$ and indexed natural transformations between them (by Lemma B2.3.13 [55]). If $\mathcal{C}$ is a small category and $\mathbb{C}$ is its internalization in a Grothendieck topos $\mathcal{E}$ then the category $[\mathbb{C}, \mathcal{E}]$ is also equivalent to the category $[\mathcal{C}, \mathcal{E}]$ (by Corollary B2.3.14 [55]). The equivalence between $[\mathbb{C}, \mathcal{E}]$ and $[\mathcal{C}, \mathcal{E}]$ restricts to an equivalence between the full subcategories **Tors**$(\mathbb{C}, \mathcal{E})$ of $\mathbb{C}$-torsors in $\mathcal{E}$ (as in section B3.2 of [55]) and **Flat**$(\mathcal{C}, \mathcal{E})$ of flat functors $\mathcal{C} \to \mathcal{E}$ (as in Chapter VII of [63]). For any internal functor between internal categories in $\mathcal{E}$ or internal diagram F in $\mathcal{E}$ we denote the corresponding $\mathcal{E}$-indexed functor by the symbol $\underline{F}_{\mathcal{E}}$. For any functor $F : \mathcal{C} \to \mathcal{E}$ we denote the internal diagram in $[\mathbb{C}, \mathcal{E}]$ corresponding to it under the equivalence $[\mathcal{C}, \mathcal{E}] \simeq [\mathbb{C}, \mathcal{E}]$ by the symbol $\overline{F}$.

The *discrete opfibration* $p : \mathbb{F} \to \mathbb{C}$ over $\mathbb{C}$ corresponding to a diagram $(f : F \to \mathbb{C}_0, \phi : \mathbb{C}_1 \times_{\mathbb{C}_0} F \to F)$ of shape $\mathbb{C}$ in $\mathcal{E}$ (where the pullback $\mathbb{C}_1 \times_{\mathbb{C}_0} F$ is taken with respect to the domain arrow $\mathbb{C}_1 \to \mathbb{C}_0$ and the arrow f) is defined as follows: $\mathbb{F}_0 = F$, $\mathbb{F}_1 = \mathbb{C}_1 \times_{\mathbb{C}_0} F$, $d_0^{\mathbb{F}} = \pi_F : \mathbb{F}_1 = \mathbb{C}_1 \times_{\mathbb{C}_0} F \to F = \mathbb{F}_0$, $d_1^{\mathbb{F}} = \phi : \mathbb{F}_1 = \mathbb{C}_1 \times_{\mathbb{C}_0} F \to F = \mathbb{F}_0$, $p_0 = f : \mathbb{F}_0 = F \to \mathbb{C}_0$, $p_1 = \pi_1 : \mathbb{F}_1 = \mathbb{C}_1 \times_{\mathbb{C}_0} F \to \mathbb{C}_1$. The discrete opfibration corresponding to a diagram $F \in [\mathbb{C}, \mathcal{E}]$ will also be denoted by $\pi_F^{\text{opf}} : \int^{\text{opf}} F \to \mathbb{C}$.

It is also natural to consider the *discrete fibration* corresponding to an internal diagram $G \in [\mathbb{C}^{\text{op}}, \mathcal{E}]$, i.e. the opposite functor $(\pi_G^{\text{opf}})^{\text{op}} : (\int^{\text{opf}} G)^{\text{op}} \to (\mathbb{C}^{\text{op}})^{\text{op}} = \mathbb{C}$. We shall denote this functor by $\pi_G^{\text{f}} : \int^{\text{f}} G \to \mathbb{C}$, or simply by $\pi_G : \int G \to \mathbb{C}$.

For any internal category $\mathbb{C}$ in a topos $\mathcal{E}$, the category $[\mathbb{C}, \mathcal{E}]$ of diagrams of shape $\mathbb{C}$ in $\mathcal{E}$ is equivalent to the category **DOpf**$/\mathbb{C}$ of *discrete opfibrations* over $\mathbb{C}$ (cf. Proposition B2.5.3 [55]); we shall denote this equivalence by

$$\tau_{\mathcal{E}}^{\mathbb{C}} : [\mathbb{C}, \mathcal{E}] \to \mathbf{DOpf}/\mathbb{C}.$$

A diagram of shape $\mathbb{C}$ in $\mathcal{E}$ lies in the subcategory **Tors**$(\mathbb{C}, \mathcal{E})$ of $[\mathbb{C}, \mathcal{E}]$ if and only if the domain of the corresponding discrete opfibration is a filtered internal category in $\mathcal{E}$.

Any internal functor $H : \mathbb{C} \to \mathbb{D}$ between internal categories $\mathbb{C}$ and $\mathbb{D}$ in a topos $\mathcal{E}$ induces a functor $[\mathbb{D}, \mathcal{E}] \to [\mathbb{C}, \mathcal{E}]$ whose action on objects is indicated by $D \to D \circ H$. This functor corresponds, under the equivalences $[\mathbb{D}, \mathcal{E}] \simeq [\underline{\mathbb{D}}_{\mathcal{E}}, \underline{\mathcal{E}}_{\mathcal{E}}]_{\mathcal{E}}$ and $[\mathbb{C}, \mathcal{E}] \simeq [\underline{\mathbb{C}}_{\mathcal{E}}, \underline{\mathcal{E}}_{\mathcal{E}}]_{\mathcal{E}}$, to the composition functor with the indexed functor associated with H, and, under the equivalences $[\mathbb{D}, \mathcal{E}] \simeq \mathbf{DOpf}/\mathbb{D}$ and

$[\mathbb{C}, \mathcal{E}] \simeq \mathbf{DOpf}/\mathbb{C}$, to the pullback along the functor H. The latter pullbacks in the category of internal categories in $\mathcal{E}$ are computed 'pointwise' as pullbacks in $\mathcal{E}$ and are preserved by the dualizing functor $\mathbb{C} \to \mathbb{C}^{\mathrm{op}}$.

Let us recall from [67] the notion of (indexed) *colimit of an $\mathcal{E}$-indexed functor*, where $\mathcal{E}$ is a cartesian category. We shall denote by $!_E$ the unique arrow from an object E of $\mathcal{E}$ to the terminal object 1 of $\mathcal{E}$. Given an $\mathcal{E}$-indexed functor $\Gamma : \underline{\mathcal{X}} \to \underline{\mathcal{A}}$ and an object A of $\mathcal{A}_1$, we denote by $\Delta_{\underline{\mathcal{A}}}^{\underline{\mathcal{X}}} A$ the $\mathcal{E}$-indexed functor $\underline{\mathcal{X}} \to \underline{\mathcal{A}}$ assigning to any $E \in \mathcal{E}$ the constant functor $\Delta(\mathcal{A}_{!_E}(A))$ on X_E with value $\mathcal{A}_{!_E}(A)$.

Let $\underline{\mathcal{X}}$ be an indexed category and $\Gamma : \underline{\mathcal{X}} \to \underline{\mathcal{A}}$ an indexed functor. An indexed cocone $\mu : \Gamma \to A$ consists of an object A in $\mathcal{A}_1$ together with an indexed natural transformation $\mu : \Gamma \to \Delta_{\underline{\mathcal{A}}}^{\underline{\mathcal{X}}} A$, i.e., for each E in $\mathcal{E}$ an ordinary cocone $\mu_E : \Gamma_E \to \Delta(\mathcal{A}_{!_E}(A))$ such that for each arrow $\alpha : E' \to E$ in $\mathcal{E}$, $\alpha^* \circ \mu_E = \mu_{E'} \circ \alpha^*$. If $\mu : \Gamma \to \Delta_{\underline{\mathcal{A}}}^{\underline{\mathcal{X}}} A$ is a universal such cocone we say that it is a *colimit* of the indexed functor Γ with vertex A. If, furthermore, for any object E of $\mathcal{E}$, the localization $\mu/E : \Gamma/E \to (\Delta_{\underline{\mathcal{A}}}^{\underline{\mathcal{X}}} A)/E = \Delta_{\underline{\mathcal{A}}/E}^{\underline{\mathcal{X}}/E}(\mathcal{A}_{!_E}(A))$ of μ at E is a colimit we say that μ is an ($\mathcal{E}$-)*indexed colimit* of Γ.

We can describe this notion more explicitly in the particular case of $\mathcal{E}$-indexed functors with values in the $\mathcal{E}$-indexed category $\underline{\mathcal{E}}_{\mathcal{E}}$. Let us suppose that $\underline{\mathcal{A}}_{\mathcal{E}}$ is an $\mathcal{E}$-indexed category and $D : \underline{\mathcal{A}}_{\mathcal{E}} \to \underline{\mathcal{E}}_{\mathcal{E}}$ is an $\mathcal{E}$-indexed functor. A cocone μ over D consists of an object U of $\mathcal{E}$ and, for each object E of $\mathcal{E}$, of a cocone $\mu_E : D_E \to \Delta(!_E^*(U))$ over the diagram $D_E : A_E \to \mathcal{E}/E$ such that for any arrow $\alpha : E' \to E$ in $\mathcal{E}$ we have $\alpha^*(\mu_E(c)) = \mu_{E'}(\mathcal{A}_\alpha(c))$ for all $c \in A_E$, that is $\alpha^* \mu_E = \mu_{E'} \mathcal{A}_\alpha$ as arrows $D_{E'}(\mathcal{A}_\alpha(c)) = \alpha^*(D_E(c)) \to !_{E'}^*(U) = \alpha^*(!_E^*(U))$ in $\mathcal{E}/E'$, where $\alpha^* : \mathcal{E}/E \to \mathcal{E}/E'$ is the pullback functor (notice that $D_{E'} \circ \mathcal{A}_\alpha = \alpha^* \circ D_E$ since D is an $\mathcal{E}$-indexed functor).

If $\underline{\mathcal{A}}_{\mathcal{E}}$ is the $\mathcal{E}$-externalization of an internal category $\mathbb{C}$ in $\mathcal{E}$ and $D : \underline{\mathcal{A}}_{\mathcal{E}} \to \underline{\mathcal{E}}_{\mathcal{E}}$ is an $\mathcal{E}$-indexed functor corresponding to an internal diagram $D \in [\mathbb{C}, \mathcal{E}]$, the discrete opfibration associated with the internal diagram in $[!_E^*(\mathbb{C}), \mathcal{E}/E]$ corresponding to the localization $D/E : \underline{\mathcal{A}}_{\mathcal{E}}/E \to \underline{\mathcal{E}}_{\mathcal{E}}/E \cong \underline{\mathcal{E}/E}_{\mathcal{E}/E}$ is the image of the discrete opfibration associated with D under the pullback functor $!_E^* : \mathcal{E} \to \mathcal{E}/E$ along the unique arrow $!_E : E \to 1_{\mathcal{E}}$.

The colimit $\mathrm{colim}_\mathcal{E}(D)$ in $\mathcal{E}$ of an internal diagram $D \in [\mathbb{C}, \mathcal{E}]$, where $\mathbb{C}$ is an internal category in $\mathcal{E}$, is by definition be the coequalizer of the two arrows $d_0^{\int^{\mathrm{opf}} D}, d_1^{\int^{\mathrm{opf}} D}$ (recall that $\int^{\mathrm{opf}} D$ is the domain of the discrete opfibration over $\mathbb{C}$ corresponding to the diagram D).

For any internal diagram $G \in [\mathbb{F}, \mathcal{E}]$, $\mathrm{colim}_\mathcal{E}(G)$ is isomorphic to $\underline{\mathrm{colim}}_\mathcal{E}(\underline{G}_\mathcal{E})$, where $\underline{G}_\mathcal{E} : \underline{\mathbb{F}}_\mathcal{E} \to \underline{\mathcal{E}}_\mathcal{E}$ is the $\mathcal{E}$-indexed functor corresponding to G under the equivalence $[\mathbb{F}, \mathcal{E}] \simeq [\underline{\mathbb{F}}_\mathcal{E}, \underline{\mathcal{E}}_\mathcal{E}]_\mathcal{E}$.

If $D \in [\mathbb{C}, \mathcal{E}]$ is an internal diagram in $\mathcal{E}$, the colimiting $\mathcal{E}$-indexed cocone $(\mathrm{colim}_\mathcal{E}(D) = \mathrm{coeq}(d_1^{\int^{\mathrm{opf}} D}, d_1^{\int^{\mathrm{opf}} D}), \mu)$ of the corresponding $\mathcal{E}$-indexed functor $\overline{D}_\mathcal{E} : \underline{\mathbb{C}}_\mathcal{E} \to \underline{\mathcal{E}}_\mathcal{E}$ can be described as follows. Let us denote by c the canonical arrow

$(\int^{\mathrm{opf}} D)_0 \to \mathrm{coeq}(d_0^{\int^{\mathrm{opf}} D}, d_1^{\int^{\mathrm{opf}} D})$ in $\mathcal{E}$. For any objects E of $\mathcal{E}$ and $x : E \to \mathbb{C}_0$ of $\mathbb{C}_E$, the arrow $\mu_E(x) : \overline{D}_{\mathcal{E}} = r_x \to !_E^*(\mathrm{colim}_{\mathcal{E}}(D)) = \mathrm{coeq}(d_0^{\int^{\mathrm{opf}} D}, d_1^{\int^{\mathrm{opf}} D}) \times E$ is equal to $< c \circ z_x, r_x >$ where the arrows r_x and z_x are defined by the following pullback square:

$$\begin{array}{ccc} R_x & \xrightarrow{z_x} & (\int^{\mathrm{opf}} D)_0 \\ \downarrow{\scriptstyle r_x} & & \downarrow{\scriptstyle (\pi_D^{\mathrm{opf}})_0} \\ E & \xrightarrow{x} & \mathbb{C}_0. \end{array}$$

Given an $\mathcal{E}$-indexed category $\mathbb{D}$ and an object $E \in \mathcal{E}$, we have an $\mathcal{E}/E$-indexed category $\mathbb{D}/E$ (defined in the obvious way) called the *localization of* $\mathbb{D}$ *at* E.

Given a topos $\mathcal{E}$, we denote by $!_E^* : \mathcal{E} \to \mathcal{E}/E$ the pullback functor along the unique arrow $!_E : E \to 1$, that is the inverse image functor of the local homeomorphism $\mathcal{E}/E \to \mathcal{E}$.

An $\mathcal{E}$-indexed category $\underline{A}_{\mathcal{E}}$ is said to have *small homs at 1* if for any objects F, G in $\mathcal{A}_1$ there exists an object $\mathrm{Hom}_{\underline{A}_{\mathcal{E}}}^{\mathcal{E}}(F, G)$ in $\mathcal{E}$ (also denoted by $\mathrm{Hom}_{\underline{A}}^{\mathcal{E}}(F, G)$) satisfying the universal property that, for every E in $\mathcal{E}$, there is a bijection between the arrows $E \to \mathrm{Hom}_{\underline{A}_{\mathcal{E}}}^{\mathcal{E}}(F, G)$ in $\mathcal{E}$ and the arrows $\mathcal{A}_{!_E^*}(F) \to \mathcal{A}_{!_E^*}(G)$ in $\mathcal{A}_E$ which natural in E. An $\mathcal{E}$-indexed category $\underline{A}_{\mathcal{E}}$ is said to be *locally small* if, for every object E of $\mathcal{E}$, the $\mathcal{E}/E$-indexed category $\underline{A}_{\mathcal{E}}/E$ has small homs at 1.

For any Grothendieck topos $\mathcal{E}$ and any internal category $\mathbb{C}$ in $\mathcal{E}$, the $\mathcal{E}$-indexed category $[\mathbb{C}, \mathcal{E}]$ is locally small (by Lemma B2.3.15 [55]); from this it follows that there exists an $\mathcal{E}$-indexed hom functor

$$\mathrm{Hom}_{[\mathbb{C}, \mathcal{E}]}^{\mathcal{E}} : [\mathbb{C}, \mathcal{E}]^{\mathrm{op}} \times [\mathbb{C}, \mathcal{E}] \to \underline{\mathcal{E}}_{\mathcal{E}}$$

whose underlying functor assigns to a pair of diagrams F and G in $[\mathbb{C}, \mathcal{E}]$ an object $\mathrm{Hom}_{[\mathbb{C}, \mathcal{E}]}^{\mathcal{E}}(F, G)$ of $\mathcal{E}$, which we call the *object of morphisms* from F to G in $[\mathbb{C}, \mathcal{E}]$; also, there is a Yoneda $\mathcal{E}$-indexed functor $\mathbb{Y} : \mathbb{C} \to [\mathbb{C}^{\mathrm{op}}, \mathcal{E}]$ which plays in this context the same role as that of a Yoneda functor in ordinary category theory. If $\mathbb{C}$ is the internalization in a Grothendieck topos $\mathcal{E}$ of a small category $\mathcal{C}$ then, since $[\mathcal{C}, \mathcal{E}] \simeq [\mathbb{C}, \mathcal{E}]$, we have a hom functor

$$\mathrm{Hom}_{[\mathcal{C}, \mathcal{E}]}^{\mathcal{E}} : [\mathcal{C}, \mathcal{E}]^{\mathrm{op}} \times [\mathcal{C}, \mathcal{E}] \to \mathcal{E}$$

which assigns to each pair of functors $F, G \in [\mathcal{C}, \mathcal{E}]$ an object $\mathrm{Hom}_{[\mathcal{C}, \mathcal{E}]}^{\mathcal{E}}(F, G)$ of $\mathcal{E}$ such that for each $E \in \mathcal{E}$ the arrows $E \to \mathrm{Hom}_{[\mathcal{C}, \mathcal{E}]}^{\mathcal{E}}(F, G)$ in $\mathcal{E}$ are in natural bijection with the arrows in $[\mathcal{C}, \mathcal{E}/E]$ from $!_E^* \circ F$ to $!_E^* \circ G$, that is with the natural transformations $!_E^* \circ F \Rightarrow !_E^* \circ G$. Note that, since $\mathbf{Flat}(\mathcal{C}, \mathcal{E})$ is a full subcategory of $[\mathcal{C}, \mathcal{E}]$, we may use the objects $\mathrm{Hom}_{[\mathcal{C}, \mathcal{E}]}^{\mathcal{E}}(F, G)$ for $F, G \in \mathbf{Flat}(\mathcal{C}, \mathcal{E})$ as the objects of morphisms from F to G, so as to obtain a locally small $\mathcal{E}$-indexed category $\mathbf{Flat}(\mathcal{C}, \mathcal{E})$.

5.1.2 $\mathcal{E}$-filtered indexed categories

The following definition will be important in the sequel:

Definition 5.1.1 Let $\mathcal{E}$ be a Grothendieck topos and $\underline{\mathcal{A}}$ an $\mathcal{E}$-indexed category. We say that $\underline{\mathcal{A}}$ is $\mathcal{E}$-*filtered* if the following conditions are satisfied:

(a) For any object E of $\mathcal{E}$, there exist an epimorphic family $\{e_i : E_i \to E \mid i \in I\}$ in $\mathcal{E}$ and for each $i \in I$ an object b_i of the category $\mathcal{A}_{E_i}$.

(b) For any $E \in \mathcal{E}$ and any objects a and b of the category $\mathcal{A}_E$, there exist an epimorphic family $\{e_i : E_i \to E \mid i \in I\}$ in $\mathcal{E}$ and for each $i \in I$ an object c_i of the category $\mathcal{A}_{E_i}$ and arrows $f_i : \mathcal{A}_{e_i}(a) \to c_i$ and $g_i : \mathcal{A}_{e_i}(b) \to c_i$ in $\mathcal{A}_{E_i}$.

(c) For any object E of $\mathcal{E}$ and any arrows $u, v : a \to b$ in the category $\mathcal{A}_E$, there exist an epimorphic family $\{e_i : E_i \to E \mid i \in I\}$ in $\mathcal{E}$ and for each $i \in I$ an object c_i of the category $\mathcal{A}_{E_i}$ and an arrow $w_i : \mathcal{A}_{e_i}(b) \to c_i$ in $\mathcal{A}_{E_i}$ such that $w_i \circ \mathcal{A}_{e_i}(u) = w_i \circ \mathcal{A}_{e_i}(v)$.

The $\mathcal{E}$-externalization of any internal filtered category in $\mathcal{E}$ is $\mathcal{E}$-filtered, but if the externalization of an internal category $\mathbb{C}$ in $\mathcal{E}$ is $\mathcal{E}$-filtered then $\mathbb{C}$ is not necessarily filtered as an internal category in $\mathcal{E}$.

A standard source of indexed, $\mathcal{E}$-filtered categories is provided by the indexed categories of elements of flat functors with values in $\mathcal{E}$ in the sense of the following definition:

Definition 5.1.2 Let $P : \mathcal{C}^{\mathrm{op}} \to \mathcal{E}$ be a functor. The $\mathcal{E}$-*indexed category* $\int P_{\mathcal{E}}$ *of elements* of P assigns

- to any object E of $\mathcal{E}$ the category $\int P_E$ whose objects are the pairs (c, x) where c is an object of $\mathcal{C}$ and $x : E \to F(c)$ is an arrow in $\mathcal{E}$ and whose arrows $(c, x) \to (d, y)$ are the arrows $f : c \to d$ in $\mathcal{C}$ such that $F(f) \circ y = x$;
- to any arrow $e : E' \to E$ in $\mathcal{E}$ the functor $\int P_e : \int P_E \to \int P_{E'}$ sending any object (c, x) of $\int P_E$ to the object $(c, x \circ e)$ of $\int P_{E'}$ and acting accordingly on arrows.

Proposition 5.1.3 Let $F : \mathcal{C}^{\mathrm{op}} \to \mathcal{E}$ be a flat functor. Then the indexed category $\int F_{\mathcal{E}}$ is $\mathcal{E}$-filtered.

Proof Straightforward from the characterization of flat functors as filtering functors given in Proposition 1.1.32. $\square$

In the sequel we shall work both with the $\mathcal{E}$-indexed category $\int F_{\mathcal{E}}$ (for a flat functor $F : \mathcal{C}^{\mathrm{op}} \to \mathcal{E}$) and with the $\mathcal{E}$-externalization $\int^{\mathrm{f}} \overline{F}_{\mathcal{E}}$ of the domain $\int^{\mathrm{f}} \overline{F}$ of the discrete fibration corresponding to the $\mathbb{C}$-torsor $\overline{F}$, which is $\mathcal{E}$-filtered since $\int^{\mathrm{f}} \overline{F}$ is filtered as an internal category in $\mathcal{E}$.

5.1.3 Indexation of internal diagrams

Given an internal diagram $D \in [\mathbb{C}, \mathcal{E}]$, the corresponding $\mathcal{E}$-indexed functor

$$\overline{D}_\mathcal{E} : \underline{\mathbb{C}}_\mathcal{E} \to \underline{\mathcal{E}}_\mathcal{E}$$

can be described as follows. For any object E of $\mathcal{E}$, $\overline{D}_\mathcal{E} : \mathbb{C}_E \to \mathcal{E}/E$ sends any object $x : E \to \mathbb{C}_0$ of $\mathbb{C}_E$ to the object $r_x : R_x \to E$ of $\mathcal{E}/E$ obtained by pulling $(\pi_D^{\mathrm{opf}})_0 : (\int^{\mathrm{opf}} D)_0 \to \mathbb{C}_0$ back along x and any arrow $h : E \to \mathbb{C}_1$ of $\mathbb{C}_E$ from $x : E \to \mathbb{C}_0$ to $x' : E \to \mathbb{C}_0$ to the arrow $\overline{D}_\mathcal{E}(h) : r_x \to r_{x'}$ in $\mathcal{E}/E$ defined as follows. Consider the pullback squares

$$\begin{array}{ccc} S & \xrightarrow{u} & (\int^{\mathrm{opf}} D)_1 \\ \downarrow{\scriptstyle s} & & \downarrow{\scriptstyle (\pi_D^{\mathrm{opf}})_1} \\ E & \xrightarrow{h} & \mathbb{C}_1 \end{array}$$

and

$$\begin{array}{ccc} R_{x'} & \xrightarrow{z_{x'}} & (\int^{\mathrm{opf}} D)_0 \\ \downarrow{\scriptstyle r_{x'}} & & \downarrow{\scriptstyle (\pi_D^{\mathrm{opf}})_0} \\ E & \xrightarrow{x'} & \mathbb{C}_0. \end{array}$$

Since $d_1^\mathbb{C} \circ h = x'$ we have that $(\pi_D^{\mathrm{opf}})_0 \circ d_1^{\int^{\mathrm{opf}} D} \circ u = x' \circ s$; thus, by the universal property of the first pullback square, there exists a unique arrow $\beta : S \to R_{x'}$ such that $r_{x'} \circ \beta = s$ and $z_{x'} \circ \beta = d_1^{\int^{\mathrm{opf}} D} \circ u$.

As $\pi_F^{\mathrm{opf}} : \int^{\mathrm{opf}} F \to \mathbb{C}$ is a discrete opfibration, the diagram

$$\begin{array}{ccc} (\int^{\mathrm{opf}} D)_1 & \xrightarrow{d_0^{\int^{\mathrm{opf}} D}} & (\int^{\mathrm{opf}} D)_0 \\ \downarrow{\scriptstyle (\pi_D^{\mathrm{opf}})_1} & & \downarrow{\scriptstyle (\pi_D^{\mathrm{opf}})_0} \\ \mathbb{C}_1 & \xrightarrow{d_0^\mathbb{C}} & \mathbb{C}_0 \end{array}$$

is a pullback. 'Composing' it with the first of the above pullback squares thus yields a pullback square

$$\begin{array}{ccc} S & \xrightarrow{d_0^{\int^{\mathrm{opf}} D} \circ u} & (\int^{\mathrm{opf}} D)_0 \\ \downarrow{\scriptstyle s} & & \downarrow{\scriptstyle (\pi_D^{\mathrm{opf}})_0} \\ E & \xrightarrow{d_0^\mathbb{C} \circ h} & \mathbb{C}_0. \end{array}$$

Now, since $(\pi_D^{\mathrm{opf}})_0 \circ z_x = x \circ r_x = d_0^\mathbb{C} \circ h \circ r_x$, the universal property of this pullback square provides a unique arrow $\gamma : R_x \to S$ such that $s \circ \gamma = r_x$ and

$d_0^{\int^{\mathrm{opf}} D} \circ u \circ \gamma = z_x$. We define $\overline{D}_{\mathcal{E}}(h) : r_x \to r_{x'}$ in $\mathcal{E}/E$ to be equal to the composite arrow $\beta \circ \gamma : R_x \to R_{x'}$.

5.1.4 Colimits and tensor products

For any internal category $\mathbb{F}$ in $\mathcal{E}$, we denote by $\mathrm{coeq}(\mathbb{F})$ the coequalizer of the two arrows $d_0^{\mathbb{F}}$ and $d_1^{\mathbb{F}}$.

From the above discussion it follows that for any $D, P \in [\mathbb{C}, \mathcal{E}]$, $\mathrm{colim}_{\mathcal{E}}(P \circ \pi_D^{\mathrm{opf}}) \cong \mathrm{colim}_{\mathcal{E}}(D \circ \pi_P^{\mathrm{opf}})$. Indeed, if we consider the pullback square

$$\begin{array}{ccc} \mathbb{R} & \xrightarrow{q^P} & \int^{\mathrm{opf}} P \\ {\scriptstyle q^D}\downarrow & & \downarrow{\scriptstyle \pi_P^{\mathrm{opf}}} \\ \int^{\mathrm{opf}} D & \xrightarrow{\pi_D^{\mathrm{opf}}} & \mathbb{C} \end{array}$$

in the category $\mathbf{cat}(\mathcal{E})$ of internal categories in $\mathcal{E}$ we have that $\mathrm{colim}_{\mathcal{E}}(P \circ \pi_D^{\mathrm{opf}}) \cong \mathrm{coeq}(\mathrm{dom}(q^D)) = \mathrm{coeq}(\mathbb{R}) = \mathrm{coeq}(\mathrm{dom}(q^P)) \cong \mathrm{colim}_{\mathcal{E}}(D \circ \pi_P^{\mathrm{opf}})$.

Similarly, by exploiting the fact that the operation $\mathbb{F} \to \mathrm{coeq}(\mathbb{F})$ on internal categories in $\mathcal{E}$ is invariant under the dualization functor $\mathbb{F} \to \mathbb{F}^{\mathrm{op}}$, we obtain another commutativity result, which we shall use in the sequel: for any internal diagrams $F \in [\mathbb{C}^{\mathrm{op}}, \mathcal{E}]$ and $P \in [\mathbb{C}, \mathcal{E}]$, we have a natural isomorphism

$$\mathrm{colim}_{\mathcal{E}}(P \circ \pi_F^{\mathrm{f}}) \cong \mathrm{colim}_{\mathcal{E}}(F \circ \pi_P^{\mathrm{f}}).$$

To prove this, we consider the following pullback squares:

$$\begin{array}{ccc} \mathbb{R} & \xrightarrow{q^P} & \int^{\mathrm{opf}} P \\ {\scriptstyle q^F}\downarrow & & \downarrow{\scriptstyle \pi_P^{\mathrm{opf}}} \\ \int^{\mathrm{f}} F & \xrightarrow{\pi_F^{\mathrm{f}}} & \mathbb{C}, \end{array} \qquad \begin{array}{ccc} \mathbb{S} & \xrightarrow{r^F} & \int^{\mathrm{opf}} F \\ {\scriptstyle r^P}\downarrow & & \downarrow{\scriptstyle \pi_F^{\mathrm{opf}}} \\ \int^{\mathrm{f}} P & \xrightarrow{\pi_P^{\mathrm{f}}} & \mathbb{C}^{\mathrm{op}}. \end{array}$$

We have that

$$\mathrm{colim}_{\mathcal{E}}(P \circ \pi_F^{\mathrm{f}}) \cong \mathrm{coeq}(\mathrm{dom}(q^F)) = \mathrm{coeq}(\mathbb{R}),$$

while

$$\mathrm{colim}_{\mathcal{E}}(F \circ \pi_P^{\mathrm{f}}) = \mathrm{coeq}(\mathrm{dom}(r^P)) = \mathrm{coeq}(\mathbb{S}).$$

But the fact that the dualization functor preserves pullbacks in $\mathbf{cat}(\mathcal{E})$ implies that $\mathbb{S} \simeq \mathbb{R}^{\mathrm{op}}$, whence $\mathrm{coeq}(\mathbb{R}) \cong \mathrm{coeq}(\mathbb{S})$, as required.

Summarizing, we have the following result:

Proposition 5.1.4 *Let $\mathbb{C}$ be an internal category in a Grothendieck topos $\mathcal{E}$, P and D internal diagrams in $[\mathbb{C}, \mathcal{E}]$ and F an internal diagram in $[\mathbb{C}^{\mathrm{op}}, \mathcal{E}]$. Then we have natural isomorphisms*

(i) $\mathrm{colim}_{\mathcal{E}}(P \circ \pi_D^{\mathrm{opf}}) \cong \mathrm{colim}_{\mathcal{E}}(D \circ \pi_P^{\mathrm{opf}})$;

(ii) $\mathrm{colim}_{\mathcal{E}}(P \circ \pi_F^{\mathrm{f}}) \cong \mathrm{colim}_{\mathcal{E}}(F \circ \pi_P^{\mathrm{f}})$. □

Let us now proceed to apply this proposition in the context of a functor $F : \mathcal{C}^{\mathrm{op}} \to \mathcal{E}$, where $\mathcal{C}$ is a small category, $\mathcal{E}$ is a Grothendieck topos and P is a functor $\mathcal{C} \to \mathbf{Set}$. To this end, we explicitly describe the discrete opfibration corresponding to the diagram $(g : \overline{G} \to \mathbb{C}_0, \phi : \mathbb{C}_1 \times_{\mathbb{C}_0} \overline{G} \to \overline{G})$ of shape $\mathbb{C}$ in $\mathcal{E}$ associated with a functor $G : \mathcal{C} \to \mathcal{E}$, where $\mathbb{C}$ is the internalization of $\mathcal{C}$ in $\mathcal{E}$.

Let $\mathrm{Ob}(\mathcal{C})$ denote the set of objects of $\mathcal{C}$ and $\mathrm{Arr}(\mathcal{C})$ the set of arrows of $\mathcal{C}$. We have that $\overline{G} = \coprod_{c \in \mathrm{Ob}(\mathcal{C})} G(c)$ and that $g : \overline{G} = \coprod_{c \in \mathrm{Ob}(\mathcal{C})} G(c) \to \mathbb{C}_0 = \coprod_{c \in \mathrm{Ob}(\mathcal{C})} 1_{\mathcal{E}}$ is equal to $\coprod_{c \in \mathrm{Ob}(\mathcal{C})} !_{G(c)}$, where $!_{G(c)}$ is the unique arrow $G(c) \to 1_{\mathcal{E}}$ in $\mathcal{E}$ (for any $c \in \mathcal{C}$).

Let $J_f : G(\mathrm{dom}(f)) \to \coprod_{f \in \mathrm{Arr}(\mathcal{C})} G(\mathrm{dom}(f))$, $\mu_c : G(c) \to \coprod_{c \in \mathrm{Ob}(\mathcal{C})} G(c)$, $\kappa_f : 1_{\mathcal{E}} \to \coprod_{f \in \mathrm{Arr}(\mathcal{C})} 1_{\mathcal{E}}$, and $\lambda_c : 1_{\mathcal{E}} \to \coprod_{c \in \mathrm{Ob}(\mathcal{C})} 1_{\mathcal{E}}$ be the canonical coproduct arrows.

First, let us show that the diagram

$$\begin{array}{ccc}
\coprod_{f \in \mathrm{Arr}(\mathcal{C})} G(\mathrm{dom}(f)) & \xrightarrow{d_0^G} & \coprod_{c \in \mathrm{Ob}(\mathcal{C})} G(c) \\
{\scriptstyle \coprod_{f \in \mathrm{Arr}(\mathcal{C})} !_{G(\mathrm{dom}(f))}} \downarrow & & \downarrow {\scriptstyle \coprod_{c \in \mathrm{Ob}(\mathcal{C})} !_{G(c)}} \\
\coprod_{f \in \mathrm{Arr}(\mathcal{C})} 1_{\mathcal{E}} & \xrightarrow{d_0^{\mathcal{C}}} & \coprod_{c \in \mathrm{Ob}(\mathcal{C})} 1_{\mathcal{E}}
\end{array}$$

is a pullback, where the arrow d_0^G is defined by setting $d_0^G \circ J_f = \mu_{\mathrm{dom}(f)}$ for any $f \in \mathrm{Arr}(\mathcal{C})$. We have to prove that, for any object E of $\mathcal{E}$ and arrows $\alpha : E \to \coprod_{f \in \mathrm{Arr}(\mathcal{C})} 1_{\mathcal{E}}$ and $\beta : E \to \coprod_{c \in \mathrm{Ob}(\mathcal{C})} G(c)$ such that $\coprod_{c \in \mathrm{Ob}(\mathcal{C})} !_{G(c)} \circ \beta = d_0^{\mathcal{C}} \circ \alpha$, there exists a unique arrow $\gamma : E \to \coprod_{f \in \mathrm{Arr}(\mathcal{C})} !_{G(\mathrm{dom}(f))}$ such that $\alpha = \coprod_{f \in \mathrm{Arr}(\mathcal{C})} !_{G(\mathrm{dom}(f))} \circ \gamma$ and $\beta = d_0^G \circ \gamma$. To this end, consider, for any $c \in \mathrm{Ob}(\mathcal{C})$ and $f \in \mathrm{Arr}(\mathcal{C})$, the commutative diagram

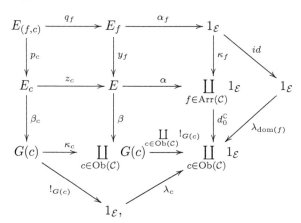

where all the squares except for the lower right one are pullbacks. The commutativity of the diagram, combined with the fact that distinct coproduct arrows are disjoint from each other, immediately implies that for any pair (f,c) such that $c \neq \text{dom}(f)$ we have $E_{(f,c)} \cong 0_{\mathcal{E}}$. On the other hand, the stability of coproducts under pullbacks implies that $E \cong \coprod_{c \in \text{Ob}(\mathcal{C})} E_c$ and $E \cong \coprod_{f \in \text{Arr}(\mathcal{C})} E_f$, whence $E \cong \coprod_{(f,c) \in \text{Arr}(\mathcal{C}) \times \text{Ob}(\mathcal{C})} E_{(f,c)} \cong \coprod_{(f,c) \mid \text{dom}(f)=c} E_{(f,c)}$ with canonical coproduct arrows $\xi_{(f,c)} = y_f \circ q_f = z_c \circ p_c : E_{(f,c)} \to E$. We define, for each pair (f,c) such that $c = \text{dom}(f)$, the arrow $\gamma_{(f,c)} : E_{(f,c)} \to \coprod_{f \in \text{Arr}(\mathcal{C})} G(\text{dom}(f))$ as the composite $J_f \circ \beta_c \circ p_c$, and set γ equal to the arrow $\coprod_{(f,c) \mid \text{dom}(f)=c} \gamma_{(f,c)} : E \to \coprod_{f \in \text{Arr}(\mathcal{C})} G(\text{dom}(f))$. We have to verify that $\alpha = \coprod_{f \in \text{Arr}(\mathcal{C})} !_{G(\text{dom}(f))} \circ \gamma$ and $\beta = d_0^G \circ \gamma$ or, equivalently, that for any pair (f,c) such that $\text{dom}(f) = c$ we have:

(1) $\alpha \circ \xi_{(f,c)} = \coprod_{f \in \text{Arr}(\mathcal{C})} !_{G(\text{dom}(f))} \circ \gamma_{(f,c)}$ and

(2) $\beta \circ \xi_{(f,c)} = d_0^G \circ \gamma_{(f,c)}$.

To prove (1) it suffices to observe that for any pair (f,c) such that $c = \text{dom}(f)$, we have $\alpha_f \circ q_f = !_{G(c)} \circ \beta_c \circ p_c$, whence $\alpha \circ \xi_{(f,c)} = \alpha \circ y_f \circ q_f = \kappa_f \circ \alpha_f \circ q_f = \kappa_f \circ !_{G(c)} \circ \beta_c \circ p_c = \coprod_{f \in \text{Arr}(\mathcal{C})} !_{G(\text{dom}(f))} \circ \gamma_{(f,c)}$. Further, $\beta \circ \xi_{(f,c)} = \beta \circ z_c \circ p_c = \mu_c \circ \beta_c \circ p_c = \mu_{\text{dom}(f)} \circ \beta_c \circ p_c = d_0^G \circ J_f \circ \beta_c \circ p_c = d_0^G \circ \gamma_{(f,c)}$, which proves (2).

Let us define an arrow
$$d_1^G : \coprod_{f \in \text{Arr}(\mathcal{C})} G(\text{dom}(f)) \to \coprod_{c \in \text{Ob}(\mathcal{C})} G(c)$$
by setting, for each $f \in \text{Arr}(\mathcal{C})$, $d_1^G \circ J_f = \mu_{\text{cod}(f)} \circ G(f)$.

The discrete opfibration $p : \mathbb{F} \to \mathbb{C}$ corresponding to G can be described as follows: $\mathbb{F}_0 = \coprod_{c \in \text{Ob}(\mathcal{C})} G(c)$, $\mathbb{F}_1 = \coprod_{f \in \text{Arr}(\mathcal{C})} G(\text{dom}(f))$, the domain and codomain arrows $d_0^{\mathbb{F}}, d_1^{\mathbb{F}} : \mathbb{F}_1 \to \mathbb{F}_0$ are respectively equal to d_0^G and to d_1^G, $p_0 : \mathbb{F}_0 \to \mathbb{C}_0$ is equal to $\coprod_{c \in \text{Ob}(\mathcal{C})} !_{G(c)}$, $p_1 : \mathbb{F}_1 \to \mathbb{C}_1$ is equal to $\coprod_{f \in \text{Arr}(\mathcal{C})} !_{G(\text{dom}(f))}$ and the composition law in the internal category $\mathbb{F}$ is defined in the obvious way.

We leave to the reader the straightforward task of verifying that this is indeed the discrete opfibration corresponding to the functor G via the composite of the equivalence $[\mathcal{C}, \mathcal{E}] \simeq [\mathbb{C}, \mathcal{E}]$ with the equivalence $\tau_{\mathcal{E}}^{\mathbb{C}} : [\mathbb{C}, \mathcal{E}] \to \mathbf{DOpf}/\mathbb{C}$. Recalling that for any internal diagram $F : \mathbb{C}^{\text{op}} \to \mathcal{E}$ the discrete fibration $\pi_F^{\text{f}} : \int^{\text{f}} F \to \mathbb{C}$ associated with it is equal to $(\pi_F^{\text{opf}})^{\text{op}} : (\int^{\text{opf}} F)^{\text{op}} \to (\mathbb{C}^{\text{op}})^{\text{op}} = \mathbb{C}$, we deduce the following explicit description of the discrete fibration $\pi_F^{\text{f}} : \int^{\text{f}} F \to \mathbb{C}$ associated with a functor $F : \mathcal{C}^{\text{op}} \to \mathcal{E}$: $(\int^{\text{f}} F)_0 = \coprod_{c \in \text{Ob}(\mathcal{C})} F(c)$, $(\int^{\text{f}} F)_1 = \coprod_{f \in \text{Arr}(\mathcal{C})} F(\text{cod}(f))$, the domain and codomain arrows $d_0^F, d_1^F : (\int^f F)_1 \to (\int^f F)_0$ are defined by the conditions $d_0^F \circ J_f = \mu_{\text{cod}(f)}$ and $d_1^F \circ J_f = \mu_{\text{dom}(f)} \circ F(f)$ for all $f \in \text{Arr}(\mathcal{C})$ (where $\mu_c : F(c) \to \coprod_{c \in \text{Ob}(\mathcal{C})} F(c)$ and $J_f : F(\text{cod}(f)) \to \coprod_{f \in \text{Arr}(\mathcal{C})} F(\text{cod}(f))$ are the canonical coproduct arrows), $(\pi_F^f)_0 : (\int^f F)_0 \to \mathbb{C}_0$ is equal to $\coprod_{c \in \text{Ob}(\mathcal{C})} !_{F(c)}$ and $(\pi_F^f)_1 : (\int^f F)_1 \to \mathbb{C}_1$ is equal to $\coprod_{f \in \text{Arr}(\mathcal{C})} !_{F(\text{cod}(f))}$. The composition law in the internal category $\int^f F$ is defined in the obvious way.

Theorem 5.1.5 *Let $\mathcal{C}$ be a small category, $\mathcal{E}$ a Grothendieck topos and $F : \mathcal{C}^{\text{op}} \to \mathcal{E}, P : \mathcal{C} \to \mathbf{Set}$ two functors. Then the following three objects are naturally isomorphic:*

(i) $\text{colim}(F \circ \pi_P^f)$
(ii) $\text{colim}_{\mathcal{E}}(\overline{F} \circ \pi_{P_{\mathcal{E}}}^f) \cong \underline{\text{colim}}_{\mathcal{E}}(P_{\mathcal{E}} \circ \pi_{\overline{F}}^f)$ *(cf. Proposition 5.1.4)*
(iii) $\underline{\text{colim}}_{\mathcal{E}}(P_{\mathcal{E}_{\mathcal{E}}} \circ \pi_{\overline{F}_{\mathcal{E}}}^f)$

where $\mathbb{C}$ is the internalization of $\mathcal{C}$ in $\mathcal{E}$ and $P_{\mathcal{E}}$ is the internal diagram in $[\mathbb{C}, \mathcal{E}]$ given by $\overline{\gamma_{\mathcal{E}}^* \circ P}$.

Proof The isomorphism between $\text{colim}_{\mathcal{E}}(P_{\mathcal{E}} \circ \pi_{\overline{F}}^f)$ and $\underline{\text{colim}}_{\mathcal{E}}(P_{\mathcal{E}_{\mathcal{E}}} \circ \pi_{\overline{F}_{\mathcal{E}}}^f)$ follows from the general fact that for any internal diagram $G \in [\mathbb{D}, \mathcal{E}]$, its colimit $\text{colim}_{\mathcal{E}}(G)$ is isomorphic to the $\mathcal{E}$-indexed colimit $\underline{\text{colim}}_{\mathcal{E}}(G_{\mathcal{E}})$. It thus remains to prove the isomorphism between $\text{colim}(F \circ \pi_P^f)$ and $\text{colim}_{\mathcal{E}}(\overline{F} \circ \pi_{P_{\mathcal{E}}}^f)$. To this end, we recall the following three general facts:

(1) For any functor $G : \mathcal{D} \to \mathcal{E}$, its colimit $\text{colim}(G)$ is naturally isomorphic to the colimit $\text{colim}_{\mathcal{E}}(\overline{G})$.
(2) For any functor $H : \mathcal{D} \to \mathcal{C}$ between small categories $\mathcal{C}$ and $\mathcal{D}$ and any functor $F : \mathcal{C} \to \mathcal{E}$, we have a natural isomorphism $\overline{F \circ H} \cong \overline{F} \circ \gamma_{\mathcal{E}}^*(H)$.
(3) For any geometric morphism $f : \mathcal{F} \to \mathcal{E}$, the diagram

$$\begin{array}{ccc} [\mathbb{C}, \mathcal{E}] & \xrightarrow{\tau_{\mathcal{E}}^{\mathbb{C}}} & \mathbf{DOpf}/\mathbb{C} \\ {\scriptstyle f^*(-)} \downarrow & & \downarrow {\scriptstyle f^*(-)} \\ [f^*(\mathbb{C}), \mathcal{F}] & \xrightarrow{\tau_{\mathcal{F}}^{f^*(\mathbb{C})}} & \mathbf{DOpf}/f^*(\mathbb{C}) \end{array}$$

commutes.

We therefore have that

$$\text{colim}(F \circ \pi_P^f) \cong \text{colim}_{\mathcal{E}}(\overline{F \circ \pi_P^f}) \cong \text{colim}_{\mathcal{E}}(\overline{F} \circ \gamma_{\mathcal{E}}^*(\pi_P^f)) \cong \text{colim}_{\mathcal{E}}(\overline{F} \circ \pi_{P_{\mathcal{E}}}^f),$$

where the first isomorphism follows from (1), the second from (2) and the third from (3). □

We shall denote the three isomorphic objects of Theorem 5.1.5 by the symbol $F \otimes_{\mathcal{C}} P$.

The following lemma is essentially contained in the proof of Giraud's theorem (cf. [42] or, for instance, the Appendix of [63]); we provide a proof of it for the reader's convenience.

Lemma 5.1.6 *Let $\mathcal{E}$ be a Grothendieck topos and $\{e_i : E_i \to E \mid i \in I\}$ an epimorphic family in $\mathcal{E}$. Then the arrow $\coprod_{i \in I} e_i : \coprod_{i \in I} E_i \to E$ yields an isomorphism $(\coprod_{i \in I} E_i)/R \cong E$, where R is the equivalence relation in $\mathcal{E}$ on $\coprod_{i \in I} E_i$ given by the subobject $\coprod_{(i,j) \in I \times I} E_i \times_E E_j \rightarrowtail \coprod_{(i,j) \in I \times I} E_i \times E_j \cong \coprod_{i \in I} E_i \times \coprod_{j \in I} E_j$; in particular, the arrow $\coprod_{i \in I} e_i : \coprod_{i \in I} E_i \to E$ is the coequalizer in $\mathcal{E}$ of the two canonical arrows $\coprod_{(i,j) \in I \times I} E_i \times_E E_j \to \coprod_{i \in I} E_i$.*

Proof Let R be the kernel pair of the epimorphism $\coprod_{i \in I} e_i$, that is the pullback of this arrow along itself; then, by the well-known exactness properties of Grothendieck toposes, R is an equivalence relation on $\coprod_{i \in I} E_i$ such that the coequalizer in $\mathcal{E}$ of the two associated projections is isomorphic to $\coprod_{i \in I} e_i$. Now, the fact that pullbacks preserve coproducts in a Grothendieck topos implies that R is isomorphic to the subobject $\coprod_{(i,j) \in I \times I} E_i \times_E E_j \rightarrowtail \coprod_{(i,j) \in I \times I} E_i \times E_j \cong \coprod_{i \in I} E_i \times \coprod_{j \in I} E_j$, as required. □

Corollary 5.1.7 *Let $a : A \to E$ and $l : L \to E$ be objects of the slice topos $\mathcal{E}/E$, and $\{e_i : E_i \to E \mid i \in I\}$ an epimorphic family in $\mathcal{E}$. Then a family of arrows $\{f_i : e_i^*(a) \to e_i^*(l) \mid i \in I\}$ in the toposes $\mathcal{E}/E_i$ defines a (unique) arrow $f : a \to l$ in $\mathcal{E}/E$ such that $e_i^*(f) = f_i$ for all $i \in I$ if and only if for every $i, j \in I$, $q_i^*(f_i) = q_j^*(f_j)$, where the arrows q_i and q_j are defined by the following pullback square:*

$$\begin{array}{ccc} E_{i,j} & \xrightarrow{q_i} & E_i \\ {\scriptstyle q_j} \downarrow & & \downarrow {\scriptstyle e_i} \\ E_j & \xrightarrow{e_j} & E. \end{array}$$

□

5.1.5 $\mathcal{E}$-final functors

The notions introduced in the following definition will be important in the sequel.

Definition 5.1.8 Let $\underline{A}_{\mathcal{E}}$ be an $\mathcal{E}$-indexed category and $i : \underline{B}_{\mathcal{E}} \to \underline{A}_{\mathcal{E}}$ an $\mathcal{E}$-indexed functor.

(a) We say that i is $\mathcal{E}$-*final* if, for every $E \in \mathcal{E}$ and every $x \in \mathcal{A}_E$, there exists a non-empty set $\mathcal{E}_E^x$ of triples $\{(e_i, b_i, f_i) \mid i \in I\}$, where the family $\{e_i : E_i \to E \mid i \in I\}$ is epimorphic, b_i is an object of $\mathcal{B}_{E_i}$ and $f_i : \mathcal{A}_{e_i}(x) \to i_{E_i}(b_i)$ is an arrow in $\mathcal{A}_{E_i}$ (for each $i \in I$), with the property that for any triples $\{(e_i, b_i, f_i) \mid i \in I\}$ and $\{(e'_j, c_j, f'_j) \mid j \in J\}$ in $\mathcal{E}_E^x$ there exist an epimorphic family $\{g_k^{i,j} : E_k^{i,j} \to E_{i,j} \mid k \in K_{i,j}\}$ in $\mathcal{E}$ and for each $k \in K_{i,j}$ an object $d_k^{i,j}$ of $\mathcal{B}_{E_k^{i,j}}$ and arrows $r_k^{i,j} : \mathcal{B}_{q_i \circ g_k^{i,j}}(b_i) \to d_k^{i,j}$ and $s_k^{i,j} : \mathcal{B}_{q_j \circ g_k^{i,j}}(c_j) \to d_k^{i,j}$ in $\mathcal{B}_{E_k^{i,j}}$ such that $i_{E_k^{i,j}}(r_k^{i,j}) \circ \mathcal{A}_{q_i}(f_i) = i_{E_k^{i,j}}(s_k^{i,j}) \circ \mathcal{A}_{q_j}(f'_j)$.

(b) We say that an $\mathcal{E}$-indexed subcategory $\underline{\mathcal{B}}_\mathcal{E}$ of an $\mathcal{E}$-indexed category $\underline{\mathcal{A}}_\mathcal{E}$ is an $\mathcal{E}$-*final subcategory* of $\underline{\mathcal{A}}_\mathcal{E}$ if the canonical $\mathcal{E}$-indexed inclusion functor $i : \underline{\mathcal{B}}_\mathcal{E} \hookrightarrow \underline{\mathcal{A}}_\mathcal{E}$ is $\mathcal{E}$-final.

(c) We say that an $\mathcal{E}$-indexed subcategory $\underline{\mathcal{B}}_\mathcal{E}$ of an $\mathcal{E}$-indexed category $\underline{\mathcal{A}}_\mathcal{E}$ is an $\mathcal{E}$-*strictly final subcategory* of $\underline{\mathcal{A}}_\mathcal{E}$ if for every arrow $f : a \to b$ in $\mathcal{A}_E$ there exists an epimorphic family $\{e_i : E_i \to E \mid i \in I\}$ in $\mathcal{E}$ such that the arrow $\mathcal{A}_{e_i}(f) : \mathcal{A}_{e_i}(a) \to \mathcal{A}_{e_i}(b)$ lies in $\mathcal{B}_{E_i}$ for every $i \in I$.

(d) We say that an $\mathcal{E}$-indexed subcategory $\underline{\mathcal{B}}_\mathcal{E}$ of an $\mathcal{E}$-indexed category $\underline{\mathcal{A}}_\mathcal{E}$ is $\mathcal{E}$-*full* if for every arrow $f : b \to b'$ in $\mathcal{A}_E$, where b and b' are objects of $\mathcal{B}_E$, there exists an epimorphic family $\{e_i : E_i \to E \mid i \in I\}$ in $\mathcal{E}$ such that the arrow $\mathcal{A}_{e_i}(f)$ lies in $\mathcal{B}_{E_i}$ for every $i \in I$.

Remarks 5.1.9 (a) Let $\underline{\mathcal{B}}_\mathcal{E}$ be an $\mathcal{E}$-indexed subcategory of $\underline{\mathcal{A}}_\mathcal{E}$ with the property that for every $E \in \mathcal{E}$ and $a \in \mathcal{A}_E$ there exists an epimorphic family $\{e_i : E_i \to E \mid i \in I\}$ in $\mathcal{E}$ with the property that for every $i \in I$ the object $\mathcal{A}_{e_i}(a)$ lies in $\mathcal{B}_{E_i}$ and for every arrow $f : a \to b$ in $\mathcal{A}_E$, where $a, b \in \mathcal{B}_E$, there exists an epimorphic family $\{e_i : E_i \to E \mid i \in I\}$ in $\mathcal{E}$ such that for every $i \in I$ the arrow $\mathcal{A}_{e_i}(f) : \mathcal{A}_{e_i}(a) \to \mathcal{A}_{e_i}(b)$ lies in $\mathcal{B}_{E_i}$. Then $\underline{\mathcal{B}}_\mathcal{E}$ is an $\mathcal{E}$-strictly final subcategory of $\underline{\mathcal{A}}_\mathcal{E}$.

(b) Every $\mathcal{E}$-strictly final subcategory is an $\mathcal{E}$-final subcategory.

(c) If i is an $\mathcal{E}$-full embedding of an $\mathcal{E}$-indexed category $\underline{\mathcal{B}}_\mathcal{E}$ into an $\mathcal{E}$-filtered $\mathcal{E}$-indexed category $\underline{\mathcal{A}}_\mathcal{E}$ then i is $\mathcal{E}$-final if and only if for every $E \in \mathcal{E}$ and $x \in \mathcal{A}_E$ there exist a non-empty epimorphic family $\{e_i : E_i \to E \mid i \in I\}$ in $\mathcal{E}$ and for each $i \in I$ an object b_i of $\mathcal{B}_{E_i}$ and an arrow $f_i : \mathcal{A}_{e_i}(x) \to b_i$ of $\mathcal{A}_{E_i}$.

Proposition 5.1.10 *Let i be the embedding of an $\mathcal{E}$-full and $\mathcal{E}$-final subcategory $\underline{\mathcal{B}}_\mathcal{E}$ of an $\mathcal{E}$-indexed category $\underline{\mathcal{A}}_\mathcal{E}$ into $\underline{\mathcal{A}}_\mathcal{E}$. Then $\underline{\mathcal{B}}_\mathcal{E}$ is $\mathcal{E}$-filtered if and only if $\underline{\mathcal{A}}_\mathcal{E}$ is $\mathcal{E}$-filtered.*

Proof The proof is entirely analogous to that of the classical result in ordinary category theory and is left to the reader. □

The following result represents an $\mathcal{E}$-indexed version of the classical theorem formalizing the behaviour of colimits with respect to final functors:

Theorem 5.1.11 *Let $i : \mathcal{B}_\mathcal{E} \to \mathcal{A}_\mathcal{E}$ be an $\mathcal{E}$-final functor and $D : \mathcal{A}_\mathcal{E} \to \mathcal{E}_\mathcal{E}$ an $\mathcal{E}$-indexed functor. Then D admits a colimit (resp. an $\mathcal{E}$-indexed colimit) if and only if $D \circ i$ admits a colimit (resp. an $\mathcal{E}$-indexed colimit) and the two colimits are equal.*

Proof First, let us show that any cocone λ over $D \circ i$ with vertex V can be (uniquely) extended to a cocone $\tilde{\lambda}$ over D. For any $E \in \mathcal{E}$ and $a \in \mathcal{A}_E$, we have to define an arrow $\tilde{\lambda}_E(a) : D_E(a) \to !_E^*(V)$ in $\mathcal{E}/E$. By our hypotheses there exists a non-empty set $\mathcal{E}_E^a$ of triples $\{(e_i, b_i, f_i) \mid i \in I\}$, where the family $\{e_i : E_i \to E \mid i \in I\}$ is epimorphic, b_i is an object of $\mathcal{B}_{E_i}$ and $f_i : \mathcal{A}_{e_i}(a) \to i_{E_i}(b_i)$ is an arrow in $\mathcal{A}_{E_i}$ (for each $i \in I$), with the property that for any triples $\{(e_i, b_i, f_i) \mid i \in I\}$ and $\{(e'_j, c_j, f'_j) \mid j \in J\}$ in $\mathcal{E}_E^a$, there exist an epimorphic family $\{g_k^{i,j} : E_k^{i,j} \to E_{i,j} \mid k \in K_{i,j}\}$ and for each $k \in K_{i,j}$ an object $d_k^{i,j}$ of $\mathcal{B}_{E_k^{i,j}}$ and arrows $r_k^{i,j} : \mathcal{B}_{q_i \circ g_k^{i,j}}(b_i) \to d_k^{i,j}$ and $s_k^{i,j} : \mathcal{B}_{q_j \circ g_k^{i,j}}(c_j) \to d_k^{i,j}$ in $\mathcal{B}_{E_k^{i,j}}$ such that $i_{E_k^{i,j}}(r_k^{i,j}) \circ \mathcal{A}_{q_i}(f_i) = i_{E_k^{i,j}}(s_k^{i,j}) \circ \mathcal{A}_{q_j}(f'_j)$.

Consider, for each $i \in I$, the arrow $\lambda_{E_i}(b_i) \circ D_{E_i}(f_i) : D_{E_i}(\mathcal{A}_{e_i}(a)) = e_i^*(D_E(a)) \to !_{E_i}^*(V) = e_i^*(!_E^*(V))$. By applying the above property to the pair $((e_i, f_i, b_i), (e_i, f_i, b_i))$ and exploiting the fact that λ is a cocone over $D \circ i$, we obtain that $q_i^*(\lambda_{E_i}(b_i) \circ D_{E_i}(f_i)) = q_j^*(\lambda_{E_j}(b_j) \circ D_{E_j}(f_j))$ for any $i, j \in I$ (where the notation is that of Corollary 5.1.7). By Corollary 5.1.7, the family of arrows $\lambda_{E_i}(\mathcal{A}_{e_i}(a))$ (for $i \in I$) thus induces a unique arrow $u_\mathcal{E} = D_E(a) \to !_E^*(V)$. To be able to set $\tilde{\lambda}_E(a)$ equal to this arrow, we have to show that such a definition does not depend on the choice of the family $\{(e_i, b_i, f_i) \mid i \in I\}$. But this follows similarly to above, by applying the condition in the definition of an $\mathcal{E}$-final functor and recalling that λ is a cocone over $D \circ i$.

Notice that if $b \in \mathcal{B}_E$ then $\tilde{\lambda}_E(i_E(b)) = \lambda_E(b)$, and that for any cocone ξ over D, $\xi = \widetilde{\xi \circ i}$.

Now that we have shown that for any $a \in \mathcal{A}_E$ the definition of the arrow $\tilde{\lambda}_E(a) : D_E(a) \to !_E^*(V)$ in the topos $\mathcal{E}/E$ is well-posed, it remains to prove that the assignment $a \to \tilde{\lambda}_E(a)$ defines a cocone over the diagram D_E with vertex $!_E^*(V)$ i.e. that for any arrow $f : a \to b$ in $\mathcal{A}_E$, $\tilde{\lambda}_E(b) \circ D_E(f) = \tilde{\lambda}_E(a)$ as arrows in $\mathcal{E}/E$. Further, we have to show that the assignment $E \to \tilde{\lambda}_E$ defines an $\mathcal{E}$-indexed cocone on the $\mathcal{E}$-indexed functor D, i.e. that for any arrow $\alpha : E' \to E$ in $\mathcal{E}$ and any $a \in \mathcal{A}_E$ we have $\alpha^*(\tilde{\lambda}_E(a)) = \tilde{\lambda}_{E'}(\mathcal{A}_\alpha(a))$. But this can be easily deduced from the fact that λ is an $\mathcal{E}$-indexed cocone over $D \circ i$.

The second part of the theorem, for indexed colimits, follows from the first part by noticing that any localization of an indexed $\mathcal{E}$-final functor with respect to a localization functor $\mathcal{E} \to \mathcal{E}/E$ is an $(\mathcal{E}/E)$-final functor. So it will be sufficient to show the first part.

To complete the proof of the theorem, it thus remains to show that for any cocone (U, μ) over D, (U, μ) is colimiting over D if and only if $(U, \mu \circ i)$ is colimiting over D.

Suppose that D has a colimiting cocone μ with vertex U. We want to prove that $\mu \circ i$ is a colimiting cocone over $D \circ i$. Let λ be a cocone over $D \circ i$ with vertex V; then, as we have just proved, $\tilde{\lambda}$ is a cocone over D with vertex V and hence $(V, \tilde{\lambda})$ factors through (U, μ), say by an arrow $z : U \to V$ in $\mathcal{E}$. Clearly, z yields a factorization of (V, λ) across $(U, \mu \circ i)$. The uniqueness of the factorization of (V, λ) across $(U, \mu \circ i)$ follows from the fact that for any two cocones ξ and χ over $D \circ i$, if ξ factors through χ by an arrow w then $\tilde{\xi}$ factors through $\tilde{\chi}$ by the same arrow. This proves that if (U, μ) is a colimiting cocone over D then $(U, \mu \circ i)$ is a colimiting cocone over $D \circ i$.

Conversely, suppose that $(U, \mu \circ i)$ is a colimiting cocone over $D \circ i$. For any cocone (Z, χ) over D, we have that $(Z, \chi \circ i)$ is a cocone over $D \circ i$; therefore $(Z, \chi \circ i)$ factors uniquely through $(U, \mu \circ i)$ or equivalently (by arguments similar to the above ones), (Z, χ) factors uniquely through (U, μ). Therefore (U, μ) is a colimiting cocone over D, as required. □

Let $\underline{\mathbf{Set}}_{\mathcal{E}}$ be the $\mathcal{E}$-indexed category given by $\mathbf{Set}_E = \mathbf{Set}$ for all $E \in \mathcal{E}$ and $\mathbf{Set}_\alpha = 1_{\mathbf{Set}}$ (where $1_{\mathbf{Set}}$ is the identical functor on $\mathbf{Set}$) for all arrows α in $\mathcal{E}$. There is an $\mathcal{E}$-indexed functor

$$\gamma_{\mathcal{E}} : \underline{\mathbf{Set}}_{\mathcal{E}} \to \underline{\mathcal{E}}_{\mathcal{E}}$$

defined by setting $\gamma_{\mathcal{E}\, E} = \gamma^*_{\mathcal{E}/E} : \mathbf{Set} \to \mathcal{E}/E$ (for any object E of $\mathcal{E}$).

We shall consider $\mathcal{E}$-indexed functors obtained by composing the $\mathcal{E}$-indexed functor $\gamma_{\mathcal{E}} : \underline{\mathbf{Set}}_{\mathcal{E}} \to \underline{\mathcal{E}}_{\mathcal{E}}$ with an $\mathcal{E}$-indexed functor $D : \underline{\mathcal{A}}_{\mathcal{E}} \to \underline{\mathbf{Set}}_{\mathcal{E}}$. Notice that an $\mathcal{E}$-indexed functor $D : \underline{\mathcal{A}}_{\mathcal{E}} \to \underline{\mathbf{Set}}_{\mathcal{E}}$ consists of a functor $D_E : \mathcal{A}_E \to \mathbf{Set}$ for each object E of $\mathcal{E}$ such that, for any arrow $\alpha : E' \to E$ in $\mathcal{E}$, $D_{E'} \circ \mathcal{A}_\alpha = D_E$.

Let us give an explicit description of cocones over the $\mathcal{E}$-indexed functor $\gamma_{\mathcal{E}} \circ D$. Specializing the general definition, we obtain that a cocone (U, μ) over $\gamma_{\mathcal{E}} \circ D$ consists of an object U of $\mathcal{E}$ and of an arrow $\mu_E(a) : \gamma^*_{\mathcal{E}/E}(D_E(a)) \to !^*_E(U)$ in $\mathcal{E}/E$ (for any objects $E \in \mathcal{E}$ and $a \in \mathcal{A}_E$) such that, for any arrow $f : a \to b$ in $\mathcal{A}_E$, $\mu_E(b) \circ \gamma^*_{\mathcal{E}/E}(D_E(f)) = \mu_E(a)$ and, for any arrow $\alpha : E' \to E$ in $\mathcal{E}$, $\alpha^*(\mu_E(a)) = \mu_{E'}(\mathcal{A}_\alpha(a))$.

By using the well-known adjunction between $\gamma^*_{\mathcal{E}}$ and the global sections functor $\mathcal{E} \to \mathbf{Set}$, we can alternatively present the above set of data as follows: a cocone (U, μ) over $\gamma_{\mathcal{E}} \circ D$ consists of an object U of $\mathcal{E}$ and of a function $\mu_E(a) : D_E(a) \to \mathrm{Hom}_{\mathcal{E}}(E, U)$ (for any objects E of $\mathcal{E}$ and a of $\mathcal{A}_E$) such that, for any arrow $f : a \to b$ in $\mathcal{A}_E$, $\mu_E(b) \circ D_E(f) = \mu_E(a)$ and, for any arrow $\alpha : E' \to E$ in $\mathcal{E}$, $\mathrm{Hom}_{\mathcal{E}}(\alpha, U) \circ \mu_E(a) = \mu_{E'}(\mathcal{A}_\alpha(a))$ (notice that the domains of these two arrows are the same since $D_E(a) = D_{E'}(\mathcal{A}_\alpha(a))$).

Theorem 5.1.12 *Let $\mathcal{C}$ be a small category, $\mathcal{E}$ a Grothendieck topos and $F : \mathcal{C}^{\mathrm{op}} \to \mathcal{E}$ a functor. Then the $\mathcal{E}$-indexed subcategory $\int \underline{F}_{\mathcal{E}}$ of $\int \underline{F}^{\mathrm{f}}_{\mathcal{E}}$ is $\mathcal{E}$-strictly final.*

Proof By definition of $\int \overline{F}^f_{\mathcal{E}}$, for any object E of $\mathcal{E}$ the category $(\int \overline{F}^f)_E$ has as objects the arrows $x : E \to \coprod_{c \in \mathrm{Ob}(\mathcal{C})} F(c)$ and as arrows $x \to x'$ the arrows $z : E \to \coprod_{f \in \mathrm{Arr}(\mathcal{C})} F(\mathrm{cod}(f))$ such that $d_0^F \circ z = x$ and $d_1^F \circ z = x'$. For any $E \in \mathcal{E}$, the category $(\int F)_E$ thus identifies with a subcategory of the category $(\int \overline{F}^f)_E$, through the assignment sending any pair (c, x), where $x : E \to F(c)$, to the arrow $\mu_c \circ x : E \to \coprod_{c \in \mathrm{Ob}(\mathcal{C})} F(c)$ and any arrow $(c, x) \to (c', x')$ to the arrow $J_f \circ x' : E \to \coprod_{f \in \mathrm{Arr}(\mathcal{C})} F(\mathrm{cod}(f))$ (where the notation is that of section 5.1.4).

These assignments clearly make $\int F_{\mathcal{E}}$ into an $\mathcal{E}$-indexed subcategory of $\int \overline{F}^f_{\mathcal{E}}$.

To prove our thesis, we shall apply the criterion of Remark 5.1.9(a). To show that the $\mathcal{E}$-indexed subcategory $\int F_{\mathcal{E}}$ satisfies the first condition in the Remark, we observe that for any object $x : E \to \coprod_{c \in \mathrm{Ob}(\mathcal{C})} F(c)$ of the category $(\int \overline{F}^f)_E$, if we consider the pullbacks

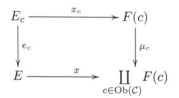

of x along the coproduct arrows $\mu_c : F(c) \to \coprod_{c \in \mathrm{Ob}(\mathcal{C})} F(c)$, we obtain an epimorphic family $\{e_c : E_c \to E \mid c \in \mathrm{Ob}(\mathcal{C})\}$ such that $(\int \overline{F}^f)_{e_c}(x)$ lies in the subcategory $(\int F)_{E_c}$ for each $c \in \mathrm{Ob}(\mathcal{C})$.

It remains to prove that the second condition of Remark 5.1.9(a) is satisfied. Suppose that (c, x) and (c', x') are objects of the subcategory $(\int F)_E$ and that $\alpha : E \to \coprod_{f \in \mathrm{Arr}(\mathcal{C})} F(\mathrm{cod}(f))$ is an arrow such that $d_0^F \circ \alpha = \mu_c \circ x$ and $d_1^F \circ \alpha = \mu_{c'} \circ x'$:

$$\begin{array}{ccc} E & \xrightarrow{\alpha} & \coprod_{f \in \mathrm{Arr}(\mathcal{C})} F(\mathrm{cod}(f)) \\ \downarrow x & & \downarrow d_0^F \\ F(c) & \xrightarrow{\mu_c} & \coprod_{c \in \mathrm{Ob}(\mathcal{C})} F(c), \end{array} \qquad \begin{array}{ccc} E & \xrightarrow{\alpha} & \coprod_{f \in \mathrm{Arr}(\mathcal{C})} F(\mathrm{cod}(f)) \\ \downarrow x' & & \downarrow d_1^F \\ F(c') & \xrightarrow{\mu_{c'}} & \coprod_{c \in \mathrm{Ob}(\mathcal{C})} F(c). \end{array}$$

Consider, for each $f \in \mathrm{Arr}(\mathcal{C})$, the following pullback square:

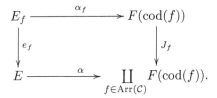

The commutativity of the three diagrams above, combined with the fact that distinct coproduct arrows are disjoint, implies that for any arrow f of $\mathcal{C}$ such that $\mathrm{dom}(f) \neq c$ or $\mathrm{cod}(f) \neq c'$ we have $E_f \cong 0_{\mathcal{E}}$. We can therefore restrict our attention to the arrows f such that $\mathrm{dom}(f) = c$ and $\mathrm{cod}(f) = c'$. For any such arrow f, we have $\alpha_f = x' \circ e_f$ (equivalently, since $\mu_{c'} = \mu_{\mathrm{cod}(f)}$ is monic, $\mu_{c'} \circ \alpha_f = \mu_{c'} \circ x' \circ e_f$). Indeed, $\mu_{c'} \circ \alpha_f = d_1^F \circ J_f \circ \alpha_f = d_1^F \circ \alpha \circ e_f = \mu_{c'} \circ x' \circ e_f$. Therefore $\alpha \circ e_f = J_f \circ x' \circ e_f$, since $\alpha \circ e_f = J_f \circ \alpha_f = J_f \circ x' \circ e_f$. Lastly, we observe that f defines an arrow $(c, x \circ e_f) \to (c', x' \circ e_f)$ in the subcategory $(\int F)_{E_f}$, i.e. $F(f) \circ x' \circ e_f = x \circ e_f$. Indeed, the arrow $\mu_{\mathrm{dom}(f)} = \mu_c$ is monic and we have $\mu_{\mathrm{dom}(f)} \circ F(f) \circ x' \circ e_f = d_0^F \circ J_f \circ x' \circ e_f = J_f \circ \alpha \circ e_f = \mu_c \circ x \circ e_f$.

The epimorphic family $\{e_f : E_f \to E \mid \mathrm{dom}(f) = c \text{ and } \mathrm{cod}(f) = c'\}$ thus satisfies the property that the arrow $(\int \overline{F}^{\mathrm{f}})_{e_f}(\alpha)$ lies in the subcategory $(\int F)_{E_f}$ for each $f : c \to c'$. This completes our proof. □

Theorem 5.1.13 *Let $\mathcal{C}$ be a small category, $\mathcal{E}$ a Grothendieck topos, $F : \mathcal{C}^{\mathrm{op}} \to \mathcal{E}$ and $P : \mathcal{C} \to \mathbf{Set}$ functors and $P_{\mathcal{E}}$ the internal diagram in $[\mathbb{C}, \mathcal{E}]$ given by $\gamma_{\mathcal{E}}^* \circ P$. Then the restriction of the $\mathcal{E}$-indexed functor $(\underline{P_{\mathcal{E}_{\mathcal{E}}}} \circ \pi_{\overline{F}_{\mathcal{E}}}^{\mathrm{f}}) \cong \underline{P_{\mathcal{E}}} \circ \pi_{\overline{F}_{\mathcal{E}}}^{\mathrm{f}}$ to the $\mathcal{E}$-indexed subcategory $\int F_{\mathcal{E}}$ of $\int \overline{F}_{\mathcal{E}}^{\mathrm{f}}$ is naturally isomorphic to $\gamma_{\mathcal{E}} \circ z_{\mathcal{E}}$, where $z_{\mathcal{E}} : \int F_{\mathcal{E}} \to \mathbf{Set}_{\mathcal{E}}$ is the $\mathcal{E}$-indexed functor defined by setting $(z_{\mathcal{E}})_E((c, x)) = P(c)$ and $(z_{\mathcal{E}})_E(f) = P(f)$ (for any $E \in \mathcal{E}$, object (c, x) and arrow f of the category $(\int F)_E$).*

Proof We shall exhibit an isomorphism

$$(\underline{P_{\mathcal{E}} \circ \pi_{\overline{F}_{\mathcal{E}}}^{\mathrm{f}}})_E((c, x)) \cong (\gamma_{\mathcal{E}} \circ z_{\mathcal{E}})_E((c, x))$$

natural in $E \in \mathcal{E}$ and $(c, x) \in (\int F)_E$.

Consider the following pullback square in $\mathbf{cat}(\mathcal{E})$:

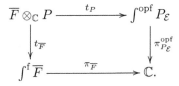

Since coproducts are stable under pullback in any Grothendieck topos, we have that

$$(\overline{F} \otimes_{\mathcal{C}} P)_0 = \coprod_{c \in \mathrm{Ob}(\mathcal{C})} F(c) \times \gamma_{\mathcal{E}}^*(P(c)),$$

$$(\overline{F} \otimes_{\mathcal{C}} P)_1 = \coprod_{f \in \mathrm{Arr}(\mathcal{C})} F(\mathrm{cod}(f)) \times \gamma_{\mathcal{E}}^*(P(\mathrm{dom}(f))),$$

$$(t_{\overline{F}})_0 : (\overline{F} \otimes_{\mathcal{C}} P)_0 = \coprod_{c \in \mathrm{Ob}(\mathcal{C})} F(c) \times \gamma_{\mathcal{E}}^*(P(c)) \to \left(\int^f \overline{F}\right)_0 = \coprod_{c \in \mathrm{Ob}(\mathcal{C})} F(c)$$

is equal to the arrow $\coprod_{c \in \mathrm{Ob}(\mathcal{C})} \pi_{F(c)}$, where $\pi_{F(c)}$ is the canonical projection $F(c) \times \gamma_{\mathcal{E}}^*(P(c)) \to F(c)$, $(t_{\overline{F}})_1 : (\overline{F} \otimes_{\mathcal{C}} P)_1 = \coprod_{f \in \mathrm{Arr}(\mathcal{C})} F(\mathrm{cod}(f)) \times \gamma_{\mathcal{E}}^*(P(\mathrm{dom}(f))) \to \left(\int^f \overline{F}\right)_1 = \coprod_{f \in \mathrm{Arr}(\mathcal{C})} F(\mathrm{cod}(f))$ is equal to the arrow $\coprod_{f \in \mathrm{Arr}(\mathcal{C})} \pi'_{F(\mathrm{cod}(f))}$, where $\pi'_{F(\mathrm{cod}(f))}$ is the canonical projection $F(\mathrm{cod}(f)) \times \gamma_{\mathcal{E}}^*(P(\mathrm{dom}(f))) \to F(\mathrm{cod}(f))$, and the domain and codomain arrows

$$d_0^{\overline{F} \otimes_{\mathcal{C}} P}, d_1^{\overline{F} \otimes_{\mathcal{C}} P} : (\overline{F} \otimes_{\mathcal{C}} P)_1 \to (\overline{F} \otimes_{\mathcal{C}} P)_0$$

are defined by the following conditions:

$$d_0^{\overline{F} \otimes_{\mathcal{C}} P} \circ W_f = Z_{\mathrm{cod}(f)} \circ <\pi'_{F(\mathrm{cod}(f))}, \gamma_{\mathcal{E}}^*(P(f)) \circ \mu_{\gamma_{\mathcal{E}}^*(P(\mathrm{dom}(f)))}>$$

and

$$d_1^{\overline{F} \otimes_{\mathcal{C}} P} \circ W_f = Z_{\mathrm{dom}(f)} \circ < F(f) \circ \pi'_{F(\mathrm{cod}(f))}, \mu_{\gamma_{\mathcal{E}}^*(P(\mathrm{dom}(f)))}>,$$

where $\mu_{\gamma_{\mathcal{E}}^*(P(\mathrm{dom}(f)))}$ is the canonical projection

$$F(\mathrm{cod}(f)) \times \gamma_{\mathcal{E}}^*(P(\mathrm{dom}(f))) \to \gamma_{\mathcal{E}}^*(P(\mathrm{dom}(f)))$$

and

$$W_f : F(\mathrm{cod}(f)) \times \gamma_{\mathcal{E}}^*(P(\mathrm{dom}(f))) \to \coprod_{f \in \mathrm{Arr}(\mathcal{C})} F(\mathrm{cod}(f)) \times \gamma_{\mathcal{E}}^*(P(\mathrm{dom}(f))),$$

$$Z_c : F(c) \times \gamma_{\mathcal{E}}^*(P(c)) \to \coprod_{c \in \mathrm{Ob}(\mathcal{C})} F(c) \times \gamma_{\mathcal{E}}^*(P(c))$$

are the canonical coproduct arrows (resp. for $f \in \mathrm{Arr}(\mathcal{C})$ and $c \in \mathrm{Ob}(\mathcal{C})$).

Now, the internal diagram $P_{\mathcal{E}} \circ \pi_{\overline{F}}^{\mathrm{f}}$ corresponds precisely to the discrete opfibration $t_{\overline{F}}$ and hence the $\mathcal{E}$-indexed functor $P_{\mathcal{E}} \circ \pi_{\overline{F}_{\mathcal{E}}}^{\mathrm{f}}$ sends any generalized element $x : E \to \coprod_{c \in \mathrm{Ob}(\mathcal{C})} F(c)$ to the object of $\mathcal{E}/E$ given by the pullback of $(t_{\overline{F}})_0$ along it. In particular, since in a Grothendieck topos any diagram of the form

$$\begin{array}{ccc} A_i & \longrightarrow & \coprod_{i \in I} A_i \\ \downarrow f_i & & \downarrow \coprod_{i \in I} f_i \\ B_i & \longrightarrow & \coprod_{i \in I} B_i, \end{array}$$

where the horizontal arrows are the canonical coproduct arrows, is a pullback, and products commute with pullbacks, the functor $(P_{\mathcal{E}} \circ \pi_{\overline{F}}^{\mathrm{f}})_E$ sends any object (c,x) of the category $(\int F)_E$ to the object given by the canonical projection $E \times \gamma_{\mathcal{E}}^*(P(c)) \to E$. But this object is canonically isomorphic to $\gamma_{\mathcal{E}/E}^*(P(c))$ in $\mathcal{E}/E$, that is to the value at (c,x) of the functor $(\gamma_{\mathcal{E}} \circ z_{\mathcal{E}})_E$. □

Let us now explicitly describe the $\mathcal{E}$-indexed colimiting cocone on the restriction to the $\mathcal{E}$-indexed subcategory $\int \overline{F}_{\mathcal{E}}$ of the $\mathcal{E}$-indexed diagram $P_{\mathcal{E}} \circ \pi_{\overline{F}_{\mathcal{E}}}^{\mathrm{f}}$. The vertex of this colimiting cocone is the codomain of the coequalizer w: $(\overline{F} \otimes_{\mathbf{C}} P)_0 \to \underline{\mathrm{colim}}_{\mathcal{E}}(\underline{P_{\mathcal{E}_{\mathcal{E}}}} \circ \pi_{\overline{F}_{\mathcal{E}}}^{\mathrm{f}})$ of the pair of arrows $d_0^{\overline{F} \otimes_{\mathbf{C}} P}, d_1^{\overline{F} \otimes_{\mathbf{C}} P} : (\overline{F} \otimes_{\mathbf{C}} P)_1 \to (\overline{F} \otimes_{\mathbf{C}} P)_0$. For any object (c,x) of the category $(\int \overline{F})_E$, the colimit arrow $(P_{\mathcal{E}} \circ \pi_{\overline{F}}^{\mathrm{f}})_E)((c,x)) \cong \gamma_{\mathcal{E}/E}^*(P(c)) \to \underline{\mathrm{colim}}_{\mathcal{E}}(\underline{P_{\mathcal{E}_{\mathcal{E}}}} \circ \pi_{\overline{F}_{\mathcal{E}}}^{\mathrm{f}})$ is given by the composite $!_E^*(w) \circ h_{(c,x)}$, where $h_{(c,x)} = !_E^*(Z_c) \circ < x \times 1_{\gamma_{\mathcal{E}}^*(P(c))}, \pi_E > : \gamma_{\mathcal{E}/E}^*(P(c)) \cong E \times \gamma_{\mathcal{E}}^*(P(c)) \to \coprod_{c \in \mathrm{Ob}(\mathcal{C})} F(c) \times \gamma_{\mathcal{E}}^*(P(c)) \times E \cong !_E^*(\coprod_{c \in \mathrm{Ob}(\mathcal{C})} F(c) \times \gamma_{\mathcal{E}}^*(P(c)))$; in other words, the function $\xi_{(c,x)} : P(c) \to \mathrm{Hom}_{\mathcal{E}}(E, \underline{\mathrm{colim}}_{\mathcal{E}}(\underline{P_{\mathcal{E}_{\mathcal{E}}}} \circ \pi_{\overline{F}_{\mathcal{E}}}^{\mathrm{f}}))$ corresponding to it assigns to every element $a \in P(c)$ the arrow $w \circ Z_c \circ < x, r_a \circ !_E >$, where $r_a : 1_{\mathcal{E}} \to \gamma_{\mathcal{E}}^*(P(c)) \cong \coprod_{a \in P(c)} 1_{\mathcal{E}}$ is the coproduct arrow corresponding to the element a.

It is immediate to verify that, under the isomorphism

$$\underline{\mathrm{colim}}_{\mathcal{E}}(\underline{P_{\mathcal{E}_{\mathcal{E}}}} \circ \pi_{\overline{F}_{\mathcal{E}}}^{\mathrm{f}}) \cong \mathrm{colim}(F \circ \pi_P^{\mathrm{f}})$$

of Theorem 5.1.5, the functions

$$\xi_{(c,x)} : P(c) \to \mathrm{Hom}_{\mathcal{E}}(E, \underline{\mathrm{colim}}_{\mathcal{E}}(\underline{P_{\mathcal{E}_{\mathcal{E}}}} \circ \pi_{\overline{F}_{\mathcal{E}}}^{\mathrm{f}})) \cong \mathrm{Hom}_{\mathcal{E}}(E, \mathrm{colim}(F \circ \pi_P^{\mathrm{f}}))$$

admit the following description in terms of the colimit arrows $\chi_{(c,a)} : F(c) \to \mathrm{colim}(F \circ \pi_P^{\mathrm{f}})$: for any $a \in P(c)$, $\xi_{(c,x)}(a) = \chi_{(c,a)} \circ x$.

5.1.6 A characterization of $\mathcal{E}$-indexed colimits

In this section we establish, for a Grothendieck topos $\mathcal{E}$, necessary and sufficient conditions for an $\mathcal{E}$-indexed cocone on a small diagram in $\mathcal{E}$ to be an ($\mathcal{E}$-indexed) colimit.

Before stating and proving the main theorem of this section we need to introduce some relevant definitions and a few technical lemmas.

Given an $\mathcal{E}$-indexed functor $D : \underline{\mathcal{A}_{\mathcal{E}}} \to \underline{\mathcal{E}_{\mathcal{E}}}$ and an object R of $\mathcal{E}$, we denote by $\mathcal{I}_R^D$ the collection of pairs of the form (x,y), where x is an object of the category $\mathcal{A}_R$ and y is an arrow $1_{\mathcal{E}/R} \to D_R(x)$ in the topos $\mathcal{E}/R$. On such a set we consider the equivalence relation $\mathcal{R}_R^D$ generated by the pairs $((x,y),(x',y'))$ with the property that there exists an arrow $f : x \to x'$ in the category $\mathcal{A}_R$ such that $y' = D_R(f) \circ y$.

The following remarks are useful in connection with the application of the localization technique.

Remarks 5.1.14 (a) For any $E \in \mathcal{E}$ we have a natural bijection $\mathcal{I}_E^D \cong \mathcal{I}_{1_{\mathcal{E}/E}}^{D/E}$ and the relation $\mathcal{R}_E^D$ corresponds to the relation $\mathcal{R}_{1_{\mathcal{E}/E}}^{D/E}$ under this bijection.

(b) For any arrow $f : F \to E$ in $\mathcal{E}$ and any object $x \in \mathcal{A}_E$, $D_F(\mathcal{A}_f(x)) = f^*(D_E(x))$ (since D is an $\mathcal{E}$-indexed functor). We shall denote the canonical arrow $\mathrm{dom}(D_F(\mathcal{A}_f(x))) \to \mathrm{dom}(D_E(x))$ in $\mathcal{E}$ by the symbol r_f.

(c) For any object x of $\mathcal{A}_E$ and any arrow $y : (f : F \to E) \to D_E(x)$ in the topos $\mathcal{E}/E$, there exists a unique arrow $y_f : (1_F : F \to F) \to D_F(\mathcal{A}_f(x))$ in the topos $\mathcal{E}/F$ such that $r_f \circ y_f = y$ (where r_f is regarded here as an arrow in $\mathcal{E}/E$ in the obvious way).

(d) For any $\mathcal{E}$-indexed cocone μ over D with vertex U, any object x of $\mathcal{A}_E$ and any arrow $y : (f : F \to E) \to D_E(x)$ in the topos $\mathcal{E}/E$, we have $\mu_E(x) \circ y = (1_U \times f) \circ \mu_F(\mathcal{A}_f(x)) \circ y_f$. Moreover, we have pullback squares

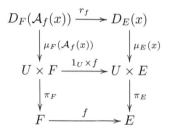

in $\mathcal{E}$, where π_E and π_F are the obvious canonical projections. In particular, for any other such pair (x', y'), $\mu_E(x) \circ y = \mu_E(x') \circ y'$ if and only if $\mu_F(\mathcal{A}_f(x)) \circ y_f = \mu_F(\mathcal{A}_f(x')) \circ y_f$ (where y_f is the arrow defined in Remark 5.1.14(c)).

(e) If $\mathcal{A}_\mathcal{E}$ is the $\mathcal{E}$-externalization of an internal category $\mathbb{C}$ in $\mathcal{E}$, x is a generalized element $E \to \mathbb{C}_0$ and y is an arrow $(f : F \to E) \to D_E(x)$ in $\mathcal{E}/E$, there exists a unique arrow $y_f : (1_F : F \to F) \to D_F(x \circ f)$ in $\mathcal{E}/F$ (provided by the universal property of the pullback square defining $D_F(x \circ f)$) such that $z_{x \circ f} \circ y_f = z_x \circ y$ (cf. section 5.1.3 for the notation), which coincides with the arrow y_f defined in Remark 5.1.14(c).

(f) If a pair $((x, y), (x', y'))$ belongs to $\mathcal{R}_R^D$ then, for any $\mathcal{E}$-indexed cocone λ over D with vertex V, $\lambda_R(x) \circ y = \lambda_R(x') \circ y'$. Indeed, for any pair $((x, y), (x', y'))$ with the property that there exists an arrow $f : x \to x'$ in the category $\mathcal{A}_R$ such that $y' = D_E(f) \circ y$, we have that $\lambda_R(x') \circ y' = \lambda_R(x') \circ D_E(f) \circ y = \lambda_R(x) \circ y$.

Lemma 5.1.15 *Let $\mathcal{E}$ be a Grothendieck topos, $\mathbb{C}$ an internal category in $\mathcal{E}$, $D \in [\mathbb{C}, \mathcal{E}]$ an internal diagram, $x : 1_\mathcal{E} \to \mathbb{C}_0$ and $f : 1_\mathcal{E} \to \mathbb{C}_1$ generalized elements such that $d_0^\mathbb{C} \circ f = x$ and $m : F \to D_{1_\mathcal{E}}(x)$ an arrow in $\mathcal{E}$. Then the unique arrow $\chi_{m,f} : F \to (\int^{\mathrm{opf}} D)_1$ in $\mathcal{E}$ such that $d_0^{\int^{\mathrm{opf}} D} \circ \chi_{m,f} = z_x \circ m$ and $(\pi_D^{\mathrm{opf}})_1 \circ \chi_{m,f} = f \circ !_F$ (provided by the universal property of the pullback square arising in the definition of the discrete opfibration associated with D – cf.*

Preliminary results on indexed colimits in toposes 159

section 5.1.3 for the notation) satisfies the property that $z_x \circ m = d_0^{\int^{\text{opf}} D} \circ \chi_{m,f}$ (by definition) and $z_{x'} \circ D_1(f) \circ m = d_1^{\int^{\text{opf}} D} \circ \chi_{m,f}$, where x' is equal to $d_1^{\mathbb{C}} \circ f$ and $d_0^{\int^{\text{opf}} D}, d_1^{\int^{\text{opf}} D}$ are respectively the domain and codomain arrows $(\int^{\text{opf}} D)_1 \to (\int^{\text{opf}} D)_0$.

Proof By definition of $D_1(f)$ we have that $z_{x'} \circ D_1(f) = d_1^{\int^{\text{opf}} D} \circ r \circ \xi$, where r is defined by the pullback square

$$\begin{array}{ccc} H & \xrightarrow{r} & (\int^{\text{opf}} D)_1 \\ {\scriptstyle !_H} \downarrow & & \downarrow {\scriptstyle (\pi_D^{\text{opf}})_1} \\ 1_{\mathcal{E}} & \xrightarrow{f} & \mathbb{C}_1 \end{array}$$

and ξ is the unique arrow $D_1(x) \to H$ such that $z_x = d_0^{\int^{\text{opf}} D} \circ r \circ \xi$, provided by the universal property of the pullback square

$$\begin{array}{ccc} H & \xrightarrow{d_0^{\int^{\text{opf}} D} \circ r} & (\int^{\text{opf}} D)_0 \\ {\scriptstyle s} \downarrow & & \downarrow {\scriptstyle (\pi_D^{\text{opf}})_0} \\ 1_{\mathcal{E}} & \xrightarrow{d_0^{\mathbb{C}} \circ f} & \mathbb{C}_0. \end{array}$$

It thus remains to show that $\chi_{m,f} = r \circ \xi \circ m$. But this immediately follows from the universal property of the pullback square

$$\begin{array}{ccc} (\int^{\text{opf}} D)_1 & \xrightarrow{d_0^{\int^{\text{opf}} D}} & (\int^{\text{opf}} D)_0 \\ {\scriptstyle (\pi_D^{\text{opf}})_1} \downarrow & & \downarrow {\scriptstyle (\pi_D^{\text{opf}})_0} \\ \mathbb{C}_1 & \xrightarrow{d_0^{\mathbb{C}}} & \mathbb{C}_0, \end{array}$$

since $d_0^{\int^{\text{opf}} D} \circ \chi_{m,f} = d_0^{\int^{\text{opf}} D} \circ (r \circ \xi \circ m)$ and $(\pi_D^{\text{opf}})_1 \circ \chi_{m,f} = (\pi_D^{\text{opf}})_1 \circ (r \circ \xi \circ m)$. □

Lemma 5.1.16 *Let $\mathcal{E}$ be a Grothendieck topos, $\mathbb{C}$ an internal category in $\mathcal{E}$, D an internal diagram in $[\mathbb{C}, \mathcal{E}]$ and χ an arrow $1_{\mathcal{E}} \to (\int^{\text{opf}} D)_1$. Let $x = (\pi_D^{\text{opf}})_0 \circ d_0^{\int^{\text{opf}} D} \circ \chi$, $x' = (\pi_D^{\text{opf}})_0 \circ d_1^{\int^{\text{opf}} D} \circ \chi$ and $f = (\pi_D^{\text{opf}})_1 \circ \chi$. Then $d_0^{\mathbb{C}} \circ f = x$, $d_1^{\mathbb{C}} \circ f = x'$, and $D_1(f) \circ \chi_0 = \chi_1$, where χ_0 is the unique arrow $1_{\mathcal{E}} \to D_1(x)$ such that $z_x \circ \chi_0 = d_0^{\int^{\text{opf}} D} \circ \chi$ (which exists by the universal property of the pullback square defining $D_1(x)$) and χ_1 is the unique arrow $1_{\mathcal{E}} \to D_1(x')$ such that $z_{x'} \circ \chi_1 = d_1^{\int^{\text{opf}} D} \circ \chi$ (which exists by the universal property of the pullback square defining $D_1(x')$).*

Proof The first two identities follow immediately from the definition of π_D^{opf}. It remains to prove that $D_1(f) \circ \chi_0 = \chi_1$. Let us employ the notation of the proof of Lemma 5.1.15. By the universal property of the pullback square defining $D_1(x')$, it is equivalent to verify that $z_{x'} \circ D_1(f) \circ \chi_0 = z_{x'} \circ \chi_1$. But $z_{x'} \circ D_1(f) \circ \chi_0 = d_1^{\int^{\text{opf}} D} \circ r \circ \xi \circ \chi_0$, while $z_{x'} \circ \chi_1 = d_1^{\int^{\text{opf}} D} \circ \chi$; so to prove the desired equality it suffices to show that $r \circ \xi \circ \chi_0 = \chi$. This follows from the fact that $d_0^{\int^{\text{opf}} D} \circ r \circ \xi \circ \chi_0 = z_x \circ \chi_0 = d_0^{\int^{\text{opf}} D} \circ \chi$ by virtue of the universal property of the pullback square given by the domain of the discrete opfibration associated with D. □

Proposition 5.1.17 *Let $\mathcal{E}$ be a Grothendieck topos, $\mathbb{C}$ an internal category in $\mathcal{E}$, $D \in [\mathbb{C}, \mathcal{E}]$ an internal diagram, $x, x' : 1_\mathcal{E} \to \mathbb{C}_0$ generalized elements and $y : 1_\mathcal{E} \to D_1(x)$, $y' : 1_\mathcal{E} \to D_1(x')$ arrows in $\mathcal{E}$. Then $(z_x \circ y, z_{x'} \circ y')$ belongs to the equivalence relation on $\text{Hom}_\mathcal{E}(1_\mathcal{E}, (\int^{\text{opf}} D)_0)$ generated by the pairs of the form $(d_0^{\int^{\text{opf}} D} \circ a, d_1^{\int^{\text{opf}} D} \circ a)$ for some arrow $a : 1_\mathcal{E} \to (\int^{\text{opf}} D)_1$ if and only if $((x,y),(x',y'))$ belongs to $\mathcal{R}_{1_\mathcal{E}}^D$.*

Proof Let us first prove the 'if' direction. We have that $((x,y),(x',y'))$ belongs to $\mathcal{R}_{1_\mathcal{E}}^D$ if and only if there exists a finite sequence of pairs

$$(x_0, y_0) = (x, y), \ldots, (x_n, y_n) = (x', y')$$

such that for any $i \in \{0, \ldots, n-1\}$

(1) either there exists $f_i : 1_\mathcal{E} \to \mathbb{C}_1$ such that $d_0^\mathbb{C} \circ f_i = x_i$, $d_1^\mathbb{C} \circ f_i = x_{i+1}$ and $D_1(f_i) \circ y_i = y_{i+1}$ or

(2) there exists $g_i : 1_\mathcal{E} \to \mathbb{C}_1$ such that $d_0^\mathbb{C} \circ g_i = x_{i+1}$, $d_1^\mathbb{C} \circ g_i = x_i$ and $D_1(g_i) \circ y_{i+1} = y_i$.

In case (1), Lemma 5.1.15 implies (by taking m to be y_i and f to be f_i) that $z_{x_i} \circ y_i = d_0^{\int^{\text{opf}} D} \circ \chi_{y_i, f_i}$ and $z_{x_{i+1}} \circ y_{i+1} = d_1^{\int^{\text{opf}} D} \circ \chi_{y_i, f_i}$ (here the notation is that of the Lemma). In case (2), Lemma 5.1.15 implies (by taking m to be y_{i+1} and f to be g_i) that $z_{x_{i+1}} \circ y_{i+1} = d_0^{\int^{\text{opf}} D} \circ \chi_{y_{i+1}, g_i}$ and $z_{x_i} \circ y_i = d_1^{\int^{\text{opf}} D} \circ \chi_{y_{i+1}, g_i}$.

From this it clearly follows that $(z_x \circ y, z_{x'} \circ y')$ belongs to the equivalence relation T on $\text{Hom}_\mathcal{E}(1_\mathcal{E}, (\int^{\text{opf}} D)_0)$ generated by the pairs of the form $(d_0^{\int^{\text{opf}} D} \circ a, d_1^{\int^{\text{opf}} D} \circ a)$ for some arrow $a : 1_\mathcal{E} \to (\int^{\text{opf}} D)_1$, as required.

Let us now prove the 'only if' direction. If $(z_x \circ y, z_{x'} \circ y')$ belongs to T, there exists a finite sequence $\xi_0, \ldots, \xi_n$ of arrows $1_\mathcal{E} \to (\int^{\text{opf}} D)_1$ and of arrows $e_0, \ldots, e_n$ (which are either $d_0^{\int^{\text{opf}} D}$ or $d_1^{\int^{\text{opf}} D}$ – we denote by e_i^{op} the arrow $d_1^{\int^{\text{opf}} D}$ if e_i is $d_0^{\int^{\text{opf}} D}$ and the arrow $d_0^{\int^{\text{opf}} D}$ if e_i is $d_1^{\int^{\text{opf}} D}$) such that $e_0 \circ \xi_0 = z_x \circ y$, $e_{i+1} \circ \xi_{i+1} = e_i^{\text{op}} \circ \xi_i$ for all $i \in \{0, \ldots, n-1\}$ and $e_n \circ \xi_n = z_{x'} \circ y'$.

To deduce our thesis, it suffices to apply Lemma 5.1.16 noticing that if $e_i \circ \chi = z_{x''} \circ y''$ then $\chi_0 = y''$ if e_i is $d_0^{\int^{\text{opf}} D}$ and $\chi_1 = y''$ if e_i is $d_1^{\int^{\text{opf}} D}$ (where χ_0 and

χ_1 are the arrows defined in the statement of Lemma 5.1.16) since $z_{x''} \circ \chi_0 = d_0^{\int^{\mathrm{opf}} D} \circ \chi = z_{x''} \circ y''$ (if e_i is $d_0^{\int^{\mathrm{opf}} D}$) and $z_{x''} \circ \chi_1 = d_1^{\int^{\mathrm{opf}} D} \circ \chi = z_{x''} \circ y''$ (if e_i is $d_1^{\int^{\mathrm{opf}} D}$). □

Theorem 5.1.18 *Let $D : \underline{\mathcal{A}_\mathcal{E}} \to \underline{\mathcal{E}_\mathcal{E}}$ be an $\mathcal{E}$-indexed functor, where $\underline{\mathcal{A}_\mathcal{E}}$ is equivalent to the $\mathcal{E}$-externalization of an internal category in $\mathcal{E}$. Then an $\mathcal{E}$-indexed cocone μ over D with vertex U is an indexed colimiting cocone for D if and only if the following conditions are satisfied:*

(i) *For any arrow $h : F \to U$ in $\mathcal{E}$, there exist an epimorphic family $\{f_i : F_i \to F \mid i \in I\}$ in $\mathcal{E}$ and for each $i \in I$ an object $x_i \in \mathcal{A}_{F_i}$ and an arrow $\alpha_i : 1_{\mathcal{E}/F_i} \to D_{F_i}(x_i)$ in the topos $\mathcal{E}/F_i$ such that $<h \circ f_i, 1_{F_i}> = \mu_{F_i}(x_i) \circ \alpha_i$ as arrows $1_{\mathcal{E}/F_i} \to F_i^*(U)$ in $\mathcal{E}/F_i$.*

(ii) *For any pairs (x, y) and (x', y'), where x and x' are objects of $\mathcal{A}_E$, y is an arrow $(f : F \to E) \to D_E(x)$ in $\mathcal{E}/E$ and y' is an arrow $(f : F \to E) \to D_E(x')$ in $\mathcal{E}/E$, we have $\mu_E(x) \circ y = \mu_E(x') \circ y'$ if and only if there exists an epimorphic family $\{f_i : F_i \to F \mid i \in I\}$ in $\mathcal{E}$ such that the pair $((\mathcal{A}_{f \circ f_i}(x), f_i^*(y)), (\mathcal{A}_{f \circ f_i}(x'), f_i^*(y')))$ belongs to the relation $\mathcal{R}_{F_i}^D$.*

Proof First, let us prove that, under the assumption that $\underline{\mathcal{A}_\mathcal{E}}$ is the $\mathcal{E}$-externalization of an internal category $\mathbb{C}$ in $\mathcal{E}$, any colimiting cocone μ for D satisfies the two conditions of the theorem.

Under this assumption, condition (i) for μ rewrites as follows: for any object F of $\mathcal{E}$ and arrow $h : F \to \mathrm{colim}(D)$ in $\mathcal{E}$, there exist an epimorphic family $\{f_i : F_i \to F \mid i \in I\}$ in $\mathcal{E}$ and for each $i \in I$ a generalized element $x_i : F_i \to \mathbb{C}_0$ and an arrow $\alpha_i : 1_{\mathcal{E}/F_i} \to D_{F_i}(x_i)$ in the topos $\mathcal{E}/F_i$ such that $<h \circ f_i, 1_{F_i}> = c \circ z_{x_i} \circ \alpha_i$ as arrows $1_{\mathcal{E}/F_i} \to !_{F_i}^*(\mathrm{colim}(D))$ in $\mathcal{E}/F_i$ (where the notation is that of section 5.1.1).

Consider the following pullback square:

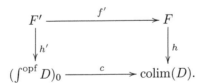

Since c is an epimorphism, $f' : F' \to F$ is also an epimorphism. This arrow will form, by itself, the unique element of an epimorphic family satisfying condition (i). We set x' equal to the composite $(\pi_D^{\mathrm{opf}})_0 \circ h'$ (recall that $(\pi_D^{\mathrm{opf}})_0 : (\int^{\mathrm{opf}} D)_0 \to \mathbb{C}_0$ is the object component of the discrete opfibration associated with D). Consider the following pullback square defining $D_{F'}(x')$:

$$\begin{array}{ccc} D_{F'}(x') & \xrightarrow{u} & (\int^{\mathrm{opf}} D)_0 \\ {\scriptstyle v}\downarrow & & \downarrow{\scriptstyle (\pi_D^{\mathrm{opf}})_0} \\ F' & \xrightarrow{x'} & \mathbb{C}_0. \end{array}$$

By the universal property of this pullback diagram, there exists a unique arrow $\alpha_{x'} : F' \to D_{F'}(x')$ in $\mathcal{E}$ such that $u \circ \alpha_{x'} = h'$ and $v \circ \alpha_{x'} = 1_{F'}$. So $\alpha_{x'}$ is an arrow $1_{\mathcal{E}/F'} \to D_{F'}(x')$ in the topos $\mathcal{E}/F'$ and $<h \circ f', 1_{F'}> = <c \circ u \circ \alpha_{x'}, v \circ \alpha_{x'}>$, as required.

Under the assumption that $\underline{A}_{\mathcal{E}}$ is the $\mathcal{E}$-externalization of an internal category $\mathbb{C}$ in $\mathcal{E}$, condition (ii) for μ rewrites as follows: for any pairs (x, y) and (x', y'), where x and x' are generalized elements $E \to \mathbb{C}_0$, y is an arrow $(f : F \to E) \to D_E(x)$ in $\mathcal{E}/E$ and y' is an arrow $(f : F \to E) \to D_E(x')$ in $\mathcal{E}/E$, we have $c \circ z_x \circ y = c \circ z_{x'} \circ y'$ if and only if there exists an epimorphic family $\{f_i : F_i \to F \mid i \in I\}$ in $\mathcal{E}$ such that the pair $((x \circ f \circ f_i, y \circ f_i), (x' \circ f \circ f_i, y' \circ f_i))$ belongs to the relation $\mathcal{R}^D_{F_i}$ for each $i \in I$.

Thanks to the localization technique, we can suppose $E = 1_{\mathcal{E}}$ without loss of generality. Indeed, condition (ii) for the diagram D, the cocone μ and the object E is equivalent to condition (ii) for the diagram D/E, the cocone μ/E and the object $1_{\mathcal{E}/E}$, and μ is, by our assumption, an indexed colimiting cocone and hence stable under localization.

We have that $c \circ z_x \circ y = c \circ z_{x'} \circ y'$ if and only if there exists an epimorphic family $\{f_i : F_i \to F \mid i \in I\}$ in $\mathcal{E}$ such that, for each $i \in I$, $(z_x \circ y \circ f_i, z_{x'} \circ y' \circ f_i)$ belongs to the equivalence relation on the set $\mathrm{Hom}_{\mathcal{E}}(F_i, (\int^{\mathrm{opf}} D)_0)$ generated by the set of pairs of the form $(d_0^{\int^{\mathrm{opf}} D} \circ a, d_1^{\int^{\mathrm{opf}} D} \circ a)$ for a generalized element $a : F_i \to (\int^{\mathrm{opf}} D)_0$ in $\mathcal{E}$ (cf. Lemma 5.1.28 below).

By Remark 5.1.14(e), we have that $(z_x \circ y \circ f_i, z_{x'} \circ y' \circ f_i) = (z_{x \circ f \circ f_i} \circ (y \circ f_i), z_{x' \circ f \circ f_i} \circ (y' \circ f_i))$. Our thesis thus follows from Proposition 5.1.17, applied to the toposes $\mathcal{E}/F_i$ and the pairs $((x \circ f \circ f_i, y \circ f_i), (x' \circ f \circ f_i, y' \circ f_i))$.

Let us now prove that the conditions of the theorem are sufficient for an indexed cocone μ over D to be an indexed colimiting cocone. For this part of the theorem, we shall not need to assume the $\mathcal{E}$-indexed category $\underline{A}_{\mathcal{E}}$ to be small.

Since conditions (i) and (ii) are both stable under localization, it suffices to prove that μ is a colimiting cocone over the diagram D. Suppose that λ is an indexed cocone over D with vertex V. We have to prove that there exists a unique arrow $l : U \to V$ in $\mathcal{E}$ such that for any $E \in \mathcal{E}$ and $x \in A_E$, $(l \times 1_E) \circ \mu_E(x) = \lambda_E(x)$. To define the arrow l we shall define a function $L : \mathrm{Hom}_{\mathcal{E}}(E, U) \to \mathrm{Hom}_{\mathcal{E}}(E, V)$ natural in $E \in \mathcal{E}$. Given $h \in \mathrm{Hom}_{\mathcal{E}}(E, U)$, by condition (i) in the statement of the theorem, there exist an epimorphic family $\mathcal{I} = \{e_i : E_i \to E \mid i \in I\}$ in $\mathcal{E}$ and for each $i \in I$ an object x_i of A_{E_i} and an arrow $\alpha_i : 1_{\mathcal{E}/E_i} \to D_{E_i}(x_i)$ in $\mathcal{E}/E_i$ such that $<h \circ e_i, 1_{E_i}> = \mu_{E_i}(x_i) \circ \alpha_i$. The family of arrows $\{\lambda_{E_i}(x_i) \circ \alpha_i \mid i \in I\}$ satisfies the hypotheses of Corollary 5.1.7. Indeed, using the notation of the corollary, we have that $q_i^*(\lambda_{E_i}(x_i) \circ \alpha_i) = q_j^*(\lambda_{E_j}(x_j) \circ \alpha_j)$ for all $i, j \in I$; this can be easily proved by using condition (ii) of the theorem, the fact that the same identity holds for the cocone μ (in place of λ) and Remark 5.1.14(d). Therefore there exists a unique arrow $L_h^{\mathcal{I}} : E \to V$ such that $<L_h^{\mathcal{I}} \circ e_i, 1_{E_i}> = \lambda_{E_i}(x_i) \circ \alpha_i$ for all $i \in I$. In order to be able to define $L(h)$ as equal to $L_h^{\mathcal{I}}$ we need to check that for any other epimorphic family

$\mathcal{J} = \{e'_j : E'_j \to E \mid j \in J\}$ in $\mathcal{E}$, we have $L_h^{\mathcal{I}} = L_h^{\mathcal{J}}$. To this end, consider the fibred product of the epimorphic families $\mathcal{I}$ and $\mathcal{J}$, that is the family of arrows $p_{i,j} = e_i \circ f_i = e'_j \circ f'_j : F_{i,j} \to E$ (for $(i,j) \in I \times J$), where the arrows f_i and f'_j are defined by the following pullback square:

$$\begin{array}{ccc} F_{i,j} & \xrightarrow{f_i} & E_i \\ {\scriptstyle f'_j}\downarrow & & \downarrow{\scriptstyle e_i} \\ E'_j & \xrightarrow{e'_j} & E. \end{array}$$

It clearly suffices to verify that for any $i \in I, j \in J$, we have $L_h^{\mathcal{I}} \circ p_{i,j} = L_h^{\mathcal{J}} \circ p_{i,j}$. Now, we have that $< L_h^{\mathcal{I}} \circ e_i, 1_{E_i} >= \lambda_{E_i}(x_i) \circ \alpha_i$, while $< L_h^{\mathcal{J}} \circ e'_j, 1_{E'_j} >= \lambda_{E'_j}(x'_j) \circ \beta_j$. Applying respectively f_i^* and $f_j'^*$ to these two identities, we obtain that $< L_h^{\mathcal{I}} \circ p_{i,j}, 1_{F_{i,j}} >= \lambda_{F_{i,j}}(x_{i,j}) \circ f_i^*(\alpha_i)$ and $< L_h^{\mathcal{J}} \circ p_{i,j}, 1_{F_{i,j}} >= \lambda_{F_{i,j}}(x'_{i,j}) \circ f_j'^*(\beta_j)$, where $x_{i,j} = \mathcal{A}_{f_i}(x_i)$ and $x'_{i,j} = \mathcal{A}_{f'_j}(x'_j)$, whence the condition $L_h^{\mathcal{I}} \circ p_{i,j} = L_h^{\mathcal{J}} \circ p_{i,j}$ is equivalent to the condition $\lambda_{F_{i,j}}(x_{i,j}) \circ f_i^*(\alpha_i) = \lambda_{F_{i,j}}(x'_{i,j}) \circ f_j'^*(\beta_j)$. This equality can be proved by using condition (ii) of the theorem, by using the fact that the same identity holds with the cocone μ in place of λ and Remark 5.1.14(d). More specifically, the identities $< h \circ e_i, 1_{E_i} >= \mu_{E_i}(x_i) \circ \alpha_i$ and $< h \circ e'_j, 1_{E'_j} >= \mu_{E'_j}(x'_j) \circ \beta_i$ imply, applying respectively f_i^* and $f_j'^*$ to them, the identity $\mu_{F_{i,j}}(x_{i,j}) \circ f_i^*(\alpha_i) = \mu_{F_{i,j}}(x'_{i,j}) \circ f_j'^*(\beta_j)$. From condition (ii) of the theorem, it thus follows that for any $(i,j) \in I \times J$, there exists an epimorphic family $\{g_k^{i,j} : G_k^{i,j} \to F_{i,j} \mid k \in K_{i,j}\}$ in $\mathcal{E}$ such that the pair $((\mathcal{A}_{g_k^{i,j}}(x_{i,j}), f_i^*(\alpha_i) \circ g_k^{i,j}), (\mathcal{A}_{g_k^{i,j}}(x'_{i,j}), f_j'^*(\beta_j) \circ g_k^{i,j}))$ belongs to the relation $\mathcal{R}_{G_k^{i,j}}^D$. This in turn implies (by Remark 5.1.14(d)) that $\lambda_{G_k^{i,j}}(\mathcal{A}_{g_k^{i,j}}(x_{i,j})) \circ (f_i^*(\alpha_i) \circ g_k^{i,j}) = \lambda_{G_k^{i,j}}(\mathcal{A}_{g_k^{i,j}}(x'_{i,j})) \circ (f_j'^*(\beta_j) \circ g_k^{i,j})$. But $\lambda_{G_k^{i,j}}(\mathcal{A}_{g_k^{i,j}}(x_{i,j})) \circ (f_i^*(\alpha_i) \circ g_k^{i,j}) = (1_V \times g_k^{i,j})^*(\lambda_{F_{i,j}}(x_{i,j}) \circ f_i^*(\alpha_i))$ and $\lambda_{G_k^{i,j}}(\mathcal{A}_{g_k^{i,j}}(x'_{i,j})) \circ (f_j'^*(\beta_j) \circ g_k^{i,j}) = (1_V \times g_k^{i,j})^*(\lambda_{F_{i,j}}(x'_{i,j}) \circ f_j'^*(\beta_j))$, whence $(1_V \times g_k^{i,j})^*(\lambda_{F_{i,j}}(x_{i,j}) \circ f_i^*(\alpha_i)) = (1_V \times g_k^{i,j})^*(\lambda_{F_{i,j}}(x'_{i,j}) \circ f_j'^*(\beta_j))$, equivalently (since the $g_k^{i,j}$ are jointly epimorphic) $\lambda_{F_{i,j}}(x_{i,j}) \circ f_i^*(\alpha_i) = \lambda_{F_{i,j}}(x'_{i,j}) \circ f_j'^*(\beta_j)$, as required.

It is clear that the assignment $h \to L(h)$ defined above is natural in $E \in \mathcal{E}$; it therefore remains to show that the resulting arrow $l : U \to V$ satisfies the required property that for any $E \in \mathcal{E}$ and any $x \in \mathcal{A}_E$, $(l \times 1_E) \circ \mu_E(x) = \lambda_E(x)$. Take s to be the canonical arrow $s : D_E(x) \to E$ in $\mathcal{E}$ and set $x' = \mathcal{A}_s(x)$, $E' = D_E(x)$; then $x' \in \mathcal{A}_{E'}$ and, considering the pullback square

$$\begin{array}{ccc} D_{E'}(x') & \xrightarrow{r} & D_E(x) \\ {\scriptstyle t}\downarrow & & \downarrow{\scriptstyle s} \\ E' & \xrightarrow{s} & E, \end{array}$$

the unique arrow $\alpha : E' \to D_{E'}(x')$ such that $r \circ \alpha = 1_{E'}$ and $t \circ \alpha = 1_{E'}$ satisfies the property that $< h, 1_{E'} > = \mu_{E'}(x') \circ \alpha$. So the epimorphic family $\{1_{E'} : E' \to E'\}$ satisfies condition (i) of the theorem with respect to the arrow $h = \pi_U \circ \mu_E(x)$, where $\pi_U : U \times E \to U$ is the canonical projection, and hence $l \circ h = L(h) = \lambda_{E'}(x') \circ \alpha$; therefore $(l \times 1_E) \circ \mu_E(x) = \lambda_E(x)$, as required. □

Remarks 5.1.19 (a) The proof of the theorem shows that the sufficiency of its conditions holds more generally for any (i.e. not necessarily small) $\mathcal{E}$-indexed category $\underline{\mathcal{A}}_\mathcal{E}$.

(b) The conditions in the statement of the theorem are both stable under localization; that is, if the cocone μ over the diagram D satisfies them then, for any $E \in \mathcal{E}$, the cocone μ/E over the diagram D/E does so too.

Proposition 5.1.20 *Let $\underline{\mathcal{B}}_\mathcal{E}$ be an $\mathcal{E}$-strictly final subcategory of an indexed category $\underline{\mathcal{A}}_\mathcal{E}$, $D : \underline{\mathcal{A}}_\mathcal{E} \to \underline{\mathcal{E}}_\mathcal{E}$ an $\mathcal{E}$-indexed functor and μ an $\mathcal{E}$-indexed cocone over D. Then μ satisfies the conditions of Theorem 5.1.18 with respect to the diagram D if and only if $\mu \circ i$ satisfies them with respect to the diagram $D \circ i$, where i is the canonical embedding $\underline{\mathcal{B}}_\mathcal{E} \hookrightarrow \underline{\mathcal{A}}_\mathcal{E}$.*

Proof Suppose that μ satisfies the conditions of Theorem 5.1.18 with respect to the diagram D. The fact that $\mu \circ i$ satisfies condition (i) of the theorem with respect to the diagram $D \circ i$ immediately follows from the fact that μ does so with respect to the diagram D, by using the fact that for any object $x_i \in \mathcal{A}_{F_i}$ there exists an epimorphic family $\{g_k^i : G_k^i \to F_i \mid k \in K_i\}$ such that $\mathcal{A}_{g_k^i}(x_i)$ lies in $\mathcal{B}_{G_k^i}$ for each $k \in K_i$. Let us now show that $\mu \circ i$ satisfies condition (ii) with respect to the diagram $D \circ i$.

Let us first establish the following fact $(*)$: for any pairs (x, y) and (x', y') in $\mathcal{I}_R^D$, if $((x, y), (x', y')) \in \mathcal{R}_R^D$, there exists an epimorphic family $\{r_i : R_i \to R \mid i \in I\}$ in $\mathcal{E}$ such that $((\mathcal{A}_{r_i}(x), r_i^*(y)), (\mathcal{A}_{r_i}(x'), r_i^*(y'))) \in \mathcal{R}_{R_i}^{D \circ i}$ for each $i \in I$.

If $((x, y), (x', y')) \in \mathcal{R}_R^D$, there exists a finite sequence

$$(x, y) = (x_0, y_0), \ldots, (x_n, y_n) = (x', y')$$

of pairs in $\mathcal{I}_R^D$ such that for any $j \in \{0, \ldots, n-1\}$, either there exists an arrow $f_j : x_j \to x_{j+1}$ in $\mathcal{A}_R$ such that $y_{j+1} = D_R(f_j) \circ y_j$ or there exists an arrow $f_j : x_{j+1} \to x_j$ in $\mathcal{A}_R$ such that $y_j = D_R(f_j) \circ y_{j+1}$. By applying the definition of an $\mathcal{E}$-strictly final subcategory a finite number of times (using the fact that the fibred product of epimorphic families is again an epimorphic family), we can find an epimorphic family $\{r_i : R_i \to R \mid i \in I\}$ in $\mathcal{E}$ such that for every $i \in I$ and every $j \in \{0, \ldots, n-1\}$ the arrow $\mathcal{A}_{r_i}(f_j)$ lies in $\mathcal{B}_{R_i}$. From this it immediately follows that $((\mathcal{A}_{r_i}(x), r_i^*(y)), (\mathcal{A}_{r_i}(x'), r_i^*(y'))) \in \mathcal{R}_{R_i}^{D \circ i}$, as required.

Using $(*)$, the proof of the fact that $\mu \circ i$ satisfies condition (ii) of the theorem with respect to the diagram $D \circ i$ follows readily from the fact that μ satisfies it with respect to the diagram D.

Conversely, let us suppose that $\mu \circ i$ satisfies the conditions of the theorem with respect to $D \circ i$ and deduce that μ satisfies them with respect to D. The fact that the validity of condition (i) for $\mu \circ i$ with respect to $D \circ i$ implies the validity of condition (i) for μ with respect to D is obvious. Concerning condition (ii) for (μ, D), this can be deduced from condition (ii) for $(\mu \circ i, D \circ i)$ by using the fact that, given (x, y) and (x', y') as in the statement of the condition for (μ, D), there exists an epimorphic family $\{e_j : E_j \to E \mid j \in J\}$ in $\mathcal{E}$ such that $\mathcal{A}_{e_j}(x)$ and $\mathcal{A}_{e_j}(x')$ lie in $\mathcal{B}_{E_j}$ for all $j \in J$. □

Remark 5.1.21 The proposition shows, in view of Theorem 5.1.18, that for any $\mathcal{E}$-strictly final subcategory $\underline{\mathcal{B}}_\mathcal{E}$ of a small indexed category $\underline{\mathcal{A}}_\mathcal{E}$ (in particular, any $\mathcal{E}$-indexed subcategory of the form $\int F_\mathcal{E} \hookrightarrow \int \overline{F}^{\mathrm{f}}_\mathcal{E}$ for a functor $F : \mathcal{C}^{\mathrm{op}} \to \mathcal{E}$ as in Theorem 5.1.12), any $\mathcal{E}$-indexed functor $D : \underline{\mathcal{A}}_\mathcal{E} \to \underline{\mathcal{E}}_\mathcal{E}$ and any $\mathcal{E}$-indexed cocone μ over D, μ is an indexed colimiting cocone for D if and only if $\mu \circ i$ satisfies the conditions of Theorem 5.1.18 with respect to the diagram $D \circ i$, where i is the canonical embedding $\underline{\mathcal{B}}_\mathcal{E} \hookrightarrow \underline{\mathcal{A}}_\mathcal{E}$.

Proposition 5.1.22 Let $D : \underline{\mathcal{A}}_\mathcal{E} \to \underline{\mathcal{E}}_\mathcal{E}$ be an indexed functor, where $\underline{\mathcal{A}}_\mathcal{E}$ is an $\mathcal{E}$-filtered category, R an object of $\mathcal{E}$ and $(x, y), (x', y')$ pairs in $\mathcal{I}^D_R$. Then there exists an epimorphic family $\{r_i : R_i \to R \mid i \in I\}$ in $\mathcal{E}$ such that, for every $i \in I$, $((\mathcal{A}_{r_i}(x), r_i^*(y)), (\mathcal{A}_{r_i}(x'), r_i^*(y'))) \in \mathcal{R}^D_{R_i}$ if and only if there exists an epimorphic family $\{e_j : E_j \to R \mid j \in J\}$ in $\mathcal{E}$ with the property that for any $j \in J$ there exist arrows $f_j : \mathcal{A}_{e_j}(x) \to z$ and $g_j : \mathcal{A}_{e_j}(x') \to z$ in the category $\mathcal{A}_{E_j}$ such that $D_{E_j}(f_j) \circ e_j^*(y) = D_{E_j}(g_j) \circ e_j^*(y')$.

Proof The 'if' direction is obvious, so it remains to prove the 'only if' one. It clearly suffices to show, by induction on the length of a sequence $(x_1, y_1), \ldots, (x_n, y_n)$ of pairs in $\mathcal{I}^D_R$ with the property that for any $i \in \{1, \ldots, n-1\}$ either there exists an arrow $f_j : x_j \to x_{j+1}$ in $\mathcal{A}_R$ such that $y_{j+1} = D_R(f_j) \circ y_j$ or there exists an arrow $f_j : x_{j+1} \to x_j$ in $\mathcal{A}_R$ such that $y_j = D_R(f_j) \circ y_{j+1}$, that there exists an epimorphic family $\{r_i : R_i \to R \mid i \in I\}$ in $\mathcal{E}$ such that for any $i \in I$ there are arrows $f_i : \mathcal{A}_{r_i}(x_1) \to z$ and $g_i : \mathcal{A}_{r_i}(x_n) \to z$ in the category $\mathcal{A}_{R_i}$ satisfying $D_{R_i}(f_i) \circ r_i^*(y_1) = D_{R_i}(g_i) \circ r_i^*(y_n)$. The case $n = 1$ is obvious. Suppose now that the claim is valid for n and show that it holds for $n + 1$. There exists an epimorphic family $\{e_j : E_j \to R \mid j \in J\}$ in $\mathcal{E}$ such that for any $j \in J$ there are arrows $f_j : \mathcal{A}_{e_j}(x_1) \to z_j$ and $g_j : \mathcal{A}_{e_j}(x_n) \to z_j$ in $\mathcal{A}_{E_j}$ such that $D_{E_j}(f_j) \circ e_j^*(y_1) = D_{E_j}(g_j) \circ e_j^*(y_n)$. On the other hand, there exists either an arrow $g : x_n \to x_{n+1}$ in $\mathcal{A}_R$ such that $D_R(g) \circ e_j^*(y_n) = e_j^*(y_{n+1})$ or an arrow $h : x_{n+1} \to x_n$ in $\mathcal{A}_R$ such that $D_R(h) \circ e_j^*(y_{n+1}) = e_j^*(y_n)$. In the latter case, our thesis follows at once; so we can concentrate on the case where there exists an arrow $g : x_n \to x_{n+1}$ in $\mathcal{A}_R$ such that $D_R(g) \circ e_j^*(y_n) = e_j^*(y_{n+1})$.

By definition of an $\mathcal{E}$-filtered category, for each $j \in J$ there exist an epimorphic family $\{f^j_k : F^j_k \to E_j \mid k \in K_j\}$ in $\mathcal{E}$ and for each $k \in K_j$ arrows $m^j_k : \mathcal{A}_{f^j_k}(z) \to w^j_k$ and $n^j_k : \mathcal{A}_{f^j_k}(\mathcal{A}_{e_j}(x_{n+1})) \to w^j_k$ such that $D_{F^j_k}(m^j_k) \circ D_{F^j_k}(\mathcal{A}_{f^j_k}(g_j)) \circ (f^j_k \circ e_j)^*(y_n) = D_{F^j_k}(n^j_k) \circ D_{F^j_k}(\mathcal{A}_{f^j_k}(g)) \circ (f^j_k \circ e_j)^*(y_n)$. Again, by definition of an $\mathcal{E}$-filtered category, for each $j \in J$ and $k \in K_j$ there exist an epimorphic family $\{g_l^{k,j} : G_l^{k,j} \to F_k^j \mid l \in L_{k,j}\}$ in $\mathcal{E}$ and for each $l \in L_{k,j}$ an arrow $p_l^{k,l} : \mathcal{A}_{g_l^{k,j}}(w_k^j) \to v_l^{k,j}$ such that $p_l^{k,l} \circ \mathcal{A}_{g_l^{k,j}}(m_k^j \circ \mathcal{A}_{f_k^j}(g_j)) = p_l^{k,j} \circ \mathcal{A}_{g_l^{k,j}}(n_k^j \circ \mathcal{A}_{f_k^j}(\mathcal{A}_{e_j}(g)))$. For any $j \in J, k \in K_j$ and $l \in L_{k,j}$, set $a_{j,k,l} : \mathcal{A}_{e_j \circ f_k^j \circ g_l^{k,j}}(x_1) \to v_l^{k,j}$ equal to $p_l^{k,j} \circ \mathcal{A}_{g_l^{j,k}}(m_k^j) \circ \mathcal{A}_{f_k^j \circ g_l^{k,j}}(f_j)$ and $b_{j,k,l} : \mathcal{A}_{e_j \circ f_k^j \circ g_l^{k,j}}(x_{n+1}) \to v_l^{k,j}$ equal to $p_l^{k,j} \circ \mathcal{A}_{g_l^{j,k}}(n_k^j)$. It is readily seen that $D_{G_l^{k,j}}(a_{j,k,l}) \circ (e_j \circ f_k^j \circ g_l^{k,j})^*(y_1) = D_{G_l^{k,j}}(b_{j,k,l}) \circ (e_j \circ f_k^j \circ g_l^{k,j})^*(y_{n+1})$, whence the required condition is satisfied (by taking as epimorphic family $\{e_j \circ f_k^j \circ g_l^{k,j} \mid j \in J, k \in K_j, l \in L_{k,j}\}$). □

Combining Theorem 5.1.18 with Proposition 5.1.22 in view of Remark 5.1.21, we immediately obtain the following result:

Corollary 5.1.23 *Let $D : \underline{\mathcal{A}_{\mathcal{E}}} \to \underline{\mathcal{E}_{\mathcal{E}}}$ be an indexed functor, where $\underline{\mathcal{A}_{\mathcal{E}}}$ is an $\mathcal{E}$-filtered, $\mathcal{E}$-strictly final subcategory of a small $\mathcal{E}$-indexed category. Then an $\mathcal{E}$-indexed cocone μ over D with vertex U is an indexed colimiting cocone for D if and only if the following conditions are satisfied:*

(i) *For any arrow $h : F \to U$ in $\mathcal{E}$, there exist an epimorphic family $\{f_i : F_i \to F \mid i \in I\}$ in $\mathcal{E}$ and for each $i \in I$ an object $x_i \in \mathcal{A}_{F_i}$ and an arrow $\alpha_i : 1_{\mathcal{E}/F_i} \to D_{F_i}(x_i)$ in $\mathcal{E}/F_i$ such that $< h \circ f_i, 1_{F_i} > = \mu_{F_i}(x_i) \circ \alpha_i$ as arrows $1_{\mathcal{E}/F_i} \to F_i^*(U)$ in $\mathcal{E}/F_i$.*

(ii) *For any object E of $\mathcal{E}$ and any pairs (x, y) and (x', y'), where x and x' are objects of $\mathcal{A}_E$, y is an arrow $(f : F \to E) \to D_E(x)$ in $\mathcal{E}/E$ and y' is an arrow $(f : F \to E) \to D_E(x')$ in $\mathcal{E}/E$, we have that $\mu_E(x) \circ y = \mu_E(x') \circ y'$ if and only if there exist an epimorphic family $\{f_i : F_i \to F \mid i \in I\}$ in $\mathcal{E}$ and for each $i \in I$ arrows $g_i : \mathcal{A}_{f \circ f_i}(x) \to z_i$ and $h_i : \mathcal{A}_{f \circ f_i}(x') \to z_i$ in $\mathcal{A}_{F_i}$ such that $D_{F_i}(g_i) \circ f_i^*(y) = D_{F_i}(h_i) \circ f_i^*(y')$.*

Moreover, the $\mathcal{E}$-filteredness of $\underline{\mathcal{A}_{\mathcal{E}}}$ implies the following 'joint embedding property': for any objects E of $\mathcal{E}$ and x, x' of $\mathcal{A}_E$, there exist an epimorphic family $\{e_i : E_i \to E \mid i \in I\}$ in $\mathcal{E}$ and for each $i \in I$ arrows $g_i : \mathcal{A}_{e_i}(x) \to z_i$ and $h_i : \mathcal{A}_{e_i}(x') \to z_i$ in $\mathcal{A}_{E_i}$ such that the diagram

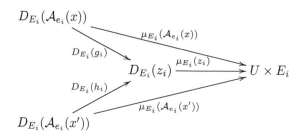

commutes. □

Remarks 5.1.24 (a) In the statement of Corollary 5.1.23, we can suppose without loss of generality the object $f : F \to E$ of the topos $\mathcal{E}/E$ to be equal to the terminal object $1_E : E \to E$. Indeed, by Remark 5.1.14(d), for any objects x and x' of $\mathcal{A}_E$ and any arrows $y : (f : F \to E) \to D_E(x)$ and $y' : (f : F \to E) \to D_E(x')$ in the topos $\mathcal{E}/E$, $\mu_E(x) \circ y = \mu_E(x') \circ y'$ if and only if $\mu_F(\mathcal{A}_f(x)) \circ y_f = \mu_F(\mathcal{A}_f(x')) \circ y_f$.

(b) For any pairs (x, y) and (x', y'), where x and x' are objects of $\mathcal{A}_E$, y is an arrow $(f : F \to E) \to D_E(x)$ in $\mathcal{E}/E$ and y' is an arrow $(f : F \to E) \to D_E(x')$ in $\mathcal{E}/E$, and any epimorphic family $\{g_j : G_j \to F \mid j \in J\}$ in $\mathcal{E}$, condition (ii) of Corollary 5.1.23 is satisfied by the pair $((x, y), (x', y'))$ if and only if it is satisfied by the pair $((x, y \circ g_j), (x', y' \circ g_j))$ for all $j \in J$.

Let us now apply Corollary 5.1.23 in the context of a flat functor $F : \mathcal{C}^{\mathrm{op}} \to \mathcal{E}$ and a functor $P : \mathcal{C} \to \mathbf{Set}$, where $\mathcal{C}$ is a small category and $\mathcal{E}$ is a Grothendieck topos. By Theorem 5.1.13, the restriction of the functor $(P_{\mathcal{E}_\mathcal{E}} \circ \pi^{\mathrm{f}}_{F_\mathcal{E}}) \cong P_\mathcal{E} \circ \pi^{\mathrm{f}}_{\overline{F_\mathcal{E}}}$ to the $\mathcal{E}$-indexed subcategory $\int F_\mathcal{E}$ of $\int F^{\mathrm{f}}_\mathcal{E}$ is naturally isomorphic to the composite $\mathcal{E}$-indexed functor $\gamma_\mathcal{E} \circ z_\mathcal{E}$, where $z_\mathcal{E} : \int F_\mathcal{E} \to \mathbf{Set}_\mathcal{E}$ is the $\mathcal{E}$-indexed functor defined by setting $(z_\mathcal{E})_E((c, x)) = P(c)$ and $(z_\mathcal{E})_E(f) = P(f)$ (for any $E \in \mathcal{E}$, object (c, x) and arrow f of the category $(\int F)_E$). By the remarks preceding the statement of Theorem 5.1.12, an $\mathcal{E}$-indexed cocone μ over $\gamma_\mathcal{E} \circ z_\mathcal{E}$ with vertex U can be identified with a family of functions $\mu_{(c,x)} : P(c) \to \mathrm{Hom}_\mathcal{E}(E, U)$ indexed by the pairs (c, x) consisting of an object c of $\mathcal{C}$ and a generalized element $x : E \to F(c)$, which satisfies the following conditions:

(i) For any generalized element $x : E \to F(c)$ and any arrow $f : d \to c$ in $\mathcal{C}$, $\mu_{(c,x)} \circ P(f) = \mu_{(c, F(f) \circ x)}$.

(ii) For any generalized element $x : E \to F(c)$ and any arrow $e : E' \to E$ in $\mathcal{E}$, $\mu_{(c, x \circ e)} = \mathrm{Hom}_\mathcal{E}(e, U) \circ \mu_{(c,x)}$.

The following theorem characterizes the cocones μ which are colimiting in terms of the functions $\mu_{(c,x)}$.

Theorem 5.1.25 *Let $\mathcal{C}$ be a small category, $\mathcal{E}$ a Grothendieck topos, $P : \mathcal{C} \to \mathbf{Set}$ a functor and $F : \mathcal{C}^{\mathrm{op}} \to \mathcal{E}$ a flat functor. Then an $\mathcal{E}$-indexed cocone $\mu = \{\mu_{(c,x)} : P(c) \to \mathrm{Hom}_\mathcal{E}(E, U) \mid c \in \mathcal{C}, x : E \to F(c) \text{ in } \mathcal{E}\}$ with vertex U over the diagram given by the restriction of the $\mathcal{E}$-indexed functor $(P_{\mathcal{E}_\mathcal{E}} \circ \pi^{\mathrm{f}}_{F_\mathcal{E}}) \cong P_\mathcal{E} \circ \pi^{\mathrm{f}}_{\overline{F_\mathcal{E}}}$ (where $P_\mathcal{E}$ is the internal diagram $[\mathbb{C}, \mathcal{E}]$ given by $\overline{\gamma^*_\mathcal{E} \circ P}$) to the $\mathcal{E}$-indexed subcategory $\int F_\mathcal{E}$ is colimiting if and only if the following conditions are satisfied:*

(i) *For any generalized element $x : E \to U$ in $\mathcal{E}$, there exist an epimorphic family $\{e_i : E_i \to E \mid i \in I\}$ in $\mathcal{E}$ and for each $i \in I$ an object c_i in $\mathcal{C}$, a generalized element $x_i : E_i \to F(c_i)$ and an element $y_i \in P(c_i)$ such that $\mu_{(c_i, x_i)}(y_i) = x \circ e_i$.*

(ii) *For any pairs (a, x) and (b, x'), where a and b are objects of $\mathcal{C}$ and $x : E \to F(a)$, $x' : E \to F(b)$ are generalized elements, and any elements $y \in P(a)$ and $y' \in P(b)$, $\mu_{(a,x)}(y) = \mu_{(b,x')}(y')$ if and only if there exist an epimorphic* family $\{e_i : E_i \to E \mid i \in I\}$ in $\mathcal{E}$ and for each $i \in I$ an object c_i of $\mathcal{C}$, arrows $f_i : a \to c_i$, $g_i : b \to c_i$ in $\mathcal{C}$ and a generalized element $x_i : E_i \to F(c_i)$ such that $<x,x'> \circ e_i = <F(f_i), F(g_i)> \circ x_i$ and $P(f_i)(y) = P(g_i)(y')$.

Moreover, the following 'joint embedding property' holds: for any pairs (a,x) and (b,x'), where a and b are objects of $\mathcal{C}$ and $x : E \to F(a)$, $x' : E \to F(b)$ are generalized elements in $\mathcal{E}$, there exist an epimorphic family $\{e_i : E_i \to E \mid i \in I\}$ in $\mathcal{E}$ and for each $i \in I$ an object c_i of $\mathcal{C}$, arrows $f_i : a \to c_i$, $g_i : b \to c_i$ in $\mathcal{C}$ and a generalized element $x_i : E_i \to F(c_i)$ such that $<x,x'> \circ e_i = <F(f_i), F(g_i)> \circ x_i$ and the following diagram commutes:

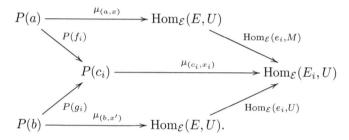

In fact, the epimorphic family $\{e_i : E_i \to E \mid i \in I\}$ can be taken to be the pullback of the family of arrows $<F(f), F(g)> : F(c) \to F(a) \times F(b)$ (indexed by the spans $(f : a \to c, g : b \to c)$ in the category $\mathcal{C}$) along the arrow $<x,x'> : E \to F(a) \times F(b)$.

Proof The theorem can be deduced from Corollary 5.1.23 by reasoning as follows.

Concerning condition (i), its equivalence with condition (i) of Corollary 5.1.23 follows from Remark 5.1.24(a) and the fact that for any object (c, z) of the category $(\int F)_E$ and any arrow $\alpha : 1_{\mathcal{E}/E} \to D_E((c,z))$ in the topos $\mathcal{E}/E$, denoting by $!_{f_r} : f_r : F_r \to E \to 1_{\mathcal{E}/E}$ the pullback along α of the coproduct arrow $u_r : 1_{\mathcal{E}/E} \to \gamma^*_{\mathcal{E}/E}(P(c)) \cong D_E((c,z))$ in $\mathcal{E}/E$ (for each $r \in P(c)$), the arrows $!_{f_r}$ define, for $r \in P(c)$, an epimorphic family of arrows in $\mathcal{E}/E$ with codomain $1_{\mathcal{E}/E}$. Indeed, this observation shows that we can suppose without loss of generality α to factor through one of the coproduct arrows u_r and hence to correspond to an element $y \in P(c)$.

Concerning condition (ii), we note that, by Remark 5.1.24, we can suppose without loss of generality that the pairs in condition (ii) of Corollary 5.1.23 have respectively the form $((x, a), y)$, where $y = u_r \circ !_{f_r}$, and $((x', b), y')$, where $y' = u_{r'} \circ !_{f_{r'}}$ (with the above notations), for some $r, r' \in P(c)$. Thus we have that $\mu_E(x) \circ y = \mu_E(x') \circ y'$ if and only if $(\mathrm{Hom}_{\mathcal{E}}(f_r, U) \circ \mu_{(x,a)})(r) = (\mathrm{Hom}_{\mathcal{E}}(f_{r'}, U) \circ \mu_{(x',b)})(r')$; but $(\mathrm{Hom}_{\mathcal{E}}(f_r, U) \circ \mu_{(x,a)})(r) = \mu_{(a, x \circ f_r)}(r)$ and $(\mathrm{Hom}_{\mathcal{E}}(f_{r'}, U) \circ \mu_{(x',b)})(r') = \mu_{(b, x' \circ f_{r'})}(r')$, whence $(\mathrm{Hom}_{\mathcal{E}}(f_r, U) \circ \mu_{(x,a)})(r) = (\mathrm{Hom}_{\mathcal{E}}(f_{r'}, U) \circ \mu_{(x',b)})(r')$ if and only if $\mu_{(a, x \circ f_r)}(r) = \mu_{(b, x' \circ f_{r'})}(r')$.

Finally, the 'joint embedding property' of the theorem is just an instance of that of Corollary 5.1.23. □

5.1.7 Explicit calculation of set-indexed colimits

It is well known (cf. for instance p. 355 of [62]) that the colimit of a functor $H : \mathcal{I} \to \mathcal{E}$ defined over a small category $\mathcal{I}$ with values in a Grothendieck topos $\mathcal{E}$ can always be realized as the coequalizer $q : \coprod_{i \in I} H(i) \to \mathrm{colim}(H)$ of the pair of arrows

$$a, b : \coprod_{u : i \to j \text{ in } \mathcal{I}} H(i) \to \coprod_{i \in I} H(i)$$

defined by the conditions

$$a \circ \lambda_u = \kappa_i$$

and

$$b \circ \lambda_u = \kappa_j \circ H(u)$$

for every arrow $u : i \to j$ in $\mathcal{I}$, where $\lambda_u : H(\mathrm{dom}(u)) \to \coprod_{u : i \to j \text{ in } \mathcal{I}} H(i)$ and $\kappa_i : H(i) \to \coprod_{i \in I} H(i)$ are the canonical coproduct arrows.

Given two arrows $s, t : E \to \coprod_{i \in I} H(i)$ we can consider, for any objects $i, j \in \mathcal{I}$, the following pullback diagrams:

$$\begin{array}{ccc} E_i^s & \xrightarrow{a_i^s} & H(i) \\ \downarrow p_i^s & & \downarrow \kappa_i \\ E & \xrightarrow{s} & \coprod_{i \in I} H(i), \end{array} \qquad \begin{array}{ccc} E_j^t & \xrightarrow{a_j^t} & H(j) \\ \downarrow p_j^t & & \downarrow \kappa_j \\ E & \xrightarrow{t} & \coprod_{i \in I} H(i). \end{array}$$

Let us also consider, for each $i, j \in \mathcal{I}$, the pullback square

$$\begin{array}{ccc} E_{i,j}^{s,t} & \xrightarrow{q_{i,j}^s} & E_i^s \\ \downarrow q_{i,j}^t & & \downarrow p_i^s \\ E_j^t & \xrightarrow{p_j^t} & E \end{array}$$

and denote by $r_{i,j}^{s,t} : E_{i,j}^{s,t} \to E$ the arrow $p_i^s \circ q_{i,j}^s = p_j^t \circ q_{i,j}^t$. Notice that the family of arrows $\{r_{i,j}^{s,t} : E_{i,j}^{s,t} \to E \mid i, j \in \mathcal{I}\}$ is epimorphic.

Lemma 5.1.26 *Let B be an object of a topos $\mathcal{E}$. Then an equivalence relation R in $\mathcal{E}$ on B can be identified with a function $E \to R_E$ sending each object E of $\mathcal{E}$ to an equivalence relation R_E on the set $\mathrm{Hom}_{\mathcal{E}}(E, B)$ satisfying the following properties:*

(i) *For any arrow $f : E \to E'$ in $\mathcal{E}$, if $(h, k) \in R_{E'}$ then $(h \circ f, k \circ f) \in R_E$.*

(ii) *For any epimorphic family $\{e_i : E_i \to E \mid i \in I\}$ in $\mathcal{E}$ and any arrows $f, g \in \mathrm{Hom}_{\mathcal{E}}(E, B)$, $(f \circ e_i, g \circ e_i) \in R_{E_i}$ for all $i \in I$ implies $(f, g) \in R_E$.*

Proof Clearly, for any object A of $\mathcal{C}$, the subobjects of A can be identified with the $J_\mathcal{E}$-closed sieves on A, or equivalently with the $c_{J_\mathcal{E}}$-closed subobjects of the representable $\text{Hom}_\mathcal{E}(-, A)$, where $c_{J_\mathcal{E}}$ is the closure operation on subobjects corresponding to the canonical Grothendieck topology $J_\mathcal{E}$ on $\mathcal{E}$. Applying this remark to $A = B \times B$ and noticing that the concept of equivalence relation is cartesian and hence preserved and reflected by cartesian fully faithful functors, in particular by the Yoneda embedding $\mathcal{E} \to [\mathcal{E}^{\text{op}}, \textbf{Set}]$, we immediately deduce our thesis. □

Lemma 5.1.26 implies that for any generalized elements $s, t : E \to \coprod_{i \in I} H(i)$, $(s, t) \in R_E$ if and only if, for every $i, j \in \mathcal{I}$, $(\kappa_i \circ a_i^s \circ q_{i,j}^s, \kappa_j \circ a_j^t \circ q_{i,j}^t) \in R_{E_{i,j}^{s,t}}$; indeed, $s \circ r_{i,j}^{s,t} = \kappa_i \circ a_i^s \circ q_{i,j}^s$ and $t \circ r_{i,j}^{s,t} = \kappa_j \circ a_j^t \circ q_{i,j}^t$. Thus, in order to completely describe the relation R, it suffices to consider generalized elements of the form $\kappa_i \circ z$ and $\kappa_j \circ w$, where $z : E \to H(i)$ and $w : E \to H(j)$, and characterize when they belong to R_E.

The following lemma provides a characterization of the coequalizer
$$q : \coprod_{i \in I} H(i) \to \text{colim}(H),$$
which will be useful later on:

Lemma 5.1.27 *Under the above assumptions, let $p : \coprod_{i \in I} H(i) \to A$ be an epimorphism. Then p is isomorphic to the canonical map $q : \coprod_{i \in I} H(i) \to \text{colim}(H)$ if and only if, denoting by R the kernel pair of q, for any objects E of $\mathcal{E}$ and i, j of $\mathcal{I}$ and any generalized elements $z : E \to H(i)$ and $w : E \to H(j)$ in $\mathcal{E}$, $(\kappa_i \circ z, \kappa_j \circ w) \in R$ if and only if $p \circ \kappa_i \circ z = p \circ \kappa_j \circ w$.*

Proof Since in a topos every epimorphism is the coequalizer of its kernel pair, any two epimorphisms $\alpha : B \to C$ and $\beta : B \to D$ in $\mathcal{E}$ are isomorphic in $B/\mathcal{E}$ if and only if their kernel pairs are isomorphic as subobjects of $B \times B$. In particular, p is isomorphic to q if and only if the kernel pair of p is isomorphic to R.

The kernel pair S of p is defined by the property that for any generalized elements $s, t : E \to \coprod_{i \in I} H(i)$, $(s, t) \in S$ if and only if $p \circ s = p \circ t$. Our thesis thus follows from the remarks following the proof of Lemma 5.1.26. □

Lemma 5.1.28 *Let $\mathcal{E}$ be a topos, $f, g : A \to B$ arrows in $\mathcal{E}$ with coequalizer $q : B \to C$ and $R \rightarrowtail B \times B$ the kernel pair of q. Then, for any object E of $\mathcal{E}$ and any elements $h, k \in \text{Hom}_\mathcal{E}(E, B)$, $<h, k> : E \to B \times B$ factors through R if and only if there exists an epimorphic family $\{e_i : E_i \to E \mid i \in I\}$ in $\mathcal{E}$ such that, for each $i \in I$, $(h \circ e_i, k \circ e_i)$ belongs to the equivalence relation on the set $\text{Hom}_\mathcal{E}(E_i, B)$ generated by the set of pairs of the form $(f \circ a, g \circ a)$ for a generalized element $a : E_i \to A$ in $\mathcal{E}$.*

Proof For any object E of $\mathcal{E}$, consider the equivalence relation $R_E^{f,g}$ on the set $\mathrm{Hom}_{\mathcal{E}}(E,B)$ consisting of the pairs (h,k) with the property that there exists an epimorphic family $\{e_i : E_i \to E \mid i \in I\}$ in $\mathcal{E}$ such that, for each $i \in I$, $(h \circ e_i, k \circ e_i)$ belongs to the equivalence relation on the set $\mathrm{Hom}_{\mathcal{E}}(E_i, B)$ generated by the set of pairs of the form $(f \circ a, g \circ a)$ for a generalized element $a : E_i \to A$ in $\mathcal{E}$. The assignment $E \to R_E^{f,g}$ clearly satisfies the conditions of Lemma 5.1.26 and hence defines an equivalence relation $R^{f,g}$ on B in $\mathcal{E}$. To prove that $R = R^{f,g}$, we argue as follows. Denoting by $\pi_1, \pi_2 : R^{f,g} \to B$ the canonical projections, we have that for any $E \in \mathcal{E}$, $\mathrm{Hom}_{\mathcal{E}}(E, q) \circ \mathrm{Hom}_{\mathcal{E}}(E, \pi_1) = \mathrm{Hom}_{\mathcal{E}}(E, q) \circ \mathrm{Hom}_{\mathcal{E}}(E, \pi_2)$; therefore $q \circ \pi_1 = q \circ \pi_2$ in $\mathcal{E}$. To prove that q is actually the coequalizer of π_1 and π_2 in $\mathcal{E}$, we observe that any arrow $p : B \to D$ such that $p \circ \pi_1 = p \circ \pi_2$ satisfies $p \circ f = p \circ g$ and hence, by definition of q, there exists a unique factorization of p through q. As q is the coequalizer of $R^{f,g}$ and in a topos any equivalence relation is the kernel pair of its own coequalizer, we conclude that $R^{f,g} = R$, as required. $\square$

Remark 5.1.29 Under the hypotheses of Lemma 5.1.28, R can be characterized as the equivalence relation on B generated by the arrows f and g, i.e. as the smallest equivalence relation on B containing the image of the arrow $<f,g>:$ $A \to B \times B$. Indeed, clearly R contains the image of this arrow, and if T is an equivalence relation on B containing the image of the arrow $<f,g>$ then the coequalizer z of T factors through q and hence the kernel pair of q, namely R, is contained in the kernel pair of z, namely T.

By using the internal language $\Sigma_{\mathcal{E}}$ of the topos $\mathcal{E}$ (cf. section 1.3.1.1), we can reformulate Lemma 5.1.28 as follows:

Lemma 5.1.30 *Let $\mathcal{E}$ be a topos, $f, g : A \to B$ arrows in $\mathcal{E}$ with coequalizer $q : B \to C$ and R the kernel pair of q. Let $(z, z') \in G_{f,g}$ (for variables z, z' of type B) be an abbreviation for the formula*

$$z = z' \vee (\exists x)(\ulcorner f \urcorner(x) = z \wedge \ulcorner g \urcorner(x) = z') \vee (\exists x')(\ulcorner f \urcorner(x') = z' \wedge \ulcorner g \urcorner(x') = z)$$

over $\Sigma_{\mathcal{E}}$. Then the geometric bisequent

$$\left(\ulcorner R \urcorner(y, y') \dashv\vdash_{y^B, y'^B} \bigvee_{n \in \mathbb{N}} \left(\exists z_1 \ldots \exists z_n \left(z_1 = x \wedge z_n = x' \wedge \bigwedge_{1 < i < n} (z_{i-1}, z_i) \in G_{f,g}\right)\right)\right)$$

is valid in the $\Sigma_{\mathcal{E}}$-structure $S_{\mathcal{E}}$.

Proof The interpretation of the formula $(z^B, z'^B) \in G_{f,g}$ is the reflexive and symmetric closure of the relation $S_{f,g}$ on B given by the image of the arrow $<f,g>: A \to B \times B$, that is the union of the diagonal subobject $\Delta : B \to B \times B$ of $B \times B$, the subobject $S_{f,g} \rightarrowtail B \times B$ and the subobject $\tau \circ S_{f,g}$ of $B \times B$ given by the composite of $S_{f,g}$ with the exchange isomorphism $\tau : B \times B \to B \times B$. The interpretation of the formula

$$\bigvee_{n \in \mathbb{N}} \left(\exists z_1 \ldots \exists z_n \left(z_1 = x \wedge z_n = x' \wedge \bigwedge_{1 < i < n} (z_{i-1}, z_i) \in G_{f,g} \right) \right)$$

thus coincides with the equivalence relation on B generated by $S_{f,g}$, that is with R (cf. Remark 5.1.29). □

Given a functor $H : \mathcal{I} \to \mathcal{E}$ from a small category $\mathcal{I}$ to a Grothendieck topos $\mathcal{E}$ and an object E of $\mathcal{E}$, we define $\mathcal{I}_H^E$ as the set of pairs (i, x), where i is an object of $\mathcal{I}$ and x is a generalized element $E \to H(i)$ in $\mathcal{E}$. Below, abusing notation, we shall occasionally identify a generalized element $x : E \to H(i)$ with its composite $\kappa_i \circ x$ with the coproduct arrow $\kappa_i : H(i) \to \coprod_{i \in \mathcal{I}} H(i)$.

The following proposition can be established by similar means to those employed in the proof of Lemma 5.1.28.

Proposition 5.1.31 *Let $H : \mathcal{I} \to \mathcal{E}$ be a functor from a small category $\mathcal{I}$ to a Grothendieck topos $\mathcal{E}$. Then the equivalence relation R on the object $\coprod_{i \in \mathcal{I}} H(i)$ given by the kernel pair of the canonical arrow $\coprod_{i \in \mathcal{I}} H(i) \to \mathrm{colim}(H)$ is characterized by the following property: for any object E of $\mathcal{E}$ and any pairs (i, x) and (i', x') in $\mathcal{I}_H^E$, we have that $(\kappa_i \circ x, \kappa_{i'} \circ x') \in R_E$ if and only if there exists an epimorphic family $\{e_k : E_k \to E \mid k \in K\}$ in $\mathcal{E}$ such that for every $k \in K$ the pair $((i, x \circ e_k), (i', x' \circ e_k))$ belongs to the equivalence relation on the set $\mathcal{I}_H^{E_k}$ generated by the pairs of the form $((j, y), (j', H(f) \circ y))$ where $f : j \to j'$ is an arrow in $\mathcal{I}$ and y is a generalized element $E_k \to H(j)$.*

Proof Let us consider the pair of arrows $a, b : \coprod_{u : i \to j \text{ in } \mathcal{I}} H(i) \to \coprod_{i \in I} H(i)$ defined at the beginning of this section. The canonical arrow $\coprod_{i \in \mathcal{I}} H(i) \to \mathrm{colim}(H)$ is the coequalizer of a and b.

For any object E of $\mathcal{E}$, let R_E be the relation on the set $\mathrm{Hom}_\mathcal{E}(E, \coprod_{i \in \mathcal{I}} H(i))$ consisting of the pairs (x, x') with the property that for any $(i, j) \in \mathcal{I} \times \mathcal{I}$ there exists an epimorphic family $\{e_k^{(i,j)} : E_k^{(i,j)} \to E_k^{x,x'} \mid k \in K_{(i,j)}\}$ in $\mathcal{E}$ such that, for each $k \in K_{(i,j)}$, $(x \circ r_{i,j}^{x,x'} \circ e_k^{(i,j)}, x' \circ r_{i,j}^{x,x'} \circ e_k^{(i,j)})$ yields a pair of elements of $\mathcal{I}_H^{E_k^{(i,j)}}$ belonging to the equivalence relation $T_k^{(i,j)}$ on $\mathcal{I}_H^{E_k^{(i,j)}}$ generated by the set of pairs of the form $((n, y), (m, H(z) \circ y))$ (for $n \in \mathcal{I}$, $y : E_k^{(i,j)} \to H(n)$ and $z : n \to m$ in $\mathcal{I}$).

It is readily seen that the assignment $E \to R_E$ satisfies the hypotheses of Lemma 5.1.26, whence it defines an equivalence relation R in $\mathcal{E}$ on the object $\coprod_{i \in \mathcal{I}} H(i)$ such that for any generalized elements $x, x' : E \to \coprod_{i \in \mathcal{I}} H(i)$, $< x, x' >$ factors through R if and only if $(x, x') \in R_E$.

The arrow $< a, b >$ factors through R since, for every $i, j \in \mathcal{I}$, $(a \circ r_{i,j}^{s,t}, b \circ r_{i,j}^{s,t})$ belongs to the equivalence relation $T_k^{(i,j)}$. Therefore, by Remark 5.1.29, R contains the kernel pair of q. The converse inclusion follows from the fact that, for any $(x, x') \in R_E$, $q \circ x = q \circ x'$; indeed, $\{e_k^{(i,j)} : E_k^{(i,j)} \to E_{i,j}^{x,x'} \mid k \in K_{(i,j)}\}$ is an epimorphic family in $\mathcal{E}$ such that, for each $k \in K_{(i,j)}$, $(x \circ r_{i,j}^{x,x'} \circ e_k^{(i,j)}, x' \circ r_{i,j}^{x,x'} \circ e_k^{(i,j)})$ belongs to the equivalence relation $T_k^{(i,j)}$ and hence we have $q \circ x \circ r_{i,j}^{x,x'} \circ e_k^{(i,j)} = q \circ x' \circ r_{i,j}^{x,x'} \circ e_k^{(i,j)}$ for all $k \in K_{(i,j)}$, that is $q \circ x = q \circ x'$. □

Remark 5.1.32 Proposition 5.1.31 could alternatively be deduced from Lemma 5.1.28, but the proof of such a derivation would be more involved than the direct argument given above.

Let us now proceed to apply the above results in the case of a functor H of the form $F \circ \pi_P$, where F is a flat functor with values in a Grothendieck topos and π_P is the fibration associated with a set-valued functor P.

Recall from Proposition 1.1.32 that a functor $F : \mathcal{C}^{\mathrm{op}} \to \mathcal{E}$ is flat if and only if it is filtering. Notice that one can suppose $E = 1_{\mathcal{E}}$ in condition (i)(a) of the proposition without loss of generality, and that if all the arrows in the category $\mathcal{C}$ are monic then condition (i)(c) of the proposition rewrites as follows: for any two parallel arrows $u, v : d \to c$ in $\mathcal{C}$, either $u = v$ or the equalizer of $F(u)$ and $F(v)$ is zero.

Proposition 5.1.33 Let $\mathcal{C}$ be a small category, $\mathcal{E}$ a Grothendieck topos, $P : \mathcal{C} \to \mathbf{Set}$ a functor and $F : \mathcal{C}^{\mathrm{op}} \to \mathcal{E}$ a flat functor. Let $\pi_P : \int P \to \mathcal{C}^{\mathrm{op}}$ be the canonical projection from the category of elements of P to $\mathcal{C}^{\mathrm{op}}$. Then

$$\mathrm{colim}(F \circ \pi_P) \cong \left(\coprod_{(c,x) \in \int P} F(c) \right) / R,$$

where R is the equivalence relation in $\mathcal{E}$ defined by saying that, for any objects (c, x) and (d, y) of the category $\int P$, the geometric bisequent

$$\left(\ulcorner R \urcorner (\ulcorner \xi_{(c,x)} \urcorner (u), \ulcorner \xi_{(d,y)} \urcorner (v)) \dashv\vdash_{u^{F(c)}, v^{F(d)}} \bigvee_{\substack{c \xrightarrow{f} a \xleftarrow{g} d \\ P(f)(x) = P(g)(y)}} (\exists \xi^{F(a)})(\ulcorner F(f) \urcorner (\xi) = u \wedge \ulcorner F(g) \urcorner (\xi) = v) \right)$$

is valid in the $\Sigma_{\mathcal{E}}$-structure $S_{\mathcal{E}}$, where $\xi_{(c,x)} : F(c) \to \coprod_{(c,x) \in \int P} F(c)$ are the canonical coproduct arrows. In particular, for any objects $(c, x), (c, y)$ of the category $\int P$, the geometric bisequent

$$\left(\ulcorner R \urcorner (\ulcorner \xi_{(c,x)} \urcorner (u), \ulcorner \xi_{(c,y)} \urcorner (u)) \dashv\vdash_{u^{F(c)}} \bigvee_{\substack{c \xrightarrow{f} a \mid \\ P(f)(x) = P(f)(y)}} (\exists \xi^{F(a)})(\ulcorner F(f) \urcorner (\xi) = u) \right)$$

is valid in $\mathcal{S}_{\mathcal{E}}$.

Proof Let T be the equivalence relation on $\coprod_{(c,x) \in \int P} F(c)$ corresponding to the quotient $\coprod_{(c,x) \in \int P} F(c) \to \operatorname{colim}(F \circ \pi_P)$. Then $\operatorname{colim}(F \circ \pi_P) \cong \left(\coprod_{(c,x) \in \int P} F(c) \right) / T$, and by Lemma 5.1.28 we have that T is the equivalence relation on $\coprod_{(c,x) \in \int P} F(c)$ generated by the image of the arrow $<a, b>$, where $a, b : \coprod_{z:(c,x) \to (d,y) \text{ in } \int P} F(c) \to \coprod_{(c,x) \in \int P} F(c)$ are the arrows defined at the beginning of this section for the functor $H = F \circ \pi_P$.

Let $\lambda_z : F(c) \to \coprod_{z:(c,x) \to (d,y) \text{ in } \int P} F(c)$ be the canonical coproduct arrows (for each arrow $z : (c, x) \to (d, y)$ of the category $\int P$). For any objects $(c, x), (d, y)$ of $\int P$, let the expression $(u^{F(c)}, v^{F(d)}) \in E_{(c,x),(d,y)}$ be an abbreviation of the formula $\bigvee_{\substack{c \xrightarrow{f} a \xleftarrow{g} d \mid \\ P(f)(x) = P(g)(y)}} (\exists \xi^{F(a)})(\ulcorner F(f) \urcorner (\xi) = u \wedge \ulcorner F(g) \urcorner (\xi) = v)$. Notice that, for any arrow $z : (c, x) \to (d, y)$ in $\int P$, the sequent

$$(\top \vdash_{u'^{F(c)}} (\ulcorner a \urcorner (\ulcorner \lambda_z \urcorner (u')), \ulcorner b \urcorner (\ulcorner \lambda_z \urcorner (u'))) \in E_{(c,x),(d,y)})$$

is valid in $\mathcal{S}_{\mathcal{E}}$.

The fact that the right-to-left implication of the bisequent in the statement of the proposition holds in $\mathcal{S}_{\mathcal{E}}$ is clear.

The validity of the left-to-right implication will follow immediately from Lemma 5.1.30 once we have shown that for any objects $(c, x), (d, y), (e, z)$ in $\int P$ the sequent

$$((u, v) \in E_{(c,x),(d,y)} \wedge (v, w) \in E_{(d,y),(e,z)} \vdash_{u^{F(c)}, v^{F(d)}, w^{F(e)}} (u, w) \in E_{(c,x),(e,z)})$$

is valid in the $\Sigma_{\mathcal{E}}$-structure $\mathcal{S}_{\mathcal{E}}$. For simplicity we shall give a proof only in the case $\mathcal{E} = \mathbf{Set}$, but all of our arguments are formalizable in the internal logic of the topos $\mathcal{E}$ and hence are valid in general; in fact, the lift of the proof from the set-theoretic to the topos-theoretic setting is made possible by the following logical characterization of flatness (cf. Proposition 1.1.32): a functor $F : \mathcal{C}^{\mathrm{op}} \to \mathcal{E}$ is flat if and only if the following sequents are satisfied in the $\Sigma_{\mathcal{E}}$-structure $\mathcal{S}_{\mathcal{E}}$:

$$\left(\top \vdash_{\emptyset} \bigvee_{c \in \mathrm{Ob}(\mathcal{C})} (\exists x^{F(c)})(x = x) \right);$$

$$\left(\top \vdash_{x^{F(a)}, y^{F(b)}} \bigvee_{a \xrightarrow{f} c \xleftarrow{g} b} (\exists z^{F(c)})(\ulcorner F(f) \urcorner (z) = x \wedge \ulcorner F(g) \urcorner (z) = y) \right)$$

for any objects a, b of $\mathcal{C}$;

$$\left(\ulcorner F(f)\urcorner(x) = \ulcorner F(g)\urcorner(x) \vdash_{x^{F(a)}} \bigvee_{h:a\to c \mid h\circ f = h\circ g} (\exists z^{F(c)})(\ulcorner F(h)\urcorner(z) = x)\right)$$

for any pair of arrows $f, g : b \to a$ in $\mathcal{C}$ with common domain and codomain.

Now, given objects $(c, x), (d, y), (e, z) \in \int P$ and elements $u \in F(c), v \in F(d), w \in F(e)$, suppose that $(u, v) \in E_{(c,x),(d,y)}$ and $(v, w) \in E_{(d,y),(e,z)}$; we want to prove that $(u, w) \in E_{(c,x),(e,z)}$.

As $(u, v) \in E_{(c,x),(d,y)}$, there exist arrows $f : c \to a$ and $g : d \to a$ in $\mathcal{C}$ and an element $\xi \in F(a)$ such that $F(f)(\xi) = u$, $F(g)(\xi) = v$ and $P(f)(x) = P(g)(y)$. Similarly, as $(v, w) \in E_{(d,y),(e,z)}$, there exist arrows $h : d \to b$ and $k : e \to b$ in $\mathcal{C}$ and an element $\chi \in F(b)$ such that $F(h)(\chi) = v$, $F(k)(\chi) = w$ and $P(h)(y) = P(k)(z)$.

Using the second condition in the definition of a filtering functor, we obtain the existence of an object m of $\mathcal{C}$, arrows $r : a \to m$ and $t : b \to m$ in $\mathcal{C}$ and an element $\epsilon \in F(m)$ such that $F(r)(\epsilon) = \xi$ and $F(t)(\epsilon) = \chi$. Consider now the arrows $r \circ g, t \circ h : d \to m$; we have that $F(r \circ g)(\epsilon) = F(t \circ h)(\epsilon)$ and hence by the third condition in the definition of a flat functor there exist an arrow $s : m \to n$ in $\mathcal{C}$ such that $s \circ r \circ g = s \circ t \circ h$ and an element $\alpha \in F(n)$ such that $F(s)(\alpha) = \epsilon$.

The arrows $s \circ r \circ f$ and $s \circ t \circ k$ satisfy the appropriate conditions ensuring that $(u, w) \in E_{(c,x),(e,z)}$, i.e. $F(s \circ r \circ f)(\alpha) = u$, $F(s \circ t \circ k)(\alpha) = w$ and $P(s \circ r \circ f)(x) = P(s \circ t \circ k)(z)$. Indeed, $F(s \circ r \circ f)(\alpha) = F(f)(F(r)(F(s)(\alpha))) = F(f)(F(r)(\epsilon)) = F(f)(\xi) = u$, $F(s \circ t \circ k)(\alpha) = F(k)(F(t)(F(s)(\alpha))) = F(k)(F(t)(\epsilon)) = F(k)(\chi) = w$ and $P(s \circ r \circ f)(x) = P(s \circ r)(P(f)(x)) = P(s \circ r)(P(g)(y)) = P(s \circ r \circ g)(y) = P(s \circ t \circ h)(y) = P(s \circ t)(P(h)(y)) = P(s \circ t)(P(k)(z)) = P(s \circ t \circ k)(z)$, as required.

This completes the proof of the first statement of the proposition; the second statement follows from the first by an easy application of the third condition in the definition of a filtering functor. □

The following proposition represents the translation of Proposition 5.1.33 into the categorical language of generalized elements:

Proposition 5.1.34 *Let $\mathcal{C}$ be a small category, $\mathcal{E}$ a Grothendieck topos with a separating family of objects $\mathcal{S}$, $P : \mathcal{C} \to \mathbf{Set}$ a functor and $F : \mathcal{C}^{\mathrm{op}} \to \mathcal{E}$ a flat functor. Let $\pi_P : \int P \to \mathcal{C}^{\mathrm{op}}$ be the canonical projection from the category of elements of P to $\mathcal{C}^{\mathrm{op}}$. Then the equivalence relation R defined by the formula*

$$\mathrm{colim}(F \circ \pi_P) \cong \left(\coprod_{(c,x) \in \int P} F(c)\right) / R$$

admits the following characterization: for any objects $(c, x), (d, y)$ of $\int P$ and generalized elements $u : E \to F(c)$, $v : E \to F(d)$ (where $E \in \mathcal{S}$), $(\xi_{(c,x)} \circ u, \xi_{(d,y)} \circ v) \in R_E$ if and only if there exist an epimorphic family $\{e_i : E_i \to$

$E \mid i \in I, E_i \in \mathcal{S}\}$ and for each $i \in I$ an object $a_i \in \mathcal{C}$, a generalized element $h_i : E_i \to F(a_i)$ and two arrows $f_i : c \to a_i$ and $f'_i : d \to a_i$ in $\mathcal{C}$ such that $P(f_i)(x) = P(f'_i)(y)$ and $< F(f_i), F(f'_i) > \circ h_i = < u, v > \circ e_i$.

In particular, for any objects (c, x) and (c, y) of $\int P$ and any generalized element $u : E \to F(c)$ (where $E \in \mathcal{S}$), $(\xi_{(c,x)} \circ u, \xi_{(c,y)} \circ v) \in R_E$ if and only if there exist an epimorphic family $\{e_i : E_i \to E \mid i \in I, E_i \in \mathcal{S}\}$ and for each $i \in I$ an object $a_i \in \mathcal{C}$, a generalized element $h_i : E_i \to F(a_i)$ and an arrow $f_i : c \to a_i$ in $\mathcal{C}$ such that $P(f_i)(x) = P(f'_i)(y)$ and $F(f_i) \circ h_i = u \circ e_i$. □

5.2 Extensions of flat functors

In this section, we investigate the operation on flat functors induced by a geometric morphism of toposes. This will be relevant for the characterization of the class of theories classified by a presheaf topos addressed in Chapter 6.

5.2.1 General extensions

Let $(\mathcal{C}, J)$ and $(\mathcal{D}, K)$ be essentially small sites and $u : \mathbf{Sh}(\mathcal{D}, K) \to \mathbf{Sh}(\mathcal{C}, J)$ a geometric morphism. Composition on the left with this morphism induces, via Diaconescu's equivalence, a functor

$$\xi_{\mathcal{E}} : \mathbf{Flat}_K(\mathcal{D}, \mathcal{E}) \to \mathbf{Flat}_J(\mathcal{C}, \mathcal{E}).$$

This functor can be explicitly described as follows. For any flat K-continuous functor $F : \mathcal{D} \to \mathcal{E}$ with corresponding geometric morphism $f_F : \mathcal{E} \to \mathbf{Sh}(\mathcal{D}, K)$, the flat functor $\tilde{F} := \xi_{\mathcal{E}}(F) : \mathcal{C} \to \mathcal{E}$ is given by $f_F^* \circ u^* \circ l_{\mathcal{C}}$, where $l_{\mathcal{C}} : \mathcal{C} \to \mathbf{Sh}(\mathcal{C}, J)$ is the composite of the Yoneda embedding $y_{\mathcal{C}} : \mathcal{C} \to [\mathcal{C}^{\text{op}}, \mathbf{Set}]$ with the associated sheaf functor $[\mathcal{C}^{\text{op}}, \mathbf{Set}] \to \mathbf{Sh}(\mathcal{C}, J)$. For any natural transformation $\alpha : F \to G$ between flat functors in $\mathbf{Flat}_K(\mathcal{D}, \mathcal{E})$, the corresponding natural transformation $f_\alpha^* : f_F^* \to f_G^*$, applied to the functor $u^* \circ l_{\mathcal{C}}$, gives rise to a natural transformation $\tilde{\alpha} : \tilde{F} \to \tilde{G}$ which is precisely $\xi_{\mathcal{E}}(\alpha)$.

We can express $\tilde{F}$ directly in terms of F by using a colimit construction, as follows.

For any $c \in \mathcal{C}$, the K-sheaf $u^*(l_{\mathcal{C}}(c)) : \mathcal{D}^{\text{op}} \to \mathbf{Set}$, considered as an object of $[\mathcal{D}^{\text{op}}, \mathbf{Set}]$, can be canonically expressed as a colimit of representables indexed over its category of elements $\mathcal{A}_c$; specifically, $u^*(l_{\mathcal{C}}(c)) \cong \text{colim}(y_{\mathcal{D}} \circ \pi_c)$ in $[\mathcal{D}^{\text{op}}, \mathbf{Set}]$, where $\pi_c : \mathcal{A}_c \to \mathcal{D}$ is the canonical projection functor and $y_{\mathcal{D}} : \mathcal{D} \to [\mathcal{D}^{\text{op}}, \mathbf{Set}]$ is the Yoneda embedding. As the associated sheaf functor $a_K : [\mathcal{D}^{\text{op}}, \mathbf{Set}] \to \mathbf{Sh}(\mathcal{D}, K)$ preserves colimits, the functor $u^*(l_{\mathcal{C}}(c))$ is isomorphic in $\mathbf{Sh}(\mathcal{D}, K)$ to the colimit of the functor $l_{\mathcal{D}} \circ \pi_c$. Therefore $\tilde{F}(c) = (f_F^* \circ u^* \circ l_{\mathcal{C}})(c) = f_F^*(\text{colim}(l_{\mathcal{D}} \circ \pi_c)) \cong \text{colim}(f_F^* \circ l_{\mathcal{D}} \circ \pi_c) = \text{colim}(F \circ \pi_c)$.

For any $(d, z) \in \mathcal{A}_c$, we write $\kappa_{(d,z)}^F : F(d) \to \tilde{F}(c)$ for the canonical colimit arrow.

Proposition 5.2.1

(i) *For any natural transformation $\alpha : F \to G$ between functors $F, G : \mathcal{D} \to \mathcal{E}$ in $\mathbf{Flat}_K(\mathcal{D}, \mathcal{E})$, the natural transformation $\tilde{\alpha} : \tilde{F} \to \tilde{G}$ is characterized by the following condition: for any object (d, z) of the category $\mathcal{A}_c$, the diagram*

$$\begin{array}{ccc} \tilde{F}(c) & \xrightarrow{\tilde{\alpha}(c)} & \tilde{G}(c) \\ \kappa^F_{(d,z)} \uparrow & & \uparrow \kappa^G_{(d,z)} \\ F(d) & \xrightarrow{\alpha(d)} & G(d) \end{array}$$

commutes.

(ii) *The functor $\xi_{\mathcal{E}} : \mathbf{Flat}_K(\mathcal{D}, \mathcal{E}) \to \mathbf{Flat}_J(\mathcal{C}, \mathcal{E})$ is natural in $\mathcal{E}$; that is, for any geometric morphism $f : \mathcal{F} \to \mathcal{E}$, the diagram*

$$\begin{array}{ccc} \mathbf{Flat}_K(\mathcal{D}, \mathcal{E}) & \xrightarrow{\xi_{\mathcal{E}}} & \mathbf{Flat}_J(\mathcal{C}, \mathcal{E}) \\ {\scriptstyle f^* \circ -} \downarrow & & \downarrow {\scriptstyle f^* \circ -} \\ \mathbf{Flat}_K(\mathcal{D}, \mathcal{F}) & \xrightarrow{\xi_{\mathcal{F}}} & \mathbf{Flat}_J(\mathcal{C}, \mathcal{F}) \end{array}$$

commutes.

(iii) *For any arrow $f : c \to c'$ in $\mathcal{C}$ and any object (d, z) in $\mathcal{A}_c$, the diagram*

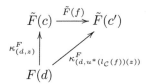

commutes.

Proof

(i) The given diagram commutes as it is the naturality square for the natural transformation $f^*_\alpha : f^*_F \to f^*_G$ with respect to the canonical colimit arrow $y_\mathcal{D}(d) \to u^*(l_\mathcal{C}(c))$ in $[\mathcal{D}^{\mathrm{op}}, \mathbf{Set}]$ corresponding to the object (d, z) of $\mathcal{A}_c$.

(ii) This follows as an immediate consequence of the fact that Diaconescu's equivalence is natural in $\mathcal{E}$.

(iii) Given a natural transformation $\beta : P \to P'$ of presheaves P and P' in $[\mathcal{D}^{\mathrm{op}}, \mathbf{Set}]$, denoting by $\kappa^P : \int P \to \mathcal{D}$ and $\kappa^{P'} : \int P' \to \mathcal{D}$ the canonical projection functors and by $\kappa^P_{(d,z)} : y_\mathcal{D}(d) \to P$, $\kappa^{P'}_{(d',z')} : y_\mathcal{D}(d') \to P$ the canonical colimit arrows in $[\mathcal{D}^{\mathrm{op}}, \mathbf{Set}]$ (for $(d, z) \in \int P$ and $(d', z') \in \int P'$), we clearly have that $\beta \circ \kappa^P_{(d,z)} = \kappa^{P'}_{(d,\beta(d)(z))}$. To obtain our thesis it suffices to apply this result to the natural transformation $\beta : u^*(l_\mathcal{C}(f)) : u^*(l_\mathcal{C}(c)) \to u^*(l_\mathcal{C}(c'))$; indeed, by taking the image under the functor f_F^* of the resulting equality we obtain precisely the identity expressing the commutativity of the diagram in the statement of point (iii) of the proposition.

$\square$

5.2.2 Extensions along embeddings of categories

Let $\mathcal{D}$ be a subcategory of a small category $\mathcal{C}$. Then the inclusion functor $i : \mathcal{D} \hookrightarrow \mathcal{C}$ induces a geometric morphism $E(i) : [\mathcal{D}, \mathbf{Set}] \to [\mathcal{C}, \mathbf{Set}]$ and hence, by Diaconescu's equivalence, a functor

$$\xi_{\mathcal{E}} : \mathbf{Flat}(\mathcal{D}^{\mathrm{op}}, \mathcal{E}) \to \mathbf{Flat}(\mathcal{C}^{\mathrm{op}}, \mathcal{E}).$$

For $F \in \mathbf{Flat}(\mathcal{D}^{\mathrm{op}}, \mathcal{E})$, we can describe $\tilde{F} = \xi_{\mathcal{E}}(F)$ directly in terms of F as follows. For any $c \in \mathcal{C}$, the functor $E(i)^*(y_{\mathcal{C}}(c)) : \mathcal{D} \to \mathbf{Set}$ coincides with the functor $\mathrm{Hom}_{\mathcal{C}}(c, i(-)) : \mathcal{D} \to \mathbf{Set}$. Its category of elements $\mathcal{A}_c$ has as objects the pairs (d, h) where d is an object of $\mathcal{D}$ and h is an arrow $c \to d$ in $\mathcal{C}$ and as arrows $(d, h) \to (d', h')$ the arrows $k : d' \to d$ in $\mathcal{D}$ such that $k \circ h' = h$ in $\mathcal{C}$. As f_F^* preserves all small colimits and $\mathrm{Hom}_{\mathcal{C}}(c, i(-))$ is the colimit in $[\mathcal{D}, \mathbf{Set}]$ of the composition of the canonical projection $\pi_c : \mathcal{A}_c \to \mathcal{D}^{\mathrm{op}}$ with the Yoneda embedding $\mathcal{D}^{\mathrm{op}} \to [\mathcal{D}, \mathbf{Set}]$, the object $\tilde{F}(c)$ can be identified with the colimit of the composite functor $F \circ \pi_c : \mathcal{A}_c \to \mathcal{E}$. For any object d of $\mathcal{D}$, the pair $(d, 1_d)$ yields an object of the category $\mathcal{A}_d$ and hence a canonical colimit arrow $\chi_d : F(d) \to \tilde{F}(d)$, also denoted by χ_d^F. For any object (d, h) of $\mathcal{A}_c$, the colimit arrow $\kappa_{(d,h)} : F(d) \to \tilde{F}(c)$ is equal to the composite $\tilde{F}(h) \circ \chi_d$.

Remarks 5.2.2 (a) The functor $\xi_{\mathcal{E}} : \mathbf{Flat}(\mathcal{D}^{\mathrm{op}}, \mathcal{E}) \to \mathbf{Flat}(\mathcal{C}^{\mathrm{op}}, \mathcal{E})$ coincides with the (restriction to the subcategories of flat functors of the) left Kan extension functor

$$[\mathcal{D}^{\mathrm{op}}, \mathcal{E}] \to [\mathcal{C}^{\mathrm{op}}, \mathcal{E}]$$

along the inclusion $\mathcal{D}^{\mathrm{op}} \hookrightarrow \mathcal{C}^{\mathrm{op}}$ (cf. Remark 5.2.13(a)).

(b) For any object $d \in \mathcal{D}$ and any natural transformation $\alpha : F \to G$ between functors in $\mathbf{Flat}(\mathcal{D}^{\mathrm{op}}, \mathcal{E})$, the diagram

$$\begin{array}{ccc} \tilde{F}(d) & \xrightarrow{\tilde{\alpha}(d)} & \tilde{G}(d) \\ \chi_d^F \uparrow & & \uparrow \chi_d^G \\ F(d) & \xrightarrow{\alpha(d)} & G(d) \end{array}$$

commutes (cf. Proposition 5.2.1(a)).

(c) For any object $d \in \mathcal{D}$, the arrow $\chi_d : F(d) \to \tilde{F}(d)$ is monic. Indeed, it can be identified with the image of the canonical subfunctor inclusion $\mathrm{Hom}_{\mathcal{D}}(d, -) \hookrightarrow \mathrm{Hom}_{\mathcal{C}}(d, i(-))$ under the inverse image f_F^* of the geometric morphism f_F. From this, in view of Remark 5.2.2(b), it follows at once that the functor $\xi_{\mathcal{E}} : \mathbf{Flat}(\mathcal{D}^{\mathrm{op}}, \mathcal{E}) \to \mathbf{Flat}(\mathcal{C}^{\mathrm{op}}, \mathcal{E})$ is faithful.

(d) For any flat functors $F, G : \mathcal{D}^{\mathrm{op}} \to \mathcal{E}$ and any natural transformation $\beta : \tilde{F} \to \tilde{G}$, $\beta = \tilde{\alpha}$ for some natural transformation $\alpha : F \to G$ if and only if for every $d \in \mathcal{D}$ the diagram

$$\begin{array}{ccc} \tilde{F}(d) & \xrightarrow{\beta(d)} & \tilde{G}(d) \\ \chi_d^F \uparrow & & \uparrow \chi_d^G \\ F(d) & \xrightarrow{\alpha(d)} & G(d) \end{array}$$

commutes. One direction follows from Remark 5.2.2(b). To prove the other one, we observe that $\beta(c) = \tilde{\alpha}(c)$ for all $c \in \mathcal{C}$ if and only if for every object $(a, z) \in \mathcal{A}_c$, $\beta(c) \circ \kappa_{(a,z)} = \tilde{\alpha}(c) \circ \kappa_{(a,z)}$. But $\beta(c) \circ \kappa_{(a,z)} = \beta(c) \circ \tilde{F}(z) \circ \chi_a^F = \tilde{G}(z) \circ \beta(a) \circ \chi_a^F = \tilde{G}(z) \circ \chi_a^G \circ \alpha(a) = \kappa_{(a,z)}^G \circ \alpha(a) = \tilde{\alpha}(c) \circ \kappa_{(a,z)}^F$ (where the third equality follows from the naturality of β and the last from Proposition 5.2.1(a)), as required.

(e) Let d be an object of $\mathcal{D}$ and $y_\mathcal{D}(d) = \mathrm{Hom}_\mathcal{D}(-, d) : \mathcal{D}^{\mathrm{op}} \to \mathbf{Set}$ the representable functor on $\mathcal{D}$ associated with d. Then $\xi_\mathcal{E}(y_\mathcal{D}(d)) = y_\mathcal{C}(d)$, where $y_\mathcal{C}(d) = \mathrm{Hom}_\mathcal{C}(-, d)$ is the representable functor on $\mathcal{C}$ associated with d (considered here as an object of $\mathcal{C}$). Indeed, the flat functor $y_\mathcal{D}(d)$ corresponds to the geometric morphism whose inverse image is the evaluation functor $\mathrm{ev}_d^\mathcal{D} : [\mathcal{D}, \mathbf{Set}] \to \mathbf{Set}$ at the object d, and the composite $\mathrm{ev}_d^\mathcal{D} \circ E(i)^*$ coincides with the evaluation functor $\mathrm{ev}_d^\mathcal{C} : [\mathcal{C}, \mathbf{Set}] \to \mathbf{Set}$ at the object d, which corresponds to the flat functor $y_\mathcal{C}(d)$.

(f) Let $\mathcal{E}$ be a Grothendieck topos and $\gamma_\mathcal{E} : \mathcal{E} \to \mathbf{Set}$ the unique (up to isomorphism) geometric morphism from $\mathcal{E}$ to $\mathbf{Set}$. Then, by Proposition 5.2.1(ii) and Remark 5.2.2(e), for any object d of $\mathcal{D}$ the functor $\xi_\mathcal{E} : \mathbf{Flat}(\mathcal{D}^{\mathrm{op}}, \mathcal{E}) \to \mathbf{Flat}(\mathcal{C}^{\mathrm{op}}, \mathcal{E})$ sends the flat functor $\gamma_\mathcal{E}^* \circ y_\mathcal{D}(d)$ to the functor $\gamma_\mathcal{E}^* \circ y_\mathcal{C}(d)$.

Proposition 5.2.3 *Under the natural equivalences*

$$e_\mathcal{C} : \mathrm{Ind}\text{-}\mathcal{C} \simeq \mathbf{Flat}(\mathcal{C}^{\mathrm{op}}, \mathbf{Set})$$

and

$$e_\mathcal{D} : \mathrm{Ind}\text{-}\mathcal{D} \simeq \mathbf{Flat}(\mathcal{D}^{\mathrm{op}}, \mathbf{Set}),$$

the functor

$$\xi_{\mathbf{Set}} : \mathbf{Flat}(\mathcal{D}^{\mathrm{op}}, \mathbf{Set}) \to \mathbf{Flat}(\mathcal{C}^{\mathrm{op}}, \mathbf{Set})$$

corresponds to the functor

$$\mathrm{Ind}\text{-}i : \mathrm{Ind}\text{-}\mathcal{D} \to \mathrm{Ind}\text{-}\mathcal{C}.$$

Proof First, let us recall the definition of the functor $\mathrm{Ind}\text{-}i : \mathrm{Ind}\text{-}\mathcal{D} \to \mathrm{Ind}\text{-}\mathcal{C}$. For any flat functor $F : \mathcal{D}^{\mathrm{op}} \to \mathbf{Set}$, $(\mathrm{Ind}\text{-}i)(F)$ is the functor given by the colimit in $\mathrm{Ind}\text{-}\mathcal{C}$ of the composite functor $i \circ \pi_F$, where $\pi_F : \int F \to \mathcal{D}$ is the canonical projection from the category of elements $\int F$ of F to $\mathcal{D}$. For any $c \in \mathcal{C}$, we thus have Ind-$i(F)(c) = \text{colim}(\text{ev}_c \circ i \circ \pi_F)$, where $\text{ev}_c : \text{Ind-}\mathcal{C} \to \mathbf{Set}$ is the evaluation functor at c.

Now, the functor $\text{ev}_c \circ i : \mathcal{D} \to \mathbf{Set}$ is equal to the functor $\text{Hom}_\mathcal{C}(c, i(-)) : \mathcal{D} \to \mathbf{Set}$ considered above, and $\tilde{F}(c) = \text{colim}(F \circ \pi_c) : \mathcal{A}_c \to \mathbf{Set}$, where $\mathcal{A}_c$ is the category of elements of $\text{ev}_c \circ i$ and $\pi_c : \mathcal{A}_c \to \mathcal{D}^{\text{op}}$ the associated canonical projection. The commutativity of the tensor product between a presheaf and a covariant set-valued functor (cf. chapter VII of [63] or Theorem 5.1.5 above) thus yields a natural isomorphism between the two sets, as required. □

Corollary 5.2.4 *Let $\mathcal{D} \hookrightarrow \mathcal{C}$ be an embedding of small categories and $F : \mathcal{D}^{\text{op}} \to \mathcal{E}$ a flat functor. With the above notation, for any object $c \in \mathcal{C}$ we have*

$$\tilde{F}(c) \cong \left(\coprod_{(a,z) \in \mathcal{A}_c} F(a) \right) / R_c,$$

where R_c is the equivalence relation in $\mathcal{E}$ defined by saying that, for any objects $(a,z), (a',z')$ of the category $\mathcal{A}_c$, the geometric bisequent

$$\left(\ulcorner R_c \urcorner (\ulcorner \xi_{(a,z)} \urcorner (x), \ulcorner \xi_{(a',z')} \urcorner (x')) \dashv\vdash_{x^{F(a)}, x'^{F(a')}} \bigvee_{\substack{a \xrightarrow{f} b \xleftarrow{g} a' \mid \\ f \circ z = g \circ z'}} (\exists y^{F(b)})(\ulcorner F(f) \urcorner (y) = x \wedge \ulcorner F(g) \urcorner (y) = x') \right)$$

is valid in the tautological $\Sigma_\mathcal{E}$-structure $S_\mathcal{E}$, where, for any object (a,z) of the category $\mathcal{A}_c$, $\xi_{(a,z)} : F(a) \to \coprod_{(a,z) \in \mathcal{A}_c} F(a)$ is the canonical coproduct arrow.
In particular, for any objects $(a,z), (a,z')$ of the category $\mathcal{A}_c$, the geometric bisequent

$$\left(\ulcorner R_c \urcorner (\ulcorner \xi_{(a,z)} \urcorner (x), \ulcorner \xi_{(a,z')} \urcorner (x)) \dashv\vdash_{x^{F(a)}} \bigvee_{\substack{a \xrightarrow{f} b \mid \\ f \circ z = f \circ z'}} (\exists y^{F(b)})(\ulcorner F(f) \urcorner (y) = x) \right)$$

is valid in $S_\mathcal{E}$.

Semantically, the relation R_c can be characterized by saying that, for any objects $(a,z), (a',z')$ of the category $\mathcal{A}_c$ and any generalized elements $x : E \to F(a)$, $x' : E \to F(a')$, $(\xi_{(a,z)} \circ x, \xi_{(a',z')} \circ x') \in R_c$ if and only if there exist an epimorphic family $\{e_i : E_i \to E \mid i \in I\}$ in $\mathcal{E}$ and for each $i \in I$ an object $b_i \in \mathcal{D}$, a generalized element $h_i : E_i \to F(b_i)$ and two arrows $f_i : a \to b_i$ and $f'_i : a' \to b_i$ in $\mathcal{D}$ such that $f'_i \circ z' = f_i \circ z$ and $< F(f_i), F(f'_i) > \circ h_i = < x, x' > \circ e_i$. In particular, for any objects (a,z) and (a,z') of $\mathcal{A}_c$ and any generalized element $x : E \to F(a)$, $(\xi_{(a,z)} \circ x, \xi_{(a,z')} \circ x) \in R_c$ if and only if there exist an epimorphic family $\{e_i : E_i \to E \mid i \in I\}$ in $\mathcal{E}$ and for each $i \in I$ an object $b_i \in \mathcal{D}$, a generalized element $h_i : E_i \to F(b_i)$ and an arrow $f_i : a \to b_i$ in $\mathcal{D}$ such that $f_i \circ z' = f_i \circ z$ and $F(f_i) \circ h_i = x \circ e_i$.

Proof Our thesis follows by applying Proposition 5.1.33 (and its categorical reformulation provided by Proposition 5.1.34) to the flat functor $F : \mathcal{D}^{\mathrm{op}} \to \mathcal{E}$ and the functor $P : \mathcal{D} \to \mathbf{Set}$ given by $\mathrm{ev}_c \circ i = \mathrm{Hom}_\mathcal{C}(c, i(-)) : \mathcal{D} \to \mathbf{Set}$, whose category of elements coincides with $\mathcal{A}_c$ and whose associated fibration π_P coincides with $\pi_c : \mathcal{A}_c \to \mathcal{D}^{\mathrm{op}}$. $\square$

5.2.3 Extensions from categories of set-based models to syntactic categories

Let $\mathbb{T}$ be a geometric theory over a signature Σ and $\mathcal{K}$ a small category of set-based $\mathbb{T}$-models. The family of geometric morphisms $\mathbf{Set} \to \mathbf{Sh}(\mathcal{C}_\mathbb{T}, J_\mathbb{T})$ corresponding to the $\mathbb{T}$-models in $\mathcal{K}$ induces a geometric morphism

$$p_\mathcal{K} : [\mathcal{K}, \mathbf{Set}] \to \mathbf{Sh}(\mathcal{C}_\mathbb{T}, J_\mathbb{T})$$

whose associated $\mathbb{T}$-model in $[\mathcal{K}, \mathbf{Set}]$ is given, at each sort, by the corresponding forgetful functor. We thus have, for each Grothendieck topos $\mathcal{E}$, an induced functor

$$u^\mathbb{T}_{(\mathcal{K}, \mathcal{E})} : \mathbf{Flat}(\mathcal{K}^{\mathrm{op}}, \mathcal{E}) \to \mathbf{Flat}_{J_\mathbb{T}}(\mathcal{C}_\mathbb{T}, \mathcal{E}),$$

which the following theorem describes explicitly. Before stating this result, we need to introduce some notation. For any object $\{\vec{x} . \phi\}$ of $\mathcal{C}_\mathbb{T}$, we write $\mathcal{A}^\mathcal{K}_{\{\vec{x} . \phi\}}$ (or simply $\mathcal{A}_{\{\vec{x} . \phi\}}$ when the category $\mathcal{K}$ can be obviously inferred from the context) for the category whose objects are the pairs (M, w), where $M \in \mathcal{K}$ and $w \in [[\vec{x} . \phi]]_M$, and whose arrows $(M, w) \to (N, z)$ are the $\mathbb{T}$-model homomorphisms $g : N \to M$ in $\mathcal{K}$ such that $g(z) = w$. We denote by $\pi^\mathcal{K}_{\{\vec{x} . \phi\}} : \mathcal{A}_{\{\vec{x} . \phi\}} \to \mathcal{K}^{\mathrm{op}}$ the canonical projection functor.

Theorem 5.2.5 *Let $\mathbb{T}$ be a geometric theory over a signature Σ and $\mathcal{K}$ a small category of set-based $\mathbb{T}$-models. Then for any flat functor $F : \mathcal{K}^{\mathrm{op}} \to \mathcal{E}$ the functor $\tilde{F} := u^\mathbb{T}_{(\mathcal{K}, \mathcal{E})}(F) : \mathcal{C}_\mathbb{T} \to \mathcal{E}$ sends any object $\{\vec{x} . \phi\}$ of $\mathcal{C}_\mathbb{T}$ to the colimit $\mathrm{colim}(F \circ \pi^\mathcal{K}_{\{\vec{x} . \phi\}})$ and acts on the arrows in the obvious way. In particular, for any formula $\{\vec{x} . \phi\}$ presenting a $\mathbb{T}$-model $M_{\{\vec{x} . \phi\}}$ in $\mathcal{K}$ (in the sense of Definition 6.1.11), $u^\mathbb{T}_{(\mathcal{K}, \mathcal{E})}(F)(\{\vec{x} . \phi\}) \cong F(M_{\{\vec{x} . \phi\}})$.*

Proof Let $g_F : \mathcal{E} \to [\mathcal{K}, \mathbf{Set}]$ be the geometric morphism corresponding, via Diaconescu's equivalence, to the flat functor F. Then the functor $u^\mathbb{T}_{(\mathcal{K}, \mathcal{E})}(F)$ is equal to the composite $g_F^* \circ p_\mathcal{K}^* \circ y$, where $y : \mathcal{C}_\mathbb{T} \to \mathbf{Sh}(\mathcal{C}_\mathbb{T}, J_\mathbb{T})$ is the Yoneda embedding.

Now, for any geometric formula $\{\vec{x} . \phi\}$ over Σ, $p_\mathcal{K}^*(y(\{\vec{x} . \phi\}))$ is the functor $F_{\{\vec{x} . \phi\}}$ sending any model $M \in \mathcal{K}$ to the set $[[\vec{x} . \phi]]_M$. This functor can be expressed as the colimit $\mathrm{colim}(y' \circ \pi^\mathcal{K}_{\{\vec{x} . \phi\}})$, where $y' : \mathcal{K}^{\mathrm{op}} \to [\mathcal{K}, \mathbf{Set}]$ is the Yoneda embedding, since the functor $\pi^\mathcal{K}_{\{\vec{x} . \phi\}}$ coincides with the canonical projection from the category of elements of the functor $F_{\{\vec{x} . \phi\}}$ to $\mathcal{K}^{\mathrm{op}}$.

Therefore we have $u^{\mathbb{T}}_{(\mathcal{K},\mathcal{E})}(F)(\{\vec{x}.\phi\}) = g_F^*(\text{colim}(y' \circ \pi_{\{\vec{x}.\phi\}})) \cong \text{colim}(g_F^* \circ y' \circ \pi^{\mathcal{K}}_{\{\vec{x}.\phi\}}) \cong \text{colim}(F \circ \pi^{\mathcal{K}}_{\{\vec{x}.\phi\}})$, as required. □

Remark 5.2.6 From the proof of Theorem 5.2.5 it is clear that the isomorphism $u^{\mathbb{T}}_{(\mathcal{K},\mathcal{E})}(F)(\{\vec{x}.\phi\}) \cong F(M_{\{\vec{x}.\phi\}})$ is natural in $\{\vec{x}.\phi\}$, that is for any geometric formulae $\{\vec{x}.\phi\}$ and $\{\vec{y}.\psi\}$ respectively presenting $\mathbb{T}$-models $M_{\{\vec{x}.\phi\}}$ and $M_{\{\vec{y}.\psi\}}$ and any $\mathbb{T}$-provably functional formula $\theta : \{\vec{x}.\phi\} \to \{\vec{y}.\psi\}$ inducing a $\mathbb{T}$-model homomorphism $M_{[\theta]} : M_{\{\vec{y}.\psi\}} \to M_{\{\vec{x}.\phi\}}$ (defined by $M_{[\theta]}(\vec{\xi_\psi}) = [[\vec{x},\vec{y}.\theta]]_{M_{\{\vec{x}.\phi\}}}(\vec{\xi_\phi})$, where $\vec{\xi_\phi}$ and $\vec{\xi_\psi}$ are respectively the tuples of generators of $M_{\{\vec{x}.\phi\}}$ and of $M_{\{\vec{y}.\psi\}}$ and $[[\vec{x},\vec{y}.\theta]]_{M_{\{\vec{x}.\phi\}}}(\vec{\xi_\phi})$ is the unique element $z \in [[\vec{y}.\psi]]_{M_{\{\vec{x}.\phi\}}}$ such that $(\vec{\xi_\phi}, z) \in [[\vec{x},\vec{y}.\theta]]_{M_{\{\vec{x}.\phi\}}}$), the arrow $u^{\mathbb{T}}_{(\mathcal{K},\mathcal{E})}(F)(\theta)$ corresponds to the arrow $F(M_{[\theta]})$ via the isomorphisms $u^{\mathbb{T}}_{(\mathcal{K},\mathcal{E})}(F)(\{\vec{x}.\phi\}) \cong F(M_{\{\vec{x}.\phi\}})$ and $u^{\mathbb{T}}_{(\mathcal{K},\mathcal{E})}(F)(\{\vec{y}.\psi\}) \cong F(M_{\{\vec{y}.\psi\}})$.

Corollary 5.2.7 *Let $\mathbb{T}$ be a geometric theory over a signature Σ, $\mathcal{K}$ a small category of set-based $\mathbb{T}$-models, $\sigma \equiv (\phi \vdash_{\vec{x}} \psi)$ a geometric sequent over Σ and $F : \mathcal{K}^{\text{op}} \to \mathcal{E}$ a flat functor. If σ is valid in every $\mathbb{T}$-model in $\mathcal{K}$ then $\tilde{F}(\{\vec{x}.\phi\}) \leq \tilde{F}(\{\vec{x}.\psi\})$ as subobjects of $\tilde{F}(\{\vec{x}.\top\})$ in $\mathcal{E}$.*

Proof By Theorem 5.2.5 we have $\tilde{F}(\{\vec{x}.\phi\}) = \text{colim}(F \circ \pi^{\mathcal{K}}_{\{\vec{x}.\phi\}})$, $\tilde{F}(\{\vec{x}.\psi\}) = \text{colim}(F \circ \pi^{\mathcal{K}}_{\{\vec{x}.\psi\}})$ and $\tilde{F}(\{\vec{x}.\top\}) = \text{colim}(F \circ \pi^{\mathcal{K}}_{\{\vec{x}.\top\}})$. Now, $\mathcal{A}_{\{\vec{x}.\phi\}}$ and $\mathcal{A}_{\{\vec{x}.\psi\}}$ canonically embed as subcategories of $\mathcal{A}_{\{\vec{x}.\top\}}$ and if σ is valid in every $\mathbb{T}$-model in $\mathcal{K}$ then we have a canonical functor $i : \mathcal{A}_{\{\vec{x}.\phi\}} \to \mathcal{A}_{\{\vec{x}.\psi\}}$ which commutes with these embeddings. It thus follows from the functoriality of colimits that $\tilde{F}(\{\vec{x}.\phi\}) \leq \tilde{F}(\{\vec{x}.\psi\})$ as subobjects of $\tilde{F}(\{\vec{x}.\top\})$ in $\mathcal{E}$, as required. □

Let us now apply Proposition 5.1.34 in the context of extensions $F \to \tilde{F}$ of flat functors induced by the geometric morphism

$$p_{\mathcal{K}} : [\mathcal{K}, \mathbf{Set}] \to \mathbf{Sh}(\mathcal{C}_{\mathbb{T}}, J_{\mathbb{T}}).$$

Proposition 5.2.8 *Let $\mathbb{T}$ be a geometric theory over a signature Σ, $\mathcal{K}$ a small category of set-based $\mathbb{T}$-models, $\mathcal{E}$ a Grothendieck topos with a separating family of objects $\mathcal{S}$ and $F : \mathcal{K}^{\text{op}} \to \mathcal{E}$ a flat functor. With the above notation, for any geometric formula-in-context $\phi(\vec{x})$ over Σ, we have*

$$\tilde{F}(\{\vec{x}.\phi\}) \cong \left(\coprod_{(M,z) \in \mathcal{A}_{\{\vec{x}.\phi\}}} F(M) \right) / R_{\{\vec{x}.\phi\}},$$

where $R_{\{\vec{x}.\phi\}}$ is the equivalence relation in $\mathcal{E}$ defined by saying that for any objects $(M,z), (N,w)$ of the category $\mathcal{A}_{\{\vec{x}.\phi\}}$ and any generalized elements $x : E \to F(M)$, $x' : E \to F(N)$ (where $E \in \mathcal{S}$), we have $(\xi_{(M,z)} \circ x, \xi_{(N,w)} \circ x') \in R_{\{\vec{x}.\phi\}}$ if and only if there exist an epimorphic family $\{e_i : E_i \to E \mid i \in I, E_i \in \mathcal{S}\}$ in $\mathcal{E}$ and for each $i \in I$ a $\mathbb{T}$-model a_i in $\mathcal{K}$, a generalized element

$h_i : E_i \to F(a_i)$ and two $\mathbb{T}$-model homomorphisms $f_i : M \to a_i$ and $f'_i : N \to a_i$ in $\mathcal{K}$ such that $f_i(z) = f'_i(w)$ and $< F(f_i), F(f'_i) > \circ h_i =< x, x' > \circ e_i$ (where for any object (M, z) of the category $\mathcal{A}_{\{\vec{x} \cdot \phi\}}$, $\xi_{(M,z)} : F(M) \to \tilde{F}(\{\vec{x} \cdot \phi\})$ is the canonical colimit arrow).

In particular, for any objects (M, z) and (M, w) of $\mathcal{A}_{\{\vec{x} \cdot \phi\}}$ and any generalized element $x : E \to F(M)$ (where $E \in \mathcal{S}$), $(\xi_{(M,z)} \circ x, \xi_{(M,w)} \circ x) \in R_{\{\vec{x} \cdot \phi\}}$ if and only if there exist an epimorphic family $\{e_i : E_i \to E \mid i \in I, E_i \in \mathcal{S}\}$ in $\mathcal{E}$ and for each $i \in I$ a $\mathbb{T}$-model a_i in $\mathcal{K}$, a generalized element $h_i : E_i \to F(a_i)$ and a $\mathbb{T}$-model homomorphism $f_i : M \to a_i$ in $\mathcal{K}$ such that $f_i(z) = f_i(w)$ and $F(f_i) \circ h_i = x \circ e_i$. $\square$

Let $M_{\mathbb{T}}$ be the universal model of $\mathbb{T}$ in its syntactic category $\mathcal{C}_{\mathbb{T}}$ (as in Definition 1.4.4). As we shall see below, for any small subcategory $\mathcal{K}$ of the category $\mathbb{T}$-mod(**Set**) and any flat functor $F : \mathcal{K}^{\mathrm{op}} \to \mathcal{E}$, we can represent the model $\tilde{F}(M_{\mathbb{T}})$ as an $\mathcal{E}$-indexed filtered colimit of 'constant' $\mathbb{T}$-models coming from $\mathcal{K}$. For simplicity, we shall first establish an explicit characterization of this colimit representation in the case $\mathcal{E} = $ **Set**, and then generalize it to an arbitrary Grothendieck topos; finally, we shall discuss the abstract characterization of $\tilde{F}(M_{\mathbb{T}})$ as an $\mathcal{E}$-indexed colimit by introducing the $\mathcal{E}$-indexed category $\underline{\mathbb{T}\text{-mod}(\mathcal{E})}$ of models of the theory $\mathbb{T}$ in slices of $\mathcal{E}$.

So let $F : \mathcal{K}^{\mathrm{op}} \to $ **Set** be a flat functor. By Theorem 5.2.5, for any sort A of Σ, $\tilde{F}(M_{\mathbb{T}})A = \tilde{F}(\{x^A . \top\}) = \mathrm{colim}(F \circ \pi_{\{x^A . \top\}})$, where $\pi_{\{x^A . \top\}}$ is the canonical projection functor $\mathcal{A}_{\{x^A . \top\}} \to \mathcal{K}^{\mathrm{op}}$ from the category of elements $\mathcal{A}_{\{x^A . \top\}}$ of the functor $P_A : \mathcal{K} \to $ **Set** which assigns any model M in $\mathcal{K}$ to the set MA and acts accordingly on the arrows. Now, it follows from the commutativity of the tensor product between a presheaf and a covariant set-valued functor (cf. Chapter VII of [63] or section 5.1.5 above) that $\mathrm{colim}(F \circ \pi_{\{x^A . \top\}}) \cong \mathrm{colim}(P_A \circ \pi_F)$, where $\pi_F : \int F \to \mathcal{K}$ is the canonical projection functor from the category of elements of the functor F to $\mathcal{K}$. Since F is flat, the category $\int F$ is filtered. Therefore, as filtered colimits in $\mathbb{T}$-mod(**Set**) are computed sortwise as in **Set**, $\tilde{F}(M_{\mathbb{T}}) \cong \mathrm{colim}(i \circ \pi_F)$, where $i : \mathcal{K} \hookrightarrow \mathbb{T}$-mod(**Set**) is the canonical inclusion functor. So for any object (c, x) of the category $\int F$ we have a $\mathbb{T}$-model homomorphism $\xi_{(c,x)} : c \to \tilde{F}(M_{\mathbb{T}})$, which can be expressed in terms of the colimit arrows $\kappa_{(a,y)} : F(c) \to \tilde{F}(\{x^A . \top\}) = \tilde{F}(M_{\mathbb{T}})A$ (for $y \in cA$ and A a sort of Σ) as follows: for any sort A of Σ, $\xi_{(c,x)_A}(y) = \kappa_{(c,y)}(x)$. The explicit description of filtered colimits in the category **Set** thus yields, for each sort A of Σ, the following characterizing properties of the colimit $\mathrm{colim}(P_A \circ \pi_F)$ in terms of the arrows $\xi_{(c,x)}$:

(i) For any element x of $\tilde{F}(M_{\mathbb{T}})A$, there exist an object (c, x) of the category $\int F$ and an element y of cA such that $\xi_{(c,x)_A}(y) = x$.

(ii) For any objects (c, x) and (c', x') of the category $\int F$ and elements $y \in cA$ and $y' \in c'A$, we have $\xi_{(c,x)_A}(y) = \xi_{(c',x')_A}(y')$ if and only if there exist an object (c'', x'') of $\int F$ and arrows $f : c \to c''$ and $g : c' \to c''$ in $\mathcal{K}$ such that $F(f)(x'') = x$, $F(g)(x'') = x'$ and $f_A(y) = g_A(y')$.

Moreover, the filteredness of the category $\int F$ implies the following 'joint embedding property': for any objects (c, x) and (c', x') of the category $\int F$, there exist an object (c'', x'') of $\int F$ and arrows $f : c \to c''$ and $g : c' \to c''$ in $\mathcal{K}$ such that $F(f)(x'') = x$, $F(g)(x'') = x'$ and $\xi_{(c'', x'')} \circ f = \xi_{(c,x)}$, $\xi_{(c'', x'')} \circ g = \xi_{(c', x')}$.

Let us now proceed to establish the $\mathcal{E}$-indexed generalization of these characterizing properties of the model $\tilde{F}(M_{\mathbb{T}})$.

Let $F : \mathcal{K}^{\mathrm{op}} \to \mathcal{E}$ be a flat functor. By Theorem 5.2.5, for any sort A of Σ we have that $\tilde{F}(M_{\mathbb{T}})A = \tilde{F}(\{x^A . \top\}) = \mathrm{colim}(F \circ \pi_{\{x^A . \top\}})$; in particular, for any object (a, y) of the category $\mathcal{A}_{\{x^A . \top\}}$ we have a colimit arrow $\kappa_{(a,y)} : F(a) \to \tilde{F}(\{x^A . \top\}) = \tilde{F}(M_{\mathbb{T}})A$ in $\mathcal{E}$. More generally, for any formula-in-context $\phi(\vec{x})$ over Σ, any model c in $\mathcal{K}$ and any element $\vec{y}$ of $[[\vec{x} . \phi]]_c$, we have a colimit arrow $\kappa_{(c, \vec{y})} : F(c) \to \tilde{F}(\{\vec{x} . \phi\})$.

Proposition 5.2.9 *Let $\mathbb{T}$ be a geometric theory over a signature Σ, $\mathcal{K}$ a small subcategory of the category $\mathbb{T}$-mod(**Set**), $\mathcal{E}$ a Grothendieck topos and $F : \mathcal{K}^{\mathrm{op}} \to \mathcal{E}$ a flat functor. With the above notation, for any pair (c, x) consisting of an object c of $\mathcal{K}$ and a generalized element $x : E \to F(c)$, there is a Σ-structure homomorphism $\xi_{(c,x)} : c \to \mathrm{Hom}_{\mathcal{E}}(E, \tilde{F}(M_{\mathbb{T}}))$ defined as follows: for any sort A over Σ, the function $\xi_{(c,x)_A} : cA \to \mathrm{Hom}_{\mathcal{E}}(E, \tilde{F}(\{x^A . \top\}))$ sends any element $y \in cA$ to the generalized element $\kappa_{(c,y)} \circ x : E \to \tilde{F}(\{x^A . \top\})$.*

Proof We have to verify that:

(1) For any function symbol $f : A_1 \cdots A_n \to B$ of Σ, the diagram

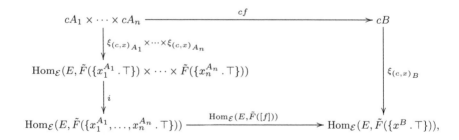

where i is the canonical isomorphism $\tilde{F}(\{x_1^{A_1} . \top\}) \times \cdots \times \tilde{F}(\{x_n^{A_n} . \top\}) \cong \tilde{F}(\{x_1^{A_1}, \ldots, x_n^{A_n} . \top\})$ induced by the preservation of finite products by $\tilde{F}$ and $[f]$ is the arrow $[x^B = f(x_1^{A_1}, \ldots, x_n^{A_n})] : \{x_1^{A_1}, \ldots, x_n^{A_n} . \top\} \to \{x^B . \top\}$, commutes;

(2) For any relation symbol $R \rightarrowtail A_1 \cdots A_n$ of Σ, we have a commutative diagram

$$\begin{CD} cR @>>> \operatorname{Hom}_{\mathcal{E}}(E, \tilde{F}(\{x_1^{A_1}, \ldots, x_n^{A_n} . R\})) \\ @VVV @VVV \\ cA_1 \times \cdots \times cA_n @>>> \operatorname{Hom}_{\mathcal{E}}(E, \tilde{F}(\{x_1^{A_1}, \ldots, x_n^{A_n} . \top\})), \end{CD}$$

where the unnamed vertical arrows are the canonical ones and the lower horizontal arrow is $\operatorname{Hom}_{\mathcal{E}}(E, i) \circ (\xi_{(c,x)_{A_1}} \times \cdots \times \xi_{(c,x)_{A_n}})$.

To prove (1), we first observe that for any n-tuple $\vec{y} = (y_1, \ldots, y_n) \in cA_1 \times \cdots \times cA_n$, $i \circ < \kappa_{(c,y_1)}, \ldots, \kappa_{(c,y_n)} >= \kappa_{(c,\vec{y})}$. Indeed, for any $i \in \{1, \ldots, n\}$, the arrow $\tilde{F}(\pi_i) \circ i$, where $\pi_i : \{x_1^{A_1}, \ldots, x_n^{A_n} . \top\} \to \{x^{A_i} . \top\}$ is the canonical projection arrow in the syntactic category $\mathcal{C}_\mathbb{T}$, is equal to the i-th product projection $\tilde{F}(\{x_1^{A_1} . \top\}) \times \cdots \times \tilde{F}(\{x_n^{A_n} . \top\}) \to \tilde{F}(\{x^{A_i} . \top\})$ by Proposition 5.2.1(iii). Therefore, to prove the required condition it is equivalent to verify that for any n-tuple $\vec{y} = (y_1, \ldots, y_n) \in cA_1 \times \cdots \times cA_n$, $\xi_{(c,x)_B}(cf(\vec{y})) = \tilde{F}([f]) \circ \kappa_{(c,\vec{y})} \circ x$. But $\xi_{(c,x)_B}(cf(\vec{y})) = \kappa_{(c,cf(\vec{y}))} \circ x$, and we have that $\tilde{F}([f]) \circ \kappa_{(c,\vec{y})} = \kappa_{(c,cf(\vec{y}))}$, again by Proposition 5.2.1(iii), as required.

To prove (2), it suffices to observe that, by Proposition 5.2.1(iii), for any n-tuple $\vec{y} = (y_1, \ldots, y_n) \in cA_1 \times \cdots \times cA_n$ in cR, $\kappa_{(c,\vec{y})}$ factors through the canonical subobject $\tilde{F}(R) \rightarrowtail \tilde{F}(\{x_1^{A_1}, \ldots, x_n^{A_n} . \top\})$. $\square$

The following lemma describes some basic properties of the homomorphisms $\xi_{(c,x)}$.

Lemma 5.2.10 *Let $\mathbb{T}$ be a geometric theory over a signature Σ, $\mathcal{K}$ a small subcategory of the category $\mathbb{T}\text{-mod}(\mathbf{Set})$ and $F : \mathcal{K}^{\operatorname{op}} \to \mathcal{E}$ a flat functor with values in a Grothendieck topos $\mathcal{E}$. Then*

(i) *For any generalized element $x : E \to F(c)$ and any arrow $f : d \to c$ in $\mathcal{K}$,*
$\xi_{(c,x)} \circ f = \xi_{(c,F(f) \circ x)}$.
(ii) *For any generalized element $x : E \to F(c)$ and any arrow $e : E' \to E$,*
$\xi_{(c,x \circ e)} = \operatorname{Hom}(e, \tilde{F}(M_\mathbb{T})) \circ \xi_{(c,x)}$.

Proof These properties can be easily proved by using the definition of the arrows $\xi_{(c,x)}$ in terms of the arrows $\kappa_{(a,y)}$ and Proposition 5.2.1(iii). $\square$

Proposition 5.2.11 *Let $\mathbb{T}$ be a geometric theory over a signature Σ, $\mathcal{K}$ a small subcategory of the category $\mathbb{T}\text{-mod}(\mathbf{Set})$ and $F : \mathcal{K}^{\operatorname{op}} \to \mathcal{E}$ a flat functor with values in a Grothendieck topos $\mathcal{E}$. Let M be the $\mathbb{T}$-model $\tilde{F}(M_\mathbb{T})$ in $\mathcal{E}$. Then, for any sort A over Σ, we have that*

(i) *For any generalized element $x : E \to MA$, there exist an epimorphic family $\{e_i : E_i \to E \mid i \in I\}$ in $\mathcal{E}$ and for each $i \in I$ a $\mathbb{T}$-model a_i in $\mathcal{K}$, a generalized element $x_i : E_i \to F(a_i)$ and an element $y_i \in a_iA$ such that $\xi_{(a_i,x_i)_A}(y_i) = x \circ e_i$.*
(ii) *For any pairs (a, x) and (b, x'), where a and b are $\mathbb{T}$-models in $\mathcal{K}$ and $x : E \to F(a)$, $x' : E \to F(b)$ are generalized elements, and any elements*

$y \in aA$ and $y' \in bA$, we have that $\xi_{(a,x)_A}(y) = \xi_{(b,x')_A}(y')$ if and only if there exist an epimorphic family $\{e_i : E_i \to E \mid i \in I\}$ in $\mathcal{E}$ and for each $i \in I$ a $\mathbb{T}$-model c_i in $\mathcal{K}$, arrows $f_i : a \to c_i$, $g_i : b \to c_i$ in $\mathcal{K}$ and a generalized element $x_i : E_i \to F(c_i)$ such that $< x, x' > \circ e_i = < F(f_i), F(g_i) > \circ x_i$ and $f_{iA}(y) = g_{iA}(y')$.

Moreover, the following 'joint embedding property' holds: for any pairs (a, x) and (b, x'), where a and b are $\mathbb{T}$-models in $\mathcal{K}$ and $x : E \to F(a)$, $x' : E \to F(b)$ are generalized elements, there exist an epimorphic family $\{e_i : E_i \to E \mid i \in I\}$ in $\mathcal{E}$ and for each index $i \in I$ a $\mathbb{T}$-model c_i in $\mathcal{K}$, arrows $f_i : a \to c_i$, $g_i : b \to c_i$ in $\mathcal{K}$ and a generalized element $x_i : E_i \to F(c_i)$ such that $< x, x' > \circ e_i = < F(f_i), F(g_i) > \circ x_i$ and (by Lemma 5.2.10) the following diagram commutes:

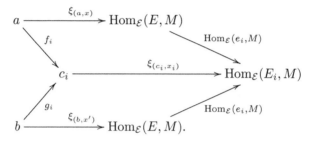

In fact, such a family can be taken to be the pullback of the family of arrows $< F(f), F(g) >: F(c) \to F(a) \times F(b)$ (indexed by the spans $(f : a \to c,\; g : b \to c)$ in the category $\mathcal{K}$) along the arrow $< x, x' >: E \to F(a) \times F(b)$.

Proof Recall from Theorem 5.2.5 that for any sort A of Σ, $\tilde{F}(M_{\mathbb{T}})A = \mathrm{colim}(F \circ \pi^{\mathcal{K}}_{\{x^A . \top\}})$. But $\pi^{\mathcal{K}}_{\{x^A . \top\}}$ is precisely the projection $\pi^{\mathrm{f}}_{P_A}$ to $\mathcal{K}^{\mathrm{op}}$ from the category of elements of the functor $P_A : \mathcal{K} \to \mathbf{Set}$ sending a model N in $\mathcal{C}$ to the set NA. The proposition thus readily follows from Theorem 5.1.25, applied to the functors $P_A : \mathcal{K} \to \mathbf{Set}$ (with A varying among the sorts of Σ), in view of Theorems 5.1.5 and 5.1.13. □

In order to express the model $\tilde{F}(M_{\mathbb{T}})$ as an $\mathcal{E}$-indexed filtered colimit of 'constant' $\mathbb{T}$-models coming from $\mathcal{K}$, we need to define the $\mathcal{E}$-indexed category $\mathbb{T}\text{-mod}(\mathcal{E})$ of models of the theory $\mathbb{T}$ in slices of $\mathcal{E}$.

For any first-order signature Σ and any Grothendieck topos $\mathcal{E}$, we have an $\mathcal{E}$-indexed category $\underline{\Sigma\text{-str}(\mathcal{E})}$ whose fibre at $E \in \mathcal{E}$ is the category $\Sigma\text{-str}(\mathcal{E}/E)$ and whose 'change of base' functors are the obvious pullback functors. For any sort A of Σ, we have an $\mathcal{E}$-indexed forgetful functor $U_A : \underline{\Sigma\text{-str}(\mathcal{E})} \to \underline{\mathcal{E}}_{\mathcal{E}}$ assigning to any Σ-structure M in the topos $\mathcal{E}/E$ the object $\overline{MA}$. It is easy to see, by adapting the classical proof for $\mathcal{E} = \mathbf{Set}$ and applying Theorem 5.1.11, that the $\mathcal{E}$-indexed functors U_A jointly create colimits of $\mathcal{E}$-indexed diagrams defined on $\mathcal{E}$-final, $\mathcal{E}$-filtered subcategories $\underline{A}_{\mathcal{E}}$ of a small $\mathcal{E}$-indexed category. Now, for any Grothendieck topos $\mathcal{E}$, every small category $\mathcal{C}$ can be made into an $\mathcal{E}$-indexed category $\underline{\mathcal{C}}_{\mathcal{E}}$ by setting $\mathcal{C}_E = \mathcal{C}$ for all $E \in \mathcal{E}$ and $\mathcal{C}_\alpha = 1_{\mathcal{C}}$ for all arrows α in $\mathcal{E}$.

To any (set-indexed) diagram $D : \mathcal{C} \to \mathcal{E}$ corresponds an $\mathcal{E}$-indexed functor $\underline{D}_\mathcal{E} : \underline{\mathcal{C}}_\mathcal{E} \to \underline{\mathcal{E}}_\mathcal{E}$ sending any object E of $\mathcal{E}$ to the functor $D_E := !_E^* \circ D$, where $!_E^* : \mathcal{E} \to \mathcal{E}/E$ is the pullback functor along the unique arrow $E \to 1_\mathcal{E}$ in $\mathcal{E}$. Since pullback functors preserve all small limits and colimits, giving a colimiting cocone (resp. a limiting cone) on the diagram D in the classical sense is equivalent to giving an $\mathcal{E}$-indexed colimiting cocone (resp. limiting cone) over the $\mathcal{E}$-indexed diagram $\underline{D}_\mathcal{E}$. Since the $\mathcal{E}$-indexed category $\underline{\mathcal{E}}_\mathcal{E}$ is complete, for any $\mathcal{E}$-indexed category $\underline{\mathcal{A}}_\mathcal{E}$, the $\mathcal{E}$-indexed functor category $[\underline{\mathcal{A}}_\mathcal{E}, \underline{\mathcal{E}}_\mathcal{E}]$ is also complete (since limits are computed pointwise); in particular, it has limits of diagrams defined on $\mathcal{E}$-indexed categories of the form $\underline{\mathcal{C}}_\mathcal{E}$. If $\underline{\mathcal{A}}_\mathcal{E}$ is an $\mathcal{E}$-final, $\mathcal{E}$-filtered subcategory of a small $\mathcal{E}$-indexed category, the cocompleteness of $\underline{\mathcal{E}}_\mathcal{E}$ as an $\mathcal{E}$-indexed category ensures that there exists a well-defined $\mathcal{E}$-indexed colimit functor $\underline{\mathrm{colim}}_\mathcal{E} : [\underline{\mathcal{A}}_\mathcal{E}, \underline{\mathcal{E}}_\mathcal{E}] \to \underline{\mathcal{E}}_\mathcal{E}$. By mimicking the classical proof of the fact that finite limits commute with filtered colimits, it is easy to see that this functor preserves limits of diagrams defined on $\mathcal{E}$-indexed categories of the form $\underline{\mathcal{C}}_\mathcal{E}$ for a finite category $\mathcal{C}$. Since it also preserves colimits, and the structure used for interpreting geometric formulae over Σ only involves set-indexed finite limits and arbitrary colimits, we can conclude that, for any geometric formula $\phi(\vec{x})$ over Σ, the $\mathcal{E}$-indexed functor $[[\vec{x} . \phi]]_- : \mathbb{T}\text{-mod}(\mathcal{E}) \to \underline{\mathcal{E}}_\mathcal{E}$ preserves $\mathcal{E}$-filtered colimits of $\mathcal{E}$-final and $\mathcal{E}$-filtered subcategories of a small $\mathcal{E}$-indexed category. This implies that, for any geometric theory $\mathbb{T}$ over Σ, the $\mathcal{E}$-indexed full subcategory $\mathbb{T}\text{-mod}(\mathcal{E})$ of $\Sigma\text{-str}(\mathcal{E})$ is closed in $\Sigma\text{-str}(\mathcal{E})$ under $\mathcal{E}$-indexed colimits of diagrams defined on $\mathcal{E}$-final, $\mathcal{E}$-filtered subcategories of a small $\mathcal{E}$-indexed category.

Now, as we observed in the proof of Proposition 5.2.11, for any sort A of Σ we have a forgetful functor $P_A : \mathcal{K} \to \mathbf{Set}$ evaluating models in $\mathcal{K}$ at A. Using the notation in the statement of Theorem 5.1.5, we have an associated internal diagram $(P_A)_\mathcal{E}$ in $[\mathbb{K}, \mathcal{E}]$ and hence an $\mathcal{E}$-indexed functor $\mathbb{K}_\mathcal{E} \to \underline{\mathcal{E}}_\mathcal{E}$ (where $\mathbb{K}$ is the internalization of $\mathcal{K}$ in $\mathcal{E}$). These functors, with A varying among the sorts of Σ, yield an $\mathcal{E}$-indexed functor $P : \mathbb{K}_\mathcal{E} \to \mathbb{T}\text{-mod}(\mathcal{E})$. We can thus conclude (in light of the proof of Proposition 5.2.11) that the model $\tilde{F}(M_\mathbb{T})$ is the $\mathcal{E}$-indexed colimit of the diagram $P \circ \pi_F^\mathrm{f}$ (or, equivalently, of its restriction to the $\mathcal{E}$-strictly final subcategory $\int \underline{F}_\mathcal{E}$ of $\int \overline{F}_\mathcal{E}^\mathrm{f}$ – cf. Theorem 5.1.12).

5.2.4 A general adjunction

Let $\mathcal{C}$ be a small category and $\mathcal{E}$ a Grothendieck topos. Recall from section 5.1.1 that the indexed category $[\mathcal{C}, \mathcal{E}]_\mathcal{E}$ is locally small, that is for any two functors $F, G : \mathcal{C} \to \mathcal{E}$ there exists an object $\mathrm{Hom}^\mathcal{E}_{[\mathcal{C}, \mathcal{E}]_\mathcal{E}}(F, G)$ of $\mathcal{E}$ satisfying the universal property that, for any object E of $\mathcal{E}$, the generalized elements $E \to \mathrm{Hom}^\mathcal{E}_{[\mathcal{C}, \mathcal{E}]_\mathcal{E}}(F, G)$ correspond bijectively, naturally in $E \in \mathcal{C}$, to the natural transformations $!_E^* \circ F \to !_E^* \circ G$.

The following theorem establishes a general adjunction between categories of $\mathcal{E}$-valued functors induced by a functor $P : \mathcal{C} \to [\mathcal{D}^{\mathrm{op}}, \mathbf{Set}]$. Before stating it, we need to introduce some notation. For any object c of $\mathcal{C}$, we denote by $\int P(c)$ the category of elements of the functor $P(c) : \mathcal{D}^{\mathrm{op}} \to \mathbf{Set}$ and by $\pi_c : \int P(c) \to \mathcal{D}$ the canonical projection functor. We denote by $y_\mathcal{D} : \mathcal{D}^{\mathrm{op}} \to [\mathcal{D}, \mathbf{Set}]$ the Yoneda embedding. For any functor $H : \mathcal{D} \to \mathcal{E}$ and any object (d, z) of the category $\int P(c)$, we denote by $\kappa^H_{(d,z)} : H(d) \to \mathrm{colim}(H \circ \pi_c)$ the canonical colimit arrow.

Theorem 5.2.12 Let $\mathcal{C}$ and $\mathcal{D}$ be small categories, $\mathcal{E}$ a Grothendieck topos and $P : \mathcal{C} \to [\mathcal{D}^{\mathrm{op}}, \mathbf{Set}]$ a functor. Let $\widetilde{(-)} : [\mathcal{D}, \mathcal{E}] \to [\mathcal{C}, \mathcal{E}]$ be the functor sending any functor $F : \mathcal{D} \to \mathcal{E}$ to the functor $\tilde{F}$ defined as follows (and acting on the arrows of $[\mathcal{D}, \mathcal{E}]$ accordingly):
for any $c \in \mathcal{C}$, $\tilde{F}(c) = \mathrm{colim}(F \circ \pi_c)$, and
for any arrow $f : c \to c'$ in $\mathcal{C}$,

$$\tilde{F}(f) : \mathrm{colim}(F \circ \pi_c) \to \mathrm{colim}(F \circ \pi_{c'})$$

is defined by the conditions $\tilde{F}(f) \circ \kappa^F_{(d,z)} = \kappa^F_{(d,P(f)(z))}$ (for any $(d,z) \in \int P(c)$).

Let $(-)_r : [\mathcal{C}, \mathcal{E}] \to [\mathcal{D}, \mathcal{E}]$ be the functor assigning to any functor $G : \mathcal{C} \to \mathcal{E}$ the functor $\mathrm{Hom}^{\mathcal{E}}_{\underline{[\mathcal{C},\mathcal{E}]}_\mathcal{E}}(\gamma^*_\mathcal{E} \circ \widetilde{y_\mathcal{D}(-)}, G)$ and acting on the arrows in the obvious way.

Then the functors

$$\widetilde{(-)} : [\mathcal{D}, \mathcal{E}] \to [\mathcal{C}, \mathcal{E}]$$

and

$$(-)_r : [\mathcal{C}, \mathcal{E}] \to [\mathcal{D}, \mathcal{E}]$$

are adjoint to each other ($\widetilde{(-)}$ on the left and $(-)_r$ on the right). The unit $\eta^F : F \to (\tilde{F})_r$ and the counit $\varepsilon^G : \widetilde{(G_r)} \to G$ are defined as follows:

- For any $d \in \mathcal{D}$, $\eta^F(d) : F(d) \to (\tilde{F})_r(d)$ is the arrow in $\mathcal{E}$ defined by means of generalized elements by saying that it sends any generalized element $E \to F(d)$, regarded as a natural transformation $\gamma^*_{\mathcal{E}/E} \circ y_\mathcal{D}(d) \to !^*_E \circ F$, to the image $\gamma^*_{\mathcal{E}/E} \circ \widetilde{y_\mathcal{D}(d)} \to !^*_E \circ \tilde{F}$ of this arrow under the functor $\widetilde{(-)}$, regarded as a generalized element $E \to \mathrm{Hom}^{\mathcal{E}}_{\underline{[\mathcal{C},\mathcal{E}]}_\mathcal{E}}(\gamma^*_\mathcal{E} \circ \widetilde{y_\mathcal{D}(d)}, \tilde{F}) = (\tilde{F})_r(d)$.

- For any $c \in \mathcal{C}$, $\varepsilon^G(c) : \widetilde{(G_r)}(c) = \mathrm{colim}(G_r \circ \pi_c) \to G(c)$ is defined by setting, for each object (d, z) of $\int P(c)$, $\varepsilon^G(c) \circ \kappa^{G_r}_{(d,z)}$ equal to the arrow $G_r(d) \to G(c)$ defined by means of generalized elements by saying that a generalized element $x : E \to G_r(d)$, corresponding to a natural transformation $\bar{x} : !^*_E \circ \gamma^*_\mathcal{E} \circ \widetilde{y_\mathcal{D}(d)} \cong \gamma^*_{\mathcal{E}/E} \circ \widetilde{y_\mathcal{D}(d)} \to !^*_E \circ \tilde{G}$, is sent to the arrow $E \to G(c)$ obtained by composing $\bar{x}(c)$ with the component of $\gamma^*_{\mathcal{E}/E}(\widetilde{y_\mathcal{D}(d)}(c))$ which corresponds to the element of $\widetilde{y_\mathcal{D}(d)}(c)$ given by the image of the identity on d under the colimit arrow $\kappa^{y_\mathcal{D}(d)}_{(d,z)} : (y_\mathcal{D}(d))(d) \to \widetilde{y_\mathcal{D}(d)}(c)$.

Proof For simplicity we shall establish the result only in the case $\mathcal{E} = \mathbf{Set}$, the proof of the general case being entirely analogous (the only care that one has to take is to use generalized elements in place of standard set-theoretic elements).

We shall define a bijective correspondence between the natural transformations $\tilde{F} \to G$ and the natural transformations $F \to G_r$ which is natural in $F \in [\mathcal{D}, \mathbf{Set}]$ and $G \in [\mathcal{C}, \mathbf{Set}]$.

Given a natural transformation $\alpha : F \to G_r$, we define $\tau(\alpha) : \tilde{F} \to G$ by setting, for each $c \in \mathcal{C}$, $\tau_\alpha(c) : \tilde{F}(c) \to G(c)$ equal to the arrow defined as follows. As $\tilde{F}(c)$ is the colimit of the cone $\{\kappa^F_{(d,z)} : F(d) \to \tilde{F}(c) \mid (d,z) \in \int P(c)\}$, it suffices to define an arrow $u_{(d,z)} : F(d) \to G(c)$ for each pair (d,z) where $d \in \mathcal{D}$ and $z \in P(c)(d)$ in such a way that whenever (d,z) and (d',z') are pairs such that $P(c)(f)(z') = z$ for an arrow $f : d \to d'$ in $\mathcal{D}$ then $u_{(d',z')} \circ F(f) = u_{(d,z)}$; indeed, by the universal property of the colimit, such a family of arrows will induce a unique arrow $\tau_\alpha(c) : \tilde{F}(c) \to G(c)$ such that $\tau_\alpha(c) \circ \kappa^F_{(d,z)} = u_{(d,z)}$ for each (d,z) in $\int P(c)$. We set $u_{(d,z)} : F(d) \to G(c)$ equal to the function sending every element $x \in F(d)$ to the element $\alpha(d)(x)(c)(\kappa^{y_\mathcal{D}(d)}_{(d,z)}(1_d))$. Given an arrow $f : d \to d'$ such that $P(c)(f)(z') = z$, we have to verify that, for every $x \in F(d)$, $\alpha(d')(F(f)(x))(c)(\kappa^{y_\mathcal{D}(d')}_{(d',z')}(1_{d'})) = \alpha(d)(x)(c)(\kappa^{y_\mathcal{D}(d)}_{(d,z)}(1_d))$. This identity immediately follows from the commutativity of the naturality diagram

$$\begin{array}{ccc} F(d) & \xrightarrow{\alpha(d)} & G_r(d) = \mathrm{Hom}_{[\mathcal{C},\mathbf{Set}]}(\widetilde{y_\mathcal{D}(d)}, G) \\ {\scriptstyle F(f)} \downarrow & & \downarrow {\scriptstyle G_r(f) = -\circ \widetilde{y_\mathcal{D}(f)}} \\ F(d') & \xrightarrow{\alpha(d')} & G_r(d') = \mathrm{Hom}_{[\mathcal{C},\mathbf{Set}]}(\widetilde{y_\mathcal{D}(d')}, G) \end{array}$$

for α with respect to the arrow f and from that of the diagram

$$\begin{array}{ccc} \widetilde{y_\mathcal{D}(d')}(c) & \xrightarrow{\widetilde{y_\mathcal{D}(f)}(c)} & \widetilde{y_\mathcal{D}(d)}(c) \\ {\scriptstyle \kappa^{y_\mathcal{D}(d')}_{(d',z')}} \uparrow & & \uparrow {\scriptstyle \kappa^{y_\mathcal{D}(d)}_{(d,z)}} \\ y_\mathcal{D}(d') & \xrightarrow{y_\mathcal{D}(f)} & y_\mathcal{D}(d), \end{array}$$

which is an instance of Proposition 5.2.1(i). Indeed,

$$\begin{aligned} \alpha(d')(F(f)(x))(c)(\kappa^{y_\mathcal{D}(d')}_{(d',z')}(1_{d'})) &= \alpha(d)(x)(c)(\widetilde{y_\mathcal{D}(f)}(c)(\kappa^{y_\mathcal{D}(d')}_{(d',z')}(1_{d'}))) \\ &= \alpha(d)(x)(c)(\kappa^{y_\mathcal{D}(d)}_{(d',z')}(f)) \\ &= \alpha(d)(x)(c)(\kappa^{y_\mathcal{D}(d)}_{(d,z)}(1_d)), \end{aligned}$$

where the first equality follows from the commutativity of the first diagram, the second follows from the commutativity of the second diagram and the last follows from the fact that the family $\{\kappa^{y_\mathcal{D}(d)}_{(d,z)} \mid (d,z) \in \int P(c)\}$ is a cocone.

To complete the definition of $\tau(\alpha)$, it remains to check that the assignment $c \to \tau(\alpha)(c)$ defines a natural transformation $\tilde{F} \to G$. We have to verify that, for any arrow $f : c \to c'$ in $\mathcal{C}$, the diagram

$$\begin{array}{ccc} \tilde{F}(c) & \xrightarrow{\tau(\alpha)(c)} & G(c) \\ \tilde{F}(f) \downarrow & & \downarrow G(f) \\ \tilde{F}(c') & \xrightarrow{\tau(\alpha)(c')} & G(c') \end{array}$$

commutes.

As, by definition of $\tilde{F}(f)$, the diagrams

$$\begin{array}{ccc} \tilde{F}(c) & \xrightarrow{\tilde{F}(f)} & \tilde{F}(c') \\ \kappa^F_{(d,z)} \uparrow & \nearrow \kappa^F_{(d,P(f)(d)(z))} & \\ F(d) & & \end{array}$$

commute and the arrows $\kappa^F_{(d,z)}$ are jointly epimorphic (as they are colimit arrows), the above diagram commutes if and only if the arrows $G(f) \circ \tau(\alpha)(c) \circ \kappa^F_{(d,z)}$ and $\tau(\alpha)(c') \circ \kappa^F_{(d,P(f)(d)(z))}$ are equal, that is if and only if they take the same values at any element $x \in F(d)$. Let us set $z' = P(f)(d)(z)$. By definition of $\tau(\alpha)$,

$$\tau(\alpha)(c)(\kappa^F_{(d,z)}(x)) = \alpha(d)(x)(c)(\kappa^{y_\mathcal{D}(d)}_{(d,z)}(1_d)),$$

while

$$\tau(\alpha)(c')(\kappa^F_{(d,z')}(x)) = \alpha(d)(x)(c')(\kappa^{y_\mathcal{D}(d)}_{(d,z')}(1_d)).$$

Now, for any $d \in \mathcal{D}$ and any $x \in F(d)$, $\alpha(d)(x)$ is a natural transformation $\widetilde{y_\mathcal{D}(d)} \to G$; in particular, the diagram

$$\begin{array}{ccc} \widetilde{y_\mathcal{D}(d)}(c) & \xrightarrow{\alpha(d)(x)(c)} & G(c) \\ \widetilde{y_\mathcal{D}(d)}(f) \downarrow & & \downarrow G(f) \\ \widetilde{y_\mathcal{D}(d)}(c') & \xrightarrow{\alpha(d)(x)(c')} & G(c') \end{array}$$

commutes.

The commutativity of this diagram, together with that of the diagram

$$\begin{array}{ccc} \widetilde{y_\mathcal{D}(d)}(c) & \xrightarrow{\widetilde{y_\mathcal{D}(d)}(f)} & \widetilde{y_\mathcal{D}(d)}(c') \\ \kappa^{y_\mathcal{D}(d)}_{(d,z)} \uparrow & \nearrow \kappa^{y_\mathcal{D}(d)}_{(d,z')} & \\ y_\mathcal{D}(d)(d) & & \end{array}$$

(which follows by definition of $\widetilde{y_\mathcal{D}(d)}(f)$), now immediately imply our thesis.

Let us now define a function χ assigning to any natural transformation $\beta : \tilde{F} \to G$ a natural transformation $\chi(\beta) : F \to G_r$. For any $d \in \mathcal{D}$, we define $\chi(\beta)(d) : F(d) \to G_r(d)$ by setting, for each $x \in F(d)$, $\chi(\beta)(d)(x)$ equal to the natural transformation $\beta \circ \widetilde{a_x} : \widetilde{y_\mathcal{D}(d)} \to G$, where $a_x : y_\mathcal{D}(d) \to F$ is the natural transformation corresponding, via the Yoneda lemma, to the element $x \in F(d)$.

To verify that $\chi(\beta)$ is well-defined, we have to check that, for every arrow $g : d \to d'$ in $\mathcal{D}$, the diagram

$$\begin{array}{ccc} F(d) & \xrightarrow{\chi(\beta)(d)} & G_r(d) \\ {\scriptstyle F(g)}\downarrow & & \downarrow{\scriptstyle G_r(g) = -\circ \widetilde{y_\mathcal{D}(g)}} \\ F(d') & \xrightarrow{\chi(\beta)(d')} & G_r(d') \end{array}$$

commutes, i.e. that, for every $x \in F(d)$, $G_r(g)(\chi(\beta)(d)(x)) = \chi(\beta)(d')(F(g)(x))$. Now, $G_r(g)(\chi(\beta)(d)(x)) = \beta \circ \widetilde{a_x} \circ \widetilde{y_\mathcal{D}(g)}$, while $\chi(\beta)(d')(F(g)(x)) = \beta \circ \widetilde{a_{F(g)(x)}}$; but $a_{F(g)(x)} = a_x \circ y_\mathcal{D}(g)$, as required.

The proof of the fact that the correspondences τ and β are natural in F and G is straightforward and left to the reader.

To conclude the proof of the theorem, it thus remains to show that τ and χ are inverse to each other. The verification of the fact that the unit and counit of the adjunction coincide with those given in the statement of the theorem is straightforward and left to the reader.

Let us show that for any natural transformation $\alpha : F \to G_r$, $\chi(\tau(\alpha)) = \alpha$. Let us set $\beta = \tau(\alpha)$. We have to prove that for any $d \in \mathcal{D}$, $x \in F(d)$ and $z \in P(c)(d')$, $\alpha(d)(x)(c) \circ \kappa_{(d',z)}^{y_\mathcal{D}(d)} = \chi(\beta)(d)(x)(c) \circ \kappa_{(d',z)}^{y_\mathcal{D}(d)}$ as functions $y_\mathcal{D}(d)(d') \to G(c)$, i.e. that for any element $g \in y_\mathcal{D}(d)(d') = \mathrm{Hom}_\mathcal{D}(d, d')$,

$$\alpha(d)(x)(c)(\kappa_{(d',z)}^{y_\mathcal{D}(d)}(g)) = (\chi(\beta)(d)(x)(c) \circ \kappa_{(d',z)}^{y_\mathcal{D}(d)})(g).$$

By definition of the functor $\widetilde{(-)}$ and of the correspondence τ, the diagram

$$\begin{array}{ccccc} \widetilde{y_\mathcal{D}(d)}(c) & \xrightarrow{\widetilde{a_x}(c)} & \tilde{F}(c) & \xrightarrow{\beta(c)} & G(c) \\ {\scriptstyle \kappa_{(d',z)}^{y_\mathcal{D}(d)}}\uparrow & & {\scriptstyle \kappa_{(d',z)}^{F}}\uparrow & \nearrow{\scriptstyle u_{(d',z)}} & \\ y_\mathcal{D}(d)(d') & \xrightarrow{a_x(d')} & F(d') & & \end{array}$$

commutes, and since $\chi(\beta)(d)(x) = \beta \circ \widetilde{a_x} : \widetilde{y_\mathcal{D}(d)} \to G$, we have that

$$(\chi(\beta)(d)(x)(c) \circ \kappa_{(d',z)}^{y_\mathcal{D}(d)})(g) = u_{(d',z)} \circ a_x(d')(g)$$
$$= \alpha(d')(a_x(d')(g))(c)(\kappa_{(d',z)}^{y_\mathcal{D}(d')}(1_{d'}))$$
$$= \alpha(d')(F(g)(x))(c)(\kappa_{(d',z)}^{y_\mathcal{D}(d')}(1_{d'})).$$

Now, the naturality diagram for α with respect to the arrow $g : d \to d'$ yields the equality $\alpha(d)(x) \circ y_{\mathcal{D}}(g) = \alpha(d')(F(g)(x))$ and hence the equality $\alpha(d)(x)(c) \circ \widetilde{y_{\mathcal{D}}(g)}(c) \circ \kappa^{y_{\mathcal{D}}(d')}_{(d',z)} = \alpha(d')(F(g)(x))(c) \circ \kappa^{y_{\mathcal{D}}(d')}_{(d',z)}$.
But the commutativity of the diagram

$$\begin{array}{ccc} \widetilde{y_{\mathcal{D}}(d')}(c) & \xrightarrow{\widetilde{y_{\mathcal{D}}(g)}(c)} & \widetilde{y_{\mathcal{D}}(d)}(c) \\ \kappa^{y_{\mathcal{D}}(d')}_{(d',z)} \uparrow & & \uparrow \kappa^{y_{\mathcal{D}}(d)}_{(d',z)} \\ y_{\mathcal{D}}(d')(d') & \xrightarrow{y_{\mathcal{D}}(g)(d')} & y_{\mathcal{D}}(d)(d') \end{array}$$

(which follows by definition of the functor $\widetilde{(-)}$) implies that $\alpha(d)(x)(c) \circ \widetilde{y_{\mathcal{D}}(g)}(c) \circ \kappa^{y_{\mathcal{D}}(d')}_{(d',z)} = \alpha(d)(x)(c) \circ \kappa^{y_{\mathcal{D}}(d)}_{(d',z)} \circ y_{\mathcal{D}}(g)(d')$. Therefore

$$\alpha(d)(x)(c) \circ \kappa^{y_{\mathcal{D}}(d)}_{(d',z)} \circ y_{\mathcal{D}}(g)(d') = \alpha(d')(F(g)(x))(c) \circ \kappa^{y_{\mathcal{D}}(d')}_{(d',z)},$$

and evaluating this at $1_{d'}$ yields the desired equality.

Finally, let us show that the composite $\tau \circ \chi$ is equal to the identity. Let $\beta : \tilde{F} \to G$ be a natural transformation. We have to show that $\beta = \tau(\chi(\beta))$. Let us set $\alpha = \chi(\beta)$. To prove that $\beta = \tau(\alpha)$, it is equivalent to verify that for every object $c \in \mathcal{C}$, any pair (d, z) with $d \in \mathcal{D}$ and $z \in P(c)(d)$ and any element $x \in F(d)$, $\beta(c)(\kappa^{F}_{(d,z)}(x)) = \tau(\alpha)(c)(\kappa^{F}_{(d,z)}(x))$. But $\tau(\alpha)(c)(\kappa^{F}_{(d,z)}(x)) = \alpha(d)(x)(c)(\kappa^{y_{\mathcal{D}}(d)}_{(d,z)}(1_d)) = \chi(\beta)(d)(x)(c)(\kappa^{y_{\mathcal{D}}(d)}_{(d,z)}(1_d)) = (\beta \circ \widetilde{a_x})(c)(\kappa^{y_{\mathcal{D}}(d)}_{(d,z)}(1_d)) = \beta(c)(\widetilde{a_x}(c)(\kappa^{y_{\mathcal{D}}(d)}_{(d,z)}(1_d)))$. Thus our thesis follows from the commutativity of the diagram

$$\begin{array}{ccc} \widetilde{y_{\mathcal{D}}(d)}(c) & \xrightarrow{\widetilde{a_x}(c)} & \tilde{F}(c) \\ \kappa^{y_{\mathcal{D}}(d)}_{(d,z)} \uparrow & & \uparrow \kappa^{F}_{(d,z)} \\ y_{\mathcal{D}}(d)(d) & \xrightarrow{a_x(d)} & F(d), \end{array}$$

which is an immediate consequence of the definition of the functor $\widetilde{(-)}$. □

Remarks 5.2.13 (a) Let $f : \mathcal{D} \to \mathcal{C}$ be a functor between two small categories, and $P : \mathcal{C} \to [\mathcal{D}^{\mathrm{op}}, \mathbf{Set}]$ be the functor $y_{\mathcal{C}}(-) \circ f^{\mathrm{op}}$, where $y_{\mathcal{C}} : \mathcal{C} \to [\mathcal{C}^{\mathrm{op}}, \mathbf{Set}]$ is the Yoneda embedding. Notice that P is the flat functor corresponding, via Diaconescu's equivalence, to the essential geometric morphism $[\mathcal{D}^{\mathrm{op}}, \mathbf{Set}] \to [\mathcal{C}^{\mathrm{op}}, \mathbf{Set}]$ induced by the functor $f^{\mathrm{op}} : \mathcal{D}^{\mathrm{op}} \to \mathcal{C}^{\mathrm{op}}$. The functor $\widetilde{(-)} : [\mathcal{D}, \mathcal{E}] \to [\mathcal{C}, \mathcal{E}]$ coincides with the left Kan extension functor along f, while the right adjoint functor $(-)_r$ coincides with the functor $- \circ f : [\mathcal{C}, \mathcal{E}] \to [\mathcal{D}, \mathcal{E}]$.

(b) Let
$$u : \mathbf{Sh}(\mathcal{D}, K) \to \mathbf{Sh}(\mathcal{C}, J)$$
be a geometric morphism. Then, for any Grothendieck topos $\mathcal{E}$, u induces as in section 5.2.1 a functor
$$\xi_\mathcal{E} : \mathbf{Flat}_K(\mathcal{D}, \mathcal{E}) \to \mathbf{Flat}_J(\mathcal{C}, \mathcal{E}).$$
Via Diaconescu's equivalence, u corresponds to a flat functor $\mathcal{C} \to \mathbf{Sh}(\mathcal{D}, K)$ which, composed with the canonical geometric inclusion
$$\mathbf{Sh}(\mathcal{D}, K) \hookrightarrow [\mathcal{D}^{\mathrm{op}}, \mathbf{Set}],$$
yields a functor $P : \mathcal{C} \to [\mathcal{D}^{\mathrm{op}}, \mathbf{Set}]$. Then $\xi_\mathcal{E}$ coincides with the restriction of the functor $\widetilde{(-)} : [\mathcal{D}, \mathcal{E}] \to [\mathcal{C}, \mathcal{E}]$ induced by P as in Theorem 5.2.12 to the full subcategories $\mathbf{Flat}_K(\mathcal{D}, \mathcal{E}) \hookrightarrow [\mathcal{D}, \mathcal{E}]$ and $\mathbf{Flat}_J(\mathcal{C}, \mathcal{E}) \hookrightarrow [\mathcal{C}, \mathcal{E}]$. In general, the right adjoint functor $(-)_r : [\mathcal{C}, \mathcal{E}] \to [\mathcal{D}, \mathcal{E}]$ does not restrict to these subcategories, but if it does, it becomes a right adjoint to the functor $\xi_\mathcal{E} : \mathbf{Flat}_K(\mathcal{D}, \mathcal{E}) \to \mathbf{Flat}_J(\mathcal{C}, \mathcal{E})$.

This notably applies in the case of the geometric morphism
$$p_\mathcal{K} : [\mathcal{K}, \mathbf{Set}] \to \mathbf{Sh}(\mathcal{C}_\mathbb{T}, J_\mathbb{T})$$
considered in section 5.2.3. In this case, $\mathcal{C} = \mathcal{C}_\mathbb{T}$, $\mathcal{D} = \mathcal{K}^{\mathrm{op}}$ and P is the functor $\mathcal{C}_\mathbb{T} \to [\mathcal{K}, \mathbf{Set}]$ sending any geometric formula $\phi(\vec{x})$ to the functor $M \to [[\vec{x}.\phi]]_M$; it is easy to see that $\widetilde{y_\mathcal{K}(-)}$ is the functor $U : \mathcal{K} \to [\mathcal{C}_\mathbb{T}, \mathbf{Set}]$ sending any model c in $\mathcal{K}$ to the functor $\{\vec{x}.\phi\} \to [[\vec{x}.\phi]]_c$, and that for any G in $[\mathcal{C}_\mathbb{T}, \mathcal{E}]$, $G_r = \mathrm{Hom}^\mathcal{E}_{[\mathcal{C}_\mathbb{T}, \mathcal{E}]_\mathcal{E}}(\gamma_\mathcal{E}^* \circ U(-), G)$.

(c) By a basic property of adjoint functors, the (left adjoint) functor
$$\widetilde{(-)} : [\mathcal{D}, \mathcal{E}] \to [\mathcal{C}, \mathcal{E}]$$
is full and faithful if and only if the unit $\eta^F : F \to (\tilde{F})_r$ is an isomorphism for every F in $[\mathcal{D}, \mathcal{E}]$. It readily follows (by purely formal considerations) that for every full subcategory $\mathcal{H}$ of $[\mathcal{D}, \mathcal{E}]$ with canonical embedding $i : \mathcal{H} \hookrightarrow [\mathcal{D}, \mathcal{E}]$, the composite functor $\widetilde{(-)} \circ i$ is full and faithful if and only if the unit $\eta^F : F \to (\tilde{F})_r$ is an isomorphism for every F in $\mathcal{H}$.

The following result will be useful in the sequel; see section 5.1.2 for the notation employed in it.

Proposition 5.2.14 *Let $f : \mathcal{D} \to \mathcal{C}$ be a functor between small categories, $\mathcal{E}$ a Grothendieck topos and F a functor $\mathcal{C}^{\mathrm{op}} \to \mathcal{E}$. Then the $\mathcal{E}$-indexed functor $\int (F \circ f^{\mathrm{op}})_\mathcal{E} \to \int F_\mathcal{E}$ sending any object (c, x) of $\int (F \circ f^{\mathrm{op}})_E$ to the object $(f(c), x)$ of $(\int F)_E$ is $\mathcal{E}$-final if and only if the $\mathcal{E}$-indexed functor $\int \varepsilon^F_\mathcal{E} : \int \widetilde{(F \circ f^{\mathrm{op}})}_\mathcal{E} \to \int \widetilde{F}_\mathcal{E}$ induced by the natural transformation $\varepsilon^F : \widetilde{F \circ f^{\mathrm{op}}} \to F$ of Theorem 5.2.12 and Remark 5.2.13(a) is $\mathcal{E}$-final.*

Proof From the general analysis of section 5.2.1 we know that for any $c \in \mathcal{C}$, $\widetilde{F \circ f^{\text{op}}}(c) = \text{colim}(F \circ f^{\text{op}} \circ \pi_c)$, where π_c is the canonical projection to $\mathcal{D}^{\text{op}}$ from the category $\mathcal{A}_c$ whose objects are the pairs (d,h), where d is an object of the category $\mathcal{D}$ and h is an arrow $c \to f(d)$ in $\mathcal{C}$, and whose arrows are the obvious ones. The natural transformation $\varepsilon^F : \widetilde{F \circ f^{\text{op}}} \to F$ is defined by the condition that, for any objects c of $\mathcal{C}$ and (d,h) of $\mathcal{A}_c$, $\varepsilon^F(c) \circ \kappa^F_{(d,h)} = F(h)$, where $\kappa^F_{(d,h)} : F(f(d)) \to \text{colim}(F \circ f^{\text{op}} \circ \pi_c)$ is the canonical colimit arrow.

Now, our thesis follows immediately from the fact that for any object of the category $\int(\widetilde{F \circ f^{\text{op}}})_E$ of the form $(c, \kappa^F_{(d,h)} \circ y)$ for a generalized element $y : E \to F(f(d))$, $(\int \varepsilon^F)_E((c, \kappa^F_{(d,h)} \circ y)) = (c, F(h) \circ y)$, by invoking the fact that the colimit arrows are jointly epimorphic. $\square$

5.3 Yoneda representations of flat functors

In this section, which is based on [17], we introduce a technique, based on ideas from indexed category theory, for representing flat functors into arbitrary Grothendieck toposes; this technique will be applied in Chapters 6 and 8 to obtain results in the theory of classifying toposes.

We shall borrow from section 5.1.1 the notation and terminology concerning indexed categories.

By the Yoneda lemma, for each $F \in [\mathcal{C}^{\text{op}}, \mathbf{Set}]$ there is a natural isomorphism of functors

$$F \cong \text{Hom}^{\mathbf{Set}}_{[\mathcal{C}^{\text{op}}, \mathbf{Set}]}(y_{\mathcal{C}}(-), F),$$

where $\text{Hom}^{\mathbf{Set}}_{[\mathcal{C}^{\text{op}}, \mathbf{Set}]}(y_{\mathcal{C}}(-), F)$ is the functor given by the composite

$$\mathcal{C}^{\text{op}} \xrightarrow{y_{\mathcal{C}}^{\text{op}} \times \Delta F} [\mathcal{C}^{\text{op}}, \mathbf{Set}]^{\text{op}} \times [\mathcal{C}^{\text{op}}, \mathbf{Set}] \xrightarrow{\text{Hom}^{\mathbf{Set}}_{[\mathcal{C}^{\text{op}}, \mathbf{Set}]}} \mathbf{Set},$$

ΔF is the constant functor equal to F and $y_{\mathcal{C}} : \mathcal{C} \to [\mathcal{C}^{\text{op}}, \mathbf{Set}]$ is the Yoneda embedding.

By exploiting the local smallness of $\mathcal{E}$-indexed diagram categories (cf. section 5.1.1), we can generalize this result to the case of functors with values in an arbitrary Grothendieck topos. More specifically, we have the following result:

Theorem 5.3.1 (cf. Theorem 2.3 [17]). *Let $\mathcal{C}$ be a small category and $\mathcal{E}$ a Grothendieck topos. Then for every functor $F : \mathcal{C}^{\text{op}} \to \mathcal{E}$ there is a natural isomorphism of functors*

$$F \cong \text{Hom}^{\mathcal{E}}_{[\mathcal{C}^{\text{op}}, \mathcal{E}]}(y^{\mathcal{E}}_{\mathcal{C}}(-), F),$$

where $y^{\mathcal{E}}_{\mathcal{C}} : \mathcal{C} \to [\mathcal{C}^{\text{op}}, \mathcal{E}]$ is the functor given by the composite

$$\mathcal{C} \xrightarrow{y_{\mathcal{C}}} [\mathcal{C}^{\text{op}}, \mathbf{Set}] \xrightarrow{\gamma^*_{\mathcal{E}} \circ -} [\mathcal{C}^{\text{op}}, \mathcal{E}]$$

and $\mathrm{Hom}^{\mathcal{E}}_{[\mathcal{C}^{\mathrm{op}},\,\mathcal{E}]}(y_{\mathcal{C}}^{\mathcal{E}}(-),F)$ is the functor given by the composite

$$\mathcal{C}^{\mathrm{op}} \xrightarrow{(y_{\mathcal{C}}^{\mathcal{E}})^{\mathrm{op}} \times \Delta F} [\mathcal{C}^{\mathrm{op}},\mathcal{E}]^{\mathrm{op}} \times [\mathcal{C}^{\mathrm{op}},\mathcal{E}] \xrightarrow{\mathrm{Hom}^{\mathcal{E}}_{[\mathcal{C}^{\mathrm{op}},\,\mathcal{E}]}} \mathcal{E}.$$

Moreover, the above isomorphism is natural in F.

In the case of flat functors, Theorem 5.3.1 specializes to the following result:

Corollary 5.3.2 *Let $\mathcal{C}$ be a small category and $\mathcal{E}$ a Grothendieck topos. Then for every flat functor $F : \mathcal{C}^{\mathrm{op}} \to \mathcal{E}$, there is a natural isomorphism of functors*

$$F \cong \mathrm{Hom}^{\mathcal{E}}_{\mathbf{Flat}(\mathcal{C}^{\mathrm{op}},\,\mathcal{E})}(y_{\mathcal{C}}^{\mathcal{E}}(-),F),$$

where $y_{\mathcal{C}}^{\mathcal{E}} : \mathcal{C} \to \mathbf{Flat}(\mathcal{C}^{\mathrm{op}},\mathcal{E})$ is the functor given by the composite

$$\mathcal{C} \xrightarrow{y_{\mathcal{C}}} \mathbf{Flat}(\mathcal{C}^{\mathrm{op}},\mathbf{Set}) \xrightarrow{\gamma_{\mathcal{E}}^{*} \circ -} \mathbf{Flat}(\mathcal{C}^{\mathrm{op}},\mathcal{E})$$

and $\mathrm{Hom}^{\mathcal{E}}_{\mathbf{Flat}(\mathcal{C},\,\mathcal{E})}(y_{\mathcal{C}}^{\mathcal{E}}(-),F)$ is the functor given by the composite

$$\mathcal{C}^{\mathrm{op}} \xrightarrow{(y_{\mathcal{C}}^{\mathcal{E}})^{\mathrm{op}} \times \Delta F} \mathbf{Flat}(\mathcal{C}^{\mathrm{op}},\mathcal{E})^{\mathrm{op}} \times \mathbf{Flat}(\mathcal{C}^{\mathrm{op}},\mathcal{E}) \xrightarrow{\mathrm{Hom}^{\mathcal{E}}_{\mathbf{Flat}(\mathcal{C}^{\mathrm{op}},\,\mathcal{E})}} \mathcal{E}.$$

Moreover, the above isomorphism is natural in F.

From now on we shall refer to this result as the *Yoneda representation of flat functors*.

5.3.1 Cauchy completion of sites

When dealing with theories classified by a presheaf topos $[\mathcal{C},\mathbf{Set}]$, it is natural to replace the category $\mathcal{C}$ with its Cauchy completion $\hat{\mathcal{C}}$, since $[\mathcal{C},\mathbf{Set}]$ and $[\hat{\mathcal{C}},\mathbf{Set}]$ are equivalent (cf. Corollary A1.1.9 [55]). Recall that $\hat{\mathcal{C}}$ can be equivalently characterized as the closure of $\mathcal{C}$ under retracts in Ind-$\mathcal{C}$.

From the equivalence of toposes

$$[\mathcal{C}^{\mathrm{op}},\mathbf{Set}] \simeq [\hat{\mathcal{C}}^{\mathrm{op}},\mathbf{Set}]$$

it follows that for any Grothendieck topology J on $\mathcal{C}$ there exists a unique Grothendieck topology $\hat{J}$ on $\hat{\mathcal{C}}$ such that subtopos

$$\mathbf{Sh}(\hat{\mathcal{C}},\hat{J}) \hookrightarrow [\hat{\mathcal{C}}^{\mathrm{op}},\mathbf{Set}]$$

corresponds to the subtopos

$$\mathbf{Sh}(\mathcal{C},J) \hookrightarrow [\mathcal{C}^{\mathrm{op}},\mathbf{Set}]$$

under the equivalence

$$[\hat{\mathcal{C}}^{\mathrm{op}},\mathbf{Set}] \simeq [\mathcal{C}^{\mathrm{op}},\mathbf{Set}].$$

We shall describe this topology explicitly in Theorem 5.3.3. For this, let us adopt the following conventions:

- If S is a sieve in $\mathcal{C}$, we denote by $\overline{S}$ the sieve in $\hat{\mathcal{C}}$ generated by the arrows in S.
- If R is a sieve in $\hat{\mathcal{C}}$, we denote by $R \cap \mathrm{Arr}(\mathcal{C})$ the sieve in $\mathcal{C}$ formed by the elements of R which are arrows in $\mathcal{C}$.
- Given an arrow $g : d \to c$ in $\mathcal{C}$ and sieves S and R on c respectively in $\mathcal{C}$ and $\hat{\mathcal{C}}$, we denote by $g^*_{\mathcal{C}}(S)$ and $g^*_{\hat{\mathcal{C}}}(R)$ the sieves obtained by pulling back S and R along g respectively in the categories $\mathcal{C}$ and $\hat{\mathcal{C}}$.

Theorem 5.3.3 (Theorem 2.15 [17]). *Let $\mathcal{C}$ be a small category and $\hat{\mathcal{C}}$ its Cauchy completion. Given a Grothendieck topology J on $\mathcal{C}$ there exists a unique Grothendieck topology $\hat{J}$ on $\hat{\mathcal{C}}$ that induces J on $\mathcal{C}$, which is defined as follows: for each sieve R on $d \in \hat{\mathcal{C}}$, $R \in \hat{J}(d)$ if and only if there exist a retract $d \overset{i}{\hookrightarrow} a \overset{r}{\to} d$ with $a \in \mathcal{C}$ and a sieve $S \in J(a)$ such that $R = i^*_{\hat{\mathcal{C}}}(\overline{S})$. Furthermore, if $d \in \mathcal{C}$ then $R \in \hat{J}(d)$ if and only if there exists a sieve S in $\mathcal{C}$ on d such that $R = \overline{S}$.*

Proof Since the full embedding $\mathcal{C} \hookrightarrow \hat{\mathcal{C}}$ is (trivially) dense with respect to every Grothendieck topology on $\hat{\mathcal{C}}$, it follows from the Comparison Lemma (Theorem 1.1.8) that there is at most one Grothendieck topology on $\hat{\mathcal{C}}$ that induces J on $\mathcal{C}$. Therefore, to prove the first part of the thesis it will be enough to show that the assignment $\hat{J}$ in the statement of the theorem is a Grothendieck topology that induces J on $\mathcal{C}$. This, as well as the second part of the thesis, can be easily deduced from the following fact (whose proof is left to the reader): given an object $c \in \mathcal{C}$, the assignments $R \to R \cap \mathrm{Arr}(\mathcal{C})$ and $S \to \overline{S}$ are inverse to each other and define a bijection between the set of sieves in $\mathcal{C}$ on c and the set of sieves in $\hat{\mathcal{C}}$ on c; moreover, these bijections are natural with respect to the operations of pullback of sieves in $\mathcal{C}$ and in $\hat{\mathcal{C}}$ along an arrow in $\mathcal{C}$.

By way of example, we provide the details of the proof that $\hat{J}$ satisfies the 'stability axiom' for Grothendieck topologies. Given $R \in \hat{J}(d)$ and $g : e \to d$ in $\hat{\mathcal{C}}$, we want to prove that $g^*(R) \in \hat{J}(e)$. Since $R \in \hat{J}(d)$, there exist a retract $d \overset{i}{\hookrightarrow} a \overset{r}{\to} d$ with $a \in \mathcal{C}$ and a sieve $S \in J(a)$ such that $R = i^*(\overline{S})$. On the other hand, there exists a retract $e \overset{j}{\hookrightarrow} b \overset{z}{\to} e$ with $b \in \mathcal{C}$. Now, $g^*(R) = g^*(i^*(\overline{S})) = (i \circ g)^*(\overline{S}) = ((i \circ g \circ z) \circ j)^*(\overline{S}) = j^*((i \circ g \circ z)^*(\overline{S})) = j^*((i \circ g \circ z)^*_{\hat{\mathcal{C}}}(\overline{S})) = j^*(\overline{(i \circ g \circ z)^*_{\mathcal{C}}(S)})$. Our claim thus follows at once from the stability axiom for J. □

6

THEORIES OF PRESHEAF TYPE: GENERAL CRITERIA

In this chapter we establish a characterization theorem providing necessary and sufficient semantic conditions for a theory to be of presheaf type (i.e. classified by a presheaf topos). This theorem subsumes the partial results previously obtained on the subject and has several corollaries which can be used in practice for testing whether a given theory is of presheaf type as well as for generating new examples of theories belonging to this class. Along the way we obtain a number of other results of independent interest, including methods for constructing theories classified by a given presheaf topos.

The contents of the chapter can be summarized as follows.

In section 6.1, in order to set up the field for the statement and proof of the characterization theorem, we identify some notable properties of theories of presheaf type. More specifically, we show that every finitely presentable model of such a theory is finitely presented, in a strong sense which we make precise in section 6.2.2, and admits an entirely syntactic description in terms of the signature of the theory and the notion of provability of geometric sequents in it (cf. section 6.1.5). We also give a semantic description of universal models of theories of presheaf type, from which we deduce a definability theorem, and establish a syntactic criterion for a theory to be classified by a presheaf topos.

In section 6.2, we show that for any geometric theory $\mathbb{T}$ and any Grothendieck topos $\mathcal{E}$ there exists, for each pair of $\mathbb{T}$-models M and N in $\mathcal{E}$, an 'object of $\mathbb{T}$-model homomorphisms' in $\mathcal{E}$ from M to N which classifies the $\mathbb{T}$-model homomorphisms between 'localizations' of M and N in slices of $\mathcal{E}$ (cf. section 6.2.1). Moreover, we introduce two notions of internal finite presentability for models of geometric theories in Grothendieck toposes, and show that the 'constant' models of a theory of presheaf type associated with its finitely presentable (set-based) models satisfy both of them.

In section 6.3, we establish our main characterization theorem providing necessary and sufficient conditions for a geometric theory to be of presheaf type. We first state the result abstractly and then proceed to obtain explicit reformulations of each of its conditions. We also derive some corollaries which allow us to verify their satisfaction in specific situations which naturally arise in practice. Lastly we show that, once expressed in the language of indexed colimits of internal diagrams in toposes, the conditions of the characterization theorem for a given geometric theory $\mathbb{T}$ amount precisely to the requirement that every model of $\mathbb{T}$ in any Grothendieck topos $\mathcal{E}$ should be a canonical $\mathcal{E}$-indexed colimit of a certain $\mathcal{E}$-filtered diagram of 'constant' models of $\mathbb{T}$ which are $\mathcal{E}$-finitely presentable.

6.1 Preliminary results

In this section we establish some results on theories of presheaf type which will be important in the sequel.

6.1.1 A canonical form for Morita equivalences

By Diaconescu's equivalence, any theory of presheaf type $\mathbb{T}$ classified by a topos $[\mathcal{C}, \mathbf{Set}]$ comes equipped with a Morita equivalence

$$\xi^{\mathcal{E}} : \mathbf{Flat}(\hat{\mathcal{C}}^{\mathrm{op}}, \mathcal{E}) \xrightarrow{\sim} \mathbb{T}\text{-mod}(\mathcal{E})$$

with the theory of flat functors on the category $\hat{\mathcal{C}}^{\mathrm{op}}$, where $\hat{\mathcal{C}}$ is the Cauchy completion of $\mathcal{C}$ (cf. section 5.3 above). Since $\hat{\mathcal{C}}$ can be recovered from $\mathrm{Ind}\text{-}\hat{\mathcal{C}} \simeq \mathbf{Flat}(\hat{\mathcal{C}}^{\mathrm{op}}, \mathbf{Set})$ as the full subcategory on the finitely presentable objects (cf. Proposition C4.2.2 [55]), we can suppose $\hat{\mathcal{C}} = \mathrm{f.p.}\mathbb{T}\text{-mod}(\mathbf{Set})$ without loss of generality, where f.p.$\mathbb{T}$-mod(**Set**) is the full subcategory of $\mathbb{T}$-mod(**Set**) on the finitely presentable objects (cf. section 6.1.4 below – this notation actually agrees with that of section 2.1.5 by Remark 6.1.14). Notice that, whilst f.p.$\mathbb{T}$-mod(**Set**) is not small in general, it is always essentially small (i.e. equivalent to a small category) by Lemma D2.3.2(i) [55]. So from now on, to avoid any size issues, we shall denote by f.p.$\mathbb{T}$-mod(**Set**) a *skeleton* of the category of finitely presentable $\mathbb{T}$-models.

Via the equivalence $\xi^{\mathcal{E}}$, the functor $y_{\hat{\mathcal{C}}}^{\mathcal{E}} : \hat{\mathcal{C}} \to \mathbf{Flat}(\hat{\mathcal{C}}^{\mathrm{op}}, \mathcal{E})$ considered in section 5.3 corresponds to the functor from f.p.$\mathbb{T}$-mod(**Set**) to $\mathbb{T}$-mod$(\mathcal{E})$ given by $\gamma_{\mathcal{E}}^{*}$; in particular, the equivalence $\xi^{\mathbf{Set}} : \mathbf{Flat}(\hat{\mathcal{C}}^{\mathrm{op}}, \mathbf{Set}) \simeq \mathbb{T}\text{-mod}(\mathbf{Set})$ restricts to a natural equivalence

$$\tau_{\xi} : \hat{\mathcal{C}} \xrightarrow{\sim} \mathrm{f.p.}\mathbb{T}\text{-mod}(\mathbf{Set}),$$

as in the following diagram:

$$\begin{array}{ccc} \hat{\mathcal{C}} & \xrightarrow{\tau_{\xi}} & \mathrm{f.p.}\mathbb{T}\text{-mod}(\mathbf{Set}) \\ {\scriptstyle y_{\hat{\mathcal{C}}}^{\mathbf{Set}}} \downarrow & & \downarrow {\scriptstyle i} \\ \mathbf{Flat}(\hat{\mathcal{C}}^{\mathrm{op}}, \mathbf{Set}) & \xrightarrow{\sim} & \mathbb{T}\text{-mod}(\mathbf{Set}). \end{array}$$

It is important to note that if $\hat{\mathcal{C}} = \mathrm{f.p.}\mathbb{T}\text{-mod}(\mathbf{Set})$, we can modify ξ so that $\tau_{\xi}(c) \cong c$ naturally in $c \in \mathrm{f.p.}\mathbb{T}\text{-mod}(\mathbf{Set})$. Indeed, it suffices to compose ξ with the equivalences

$$\mathbf{Flat}(\mathrm{f.p.}\mathbb{T}\text{-mod}(\mathbf{Set})^{\mathrm{op}}, \mathcal{E}) \to \mathbf{Flat}(\hat{\mathcal{C}}^{\mathrm{op}}, \mathcal{E})$$

(natural in $\mathcal{E} \in \mathfrak{Btop}$) induced by composition with $(\tau_{\xi}^{-1})^{\mathrm{op}}$.

We can thus assume, given a theory of presheaf type $\mathbb{T}$, that $\mathbb{T}$ comes equipped with a Morita equivalence ξ satisfying the condition $\tau_\xi \cong 1_{\text{f.p.}\mathbb{T}\text{-mod}(\mathbf{Set})}$; we call such an equivalence *canonical*, and accordingly we say that an equivalence

$$\chi^\mathcal{E} : \mathbb{T}\text{-mod}(\mathcal{E}) \simeq \mathbf{Geom}(\mathcal{E}, [\text{f.p.}\mathbb{T}\text{-mod}(\mathbf{Set}), \mathbf{Set}])$$

natural in $\mathcal{E} \in \mathfrak{Btop}$ is canonical if it is induced by a canonical equivalence

$$\mathbf{Flat}(\text{f.p.}\mathbb{T}\text{-mod}(\mathbf{Set})^{\text{op}}, \mathcal{E}) \xrightarrow{\sim} \mathbb{T}\text{-mod}(\mathcal{E})$$

by composition with Diaconescu's equivalence.

Note that 'the' canonical Morita equivalence for a theory of presheaf type $\mathbb{T}$ is induced by the canonical geometric morphism

$$[\text{f.p.}\mathbb{T}\text{-mod}(\mathbf{Set}), \mathbf{Set}] \to \mathbf{Sh}(\mathcal{C}_\mathbb{T}, J_\mathbb{T})$$

via Diaconescu's equivalence. This yields in particular a canonical representation for the classifying topos of a theory of presheaf type $\mathbb{T}$ as the topos $[\text{f.p.}\mathbb{T}\text{-mod}(\mathbf{Set}), \mathbf{Set}]$. Notice that, since

$$\mathbb{T}\text{-mod}(\mathbf{Set}) \simeq \mathbf{Flat}(\text{f.p.}\mathbb{T}\text{-mod}(\mathbf{Set})^{\text{op}}, \mathbf{Set}),$$

the category $\mathbb{T}\text{-mod}(\mathbf{Set})$ is the ind-completion of the category f.p.$\mathbb{T}$-mod($\mathbf{Set}$) (cf. section 8.1.3 below for some general results about ind-completions).

6.1.2 Universal models and definability

The following result gives an explicit description of 'the' universal model of a theory of presheaf type.

Theorem 6.1.1 (Theorem 3.1 [19]). *Let $\mathbb{T}$ be a theory of presheaf type over a signature Σ. Then the Σ-structure $N_\mathbb{T}$ in $[\text{f.p.}\mathbb{T}\text{-mod}(\mathbf{Set}), \mathbf{Set}]$ which assigns to a sort A of Σ the evaluation functor $N_\mathbb{T} A$ at A (sending any model $M \in$ f.p.$\mathbb{T}$-mod($\mathbf{Set}$) to the set MA and acting on the arrows accordingly), to a function symbol $f : A_1 \cdots A_n \to B$ of Σ the morphism $N_\mathbb{T} A_1 \times \cdots \times N_\mathbb{T} A_n \to N_\mathbb{T} B$ given by $(N_\mathbb{T} f)(M) = Mf$ and to a relation symbol $R \rightarrowtail A_1 \cdots A_n$ of Σ the subobject $N_\mathbb{T} R \rightarrowtail N_\mathbb{T} A_1 \times \cdots \times N_\mathbb{T} A_n$ given by $(N_\mathbb{T} R)(M) = MR$ (for any $M \in$ f.p.$\mathbb{T}$-mod($\mathbf{Set}$)) is a universal model for $\mathbb{T}$; moreover, for any geometric formula $\phi(\vec{x})$ over Σ, the interpretation $[[\vec{x}.\phi]]_{N_\mathbb{T}}$ of $\phi(\vec{x})$ in $N_\mathbb{T}$ is given by $[[\vec{x}.\phi]]_{N_\mathbb{T}}(M) = [[\vec{x}.\phi]]_M$ for any $M \in$ f.p.$\mathbb{T}$-mod($\mathbf{Set}$). In particular, the finitely presentable $\mathbb{T}$-models are jointly conservative for $\mathbb{T}$.*

Proof Consider a canonical Morita equivalence

$$\xi^\mathcal{E} : \mathbf{Flat}(\text{f.p.}\mathbb{T}\text{-mod}(\mathbf{Set})^{\text{op}}, \mathcal{E}) \xrightarrow{\sim} \mathbb{T}\text{-mod}(\mathcal{E})$$

for the theory of presheaf type $\mathbb{T}$. By composing it with Diaconescu's equivalence

$$\mathbf{Geom}(\mathcal{E}, [\text{f.p.}\mathbb{T}\text{-mod}(\mathbf{Set}), \mathbf{Set}]) \simeq \mathbf{Flat}(\text{f.p.}\mathbb{T}\text{-mod}(\mathbf{Set})^{\text{op}}, \mathcal{E}),$$

we get an equivalence $\tau^\mathcal{E} : \mathbf{Geom}(\mathcal{E}, [\text{f.p.}\mathbb{T}\text{-mod}(\mathbf{Set}), \mathbf{Set}]) \simeq \mathbb{T}\text{-mod}(\mathcal{E})$ natural in $\mathcal{E} \in \mathfrak{Btop}$; let us define U as the image of the identical geometric morphism on $[\text{f.p.}\mathbb{T}\text{-mod}(\mathbf{Set}), \mathbf{Set}]$ under this equivalence. Then U is a universal model of $\mathbb{T}$ and a geometric morphism $f : \mathcal{E} \to [\text{f.p.}\mathbb{T}\text{-mod}(\mathbf{Set}), \mathbf{Set}]$ is sent by $\tau^{\mathcal{E}}$ to the $\mathbb{T}$-model $f^*(U)$. Recall that, via Diaconescu's equivalence, the identity on $[\text{f.p.}\mathbb{T}\text{-mod}(\mathbf{Set}), \mathbf{Set}]$ is sent to the flat functor $y : \text{f.p.}\mathbb{T}\text{-mod}(\mathbf{Set})^{\text{op}} \to [\text{f.p.}\mathbb{T}\text{-mod}(\mathbf{Set}), \mathbf{Set}]$ given by the Yoneda embedding. So, by naturality, for any model $M \in \text{f.p.}\mathbb{T}\text{-mod}(\mathbf{Set})$, the geometric morphism

$$e_M : \mathbf{Set} \to [\text{f.p.}\mathbb{T}\text{-mod}(\mathbf{Set}), \mathbf{Set}]$$

whose inverse image is the evaluation functor $\text{ev}_M : [\text{f.p.}\mathbb{T}\text{-mod}(\mathbf{Set}), \mathbf{Set}] \to \mathbf{Set}$ at M is sent to the flat functor $(e_M)^* \circ y = \text{Hom}_{\text{f.p.}\mathbb{T}\text{-mod}(\mathbf{Set})}(-, M) : \text{f.p.}\mathbb{T}\text{-mod}(\mathbf{Set})^{\text{op}} \to \mathbf{Set}$. But, since ξ is canonical, $\xi^{\mathbf{Set}}$ sends the functor $\text{Hom}_{\text{f.p.}\mathbb{T}\text{-mod}(\mathbf{Set})}(-, M)$ to the model M and hence $\tau^{\mathbf{Set}}$ sends e_M to M (for any $M \in \text{f.p.}\mathbb{T}\text{-mod}(\mathbf{Set})$). Thus $M \cong (e_M)^*(U) = \text{ev}_M(U)$ for any $M \in \text{f.p.}\mathbb{T}\text{-mod}(\mathbf{Set})$, and hence U is (isomorphic to) the Σ-structure $N_{\mathbb{T}}$ defined in the statement of the theorem. Now, the fact that $[[\vec{x}\,.\,\phi]]_{N_{\mathbb{T}}}(M) = [[\vec{x}\,.\,\phi]]_M$ for any geometric formula $\phi(\vec{x})$ and model $M \in \text{f.p.}\mathbb{T}\text{-mod}(\mathbf{Set})$ follows from the fact that the functors ev_M are geometric (being inverse image functors of geometric morphisms). □

Remarks 6.1.2 (a) Theorem 6.1.1 specializes to Corollary D3.1.2 [55] in the case when $\mathbb{T}$ is cartesian.

(b) Since universal models are preserved by inverse image functors of geometric inclusions (Lemma 2.1 [19]), 'the' universal model of a quotient $\mathbb{T}'$ of a theory of presheaf type $\mathbb{T}$ is given by the image of the model $N_{\mathbb{T}}$ under the associated sheaf functor $\mathcal{E}_{\mathbb{T}} \to \mathcal{E}_{\mathbb{T}'}$. In particular, if for each sort A of the signature of $\mathbb{T}$ the formula $\{x^A\,.\,\top\}$ presents a $\mathbb{T}$-model then the functors $N_{\mathbb{T}}A$ are all representable and hence, if the induced $\mathbb{T}$-topology J of $\mathbb{T}$ (in the sense of section 8.1.2) is subcanonical, the assignment $A \to N_{\mathbb{T}}A$ yields a universal model of $\mathbb{T}'$ inside its classifying topos $\mathbf{Sh}(\text{f.p.}\mathbb{T}\text{-mod}(\mathbf{Set})^{\text{op}}, J) \hookrightarrow [\text{f.p.}\mathbb{T}\text{-mod}(\mathbf{Set}), \mathbf{Set}]$.

(c) The proof of Theorem 6.1.1 shows that if $\mathbb{T}$ is a geometric theory classified by a topos $[\mathcal{C}, \mathbf{Set}]$ then the $\mathbb{T}$-models corresponding to the objects of $\mathcal{C}$ are jointly conservative for $\mathbb{T}$.

As shown by the following theorem, universal models of geometric theories enjoy a strong form of logical completeness.

Theorem 6.1.3 (Theorem 2.2 [19]). *Let $\mathbb{T}$ be a geometric theory over a signature Σ and U a universal model of $\mathbb{T}$ in a Grothendieck topos $\mathcal{E}$. Then*

(i) *For any subobject $S \rightarrowtail UA_1 \times \cdots \times UA_n$ in $\mathcal{E}$, there exists a geometric formula $\phi(\vec{x}) = \phi(x_1^{A_1}, \ldots, x_n^{A_n})$ over Σ such that $S = [[\vec{x}\,.\,\phi]]_U$.*

(ii) *For any arrow $f : [[\vec{x}\,.\,\phi]]_U \to [[\vec{y}\,.\,\psi]]_U$ in $\mathcal{E}$, where $\phi(\vec{x})$ and $\psi(\vec{y})$ are geometric formulae over Σ, there exists a geometric formula $\theta(\vec{x}, \vec{y})$ over Σ such that the sequents $(\phi \vdash_{\vec{x}} (\exists \vec{y})\theta)$, $(\theta \vdash_{\vec{x}, \vec{y}} \phi \wedge \psi)$ and $(\theta \wedge \theta[\vec{y}'/\vec{y}] \vdash_{\vec{x}, \vec{y}, \vec{y}'} \vec{y} = \vec{y}')$ are provable in $\mathbb{T}$ and $[[\vec{x}, \vec{y}\,.\,\theta]]_U$ is the graph of f.*

Proof We can clearly suppose, without loss of generality, that U is the universal model $U_\mathbb{T}$ of $\mathbb{T}$ lying in its classifying topos $\mathbf{Sh}(\mathcal{C}_\mathbb{T}, J_\mathbb{T})$ described in Remark 2.1.9.

To prove (i) we observe that, given a geometric formula $\phi(x_1^{A_1}, \ldots, x_n^{A_n})$ over Σ, its interpretation in $U_\mathbb{T}$ identifies with the sieve on $\{x_1^{A_1}, \ldots, x_n^{A_n} . \top\}$ generated by the canonical monomorphism $[\phi] : \{\vec{x} . \phi\} \rightarrowtail \{x_1^{A_1}, \ldots, x_n^{A_n} . \top\}$. Now, if S is a subobject of

$$U_\mathbb{T} A_1 \times \cdots \times U_\mathbb{T} A_n \cong \mathrm{Hom}_{\mathcal{C}_\mathbb{T}}(-, \{x_1^{A_1}, \ldots, x_n^{A_n} . \top\})$$

then S, regarded as a sieve, is $J_\mathbb{T}$-closed and hence, by Proposition 3.1.6(ii), it is generated by a subobject of $\{x_1^{A_1}, \ldots, x_n^{A_n} . \top\}$ in $\mathcal{C}_\mathbb{T}$. Thus (i) follows from the characterization of subobjects in $\mathcal{C}_\mathbb{T}$ given by Lemma 1.4.2.

Let us now prove (ii). By the Yoneda lemma, any arrow

$$f : [[\vec{x} . \phi]]_{U_\mathbb{T}} \cong \mathrm{Hom}_{\mathcal{C}_\mathbb{T}}(-, \{\vec{x} . \phi\}) \to \mathrm{Hom}_{\mathcal{C}_\mathbb{T}}(-, \{\vec{y} . \psi\}) \cong [[\vec{y} . \psi]]_{U_\mathbb{T}}$$

in $\mathbf{Sh}(\mathcal{C}_\mathbb{T}, J_\mathbb{T})$ is of the form $\mathrm{Hom}_{\mathcal{C}_\mathbb{T}}(-, [\theta])$ for a unique arrow $[\theta] : \{\vec{x} . \phi\} \to \{\vec{y} . \psi\}$ in $\mathcal{C}_\mathbb{T}$; but $[[\vec{x}, \vec{y} . \theta]]_{U_\mathbb{T}}$ is the graph of the arrow $\mathrm{Hom}_{\mathcal{C}_\mathbb{T}}(-, [\theta])$, from which our thesis follows. □

Theorem 6.1.3, combined with Theorem 6.1.1, yields the following definability theorem for theories of presheaf type.

Theorem 6.1.4 (Corollary 3.2 [19]). *Let $\mathbb{T}$ be a theory of presheaf type and $A_1, \ldots, A_n$ sorts of its signature. Suppose that we are given, for every finitely presentable set-based model M of $\mathbb{T}$, a subset R_M of $M A_1 \times \cdots \times M A_n$ in such a way that every $\mathbb{T}$-model homomorphism $h : M \to N$ maps R_M into R_N. Then there exists a (unique up to $\mathbb{T}$-provable equivalence) geometric formula-in-context $\phi(\vec{x})$, where $\vec{x} = (x_1^{A_1}, \ldots, x_n^{A_n})$, such that $R_M = [[\vec{x} . \phi]]_M$ for every finitely presentable $\mathbb{T}$-model M.*

Proof The thesis immediately follows from Theorems 6.1.3 and 6.1.1 observing that the assignment $M \to R_M$ in the statement of the theorem gives rise to a subfunctor $R \rightarrowtail N_\mathbb{T}$, where $N_\mathbb{T}$ is the universal model of $\mathbb{T}$ in $[\text{f.p.}\mathbb{T}\text{-mod}(\mathbf{Set}), \mathbf{Set}]$ defined in the statement of Theorem 6.1.1. Conversely, any geometric formula $\phi(\vec{x})$ yields a covariant functorial assignment $M \to [[\vec{x} . \phi]]_M$, since any $\mathbb{T}$-model homomorphism $h : M \to N$ maps $[[\vec{x} . \phi]]_M$ to $[[\vec{x} . \phi]]_N$ (cf. Remark 1.3.21). In other words, we have the following 'bridge':

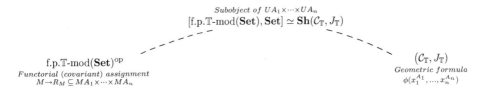

where U is 'the' universal model of $\mathbb{T}$ in its classifying topos. □

Remarks 6.1.5 (a) The proof of Theorem 6.1.4 also shows that, for any two geometric formulae $\phi(\vec{x})$ and $\psi(\vec{y})$ over the signature of $\mathbb{T}$, every assignment $M \to f_M : [[\vec{x}\,.\,\phi]]_M \to [[\vec{y}\,.\,\psi]]_M$ (for finitely presentable $\mathbb{T}$-models M) which is natural in M is definable in finitely presentable $\mathbb{T}$-models by a $\mathbb{T}$-provably functional formula $\theta(\vec{x}, \vec{y})$ from $\phi(\vec{x})$ to $\psi(\vec{y})$.

(b) If the property R of tuples $\vec{x}$ of elements of finitely presentable $\mathbb{T}$-models in the statement of Theorem 6.1.4 is extended to all set-based $\mathbb{T}$-models so as to be preserved by filtered colimits in $\mathbb{T}\text{-mod}(\mathbf{Set})$ then we have $R_M = [[\vec{x}\,.\,\phi]]_M$ for each set-based $\mathbb{T}$-model M.

(c) If in Theorem 6.1.4 the theory $\mathbb{T}$ is coherent and the property R is not only preserved but also reflected by arbitrary $\mathbb{T}$-model homomorphisms then the formula $\phi(\vec{x})$ in the statement of the theorem can be taken to be coherent and $\mathbb{T}$-Boolean (in the sense that there exists a coherent formula $\psi(\vec{x})$ in the same context such that the sequents $(\phi \wedge \psi \vdash_{\vec{x}} \bot)$ and $(\top \vdash_{\vec{x}} \phi \vee \psi)$ are provable in $\mathbb{T}$). Indeed, the theorem can be applied both to the property R and to the negation of it yielding two geometric formulae $\phi(\vec{x})$ and $\psi(\vec{x})$ defining them such that $(\phi \wedge \psi \vdash_{\vec{x}} \bot)$ and $(\top \vdash_{\vec{x}} \phi \vee \psi)$ are provable in $\mathbb{T}$. Since $\mathbb{T}$ is coherent and every geometric formula in a given context is provably equivalent to a disjunction of coherent formulae in the same context, we can suppose ϕ and ψ to be coherent without loss of generality (cf. Theorem 10.8.6(iii)).

6.1.3 A syntactic criterion for a theory to be of presheaf type

In this section we shall establish a syntactic criterion for a geometric theory to be of presheaf type.

The following notion will be central in our analysis:

Definition 6.1.6 (Definition C2.2.18 [55]) A Grothendieck topology J on a category $\mathcal{C}$ is said to be *rigid* if, for every object c of $\mathcal{C}$, the collection of arrows from J-irreducible objects of $\mathcal{C}$ (i.e. objects whose J-covering sieves are only the maximal ones) generates a J-covering sieve.

In Theorem 6.1.7 below, for a subcanonical site $(\mathcal{C}, J)$, we denote by $y : \mathcal{C} \to \mathbf{Sh}(\mathcal{C}, J)$ the factorization through $\mathbf{Sh}(\mathcal{C}, J) \hookrightarrow [\mathcal{C}^{\mathrm{op}}, \mathbf{Set}]$ of the Yoneda embedding.

Theorem 6.1.7 (Theorem 3.13 [20]). *Let $(\mathcal{C}, J)$ be a small subcanonical site such that $y(\mathcal{C})$ is closed in $\mathbf{Sh}(\mathcal{C}, J)$ under retracts. Then $\mathbf{Sh}(\mathcal{C}, J)$ is equivalent to a presheaf topos if and only if J is rigid.*

Proof The 'if' direction follows at once from the Comparison Lemma (Theorem 1.1.8). It thus remains to prove the 'only if' direction. If $\mathcal{E} = \mathbf{Sh}(\mathcal{C}, J)$ is equivalent to a presheaf topos then, by Lemma C2.2.20 [55], $\mathcal{E}$ has a separating set of indecomposable projective objects. Now, suppose A is an indecomposable projective object of $\mathcal{E}$. Then, as it is observed in the proof of Lemma C2.2.20 [55], given any epimorphic family $\{f_i : B_i \to A \mid i \in I\}$, at least one f_i must be a split epimorphism; in particular, A is $J_\mathcal{E}$-irreducible, where $J_\mathcal{E}$ is the canonical topology on $\mathcal{E}$. Hence, by taking as epimorphic family the collection of all the arrows in $\mathcal{E}$ from objects of the form $y(c)$ to A, we obtain that A is a retract in $\mathcal{E}$ of an object of the form $y(c)$ for some $c \in \mathcal{C}$. Thus, by our hypotheses, A is itself, up to isomorphism, of the form $y(c)$. Let us denote by $\mathcal{C}'$ the full subcategory of $\mathcal{C}$ on the objects c such that $y(c)$ is indecomposable and projective in $\mathcal{E}$; then the objects in $y(\mathcal{C}')$ form a separating set of objects for $\mathcal{E}$. Thus, for any object B of $\mathcal{E}$, the family of all the arrows in $\mathcal{E}$ from objects of the form $y(c)$, for $c \in \mathcal{C}'$, to B generates a $J_\mathcal{E}$-covering sieve. Now, J being subcanonical, $J = J_\mathcal{E}|_\mathcal{C}$ (by Proposition C2.2.16 [55]) and hence, for any object $c \in \mathcal{C}$, the collection of all arrows in $\mathcal{C}$ from objects of $\mathcal{C}'$ to c is J-covering; but every J-covering sieve on an object of $\mathcal{C}'$ is trivial, so J is rigid, as required. $\square$

Remark 6.1.8 Under the hypotheses of the theorem, if $\mathcal{C}$ is Cauchy-complete (in particular, if $\mathcal{C}$ is cartesian) then $y(\mathcal{C})$ is closed in $\mathcal{E} = \mathbf{Sh}(\mathcal{C}, J)$ under retracts. Indeed, let $i : A \rightarrowtail y(c)$, $r : y(c) \twoheadrightarrow A$ be a retract of A in $\mathcal{E}$, that is, $r \circ i = 1_A$. Then $i \circ r : y(c) \to y(c)$ is idempotent. Now, since y is full and faithful, $i \circ r = y(e)$ for some idempotent $e : c \to c$ in $\mathcal{C}$. Since $\mathcal{C}$ is Cauchy-complete, e splits as $s \circ t$ where $t \circ s = 1$. Then $y(s)$ and $y(t)$ form a retract of A and hence, by the uniqueness up to isomorphism of the splitting of an idempotent in a category, r is isomorphic to $y(t)$ and i is isomorphic to $y(s)$; in particular, A is isomorphic to $y(\mathrm{dom}(s))$.

By Remark 1.4.9, the regular (resp. coherent, geometric) syntactic sites of regular (resp. coherent, geometric) theories all satisfy the hypotheses of Theorem 6.1.7. This motivates the following definition:

Definition 6.1.9 Let $\mathbb{T}$ be a geometric theory over a signature Σ. A geometric formula-in-context $\{\vec{x}.\phi\}$ is said to be $\mathbb{T}$-*irreducible* if it is $J_\mathbb{T}$-irreducible as an object of the syntactic category $\mathcal{C}_\mathbb{T}$, equivalently if for any family $\{\theta_i \mid i \in I\}$ of $\mathbb{T}$-provably functional geometric formulae $\{\vec{x}_i, \vec{x}.\theta_i\}$ from $\{\vec{x}_i.\phi_i\}$ to $\{\vec{x}.\phi\}$ such that $(\phi \vdash_{\vec{x}} \bigvee_{i \in I} (\exists \vec{x}_i) \theta_i)$ is provable in $\mathbb{T}$, there exist $i \in I$ and a $\mathbb{T}$-provably functional geometric formula $\{\vec{x}, \vec{x}_i.\theta'\}$ from $\{\vec{x}.\phi\}$ to $\{\vec{x}_i.\phi_i\}$ such that the composite arrow $[\theta_i] \circ [\theta']$ in $\mathcal{C}_\mathbb{T}$ is equal to the identity on $\{\vec{x}.\phi\}$ (equivalently, the bisequent $(\phi(\vec{x}) \wedge \vec{x} = \vec{x}' \dashv\vdash_{\vec{x},\vec{x}'} (\exists \vec{x}_i)(\theta'(\vec{x}', \vec{x}_i) \wedge \theta_i(\vec{x}_i, \vec{x})))$ is provable in $\mathbb{T}$).

Theorem 6.1.7 thus yields the following syntactic criterion for a geometric theory to be of presheaf type (cf. also the proof of Theorem 10.3.4(i)):

Corollary 6.1.10 (Corollary 3.15 [20]). *Let $\mathbb{T}$ be a geometric theory over a signature Σ. Then $\mathbb{T}$ is of presheaf type if and only if every geometric formula-in-context over Σ, regarded as an object of $\mathcal{C}_\mathbb{T}$, is $J_\mathbb{T}$-covered by $\mathbb{T}$-irreducible formulae, i.e. if and only if there exists a collection $\mathcal{F}$ of geometric formulae-in-context over Σ satisfying the following properties:*

(i) *For any geometric formula $\{\vec{y}.\psi\}$ over Σ, there exist objects $\{\vec{x_i}.\phi_i\}$ in $\mathcal{F}$ (for $i \in I$) and $\mathbb{T}$-provably functional geometric formulae $\{\vec{x_i},\vec{y}.\theta_i\}$ from $\{\vec{x_i}.\phi_i\}$ to $\{\vec{y}.\psi\}$ such that $(\psi \vdash_{\vec{y}} \bigvee_{i \in I}(\exists \vec{x_i})\theta_i)$ is provable in $\mathbb{T}$.*

(ii) *Every formula $\{\vec{x}.\phi\}$ in $\mathcal{F}$ is $\mathbb{T}$-irreducible, i.e. for any family $\{\theta_i \mid i \in I\}$ of $\mathbb{T}$-provably functional geometric formulae $\{\vec{x_i},\vec{x}.\theta_i\}$ from $\{\vec{x_i}.\phi_i\}$ to $\{\vec{x}.\phi\}$ such that $(\phi \vdash_{\vec{x}} \bigvee_{i \in I}(\exists \vec{x_i})\theta_i)$ is provable in $\mathbb{T}$, there exist $i \in I$ and a $\mathbb{T}$-provably functional geometric formula $\{\vec{x},\vec{x_i}.\theta'\}$ from $\{\vec{x}.\phi\}$ to $\{\vec{x_i}.\phi_i\}$ such that the bisequent $(\phi(\vec{x}) \wedge \vec{x} = \vec{x}' \dashv\vdash_{\vec{x},\vec{x}'} (\exists \vec{x_i})(\theta'(\vec{x}',\vec{x_i}) \wedge \theta_i(\vec{x_i},\vec{x})))$ is provable in $\mathbb{T}$.*

6.1.4 Finitely presentable = finitely presented

We shall prove in this section that the semantic and syntactic notions of finite presentability for models of geometric theories coincide for all theories of presheaf type. Moreover, we shall characterize in a syntactic way the geometric formulae which present some model of a given theory of presheaf type. This generalizes the well-known result for cartesian theories (cf. section D2.4 of [55]).

First, let us recall the standard notions of finite presentability of a set-based model of a geometric theory.

Let Σ be a first-order signature. It is well known that the category $\Sigma\text{-str}(\mathbf{Set})$ of Σ-structures in $\mathbf{Set}$ has all filtered colimits, which can be constructed as follows. For any diagram $D : \mathcal{I} \to \Sigma\text{-str}(\mathbf{Set})$ defined on a filtered category $\mathcal{I}$, its colimit $\text{colim}(D)$ can be realized by taking, for any sort A of Σ, $\text{colim}(D)A$ equal to the colimit of the functor $D_A : \mathcal{I} \to \mathbf{Set}$ given by the composite of D with the forgetful functor $\Sigma\text{-str}(\mathbf{Set}) \to \mathbf{Set}$ corresponding to the sort A. By Proposition 2.13.3 [8], this can be identified with the disjoint union $\coprod_{i \in \mathcal{I}} D_A(i)$ modulo the equivalence relation $\approx_A$ defined as follows: given $x \in D_A(i)$ and $x' \in D_A(i')$, $x \approx_A x'$ if and only if there exist arrows $f : i \to i''$ and $g : i' \to i''$ in $\mathcal{I}$ such that $(D_A(f))(x) = (D_A(g))(x')$. Let us denote by $J_i : D(i) \to \text{colim}(D)$ the canonical colimit maps (for $i \in \mathcal{I}$). For any function symbol $f : A_1 \cdots A_n \to B$ in Σ, we define $\text{colim}(D)f : \text{colim}(D)A_1 \times \cdots \times \text{colim}(D)A_n \to \text{colim}(D)B$ as follows. For any $(x_1,\ldots,x_n) \in \text{colim}(D)A_1 \times \cdots \times \text{colim}(D)A_n$, we can suppose, thanks to the filteredness of $\mathcal{I}$, that there exist $i \in I$ and for each $k \in \{1,\ldots,n\}$ an element $x'_k \in D_{A_k}(i)$ such that $x_k = J_i(x'_k)$ for all k; we thus set $(\text{colim}(D)f)((x_1,\ldots,x_n))$ equal to $J_{iB}((D(i)f)((x'_1,\ldots,x'_n)))$. It is readily seen that this map is well-defined. For any relation symbol $R \rightarrowtail A_1 \cdots A_n$ in Σ, we define $\text{colim}(D)R$ as the subset of $\text{colim}(D)A_1 \times \cdots \times \text{colim}(D)A_n$ consisting of the tuples of the form $J_i((x_1,\ldots,x_n))$ where $(x_1,\ldots,x_n) \in D(i)R$ for some $i \in \mathcal{I}$. Notice that for any $i \in \mathcal{I}$ and any $x,x' \in D_A(i)$, $x \approx_A x'$ if and only if there exists an arrow $s : i \to j$ in $\mathcal{I}$ such that $D_A(s)(x) = D_A(s)(x')$, and that if $x \approx_A x'$ then for any arrow $t : i \to k$ in $\mathcal{I}$, $D_A(t)(x) \approx_A D_A(t)(x')$.

Recall from Lemma D2.4.9 [55] that for any geometric theory $\mathbb{T}$ over a signature Σ, the category $\mathbb{T}$-mod(**Set**) of $\mathbb{T}$-models in **Set** is closed in Σ-str(**Set**) under filtered colimits.

Definition 6.1.11 Let $\mathbb{T}$ be a geometric theory over a signature Σ and M a set-based $\mathbb{T}$-model. Then

(a) The model M is said to be *finitely presentable* if the representable functor

$$\mathrm{Hom}_{\mathbb{T}\text{-mod}(\mathbf{Set})}(M, -) : \mathbb{T}\text{-mod}(\mathbf{Set}) \to \mathbf{Set}$$

preserves filtered colimits.

(b) The model M is said to be *finitely presented* if there is a geometric formula $\{\vec{x} \cdot \phi\}$ over Σ and a string of elements $(\xi_1, \ldots, \xi_n) \in MA_1 \times \cdots \times MA_n$ (where $A_1, \ldots, A_n$ are the sorts of the variables in $\vec{x}$), called the *generators* of M, such that for any $\mathbb{T}$-model N in **Set** and string of elements $(b_1, \ldots, b_n) \in MA_1 \times \cdots \times MA_n$ such that $(b_1, \ldots, b_n) \in [[\vec{x} \cdot \phi]]_N$, there exists a unique arrow $f : M \to N$ in $\mathbb{T}$-mod(**Set**) such that $(f_{A_1} \times \cdots \times f_{A_n})((\xi_1, \ldots, \xi_n)) = (b_1, \ldots, b_n)$.

The full subcategory of $\mathbb{T}$-mod(**Set**) on the finitely presentable models (or a skeleton of it) will be denoted by f.p.$\mathbb{T}$-mod(**Set**). We shall denote 'the' model finitely presented by a formula $\{\vec{x} \cdot \phi\}$ by $M_{\{\vec{x} \cdot \phi\}}$, and its tuple of generators by $\vec{\xi}_\phi$.

The following lemma provides an explicit categorical characterization of the finitely presentable objects of a category with filtered colimits. By the above remark, this applies in particular to the categories $\mathbb{T}$-mod(**Set**) of set-based models of a geometric theory $\mathbb{T}$; indeed, a set-based model of a geometric theory $\mathbb{T}$ is finitely presentable in the sense of Definition 6.1.11(a) if and only if it is finitely presentable as an object of the category $\mathbb{T}$-mod(**Set**).

Lemma 6.1.12 *Let $\mathcal{D}$ be a category with filtered colimits and M an object of $\mathcal{D}$. Then M is finitely presentable in $\mathcal{D}$ if and only if for any diagram $D : \mathcal{I} \to \mathcal{D}$ defined on a small filtered category $\mathcal{I}$, every arrow $M \to \mathrm{colim}(D)$ factors through one of the canonical colimit arrows $J_i : D(i) \to \mathrm{colim}(D)$ and for any arrows $f : M \to D(i)$ and $g : M \to D(j)$ in $\mathcal{D}$ such that $J_i \circ f = J_j \circ g$ there exist $k \in \mathcal{I}$ and two arrows $s : i \to k$ and $t : j \to k$ in $\mathcal{I}$ such that $D(s) \circ f = D(t) \circ g$.*

Proof By definition, M is finitely presentable in $\mathcal{D}$ if and only if the functor $\mathrm{Hom}_\mathcal{D}(M, -) : \mathcal{D} \to \mathbf{Set}$ preserves filtered colimits, that is for any diagram $D : \mathcal{I} \to \mathcal{D}$ defined on a small filtered category $\mathcal{I}$, the canonical map ξ from the colimit in **Set** of the composite diagram $\mathrm{Hom}_\mathcal{D}(M, -) \circ D$ to $\mathrm{Hom}_\mathcal{D}(M, \mathrm{colim}(D))$ is a bijection. The former set can be described by Proposition 2.13.3 [8] as the disjoint union of the $\mathrm{Hom}_\mathcal{D}(M, D(i))$ (for $i \in \mathcal{I}$) modulo the equivalence relation $\approx$ defined as follows: for any $f : M \to D(i)$ and $g : M \to D(j)$, $f \approx g$ if and only if there exists $k \in \mathcal{I}$ and two arrows $s : i \to k$ and $t : j \to k$ such that $D(s) \circ f = D(t) \circ g$. The map ξ sends the $\approx$-equivalence class of an arrow

$f : M \to D(i)$ to the composite arrow $J_i \circ f$, where J_i is the canonical colimit arrow $D(i) \to \mathrm{colim}(D)$; the first and second conditions in the statement of the lemma thus correspond respectively to the requirements that ξ be surjective and injective. □

Our argument to establish the equivalence between the syntactic and semantic notions of finite presentability for a model of a theory of presheaf type $\mathbb{T}$ is based on the concept of irreducible object of a Grothendieck topos; in fact, this is precisely the invariant that we shall 'transfer' across the two distinct representations
$$[\mathrm{f.p.}\mathbb{T}\text{-mod}(\mathbf{Set}), \mathbf{Set}] \simeq \mathbf{Sh}(\mathcal{C}_\mathbb{T}, J_\mathbb{T})$$
of the classifying topos of $\mathbb{T}$. Recall that an object A of a Grothendieck topos $\mathcal{E}$ is said to be *irreducible* if every covering sieve on A with respect to the canonical topology on $\mathcal{E}$ is maximal, that is if any sieve in $\mathcal{E}$ containing a small epimorphic family contains the identity on A.

Theorem 6.1.13 (Theorem 4.3 [20]). *Let $\mathbb{T}$ be a theory of presheaf type over a signature Σ. Then*

(i) *Any finitely presentable $\mathbb{T}$-model in $\mathbf{Set}$ is presented by a $\mathbb{T}$-irreducible geometric formula $\phi(\vec{x})$ over Σ.*

(ii) *Conversely, any $\mathbb{T}$-irreducible geometric formula $\phi(\vec{x})$ over Σ presents a $\mathbb{T}$-model.*

In fact, the category f.p.$\mathbb{T}$-mod$(\mathbf{Set})^{\mathrm{op}}$ *is equivalent to the full subcategory* $\mathcal{C}_\mathbb{T}^{\mathrm{irr}}$ *of* $\mathcal{C}_\mathbb{T}$ *on the $\mathbb{T}$-irreducible formulae.*

Proof It is clear from the proof of Theorem 6.1.7 that a geometric formula $\{\vec{x}.\phi\}$ over the signature of $\mathbb{T}$ is $\mathbb{T}$-irreducible if and only if the object $y_\mathbb{T}(\{\vec{x}.\phi\})$, where $y_\mathbb{T}$ is the Yoneda embedding
$$y_\mathbb{T} : \mathcal{C}_\mathbb{T} \hookrightarrow \mathbf{Sh}(\mathcal{C}_\mathbb{T}, J_\mathbb{T}),$$
is irreducible in the topos $\mathbf{Sh}(\mathcal{C}_\mathbb{T}, J_\mathbb{T})$.

Since $\mathbb{T}$ is of presheaf type, we have a canonical equivalence of classifying toposes
$$\tau : \mathbf{Sh}(\mathcal{C}_\mathbb{T}, J_\mathbb{T}) \simeq [\mathrm{f.p.}\mathbb{T}\text{-mod}(\mathbf{Set}), \mathbf{Set}]$$
(in the sense of section 6.1.1).

If $\mathcal{C}_\mathbb{T}^{\mathrm{irr}}$ is the full subcategory of $\mathcal{C}_\mathbb{T}$ on the $\mathbb{T}$-irreducible formulae then, by Theorem 6.1.7, the Comparison Lemma yields an equivalence $\mathbf{Sh}(\mathcal{C}_\mathbb{T}, J_\mathbb{T}) \simeq [(\mathcal{C}_\mathbb{T}^{\mathrm{irr}})^{\mathrm{op}}, \mathbf{Set}]$.

If $\widehat{\mathcal{C}_\mathbb{T}^{\mathrm{irr}}} \hookrightarrow \mathcal{C}_\mathbb{T}$ is the Cauchy completion of $\mathcal{C}_\mathbb{T}^{\mathrm{irr}}$ (as in section 5.3.1) then $[(\mathcal{C}_\mathbb{T}^{\mathrm{irr}})^{\mathrm{op}}, \mathbf{Set}] \simeq [(\widehat{\mathcal{C}_\mathbb{T}^{\mathrm{irr}}})^{\mathrm{op}}, \mathbf{Set}]$ and the resulting equivalence
$$[\widehat{\mathcal{C}_\mathbb{T}^{\mathrm{irr}}}^{\mathrm{op}}, \mathbf{Set}] \simeq [\mathrm{f.p.}\mathbb{T}\text{-mod}(\mathbf{Set}), \mathbf{Set}]$$

restricts to an equivalence

$$l : \widehat{\mathcal{C}_{\mathbb{T}}^{\text{irr}}}^{\text{op}} \simeq \text{f.p.} \mathbb{T}\text{-mod}(\mathbf{Set})$$

between the full subcategories on the irreducible objects (cf. the proof of Theorem 6.1.7):

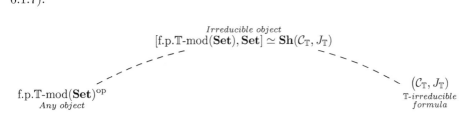

Now, given $\{\vec{x}.\phi\} \in \widehat{\mathcal{C}_{\mathbb{T}}^{\text{irr}}}$, τ sends the functor $y_{\mathbb{T}}(\{\vec{x}.\phi\}) = [[\vec{x}.\phi]]_{U_{\mathbb{T}}}$ to $y(l(\{\vec{x}.\phi\})) = [[\vec{x}.\phi]]_{N_{\mathbb{T}}}$ (where the notation is that of section 6.1.2), whence the model $l(\{\vec{x}.\phi\})$ is finitely presented by $\{\vec{x}.\phi\}$. Thus, since all representables in presheaf toposes are irreducible objects, $\phi(\vec{x})$ is $\mathbb{T}$-irreducible. So $\widehat{\mathcal{C}_{\mathbb{T}}^{\text{irr}}} \simeq \mathcal{C}_{\mathbb{T}}^{\text{irr}}$, that is, $\mathcal{C}_{\mathbb{T}}^{\text{irr}}$ is Cauchy-complete. Hence l yields an equivalence

$$(\mathcal{C}_{\mathbb{T}}^{\text{irr}})^{\text{op}} \simeq \text{f.p.} \mathbb{T}\text{-mod}(\mathbf{Set}).$$

This equivalence sends any formula $\{\vec{x}.\phi\} \in \mathcal{C}_{\mathbb{T}}^{\text{irr}}$ to the model $M_{\{\vec{x}.\phi\}}$ presented by it and any arrow $[\theta] : \{\vec{y}.\psi\} \to \{\vec{x}.\phi\}$ in $\mathcal{C}_{\mathbb{T}}^{\text{irr}}$ to the arrow $M_{[\theta]} : M_{\{\vec{x}.\phi\}} \to M_{\{\vec{y}.\psi\}}$ in f.p.$\mathbb{T}$-mod$(\mathbf{Set})$ defined by setting $M_{[\theta]}(\vec{\xi_\phi})$ equal to the unique element $z \in [[\vec{x}.\phi]]_{M_{\{\vec{y}.\psi\}}}$ such that $(\vec{\xi_\psi}, z) \in [[\vec{y}, \vec{x}.\theta]]_{M_{\{\vec{y}.\psi\}}}$. □

Remark 6.1.14 Theorem 6.1.13 generalizes Theorem 2.1.21. Indeed, for any cartesian theory $\mathbb{T}$, any formula presenting a $\mathbb{T}$-model (equivalently, any $\mathbb{T}$-irreducile formula) is isomorphic, in the syntactic category $\mathcal{C}_{\mathbb{T}}$, to a $\mathbb{T}$-cartesian formula (cf. Remark 6.1.18(b) below).

Corollary 6.1.15 *Let $\mathbb{T}$ be a theory of presheaf type and M a set-based $\mathbb{T}$-model. Then M is finitely presentable if and only if it is finitely presented.*

Proof Every finitely presented model of a geometric theory $\mathbb{T}$ is finitely presentable. Indeed, if M is presented by a formula $\{\vec{x}.\phi\}$ then the hom functor $\text{Hom}_{\mathbb{T}\text{-mod}(\mathbf{Set})}(M, -) : \mathbb{T}\text{-mod}(\mathbf{Set}) \to \mathbf{Set}$ is isomorphic to the functor $[[\vec{x}.\phi]]_-$, which preserves filtered colimits since the formula $\phi(\vec{x})$ is geometric (recall that finite limits as well as arbitrary colimits commute with filtered colimits). The converse direction follows from Theorem 6.1.13. □

Remark 6.1.16 The 'only if' part of the corollary is not true for general geometric theories; for instance, in the coherent theory of fields the model $\mathbb{Q}$ is finitely presentable but not finitely presented (cf. section 9.4 below).

6.1.5 A syntactic description of the finitely presentable models

For a theory of presheaf type $\mathbb{T}$, it is possible to give an explicit syntactic description of the finitely presentable $\mathbb{T}$-models; specifically, we have the following result:

Theorem 6.1.17 *Let $\mathbb{T}$ be a theory of presheaf type over a signature Σ and $\{\vec{x}.\phi\}$ a formula over Σ presenting a $\mathbb{T}$-model. Then this model is isomorphic to the Σ-structure $M_{\{\vec{x}.\phi\}}$ defined as follows:*

(i) *For any sort A of Σ, $M_{\{\vec{x}.\phi\}}A$ is equal to the set $\mathrm{Hom}_{\mathcal{C}_{\mathbb{T}}}(\{\vec{x}.\phi\},\{x^A.\top\})$ of $\mathbb{T}$-provable equivalence classes of $\mathbb{T}$-provably functional geometric formulae from $\{\vec{x}.\phi\}$ to $\{x^A.\top\}$.*

(ii) *For any function symbol $f: A_1 \cdots A_n \to B$ of Σ, the function*

$$M_{\{\vec{x}.\phi\}}f : \mathrm{Hom}_{\mathcal{C}_{\mathbb{T}}}(\{\vec{x}.\phi\},\{x^{A_1}.\top\}) \times \cdots \times \mathrm{Hom}_{\mathcal{C}_{\mathbb{T}}}(\{\vec{x}.\phi\},\{x^{A_n}.\top\}) \cong$$
$$\mathrm{Hom}_{\mathcal{C}_{\mathbb{T}}}(\{\vec{x}.\phi\},\{x^{A_1},\ldots,x^{A_n}.\top\}) \to \mathrm{Hom}_{\mathcal{C}_{\mathbb{T}}}(\{\vec{x}.\phi\},\{x^B.\top\})$$

is equal to $[f] \circ -$ (where $[f] : \{x^{A_1},\ldots,x^{A_n}.\top\} \to \{x^B.\top\}$ is the morphism in $\mathcal{C}_{\mathbb{T}}$ corresponding to f).

(iii) *For any relation symbol $R \rightarrowtail A_1 \cdots A_n$ of Σ, $M_{\{\vec{x}.\phi\}}R$ is the subobject of*

$$\mathrm{Hom}_{\mathcal{C}_{\mathbb{T}}}(\{\vec{x}.\phi\},\{x^{A_1}.\top\}) \times \cdots \times \mathrm{Hom}_{\mathcal{C}_{\mathbb{T}}}(\{\vec{x}.\phi\},\{x^{A_n}.\top\}) \cong$$
$$\mathrm{Hom}_{\mathcal{C}_{\mathbb{T}}}(\{\vec{x}.\phi\},\{x^{A_1},\ldots,x^{A_n}.\top\})$$

given by $[R] \circ -$, (where $[R] : \{x^{A_1},\ldots,x^{A_n}.R\} \rightarrowtail \{x^{A_1},\ldots,x^{A_n}.\top\}$ is the subobject in $\mathcal{C}_{\mathbb{T}}$ corresponding to R).

Proof Recall that we have a canonical equivalence of categories

$$\mathbf{Flat}_{J_{\mathbb{T}}}(\mathcal{C}_{\mathbb{T}},\mathbf{Set}) \simeq \mathbb{T}\text{-mod}(\mathbf{Set})$$

sending any flat $J_{\mathbb{T}}$-continuous functor $F : \mathcal{C}_{\mathbb{T}} \to \mathbf{Set}$ to the $\mathbb{T}$-model $F(M_{\mathbb{T}})$, where $M_{\mathbb{T}}$ is the universal model of $\mathbb{T}$ in $\mathcal{C}_{\mathbb{T}}$ as in Definition 1.4.4.

We know from (the proof of) Theorem 6.1.13 that for any theory of presheaf type $\mathbb{T}$ over a signature Σ, any formula-in-context $\{\vec{x}.\phi\}$ over Σ which presents a $\mathbb{T}$-model is $\mathbb{T}$-irreducible, in the sense that every $J_{\mathbb{T}}$-covering sieve on $\{\vec{x}.\phi\}$ in $\mathcal{C}_{\mathbb{T}}$ is maximal. From this it easily follows that the (flat) representable functor $\mathrm{Hom}_{\mathcal{C}_{\mathbb{T}}}(\{\vec{x}.\phi\},-) : \mathcal{C}_{\mathbb{T}} \to \mathbf{Set}$ is $J_{\mathbb{T}}$-continuous; indeed, for any $J_{\mathbb{T}}$-covering sieve S on an object $\{\vec{y}.\psi\}$ of $\mathcal{C}_{\mathbb{T}}$, any arrow $\gamma : \{\vec{x}.\phi\} \to \{\vec{y}.\psi\}$ in $\mathcal{C}_{\mathbb{T}}$ factors through one of the arrows belonging to S since the pullback of S along γ is $J_{\mathbb{T}}$-covering and hence maximal. The image $\mathrm{Hom}_{\mathcal{C}_{\mathbb{T}}}(\{\vec{x}.\phi\},M_{\mathbb{T}})$ under this functor of the universal model $M_{\mathbb{T}}$ of $\mathbb{T}$ in $\mathcal{C}_{\mathbb{T}}$, which clearly coincides with the Σ-structure $M_{\{\vec{x}.\phi\}}$ in the statement of the theorem, is therefore a $\mathbb{T}$-model. In order to deduce our thesis, it thus remains to verify that the model $M_{\{\vec{x}.\phi\}}$ satisfies the universal property of the $\mathbb{T}$-model presented by the formula $\{\vec{x}.\phi\}$, i.e. that for any $\mathbb{T}$-model N in $\mathbf{Set}$, the $\mathbb{T}$-model homomorphisms $M_{\{\vec{x}.\phi\}} = \mathrm{Hom}_{\mathcal{C}_{\mathbb{T}}}(\{\vec{x}.\phi\},M_{\mathbb{T}}) \to$

N are in natural bijection with the elements of the set $[[\vec{x}.\phi]]_N$. But the $\mathbb{T}$-model homomorphisms $\mathrm{Hom}_{\mathcal{C}_\mathbb{T}}(\{\vec{x}.\phi\}, M_\mathbb{T}) \to N$ are in natural bijection, by the equivalence $\mathbf{Flat}_{J_\mathbb{T}}(\mathcal{C}_\mathbb{T}, \mathbf{Set}) \simeq \mathbb{T}\text{-mod}(\mathbf{Set})$, with the natural transformations $\mathrm{Hom}_{\mathcal{C}_\mathbb{T}}(\{\vec{x}.\phi\}, -) \to F_N$, that is, by the Yoneda lemma, with the elements of the set $F_N(\{\vec{x}.\phi\}) = [[\vec{x}.\phi]]_N$, as required. □

Remarks 6.1.18 (a) If $\mathbb{T}$ is a universal Horn theory and $\vec{x} = (x_1^{A_1}, \ldots, x_n^{A_n})$ then the set $M_{\{\vec{x}.\phi\}} A$ can be identified with the set of equivalence classes of terms over Σ of type $A_1 \cdots A_n \to A$ modulo the equivalence relation which identifies any two such terms t_1 and t_2 precisely when the sequent $(\phi \vdash_{\vec{x}} t_1 = t_2)$ is provable in $\mathbb{T}$. Indeed, it is shown in [7] (cf. p. 120 therein) that any $\mathbb{T}$-provably functional geometric formula $\theta(\vec{x}, \vec{y})$ between Horn formulae over Σ is $\mathbb{T}$-provably equivalent to a formula of the form $\vec{y} = \vec{t}(\vec{x})$, where $\vec{t}$ is a sequence of terms of the appropriate sorts in the context $\vec{x}$.

(b) If $\mathbb{T}$ is a cartesian theory then any formula presenting a $\mathbb{T}$-model is isomorphic, in the syntactic category $\mathcal{C}_\mathbb{T}$, to a $\mathbb{T}$-cartesian formula, and for any such formula $\phi(\vec{x})$, any $\mathbb{T}$-provably functional geometric formula from $\{\vec{x}.\phi\}$ to $\{x^A.\top\}$ is $\mathbb{T}$-cartesian, up to $\mathbb{T}$-provable equivalence. This follows immediately from Theorem 6.1.3 by using the explicit description of the universal model of $\mathbb{T}$ in its classifying topos $[(\mathcal{C}_\mathbb{T}^{\mathrm{cart}})^{\mathrm{op}}, \mathbf{Set}]$ (cf. Remark 2.1.9).

(c) Let $\mathbb{T}$ be a geometric theory over a signature Σ, $\mathcal{K}$ a small category of set-based $\mathbb{T}$-models and $\phi(\vec{x})$ a geometric formula over Σ presenting a $\mathbb{T}$-model in $\mathcal{K}$. If the geometric morphism

$$p_\mathcal{K} : [\mathcal{K}, \mathbf{Set}] \to \mathbf{Sh}(\mathcal{C}_\mathbb{T}, J_\mathbb{T})$$

considered in section 5.2.3 has the property that its inverse image $p_\mathcal{K}^*$ is full and faithful (for instance, if $p_\mathcal{K}$ is hyperconnected) then $\phi(\vec{x})$ is $\mathbb{T}$-irreducible and the argument in the proof of Theorem 6.1.17 applies yielding a syntactic description of the model presented by $\phi(\vec{x})$ as in the statement of Theorem 6.1.17. Indeed, denoting by y and y' the Yoneda embeddings respectively of $\mathcal{C}_\mathbb{T}$ into $\mathbf{Sh}(\mathcal{C}_\mathbb{T}, J_\mathbb{T})$ and of $\mathcal{K}^{\mathrm{op}}$ into $[\mathcal{K}, \mathbf{Set}]$, we have that $p_\mathcal{K}^*(y\{\vec{x}.\phi\}) \cong y'(M_{\{\vec{x}.\phi\}})$. Now, as $y'(M_{\{\vec{x}.\phi\}})$ is an irreducible object of the topos $[\mathcal{K}, \mathbf{Set}]$ and the property of an object of a topos to be irreducible is reflected by full and faithful inverse images of geometric morphisms, $y\{\vec{x}.\phi\}$ is an irreducible object of the topos $\mathbf{Sh}(\mathcal{C}_\mathbb{T}, J_\mathbb{T})$, equivalently $\phi(\vec{x})$ is $\mathbb{T}$-irreducible, as required.

Corollary 6.1.19 *Let $\mathbb{T}$ be a theory of presheaf type over a signature Σ and $\phi(\vec{x})$ a geometric formula-in-context over Σ presenting a $\mathbb{T}$-model $M_{\{\vec{x}.\phi\}}$. Then, for any geometric formula $\psi(\vec{x})$ in the same context, $M_{\{\vec{x}.\phi\}} \vDash \psi(\vec{\xi_\phi})$ if and only if the sequent $(\phi \vdash_{\vec{x}} \psi)$ is provable in $\mathbb{T}$.*

Proof Let us use the syntactic description of $M_{\{\vec{x}.\phi\}}$ provided by Theorem 6.1.17. The generators $\vec{\xi_\phi} \in M_{\{\vec{x}.\phi\}} A_1 \times \cdots \times M_{\{\vec{x}.\phi\}} A_n$ are given by the canonical projections $\pi_i : \{\vec{x}.\phi\} \to \{x_i^{A_i}.\top\}$ (for $i = 1, \ldots, n$). We thus have that

$M_{\{\vec{x}.\phi\}} \vDash \psi(\vec{\xi_\phi})$ if and only if $\vec{\xi_\phi} \in [[\vec{x}.\psi]]_{M_{\{\vec{x}.\phi\}}} = \mathrm{Hom}_{\mathcal{C}_\mathbb{T}}(\{\vec{x}.\phi\}, \{\vec{x}.\psi\})$, that is if and only if $<\pi_1,\ldots,\pi_n>: \{\vec{x}.\phi\} \to \{\vec{x}.\top\}$ factors through the subobject $\{\vec{x}.\psi\} \rightarrowtail \{\vec{x}.\top\}$ in $\mathcal{C}_\mathbb{T}$; but this occurs if and only if the sequent $(\phi \vdash_{\vec{x}} \psi)$ is provable in $\mathbb{T}$. □

6.2 Internal finite presentability

In this section we introduce two notions of internal finite presentability for models of geometric theories in Grothendieck toposes, and show that the models of the form $\gamma_\mathcal{E}^*(c)$ for a finitely presentable model c of a theory of presheaf type $\mathbb{T}$ satisfy both of them.

6.2.1 Objects of homomorphisms

The following result asserts that, for any geometric theory $\mathbb{T}$ and any Grothendieck topos $\mathcal{E}$, the $\mathcal{E}$-indexed category $\underline{\mathbb{T}\text{-mod}(\mathcal{E})}$ introduced in section 5.2.3 is locally small.

Theorem 6.2.1 *Let $\mathbb{T}$ be a geometric theory. Then for any $\mathbb{T}$-models M and N in a Grothendieck topos $\mathcal{E}$ there exists an object $\mathrm{Hom}^\mathcal{E}_{\mathbb{T}\text{-mod}(\mathcal{E})}(M,N)$ of $\mathcal{E}$, called the 'object of $\mathbb{T}$-model homomorphisms from M to N', satisfying the universal property that for any object E of $\mathcal{E}$ the generalized elements $E \to \mathrm{Hom}^\mathcal{E}_{\mathbb{T}\text{-mod}(\mathcal{E})}(M,N)$ are in bijective correspondence, naturally in $E \in \mathcal{E}$, with the $\mathbb{T}$-model homomorphisms $!_E^*(M) \to !_E^*(N)$ in $\mathbb{T}\text{-mod}(\mathcal{E}/E)$.*

Proof Recall from section 5.1.1 that for any small category $\mathcal{C}$ and any Grothendieck topos $\mathcal{E}$, for any two functors $F, G : \mathcal{C} \to \mathcal{E}$ there exists an object $\mathrm{Hom}^\mathcal{E}_{[\mathcal{C},\mathcal{E}]}(F,G)$ satisfying the property that for any object E of $\mathcal{E}$ the generalized elements $E \to \mathrm{Hom}^\mathcal{E}_{[\mathcal{C},\mathcal{E}]}(F,G)$ are in bijective correspondence, naturally in $E \in \mathcal{E}$, with the arrows $!_E^* \circ F \to !_E^* \circ G$ in $[\mathcal{C}, \mathcal{E}/E]$, that is, with the natural transformations $!_E^* \circ F \to !_E^* \circ G$. This implies that for any Grothendieck topos $\mathcal{E}$, the indexed category $\underline{[\mathcal{C},\mathcal{E}]}_\mathcal{E}$ of functors on $\mathcal{C}$ with values in $\mathcal{E}$ and natural transformations between them is locally small. It follows in particular that any $\mathcal{E}$-indexed full subcategory of $\underline{[\mathcal{C},\mathcal{E}]}_\mathcal{E}$, such as the indexed category of J-continuous flat functors on $\mathcal{C}$ with values in $\mathcal{E}$, is also locally small.

Now, every geometric theory $\mathbb{T}$ is Morita-equivalent to the theory of flat $J_\mathbb{T}$-continuous functors $\mathcal{C}_\mathbb{T}$ (cf. section 2.1.2); so the $\mathcal{E}$-indexed category $\underline{\mathbb{T}\text{-mod}(\mathcal{E})}$ is equivalent to the $\mathcal{E}$-indexed category of flat $J_\mathbb{T}$-continuous functors on $\mathcal{C}_\mathbb{T}$ with values in $\mathcal{E}$. Hence $\underline{\mathbb{T}\text{-mod}(\mathcal{E})}$ is locally small, as required. □

Remarks 6.2.2 Let $\mathbb{T}$ be a geometric theory over a signature Σ. Then

(a) The assignment $(M,N) \to \mathrm{Hom}^\mathcal{E}_{\mathbb{T}\text{-mod}(\mathcal{E})}(M,N)$ is functorial both in M and N; that is, any homomorphism $f : M \to M'$ in $\mathbb{T}\text{-mod}(\mathcal{E})$ (resp. any homomorphism $g : N' \to N$ in $\mathbb{T}\text{-mod}(\mathcal{E})$) induces an arrow

$$\mathrm{Hom}^\mathcal{E}_{\mathbb{T}\text{-mod}(\mathcal{E})}(f,N) : \mathrm{Hom}^\mathcal{E}_{\mathbb{T}\text{-mod}(\mathcal{E})}(M',N) \to \mathrm{Hom}^\mathcal{E}_{\mathbb{T}\text{-mod}(\mathcal{E})}(M,N)$$

(resp. an arrow
$$\mathrm{Hom}^{\mathcal{E}}_{\mathbb{T}\text{-mod}(\mathcal{E})}(M, g) : \mathrm{Hom}^{\mathcal{E}}_{\mathbb{T}\text{-mod}(\mathcal{E})}(M, N') \to \mathrm{Hom}^{\mathcal{E}}_{\mathbb{T}\text{-mod}(\mathcal{E})}(M, N))$$

functorially in f (resp. functorially in g).

(b) For any Grothendieck topos $\mathcal{E}$ and $\mathbb{T}$-models M and N in $\mathcal{E}$, we have a canonical embedding

$$\mathrm{Hom}^{\mathcal{F}}_{\mathbb{T}\text{-mod}(\mathcal{E})}(M, N) \to \prod_{A \text{ sort of } \Sigma} NA^{MA}$$

induced by arrows

$$\pi_A : \mathrm{Hom}^{\mathcal{E}}_{\mathbb{T}\text{-mod}(\mathcal{E})}(M, N) \to NA^{MA}$$

(for any sort A of Σ) defined in terms of generalized elements as follows: π_A sends any arrow $E \to \mathrm{Hom}^{\mathcal{E}}_{\mathbb{T}\text{-mod}(\mathcal{E})}(M, N)$ in $\mathcal{E}$, corresponding to a $\mathbb{T}$-model homomorphism $r :!^*_E(M) \to !^*_E(N)$ in $\mathcal{E}/E$, to the arrow $E \to NA^{MA}$ whose transpose is the arrow $r_A :!^*_E(M)A \to !^*_E(N)A$.

(c) For any $\mathbb{T}$-models M and N in a Grothendieck topos $\mathcal{E}$ and any geometric morphism $f : \mathcal{F} \to \mathcal{E}$, there is a canonical arrow

$$f^*(\mathrm{Hom}^{\mathcal{E}}_{\mathbb{T}\text{-mod}(\mathcal{E})}(M, N)) \to \mathrm{Hom}^{\mathcal{F}}_{\mathbb{T}\text{-mod}(\mathcal{F})}(f^*(M), f^*(N)).$$

Indeed, this arrow corresponds, by the universal property of the object

$$\mathrm{Hom}^{\mathcal{F}}_{\mathbb{T}\text{-mod}(\mathcal{F})}(f^*(M), f^*(N)),$$

to the $\mathbb{T}$-model homomorphism

$$f^*(\mathrm{Hom}^{\mathcal{E}}_{\mathbb{T}\text{-mod}(\mathcal{E})}(M, N)) \times f^*(M) \to f^*(\mathrm{Hom}^{\mathcal{E}}_{\mathbb{T}\text{-mod}(\mathcal{E})}(M, N)) \times f^*(N)$$

in the topos $\mathcal{F}/f^*(\mathrm{Hom}^{\mathcal{F}}_{\mathbb{T}\text{-mod}(\mathcal{E})}(M, N))$ whose first component, at any sort A of Σ, is the canonical projection and whose second component at A is the arrow

$$f^*(\mathrm{Hom}^{\mathcal{E}}_{\mathbb{T}\text{-mod}(\mathcal{E})}(M, N)) \times f^*(MA) \cong f^*(\mathrm{Hom}^{\mathcal{E}}_{\mathbb{T}\text{-mod}(\mathcal{E})}(M, N) \times MA) \to f^*(NA)$$

obtained by taking the image under f^* of the arrow

$$\mathrm{Hom}^{\mathcal{E}}_{\mathbb{T}\text{-mod}(\mathcal{E})}(M, N) \times MA \to NA$$

given by the transpose of the arrow $\pi_A : \mathrm{Hom}^{\mathcal{E}}_{\mathbb{T}\text{-mod}(\mathcal{E})}(M, N) \to NA^{MA}$ defined above.

(d) Since the $\mathcal{E}$-indexed category $\mathbb{T}\text{-mod}(\mathcal{E})$ is locally small (cf. Theorem 6.2.1), we have an $\mathcal{E}$-indexed hom functor

$$\mathrm{Hom}^{\mathcal{E}}_{\mathbb{T}\text{-mod}(\mathcal{E})}(-, -) : \mathbb{T}\text{-mod}(\mathcal{E}) \times \mathbb{T}\text{-mod}(\mathcal{E}) \to \underline{\mathcal{E}}_{\mathcal{E}}.$$

The following proposition gives an explicit description of the generalized elements of the objects of homomorphisms of Theorem 6.2.1 in a particular case of interest. In the statement of the proposition, for a $\mathbb{T}$-model M in a Grothendieck topos $\mathcal{E}$, we denote by $\mathrm{Hom}_{\mathcal{E}}(E, M)$ the Σ-structure in **Set** given by the image of M under the product-preserving functor $\mathrm{Hom}_{\mathcal{E}}(E, -) : \mathcal{E} \to \mathbf{Set}$.

Proposition 6.2.3 *Let $\mathbb{T}$ be a geometric theory over a signature Σ, M a model of $\mathbb{T}$ in a Grothendieck topos $\mathcal{E}$, c a set-based $\mathbb{T}$-model and E an object of $\mathcal{E}$. Then the generalized elements $x : E \to \mathrm{Hom}^{\mathcal{E}}_{\mathbb{T}\text{-mod}(\mathcal{E})}(\gamma^*_{\mathcal{E}}(c), M)$ correspond bijectively to the Σ-structure homomorphisms $\xi_x : c \to \mathrm{Hom}_{\mathcal{E}}(E, M)$.*

Proof By definition of $\mathrm{Hom}^{\mathcal{E}}_{\mathbb{T}\text{-mod}(\mathcal{E})}(\gamma^*_{\mathcal{E}}(c), M)$, a generalized element

$$E \to \mathrm{Hom}^{\mathcal{E}}_{\mathbb{T}\text{-mod}(\mathcal{E})}(\gamma^*_{\mathcal{E}}(c), M)$$

corresponds precisely to a $\mathbb{T}$-model homomorphism $\gamma^*_{\mathcal{E}/E}(c) \to !^*_E(M)$ in the topos $\mathcal{E}/E$. Concretely, such a $\mathbb{T}$-model homomorphism consists of a family of arrows $\tau_A : \gamma^*_{\mathcal{E}/E}(cA) \to !^*_E(MA)$ in $\mathcal{E}/E$ (indexed by the sorts A of Σ) satisfying the conditions defining the notion of Σ-structure homomorphism. Now, each of the arrows $\tau_A : \gamma^*_{\mathcal{E}/E}(cA) \to !^*_E(MA)$ corresponds, via the adjunction between $\gamma^*_{\mathcal{E}/E}$ and the global section functor on the topos $\mathcal{E}/E$, to a function $\xi_A : cA \to \mathrm{Hom}_{\mathcal{E}}(E, MA)$, and it is immediate to see that the above-mentioned conditions translate precisely into the requirement that the functions $\xi_A : cA \to \mathrm{Hom}_{\mathcal{E}}(E, MA)$ should yield a Σ-structure homomorphism $c \to \mathrm{Hom}_{\mathcal{E}}(E, M)$. □

Remark 6.2.4 If the model c is finitely presentable also as a $\mathbb{T}_c$-model, where $\mathbb{T}_c$ is the cartesianization of $\mathbb{T}$ (notice that this is always the case if Σ is finite, c is finite and $\mathbb{T}$ has only a finite number of axioms, cf. Proposition 9.1.1 below) then, for any ($\mathbb{T}_c$-cartesian) formula $\phi(\vec{x})$ over Σ which presents c as a $\mathbb{T}_c$-model, the Σ-structure homomorphisms $c \to \mathrm{Hom}_{\mathcal{E}}(E, M)$ are in natural bijection with the elements of the interpretation of the formula $\phi(\vec{x})$ in the Σ-structure $\mathrm{Hom}_{\mathcal{E}}(E, M)$ (cf. Remark 6.2.6(c) below).

6.2.2 Strong finite presentability

We shall show in this section that the finitely presentable models of a theory $\mathbb{T}$ of presheaf type enjoy a strong form of syntactic finite presentability with respect to the models of $\mathbb{T}$ in arbitrary Grothendieck toposes.

Internal finite presentability

Let $\mathbb{T}$ be a geometric theory over a signature Σ, c a set-based model of $\mathbb{T}$, $\phi(\vec{x})$ a geometric formula over Σ and $\vec{u}$ an element of $[[\vec{x}\,.\,\phi]]_c$. Then, for any model M of $\mathbb{T}$ in a Grothendieck topos $\mathcal{E}$, there is a canonical arrow

$$\epsilon^M_{(\phi(\vec{x}),\vec{u})} : \mathrm{Hom}^{\mathcal{E}}_{\mathbb{T}\text{-mod}(\mathcal{E})}(\gamma^*_{\mathcal{E}}(c), M) \to [[\vec{x}\,.\,\phi]]_M$$

in $\mathcal{E}$, defined in terms of generalized elements as follows: $\epsilon^M_{(\phi(\vec{x}),\vec{u})}$ sends a generalized element

$$E \to \mathrm{Hom}^{\mathcal{E}}_{\mathbb{T}\text{-mod}(\mathcal{E})}(\gamma^*_{\mathcal{E}}(c), M),$$

corresponding under the bijection of Proposition 6.2.3 to a Σ-structure homomorphism $f : c \to \mathrm{Hom}_{\mathcal{E}}(E, M)$, to the image of $\vec{u}$ under f; notice that this element indeed belongs to $\mathrm{Hom}_{\mathcal{E}}(E, [[\vec{x}\,.\,\phi]]_M)$ since, as f is a Σ-structure homomorphism, the image of $[[\vec{x}\,.\,\phi]]_c$ under f is contained in $[[\vec{x}\,.\,\phi]]_{\mathrm{Hom}_{\mathcal{E}}(E,M)}$, which is in turn contained in $\mathrm{Hom}_{\mathcal{E}}(E, [[\vec{x}\,.\,\phi]]_M)$ as the functor $\mathrm{Hom}_{\mathcal{E}}(E, -)$ is cartesian.

Definition 6.2.5 Let $\mathbb{T}$ be a geometric theory over a signature Σ and c a $\mathbb{T}$-model in **Set**. Then c is said to be *strongly finitely presented* if there exist a geometric formula $\phi(\vec{x})$ over Σ and a finite string of elements $\vec{u}$ of $[[\vec{x}\,.\,\phi]]_c$, called the *strong generators* of c, such that for any $\mathbb{T}$-model M in a Grothendieck topos $\mathcal{E}$ the arrow

$$\epsilon^M_{(\phi(\vec{x}),\vec{u})} : \mathrm{Hom}^{\mathcal{E}}_{\mathbb{T}\text{-mod}(\mathcal{E})}(\gamma^*_{\mathcal{E}}(c), M) \to [[\vec{x}\,.\,\phi]]_M$$

is an isomorphism, equivalently if for every $E \in \mathcal{E}$ the Σ-structure homomorphisms $f : c \to \mathrm{Hom}_{\mathcal{E}}(E, M)$ are in natural bijection with the generalized elements $E \to [[\vec{x}\,.\,\phi]]_M$ via the assignment $f \to f(\vec{u})$.

Remarks 6.2.6 (a) If the equivalent condition in Definition 6.2.5 is satisfied by all models M of $\mathbb{T}$ inside Grothendieck toposes for $E = 1_{\mathcal{E}}$ then it is true in general, by the localization principle.

(b) If the condition in Definition 6.2.5 is satisfied then

$$[[\vec{x}\,.\,\phi]]_{\mathrm{Hom}_{\mathcal{E}}(E,M)} \cong \mathrm{Hom}_{\mathcal{E}}(E, [[\vec{x}\,.\,\phi]]_M)$$

for every $\mathbb{T}$-model M in a Grothendieck topos $\mathcal{E}$ and any object E of $\mathcal{E}$.

(c) If a model of a geometric theory $\mathbb{T}$ is finitely presentable as a model of its cartesianization then it is strongly finitely presented. Indeed, any structure of the form $\mathrm{Hom}_{\mathcal{E}}(E, M)$, where M is a model of $\mathbb{T}$ in a Grothendieck topos $\mathcal{E}$, is a model of the cartesianization of $\mathbb{T}$ (as it is obtained by applying a global section functor, which is cartesian, to a model of $\mathbb{T}$), and for any cartesian formula $\phi(\vec{x})$, $[[\vec{x}\,.\,\phi]]_{\mathrm{Hom}_{\mathcal{E}}(E,M)} \cong \mathrm{Hom}_{\mathcal{E}}(E, [[\vec{x}\,.\,\phi]]_M)$.

Proposition 6.2.7 *Let $\mathbb{T}$ be a theory of presheaf type classified by the topos $[\mathcal{K}, \mathbf{Set}]$, where $\mathcal{K}$ is a full subcategory of* f.p.$\mathbb{T}$-mod$(\mathbf{Set})$. *Then for any $\mathbb{T}$-model M in a Grothendieck topos $\mathcal{E}$ and any $\mathbb{T}$-model c in $\mathcal{K}$ there is a natural bijective correspondence between the Σ-structure homomorphisms $c \to \mathrm{Hom}_{\mathcal{E}}(E, M)$ and* the elements of the set $\mathrm{Hom}_{\mathcal{E}}(E, F_M(c))$, where F_M is the flat functor $\mathcal{K}^{\mathrm{op}} \to \mathcal{E}$ corresponding to the model M via the canonical Morita equivalence for $\mathbb{T}$.

Proof We can clearly suppose without loss of generality that $E = 1_{\mathcal{E}}$. The adjunction between $\gamma_{\mathcal{E}}^*$ and the global sections functor $\Gamma_{\mathcal{E}} : \mathcal{E} \to \mathbf{Set}$ provides a natural bijective correspondence between the elements of the set $\mathrm{Hom}_{\mathcal{E}}(1_{\mathcal{E}}, F_M(c))$ and the natural transformations $\gamma_{\mathcal{E}}^* \circ y_{\mathcal{K}}(c) \to F_M$, where $y_{\mathcal{K}} : \mathcal{K} \to [\mathcal{K}^{\mathrm{op}}, \mathbf{Set}]$ is the Yoneda embedding. Via the canonical Morita equivalence

$$\tau_{\mathcal{E}} : \mathbf{Flat}(\mathcal{K}^{\mathrm{op}}, \mathcal{E}) \simeq \mathbb{T}\text{-}\mathrm{mod}(\mathcal{E})$$

for $\mathbb{T}$, such natural transformations are in natural bijective correspondence with the $\mathbb{T}$-model homomorphisms $\gamma_{\mathcal{E}}^*(c) \cong \tau_{\mathcal{E}}(\gamma_{\mathcal{E}}^* \circ y_{\mathcal{K}}(c)) \to \tau_{\mathcal{E}}(F_M) = M$. But these homomorphisms are, by Proposition 6.2.3, in natural bijective correspondence with the Σ-structure homomorphisms $c \to \mathrm{Hom}_{\mathcal{E}}(1_{\mathcal{E}}, M)$, as required. □

Corollary 6.2.8 *Let $\mathbb{T}$ be a theory of presheaf type over a signature Σ and $\phi(\vec{x})$ a geometric formula over Σ presenting a $\mathbb{T}$-model $M_{\{\vec{x}.\phi\}}$. Then $M_{\{\vec{x}.\phi\}}$ is strongly finitely presented by $\phi(\vec{x})$; that is, for any $\mathbb{T}$-model M in a Grothendieck topos $\mathcal{E}$ and any object E of $\mathcal{E}$, the Σ-structure homomorphisms $M_{\{\vec{x}.\phi\}} \to \mathrm{Hom}_{\mathcal{E}}(E, M)$ correspond bijectively to the elements of the set $\mathrm{Hom}_{\mathcal{E}}(E, [[\vec{x}.\phi]]_M)$ via the assignment sending a Σ-structure homomorphism $M_{\{\vec{x}.\phi\}} \to \mathrm{Hom}_{\mathcal{E}}(E, M)$ to the image of the strong generators of $M_{\{\vec{x}.\phi\}}$ under it.*

Proof Clearly, we can suppose without loss of generality $E = 1_{\mathcal{E}}$ in the condition of the corollary.

The canonical Morita equivalence

$$\xi^{\mathcal{E}} : \mathbf{Flat}(\text{f.p.}\mathbb{T}\text{-}\mathrm{mod}(\mathbf{Set})^{\mathrm{op}}, \mathcal{E}) \simeq \mathbb{T}\text{-}\mathrm{mod}(\mathcal{E})$$

for $\mathbb{T}$ (in the sense of section 6.1.1) sends any $\mathbb{T}$-model M to the flat functor $F_M := \mathrm{Hom}_{\mathbb{T}\text{-}\mathrm{mod}(\mathcal{E})}(\gamma_{\mathcal{E}}^*(-), M)$ (cf. the proof of Theorem 6.3.1) and any flat functor $F : \text{f.p.}\mathbb{T}\text{-}\mathrm{mod}(\mathbf{Set})^{\mathrm{op}} \to \mathcal{E}$ to the model $\tilde{F}(M_{\mathbb{T}})$, where $\tilde{F} : \mathcal{C}_{\mathbb{T}} \to \mathcal{E}$ is the extension of F to the syntactic category $\mathcal{C}_{\mathbb{T}}$ (in the sense of section 5.2.3) and $M_{\mathbb{T}}$ is the universal model of $\mathbb{T}$ in $\mathcal{C}_{\mathbb{T}}$ (as in Definition 1.4.4). Indeed, $\xi^{\mathcal{E}}$ is induced by the canonical geometric morphism (in fact, equivalence)

$$p_{\text{f.p.}\mathbb{T}\text{-}\mathrm{mod}(\mathbf{Set})} : [\text{f.p.}\mathbb{T}\text{-}\mathrm{mod}(\mathbf{Set}), \mathbf{Set}] \to \mathbf{Sh}(\mathcal{C}_{\mathbb{T}}, J_{\mathbb{T}}),$$

i.e. it is given by the composite of the induced equivalence

$$\mathbf{Flat}(\text{f.p.}\mathbb{T}\text{-}\mathrm{mod}(\mathbf{Set})^{\mathrm{op}}, \mathcal{E}) \simeq \mathbf{Flat}_{J_{\mathbb{T}}}(\mathcal{C}_{\mathbb{T}}, \mathcal{E})$$

with the canonical equivalence $\mathbf{Flat}_{J_{\mathbb{T}}}(\mathcal{C}_{\mathbb{T}}, \mathcal{E}) \simeq \mathbb{T}\text{-}\mathrm{mod}(\mathcal{E})$ sending any $J_{\mathbb{T}}$-continuous flat functor G on $\mathcal{C}_{\mathbb{T}}$ to the $\mathbb{T}$-model $G(M_{\mathbb{T}})$.

By Theorem 5.2.5, for any $\mathbb{T}$-model M in a Grothendieck topos $\mathcal{E}$, there is an isomorphism $z_{(M, \{\vec{x}.\phi\})} : F_M(M_{\{\vec{x}.\phi\}}) \cong \widetilde{F_M}(\{\vec{x}.\phi\}) = [[\vec{x}.\phi]]_M$. Thus, applying Proposition 6.2.7 to the model $c = M_{\{\vec{x}.\phi\}}$, we obtain a bijective correspondence between the Σ-structure homomorphisms $M_{\{\vec{x}.\phi\}} \to \mathrm{Hom}_{\mathcal{E}}(1_{\mathcal{E}}, M)$ and the elements of the set $\mathrm{Hom}_{\mathcal{E}}(1_{\mathcal{E}}, [[\vec{x}.\phi]]_M)$. It remains to show that this correspondence can be identified with the assignment sending a Σ-structure homomorphism $M_{\{\vec{x}.\phi\}} \to \mathrm{Hom}_{\mathcal{E}}(1_{\mathcal{E}}, M)$ to the image of the strong generators of $M_{\{\vec{x}.\phi\}}$ under it. Recall that the bijection of Proposition 6.2.7 can be described as follows: any Σ-structure homomorphism $f : M_{\{\vec{x}.\phi\}} \to \mathrm{Hom}_{\mathcal{E}}(1_{\mathcal{E}}, M)$ corresponding to a $\mathbb{T}$-model homomorphism $\gamma_{\mathcal{E}}^*(M_{\{\vec{x}.\phi\}}) \to M$ and hence, via $\xi^{\mathcal{E}}$, to a natural transformation $\gamma_{\mathcal{E}}^* \circ y(M_{\{\vec{x}.\phi\}}) \cong F_{\gamma_{\mathcal{E}}^*(M_{\{\vec{x}.\phi\}})} \to F_M$ (where y is the Yoneda embedding f.p.$\mathbb{T}$-mod$(\mathbf{Set})^{\mathrm{op}} \hookrightarrow [\text{f.p.}\mathbb{T}\text{-mod}(\mathbf{Set}), \mathbf{Set}])$, is sent to the global element of $F_M(M_{\{\vec{x}.\phi\}})$ corresponding via the Yoneda lemma to this transformation. To deduce our thesis, it thus remains to verify that the canonical isomorphism of functors $\gamma_{\mathcal{E}}^* \circ y(M_{\{\vec{x}.\phi\}}) = \gamma_{\mathcal{E}}^* \circ F_{M_{\{\vec{x}.\phi\}}} \cong F_{\gamma_{\mathcal{E}}^*(M_{\{\vec{x}.\phi\}})}$, when evaluated at $M_{\{\vec{x}.\phi\}}$ and composed with the isomorphism $z_{(\gamma_{\mathcal{E}}^*(M_{\{\vec{x}.\phi\}}),\{\vec{x}.\phi\})}$: $F_{\gamma_{\mathcal{E}}^*(M_{\{\vec{x}.\phi\}})}(M_{\{\vec{x}.\phi\}}) \cong [[\vec{x}.\phi]]_{\gamma_{\mathcal{E}}^*(M_{\{\vec{x}.\phi\}})} \cong \gamma_{\mathcal{E}}^*([[\vec{x}.\phi]]_{M_{\{\vec{x}.\phi\}}})$, sends the coproduct component of $\gamma_{\mathcal{E}}^*((y(M_{\{\vec{x}.\phi\}}))(M_{\{\vec{x}.\phi\}}))$ corresponding to the identity on $M_{\{\vec{x}.\phi\}}$ to the coproduct component of $\gamma_{\mathcal{E}}^*([[\vec{x}.\phi]]_{M_{\{\vec{x}.\phi\}}})$ corresponding to the strong generators of $M_{\{\vec{x}.\phi\}}$. By the naturality in $\mathcal{E}$ of $\xi^{\mathcal{E}}$, we can clearly suppose, without loss of generality, that $\mathcal{E}$ is equal to $\mathbf{Set}$. But from the proof of Theorem 6.1.17 it is clear that the strong generators of $M_{\{\vec{x}.\phi\}}$ correspond to the identity on $\{\vec{x}.\phi\}$ via the Yoneda lemma and hence to the identity on $M_{\{\vec{x}.\phi\}}$ via the above-mentioned bijection, as required. $\square$

Remark 6.2.9 We can express the bijective correspondence of Corollary 6.2.8 by saying that for any Grothendieck topos $\mathcal{E}$ the $\mathcal{E}$-indexed functor $[[\vec{x}.\phi]]_- :$ $\mathbb{T}\text{-mod}(\mathcal{E}) \to \mathcal{E}_{\mathcal{E}}$ assigning to any $\mathbb{T}$-model M the interpretation of $\phi(\vec{x})$ in M is represented as an $\mathcal{E}$-indexed functor by the object $\gamma_{\mathcal{E}}^*(M_{\{\vec{x}.\phi\}})$, in the sense of being isomorphic to the $\mathcal{E}$-representable functor $\mathrm{Hom}_{\mathbb{T}\text{-mod}(\mathcal{E})}^{\mathcal{E}}(\gamma_{\mathcal{E}}^*(M_{\{\vec{x}.\phi\}}), -)$.

Proposition 6.2.10 *Let $\mathbb{T}$ be a theory of presheaf type and $\mathbb{S}$ be a subtheory of $\mathbb{T}$ (that is, a theory $\mathbb{S}$ of which $\mathbb{T}$ is a quotient) such that every set-based model of $\mathbb{S}$ admits a representation as the structure $\mathrm{Hom}_{\mathcal{E}}(1_{\mathcal{E}}, M)$ of global sections of a $\mathbb{T}$-model M in a Grothendieck topos $\mathcal{E}$. Then every finitely presentable model of $\mathbb{T}$ is finitely presented as a model of $\mathbb{S}$.*

Proof If a theory $\mathbb{T}$ is of presheaf type then any finitely presentable model of $\mathbb{T}$ is strongly finitely presented (by Corollary 6.2.8) and hence, by Remark 6.2.6(b), finitely presented relatively to whatever subtheory $\mathbb{S}$ of $\mathbb{T}$ whose set-based models admit representations as sections $\mathrm{Hom}_{\mathcal{E}}(E, M)$ of models M of $\mathbb{T}$ in Grothendieck toposes. $\square$

Remarks 6.2.11 (a) Pairs of theories satisfying the hypotheses of the proposition are investigated for instance in [34] and [38].
(b) The proposition can often be profitably applied to the cartesianization of $\mathbb{T}$; however, it is not known if it is always the case that every model of it admits a representation of the above kind (cf. also Remark 8.2.2(c)).

6.2.3 Semantic $\mathcal{E}$-finite presentability

In this section, we introduce a semantic notion of $\mathcal{E}$-finite presentability of a model of a geometric theory in a Grothendieck topos $\mathcal{E}$, which generalizes the classical notion in the theory of finitely accessible categories, and show that all the 'constant' finitely presentable models of a theory of presheaf type in a Grothendieck topos $\mathcal{E}$ are $\mathcal{E}$-finitely presentable.

Recall from section 5.2.3 that, for any geometric theory $\mathbb{T}$ over a signature Σ and any Grothendieck topos $\mathcal{E}$, the $\mathcal{E}$-indexed category $\underline{\mathbb{T}\text{-mod}(\mathcal{E})}$ admits $\mathcal{E}$-indexed colimits of diagrams defined on $\mathcal{E}$-final, $\mathcal{E}$-filtered subcategories of a small $\mathcal{E}$-indexed category, which are jointly created by the forgetful functors $U_A : \underline{\mathbb{T}\text{-mod}(\mathcal{E})} \to \underline{\mathcal{E}_\mathcal{E}}$ corresponding to the sorts A of Σ. In particular, an $\mathcal{E}$-indexed cocone (M, μ) over a diagram $D : \underline{A_\mathcal{E}} \to \underline{\mathbb{T}\text{-mod}(\mathcal{E})}$ defined on a $\mathcal{E}$-final, $\mathcal{E}$-filtered subcategory of a small $\mathcal{E}$-indexed category is colimiting if and only if, for every sort A of Σ, $U_A((M, \mu))$ is a colimiting cocone over the diagram $U_A \circ D$ in $\underline{\mathcal{E}_\mathcal{E}}$.

Definition 6.2.12 Let $\mathbb{T}$ be a geometric theory and M be a model of $\mathbb{T}$ in a Grothendieck topos $\mathcal{E}$. Then M is said to be $\mathcal{E}$-finitely presentable if the $\mathcal{E}$-indexed functor $\text{Hom}^{\mathcal{E}}_{\underline{\mathbb{T}\text{-mod}(\mathcal{E})}}(M, -) : \underline{\mathbb{T}\text{-mod}(\mathcal{E})} \to \underline{\mathcal{E}_\mathcal{E}}$ of section 6.2.1 preserves $\mathcal{E}$-filtered colimits (of diagrams defined on $\mathcal{E}$-final, $\mathcal{E}$-filtered subcategories of a small $\mathcal{E}$-indexed category).

Proposition 6.2.13 Let $\mathbb{T}$ be a theory of presheaf type and c a finitely presentable $\mathbb{T}$-model. Then, for any Grothendieck topos $\mathcal{E}$, the $\mathbb{T}$-model $\gamma^*_\mathcal{E}(c)$ is $\mathcal{E}$-finitely presentable.

Proof If c is presented by a geometric formula $\phi(\vec{x})$ over the signature Σ of $\mathbb{T}$, by Corollary 6.2.8 the functor $\text{Hom}^{\mathcal{E}}_{\underline{\mathbb{T}\text{-mod}(\mathcal{E})}}(\gamma^*_\mathcal{E}(c), M)$ is naturally isomorphic to the functor $[[\vec{x}.\phi]]_- : \underline{\mathbb{T}\text{-mod}(\mathcal{E})} \to \underline{\mathcal{E}_\mathcal{E}}$, which, as we observed in section 5.2.3, preserves $\mathcal{E}$-filtered colimits of diagrams defined on $\mathcal{E}$-final and $\mathcal{E}$-filtered subcategories of a small $\mathcal{E}$-indexed category. □

Remark 6.2.14 The proof of the proposition shows that, more generally, for any strongly finitely presentable model c of a geometric theory $\mathbb{T}$ in the sense of Definition 6.2.5 (for instance, a finite model of $\mathbb{T}$ if the signature of $\mathbb{T}$ is finite – note that such a model is finitely presented with respect to the empty theory over the signature of $\mathbb{T}$ by Proposition 9.1.1, whence Remark 6.2.6(c) applies) and any Grothendieck topos $\mathcal{E}$, the $\mathbb{T}$-model $\gamma^*_\mathcal{E}(c)$ is $\mathcal{E}$-finitely presentable.

The following proposition provides an explicit characterization of the $\mathcal{E}$-finitely presentable models of a geometric theory $\mathbb{T}$.

Proposition 6.2.15 Let $\mathbb{T}$ be a geometric theory, $\mathcal{E}$ a Grothendieck topos and c a $\mathbb{T}$-model in $\mathcal{E}$. Then c is $\mathcal{E}$-finitely presentable if and only if for every $\mathcal{E}$-indexed diagram $D : \underline{A_\mathcal{E}} \to \underline{\mathbb{T}\text{-mod}(\mathcal{E})}$ defined on an $\mathcal{E}$-filtered, $\mathcal{E}$-strictly final subcategory $\underline{A_\mathcal{E}}$ of a small $\mathcal{E}$-indexed category with $\mathcal{E}$-indexed colimiting cocone (M, μ) (and colimit arrows $\mu_{(E,x)} : D_E(x) \to !^*_E(M)$), the following conditions are verified:

(i) *For any object E of $\mathcal{E}$ and $\mathbb{T}$-model homomorphism $h :!_E^*(c) \to !_E^*(M)$ in the topos $\mathcal{E}/E$, there exist an epimorphic family $\{e_i : E_i \to E \mid i \in I\}$ in $\mathcal{E}$ and for each $i \in I$ an object x_i of $\mathcal{A}_{E_i}$ and a $\mathbb{T}$-model homomorphism $\alpha_i :!_{E_i}^*(c) \to D_{E_i}(x_i)$ in the topos $\mathcal{E}/E_i$ such that $\mu_{(E_i, x_i)} \circ \alpha_i = e_i^*(h)$.*

(ii) *For any pairs (x, y) and (x', y'), where x and x' are objects of $\mathcal{A}_E$, y is a $\mathbb{T}$-model homomorphism $!_F^*(c) \to f^*(D_E(x))$ in $\mathcal{E}/F$ and y' is a $\mathbb{T}$-model homomorphism $!_F^*(c) \to f^*(D_E(x'))$ in $\mathcal{E}/F$ (for an arrow $f : F \to E$ in $\mathcal{E}$), we have $f^*(\mu_E(x)) \circ y = f^*(\mu_E(x')) \circ y'$ if and only if there exist an epimorphic family $\{f_i : F_i \to F \mid i \in I\}$ in $\mathcal{E}$ and for each $i \in I$ arrows $g_i : \mathcal{A}_{f \circ f_i}(x) \to z_i$ and $h_i : \mathcal{A}_{f \circ f_i}(x') \to z_i$ in the category $\mathcal{A}_{F_i}$ such that $D_{F_i}(g_i) \circ f_i^*(y) = D_{F_i}(h_i) \circ f_i^*(y')$.*

Proof The proposition follows as an immediate consequence of Corollary 5.1.23, applied to the composite $\mathcal{E}$-indexed functor $\underline{\mathrm{Hom}}_{\mathbb{T}\text{-mod}(\mathcal{E})}^{\mathcal{E}}(c, -) \circ D$. □

6.3 Semantic criteria for a theory to be of presheaf type

In this section we establish our main characterization theorem providing necessary and sufficient conditions for a geometric theory to be of presheaf type. These conditions are entirely expressed in terms of the models of the theory in arbitrary Grothendieck toposes and are fully constructive.

We shall first prove the theorem and then proceed to reformulate its conditions in more concrete terms so as to make them easier to check in practice.

6.3.1 The characterization theorem

Recall from section 6.1.1 that the classifying topos of a theory of presheaf type $\mathbb{T}$ can be canonically represented as the topos $[\text{f.p.}\mathbb{T}\text{-mod}(\mathbf{Set}), \mathbf{Set}]$ of set-valued functors on the category f.p.$\mathbb{T}$-mod($\mathbf{Set}$) of finitely presentable $\mathbb{T}$-models in $\mathbf{Set}$; this is not the only possible representation of the classifying topos of $\mathbb{T}$ as a presheaf topos, but for any small category $\mathcal{K}$, $\mathbb{T}$ is classified by the topos $[\mathcal{K}, \mathbf{Set}]$ if and only if the Cauchy completion of $\mathcal{K}$ is equivalent to f.p.$\mathbb{T}$-mod($\mathbf{Set}$).

Theorem 6.3.1 *Let $\mathbb{T}$ be a geometric theory over a signature Σ and $\mathcal{K}$ a full subcategory of f.p.$\mathbb{T}$-mod($\mathbf{Set}$). Then $\mathbb{T}$ is a theory of presheaf type classified by the topos $[\mathcal{K}, \mathbf{Set}]$ if and only if all of the following conditions are satisfied:*

(i) *For any $\mathbb{T}$-model M in a Grothendieck topos $\mathcal{E}$, the functor*
$$H_M := \underline{\mathrm{Hom}}_{\mathbb{T}\text{-mod}(\mathcal{E})}^{\mathcal{E}}(\gamma_{\mathcal{E}}^*(-), M) : \mathcal{K}^{\mathrm{op}} \to \mathcal{E}$$
is flat.

(ii) *For any $\mathbb{T}$-model M in a Grothendieck topos $\mathcal{E}$, if the functor $H_M : \mathcal{K}^{\mathrm{op}} \to \mathcal{E}$ in condition (i) is flat then its extension $\widetilde{H_M} : \mathcal{C}_{\mathbb{T}} \to \mathcal{E}$ to the syntactic category $\mathcal{C}_{\mathbb{T}}$ (in the sense of section 5.2.3) satisfies the property that the canonical morphism $\varepsilon^M : \widetilde{H_M}(M_{\mathbb{T}}) \to M$ (given by the counit of the adjunction of Theorem 5.2.12 via the canonical equivalence $\mathbb{T}\text{-mod}(\mathcal{E}) \simeq \mathbf{Flat}_{J_{\mathbb{T}}}(\mathcal{C}_{\mathbb{T}}, \mathcal{E})$ – cf. Remark 5.2.13(b)) is an isomorphism.*

(iii) *Any of the following (equivalent, if (i) and (ii) hold) conditions is satisfied:*
   (a) *For any model c in $\mathcal{K}$, $\mathbb{T}$-model M in a Grothendieck topos $\mathcal{E}$ and geometric morphism $f : \mathcal{F} \to \mathcal{E}$, the canonical morphism*

$$f^*(\mathrm{Hom}^{\mathcal{E}}_{\underline{\mathbb{T}\text{-mod}(\mathcal{E})}}(\gamma^*_{\mathcal{E}}(c), M)) \to \mathrm{Hom}^{\mathcal{F}}_{\underline{\mathbb{T}\text{-mod}(\mathcal{F})}}(\gamma^*_{\mathcal{F}}(c), f^*(M)),$$

   provided by Remark 6.2.2(c) via the identification $\gamma^*_{\mathcal{F}}(c) \cong f^*(\gamma^*_{\mathcal{E}}(c))$, is an isomorphism.
   (b) *For any flat functor $F : \mathcal{K}^{\mathrm{op}} \to \mathcal{E}$, the canonical natural transformation*

$$\eta^F : F \to \mathrm{Hom}^{\mathcal{E}}_{\underline{\mathbb{T}\text{-mod}(\mathcal{E})}}(\gamma^*_{\mathcal{E}}(-), \tilde{F}(M_{\mathbb{T}})) \cong \mathrm{Hom}^{\mathcal{E}}_{\mathbf{Flat}_{J_{\mathbb{T}}}(\mathcal{C}_{\mathbb{T}}, \mathcal{E})}(\gamma^*_{\mathcal{E}} \circ \widetilde{y_{\mathcal{K}}(-)}, \tilde{F})$$

   of Theorem 5.2.12 is an isomorphism, where $y_{\mathcal{K}} : \mathcal{K} \to \mathbf{Flat}(\mathcal{K}^{\mathrm{op}}, \mathbf{Set})$ is the Yoneda embedding.
   (c) *The functor*

$$u^{\mathbb{T}}_{(\mathcal{K}, \mathcal{E})} : \mathbf{Flat}(\mathcal{K}^{\mathrm{op}}, \mathcal{E}) \to \mathbf{Flat}_{J_{\mathbb{T}}}(\mathcal{C}_{\mathbb{T}}, \mathcal{E}) \simeq \mathbb{T}\text{-mod}(\mathcal{E})$$

   of section 5.2.3 is full and faithful.

Proof Let us begin by proving that conditions (i), (ii) and (iii) are all necessary for $\mathbb{T}$ to be classified by the topos $[\mathcal{K}, \mathbf{Set}]$.

As remarked in section 6.1.1, if $\mathbb{T}$ is of presheaf type classified by the topos $[\mathcal{K}, \mathbf{Set}]$ then we have a Morita equivalence

$$\tau_{\mathcal{E}} : \mathbf{Flat}(\mathcal{K}^{\mathrm{op}}, \mathcal{E}) \simeq \mathbb{T}\text{-mod}(\mathcal{E})$$

which can be supposed to be canonical without loss of generality, i.e. to send any finitely presentable $\mathbb{T}$-model c in $\mathcal{K}$ to the functor $\gamma^*_{\mathcal{E}} \circ y_{\mathcal{K}}(c)$. It follows that, for any $\mathbb{T}$-model M in a Grothendieck topos $\mathcal{E}$ corresponding to a flat functor F_M under the Morita equivalence $\tau_{\mathcal{E}}$, the object $\mathrm{Hom}^{\mathcal{E}}_{\underline{\mathbb{T}\text{-mod}(\mathcal{E})}}(\gamma^*_{\mathcal{E}}(c), M)$ is isomorphic to the object $\mathrm{Hom}^{\mathcal{E}}_{\mathbf{Flat}(\mathcal{K}^{\mathrm{op}}, \mathcal{E})}(\gamma^*_{\mathcal{E}} \circ y_{\mathcal{K}}(c), F_M) \cong F_M(c)$, naturally in M and in c (cf. Corollary 5.3.2). Therefore the functor

$$\mathrm{Hom}^{\mathcal{E}}_{\underline{\mathbb{T}\text{-mod}(\mathcal{E})}}(\gamma^*_{\mathcal{E}}(-), M) : \mathcal{K}^{\mathrm{op}} \to \mathcal{E}$$

is flat, it being isomorphic to F_M. This proves that condition (i) of the theorem is satisfied.

We observed in the proof of Corollary 6.2.8 that the functor

$$\mathbf{Flat}(\mathcal{K}^{\mathrm{op}}, \mathcal{E}) \to \mathbb{T}\text{-mod}(\mathcal{E})$$

forming the canonical Morita equivalence $\tau_{\mathcal{E}}$ for $\mathbb{T}$ sends any flat functor $F : \mathcal{K}^{\mathrm{op}} \to \mathcal{E}$, to the $\mathbb{T}$-model $\tilde{F}(M_{\mathbb{T}})$ (where $M_{\mathbb{T}}$ is the universal model of $\mathbb{T}$ in $\mathcal{C}_{\mathbb{T}}$ as in Definition 1.4.4).

The fact that $\tau_{\mathcal{E}}$ is an equivalence thus implies that the canonical morphism $\widetilde{H_M}(M_\mathbb{T}) \to M$ is an isomorphism. This shows that condition (ii) of the theorem is satisfied.

The fact that conditions (iii)(b) and (iii)(c) hold also follows at once from the fact that $\tau_{\mathcal{E}}$ is an equivalence.

We have thus proved that conditions (i), (ii), (iii)(b) and (iii)(c) of the theorem are necessary for $\mathbb{T}$ to be classified by the topos $[\mathcal{K}, \mathbf{Set}]$.

Next, we notice that condition (iii)(b) implies condition (iii)(a) under the assumption that conditions (i) and (ii) hold. Indeed, for any $\mathbb{T}$-model M in a Grothendieck topos $\mathcal{E}$, condition (i) ensures that the functor $H_M : \mathcal{K}^{\mathrm{op}} \to \mathcal{E}$ is flat; for any geometric morphism $f : \mathcal{F} \to \mathcal{E}$, applying condition (iii)(b) to the flat functor $f^* \circ H_M$ and invoking condition (ii) thus yields, in view of the naturality in $\mathcal{E}$ of the operation $\widetilde{(-)}$, a natural isomorphism between the flat functor $f^* \circ H_M$ and the functor $H_{f^*(M)}$.

Since conditions (iii)(b) and (iii)(c) are equivalent by Remarks 5.2.13(b)-(c), to complete the proof of the theorem it remains to show that conditions (i), (ii) and (iii)(a) imply all together that $\mathbb{T}$ is classified by the presheaf topos $[\mathcal{K}, \mathbf{Set}]$.

Under conditions (i), (ii) and (iii)(a), we can define functors

$$G_{\mathcal{E}} : \mathbf{Flat}(\mathcal{K}^{\mathrm{op}}, \mathcal{E}) \to \mathbb{T}\text{-mod}(\mathcal{E})$$

and

$$H_{\mathcal{E}} : \mathbb{T}\text{-mod}(\mathcal{E}) \to \mathbf{Flat}(\mathcal{K}^{\mathrm{op}}, \mathcal{E}),$$

for each Grothendieck topos $\mathcal{E}$, which are natural in $\mathcal{E}$ and inverse to each other up to isomorphism, as follows.

We set $H_{\mathcal{E}}(M)$ equal to the functor

$$H_M = \mathrm{Hom}^{\mathcal{E}}_{\mathbb{T}\text{-mod}(\mathcal{E})}(\gamma^*_{\mathcal{E}}(-), M) : \mathcal{K}^{\mathrm{op}} \to \mathcal{E}.$$

This assignment is natural in M (cf. Remark 6.2.2) and hence defines a functor $H_{\mathcal{E}} : \mathbb{T}\text{-mod}(\mathcal{E}) \to \mathbf{Flat}(\mathcal{K}^{\mathrm{op}}, \mathcal{E})$ which is natural in $\mathcal{E}$ by condition (iii)(a).

In the converse direction, for any flat functor $F : \mathcal{K}^{\mathrm{op}} \to \mathcal{E}$, we set $G_{\mathcal{E}}(F) = \tilde{F}(M_\mathbb{T})$. Clearly, this assignment is natural in F and hence defines a functor $G_{\mathcal{E}} : \mathbf{Flat}(\mathcal{K}^{\mathrm{op}}, \mathcal{E}) \to \mathbb{T}\text{-mod}(\mathcal{E})$.

The functors $G_{\mathcal{E}}$ and $H_{\mathcal{E}}$ are natural in $\mathcal{E}$ and hence respectively induce geometric morphisms

$$G : [\mathcal{K}, \mathbf{Set}] \to \mathbf{Sh}(\mathcal{C}_\mathbb{T}, J_\mathbb{T})$$

and

$$H : \mathbf{Sh}(\mathcal{C}_\mathbb{T}, J_\mathbb{T}) \to [\mathcal{K}, \mathbf{Set}].$$

Notice that G coincides with the morphism $p_{\mathcal{K}}$ canonically induced by the universal property of the classifying topos of $\mathbb{T}$ as in section 5.2.3.

We have to prove that
$$H \circ G \cong 1_{[\mathcal{K},\mathbf{Set}]},$$
or equivalently that $G^* \circ H^* \circ y_\mathcal{K} \cong y_\mathcal{K}$.

Let us consider the geometric morphisms
$$e_N : \mathbf{Set} \to [\mathcal{K},\mathbf{Set}]$$
corresponding to the objects N of $\mathcal{K}$. We shall prove that $G^* \circ H^* \circ y_\mathcal{K} \cong y_\mathcal{K}$ by showing that $e_N^* \circ G^* \circ H^* \circ y_\mathcal{K} \cong e_N^* \circ y_\mathcal{K} = \mathrm{Hom}_\mathcal{K}(-, N)$ naturally in $N \in \mathcal{K}$.

Let us denote by M_G the $\mathbb{T}$-model in $[\mathcal{K},\mathbf{Set}]$ corresponding to the geometric morphism G. Clearly, for any $\mathbb{T}$-model N in $\mathcal{K}$, $e_N^*(M_G) \cong N$.

For any $\mathbb{T}$-model P in $\mathcal{E}$, $H_\mathcal{E}(P) = f_P^* \circ H^* \circ y'_\mathcal{K}$, where $f_P : \mathcal{E} \to \mathbf{Sh}(\mathcal{C}_\mathbb{T}, J_\mathbb{T})$ is the geometric morphism corresponding to P via the universal property of the classifying topos of $\mathbb{T}$ and $y'_\mathcal{K} : \mathcal{K}^{\mathrm{op}} \to [\mathcal{K},\mathbf{Set}]$ is the Yoneda embedding.

We have $H_{[\mathcal{K},\mathbf{Set}]}(M_G) = G^* \circ H^* \circ y'_\mathcal{K}$. But
$$H_{[\mathcal{K},\mathbf{Set}]}(M_G) = \mathrm{Hom}^{[\mathcal{K},\mathbf{Set}]}_{\mathbb{T}\text{-mod}([\mathcal{K},\mathbf{Set}])}(\gamma^*_{[\mathcal{K},\mathbf{Set}]}(-), M_G)$$

and, by condition (iii)(a),
$$e_N^*(\mathrm{Hom}^{[\mathcal{K},\mathbf{Set}]}_{\mathbb{T}\text{-mod}([\mathcal{K},\mathbf{Set}])}(\gamma^*_{[\mathcal{K},\mathbf{Set}]}(-), M_G)) \cong \mathrm{Hom}^{\mathbf{Set}}_{\mathbb{T}\text{-mod}(\mathbf{Set})}(-, e_N^*(M_G))$$
$$\cong \mathrm{Hom}_\mathcal{K}(-, N),$$

as required. This proves that $H \circ G \cong 1_{[\mathcal{K},\mathbf{Set}]}$. On the other hand, by condition (ii), $G \circ H$ is isomorphic to the identity. We can thus conclude that $\mathbb{T}$ is classified by the topos $[\mathcal{K},\mathbf{Set}]$, as required. □

Remarks 6.3.2 (a) Conditions (i) and (ii) of Theorem 6.3.1 together imply that the canonical geometric morphism
$$p_\mathcal{K} : [\mathcal{K},\mathbf{Set}] \to \mathbf{Sh}(\mathcal{C}_\mathbb{T}, J_\mathbb{T})$$
(cf. section 5.2.3) is a surjection, equivalently that the models in $\mathcal{K}$ are jointly conservative for $\mathbb{T}$. Indeed, by condition (i) the functor H_U, where U is 'the' universal model of $\mathbb{T}$ lying in its classifying topos, is flat. By applying Corollary 5.2.7 to it, we obtain that for any geometric sequent $\sigma \equiv (\phi \vdash_{\vec{x}} \psi)$ over Σ which is valid in every $\mathbb{T}$-model in $\mathcal{K}$, $\widetilde{H_U}(\{\vec{x}.\phi\}) \leq \widetilde{H_U}(\{\vec{x}.\psi\})$; condition (ii), combined with the conservativity of U, thus allows us to conclude that σ is provable in $\mathbb{T}$, as required.

(b) The following condition is sufficient, together with conditions (i) and (ii) of the theorem (or equivalently, together with condition (i) and the requirement that the models in $\mathcal{K}$ be jointly conservative for $\mathbb{T}$), for $\mathbb{T}$ to be classified by the topos $[\mathcal{K},\mathbf{Set}]$, but necessary only if one assumes the axiom of choice:

(∗) There is an assignment $M \to \phi_M(\vec{x_M})$ sending a $\mathbb{T}$-model M in $\mathcal{K}$ to a geometric formula-in-context $\phi_M(\vec{x_M})$ presenting it such that every $\mathbb{T}$-model homomorphism $M \to N$ between models in $\mathcal{K}$ is induced by a $\mathbb{T}$-provably functional formula from $\phi_N(\vec{x_N})$ to $\phi_M(\vec{x_M})$.

The necessity of condition (∗) (under the axiom of choice) follows from Theorem 6.1.13. To prove its sufficiency, let us show that, under conditions (i) and (ii), or under condition (i) and the assertion that the models in $\mathcal{K}$ are jointly conservative for $\mathbb{T}$, condition (∗) implies condition (iii)(c). Under either of these assumptions, the canonical geometric morphism $p_\mathcal{K} : [\mathcal{K}, \mathbf{Set}] \to \mathbf{Sh}(\mathcal{C}_\mathbb{T}, J_\mathbb{T})$ is a surjection (cf. Remark 6.3.2(a)). From this it follows that, for any geometric formulae $\{\vec{x}.\phi\}$ and $\{\vec{y}.\psi\}$ over Σ respectively presenting models c and d in $\mathcal{K}$, there can be at most one $\mathbb{T}$-provably functional formula $\{\vec{x}.\phi\} \to \{\vec{y}.\psi\}$ over Σ, up to $\mathbb{T}$-provable equivalence, inducing a given homomorphism of $\mathbb{T}$-models $d \to c$. Condition (∗) thus yields a (full and faithful) functor $R : \mathcal{K}^{\mathrm{op}} \to \mathcal{C}_\mathbb{T}$ such that $u^\mathbb{T}_{(\mathcal{K},\mathcal{E})}(F) \circ R \cong F$ naturally in F (cf. Theorem 5.2.5 and Remark 5.2.6), making condition (iii)(c) satisfied.

(c) Conditions (ii) and (iii)(c) of Theorem 6.3.1 admit invariant formulations, which can be profitably exploited in the presence of different representations for the classifying topos of $\mathbb{T}$. Indeed, they can both be entirely reformulated in terms of the extension of flat functors operation (in the sense of section 5.2.1) along the canonical geometric morphism

$$p_\mathcal{K} : [\mathcal{K}, \mathbf{Set}] \to \mathbf{Set}[\mathbb{T}].$$

Specifically, any (essentially small) site of definition $(\mathcal{C}, J)$ for $\mathbf{Set}[\mathbb{T}]$ gives rise to a functor

$$G^{(\mathcal{C},J)}_\mathcal{E} : \mathbf{Flat}(\mathcal{K}^{\mathrm{op}}, \mathcal{E}) \to \mathbf{Flat}_J(\mathcal{C}, \mathcal{E})$$

(note that if $(\mathcal{C}, J) = (\mathcal{C}_\mathbb{T}, J_\mathbb{T})$ then $G^{(\mathcal{C},J)}_\mathcal{E}$ is the functor $u^\mathbb{T}_{(\mathcal{K},\mathcal{E})}$ defined in section 5.2.3). Condition (iii)(c) asserts that this functor, whose explicit description is given in section 5.2.1, is full and faithful, while condition (ii) asserts that, denoting by

$$u_\mathcal{E} : \mathbf{Flat}_J(\mathcal{C}, \mathcal{E}) \simeq \mathbb{T}\text{-mod}(\mathcal{E})$$

the equivalence canonically induced by the universal property of the classifying topos of $\mathbb{T}$, the canonical morphism $u_\mathcal{E}(G^{(\mathcal{C},J)}_\mathcal{E}(H_M)) \to M$ is an isomorphism for any $\mathbb{T}$-model M in a Grothendieck topos $\mathcal{E}$.

(d) Condition (ii) of Theorem 6.3.1 is satisfied by every $\mathbb{T}$-model M in $\mathcal{K}$. Indeed, for any $M \in \mathcal{K}$, the canonical morphism $u_{\mathbf{Set}}(u^\mathbb{T}_{(\mathcal{K},\mathcal{E})}(H_M)) \to M$ of Remark 6.3.2(c) is clearly an isomorphism.

(e) Under condition (i) of Theorem 6.3.1, if condition (iii)(a) is satisfied and the models in $\mathcal{K}$ are jointly conservative for $\mathbb{T}$ then $\mathbb{T}$ satisfies condition (ii). To prove this, we observe that the assertion that the models in $\mathcal{K}$ be jointly conservative for the theory $\mathbb{T}$ is equivalent to the requirement that the geometric morphism $p_\mathcal{K}$ be a surjection. By the naturality in $\mathcal{E}$ of the operation $\widetilde{(-)}$ and of the assignment $\mathcal{E} \to H_{\mathcal{E}}$ (notice that the latter follows from condition (iii)(a)), it suffices to verify, since $p_{\mathcal{K}}$ is surjective, that the canonical morphism $\widetilde{H_M(M_{\mathbb{T}})} \to M$ is an isomorphism for M equal to a model in $\mathcal{K}$; but this is true by Remark 6.3.2(d).

(f) Condition (iii)(b) can be split into two separate conditions: η^F is pointwise monic and η^F is pointwise epic. We shall refer to the first condition as condition (iii)(b)-(1) and to the second as condition (iii)(b)-(2). By Theorem 5.2.12 and Lemma 7.1.9, condition (iii)(b)-(1) is equivalent to the requirement that the functor $u^{\mathbb{T}}_{(\mathcal{K},\mathcal{E})}$ be faithful.

(g) If for every sort A of Σ the formula $\{x^A . \top\}$ strongly presents a $\mathbb{T}$-model F_A in $\mathcal{K}$ (in the sense of Definition 6.2.5) then condition (i) of the theorem is automatically satisfied; indeed, under this hypothesis, for every sort A of Σ the canonical arrow $\widetilde{H_M(M_{\mathbb{T}})}A \cong \mathrm{Hom}^{\mathcal{E}}_{\mathbb{T}\text{-mod}(\mathcal{E})}(\gamma^*_{\mathcal{E}}(F_A), M) \to MA$ is an isomorphism (cf. Theorem 5.2.5).

(h) As observed in the proof of the theorem, the equivalence of conditions (iii)(b) and (iii)(c) holds in general (that is, not only under the assumption that (i) and (ii) hold).

Proposition 6.3.3 *Condition (iii)(c) of Theorem 6.3.1 is always satisfied for $\mathcal{E} = \mathbf{Set}$.*

Proof Let us prove that for any small category $\mathcal{C}$ and any functor $F : \mathcal{C} \to \mathcal{D}$, where $\mathcal{D}$ is a category with filtered colimits, if F is full and faithful and takes values in the full subcategory of $\mathcal{D}$ on the finitely presentable objects then its canonical extension $F' : \mathrm{Ind}\text{-}\mathcal{C} = \mathbf{Flat}(\mathcal{C}^{\mathrm{op}}, \mathbf{Set}) \to \mathcal{D}$ is full and faithful. This will imply our thesis by taking F to be the inclusion functor $\mathcal{K} \hookrightarrow \mathbb{T}\text{-mod}(\mathbf{Set})$.

For any $P \in \mathbf{Flat}(\mathcal{C}^{\mathrm{op}}, \mathbf{Set})$, $F'(P) = \mathrm{colim}(F \circ \pi_P)$, where $\pi_P : \int P \to \mathcal{C}$ is the fibration associated with P.

To prove that F' is faithful, we have to show that, for any natural transformations $\alpha, \beta : P \to Q$, $F'(\alpha) = F'(\beta)$ implies $\alpha = \beta$. Since every $P \in \mathbf{Flat}(\mathcal{C}^{\mathrm{op}}, \mathbf{Set})$ can be covered by representables, we can suppose P to be a representable $y_{\mathcal{C}}(c)$ (where $y_{\mathcal{C}} : \mathcal{C} \to [\mathcal{C}^{\mathrm{op}}, \mathbf{Set}]$ is the Yoneda embedding) without loss of generality. Now, since $F(c)$ is finitely presentable in $\mathcal{D}$, by Lemma 6.1.12 we can suppose without loss of generality that Q is a representable as well. The faithfulness of F' thus follows from that of F.

To prove the fullness of F', we have to show that for any natural transformation $\beta : F'(P) \to F'(Q)$ there exists a natural transformation $\alpha : P \to Q$ such that $F'(\alpha) = \beta$. Thanks to the faithfulness of F', we can clearly suppose P to be a representable $y_{\mathcal{C}}(c)$ without loss of generality. But since $F'(y_{\mathcal{C}}(c)) \cong F(c)$ is finitely presentable in $\mathcal{D}$ and F' preserves filtered colimits, by Lemma 6.1.12 we can suppose without loss of generality Q to be a representable as well. The fullness of F' thus follows from that of F. □

6.3.2 Concrete reformulations

In this section we shall give 'concrete' reformulations of the conditions of Theorem 6.3.1 in full generality as well as in some particular cases in which they admit relevant simplifications.

Condition (i)

Theorem 6.3.4 *Let $\mathbb{T}$ be a geometric theory over a signature Σ, $\mathcal{K}$ a full subcategory of f.p.$\mathbb{T}$-mod($\mathbf{Set}$), $\mathcal{E}$ a Grothendieck topos with a separating family of objects $\mathcal{S}$ and M a $\mathbb{T}$-model in $\mathcal{E}$. Then the following conditions are equivalent:*

(i) *The functor*
$$H_M := \mathrm{Hom}^{\mathcal{E}}_{\mathbb{T}\text{-mod}(\mathcal{E})}(\gamma_{\mathcal{E}}^*(-), M) : \mathcal{K}^{\mathrm{op}} \to \mathcal{E}$$
is flat.

(ii) (a) *There exist an epimorphic family $\{E_i \to 1_{\mathcal{E}} \mid i \in I, E_i \in \mathcal{S}\}$ in $\mathcal{E}$ and for each $i \in I$ a $\mathbb{T}$-model c_i in $\mathcal{K}$ and a Σ-structure homomorphism $c_i \to \mathrm{Hom}_{\mathcal{E}}(E_i, M)$.*

(b) *For any $\mathbb{T}$-models c and d in $\mathcal{K}$ and Σ-structure homomorphisms $x : c \to \mathrm{Hom}_{\mathcal{E}}(E, M)$ and $y : d \to \mathrm{Hom}_{\mathcal{E}}(E, M)$ (where $E \in \mathcal{S}$), there exist an epimorphic family $\{e_i : E_i \to E \mid i \in I, E_i \in \mathcal{S}\}$ in $\mathcal{E}$ and for each $i \in I$ a $\mathbb{T}$-model b_i in $\mathcal{K}$, $\mathbb{T}$-model homomorphisms $u_i : c \to b_i$, $v_i : d \to b_i$ and a Σ-structure homomorphism $z_i : b_i \to \mathrm{Hom}_{\mathcal{E}}(E_i, M)$ such that $\mathrm{Hom}_{\mathcal{E}}(e_i, M) \circ x = z_i \circ u_i$ and $\mathrm{Hom}_{\mathcal{E}}(e_i, M) \circ y = z_i \circ v_i$.*

(c) *For any two homomorphisms $u, v : d \to c$ of $\mathbb{T}$-models in $\mathcal{K}$ and any Σ-structure homomorphism $x : c \to \mathrm{Hom}_{\mathcal{E}}(E, M)$ in $\mathcal{E}$ (where $E \in \mathcal{S}$) for which $x \circ u = x \circ v$, there exist an epimorphic family $\{e_i : E_i \to E \mid i \in I, E_i \in \mathcal{S}\}$ in $\mathcal{E}$ and for each $i \in I$ a homomorphism $w_i : c \to b_i$ of $\mathbb{T}$-models in $\mathcal{K}$ and a Σ-structure homomorphism $y_i : b_i \to \mathrm{Hom}_{\mathcal{E}}(E_i, M)$ such that $w_i \circ u = w_i \circ v$ and $y_i \circ w_i = \mathrm{Hom}_{\mathcal{E}}(e_i, M) \circ x$.*

Proof This follows immediately from Proposition 6.2.3 in view of the characterization of flat functors as filtering functors (cf. Proposition 1.1.32). □

Remarks 6.3.5 (a) If for any $\mathbb{T}$-model M in a Grothendieck topos $\mathcal{E}$ and any object E of $\mathcal{E}$ the Σ-structure $\mathrm{Hom}_{\mathcal{E}}(E, M)$ is a $\mathbb{T}$-model then the three conditions (ii)(a), (ii)(b) and (ii)(c) of Theorem 6.3.4 hold if they are satisfied for $\mathcal{E} = \mathbf{Set}$.

(b) Condition (ii)(c) of Theorem 6.3.4 is automatically satisfied if all the $\mathbb{T}$-model homomorphisms in any Grothendieck topos are monic. Indeed, by Proposition 6.2.3, the Σ-structure homomorphisms $x : c \to \mathrm{Hom}_{\mathcal{E}}(E, M)$ correspond precisely to the $\mathbb{T}$-model homomorphisms $\bar{x} : \gamma^*_{\mathcal{E}/E}(c) \to !^*_E(M)$. The condition $x \circ u = x \circ v$ is therefore equivalent to $\bar{x} \circ \gamma^*_{\mathcal{E}/E}(u) = \bar{x} \circ \gamma^*_{\mathcal{E}/E}(v)$ and hence, if $\bar{x}$ is monic, to the condition $\gamma^*_{\mathcal{E}/E}(u) = \gamma^*_{\mathcal{E}/E}(v)$, which is equivalent, if $E \not\cong 0$, to $u = v$ (cf. Lemma 6.3.19).

(c) Condition (ii)(a) of Theorem 6.3.4 follows from condition (ii)(a) of Theorem 6.3.8 if the signature of $\mathbb{T}$ contains at least one constant. Indeed, for any $\mathbb{T}$-model M in a Grothendieck topos $\mathcal{E}$, the interpretation of such a constant in M will be an arrow $1 \to MA$ in $\mathcal{E}$, where A is the sort of the constant; condition (ii)(a) of Theorem 6.3.8 thus provides an epimorphic family satisfying the required property.

(d) If all the $\mathbb{T}$-models in $\mathcal{K}$ are finitely generated as Σ-structures and all the Σ-structure homomorphisms of the form $c \to \text{Hom}_{\mathcal{E}}(E, M)$ (where $c \in \mathcal{K}$, M is a $\mathbb{T}$-model in $\mathcal{E}$ and $E \in \mathcal{E}$) are injective if $E \not\cong 0$, a sufficient condition for condition (ii)(b) of Theorem 6.3.4 to hold is that, for any context $\vec{x}$, the sequent $(\top \vdash_{\vec{x}} \bigvee_{\phi(\vec{x}) \in \mathcal{I}_{\mathcal{K}}^{\vec{x}}} \phi(\vec{x}))$ be provable in $\mathbb{T}$, where $\mathcal{I}_{\mathcal{K}}^{\vec{x}}$ is the set of (representatives of $\mathbb{T}$-provable-equivalence classes of) geometric formulae in the context $\vec{x}$ which strongly finitely present a $\mathbb{T}$-model in $\mathcal{K}$ (in the sense of Definition 6.2.5). Indeed, under these hypotheses, for any two Σ-structure homomorphisms $x : c \to \text{Hom}_{\mathcal{E}}(E, M)$ and $y : d \to \text{Hom}_{\mathcal{E}}(E, M)$, where c and d are finitely generated Σ-structures and $E \not\cong 0$, the substructure $r_e : e \hookrightarrow \text{Hom}_{\mathcal{E}}(E, M)$ of $\text{Hom}_{\mathcal{E}}(E, M)$ generated by the images of c under x and of d under y is finitely generated, say by elements $\xi_1, \ldots, \xi_n$. By choosing a context $\vec{x}$ of the same length as the number of generators of e, we obtain an epimorphic family $\{e_i : E_i \to E \mid i \in I, E_i \in S, E_i \not\cong 0\}$ in $\mathcal{E}$ with the property that for each $i \in I$ there exists a geometric formula $\phi_i(\vec{x})$ strongly presenting a $\mathbb{T}$-model u_i in $\mathcal{K}$ such that $(\xi_1 \circ e_i, \ldots, \xi_n \circ e_i)$ factors through $[[\vec{x} . \phi_i]]_M$. For each $i \in I$ we thus have a Σ-structure homomorphism $z_i : u_i \to \text{Hom}_{\mathcal{E}}(E_i, M)$ which sends the generators of u_i to the tuple $(\xi_1 \circ e_i, \ldots, \xi_n \circ e_i)$. Since z_i is injective by our hypothesis and its image contains a set of generators for e, we have a factorization of $\text{Hom}_{\mathcal{E}}(e_i, M) \circ r_e$ through z_i. Therefore both $\text{Hom}_{\mathcal{E}}(e_i, M) \circ x$ and $\text{Hom}_{\mathcal{E}}(e_i, M) \circ y$ factor through z_i. This shows that condition (ii)(b) is satisfied.

Let us suppose that every $\mathbb{T}$-model in $\mathcal{K}$ is strongly finitely presented by a formula over Σ (in the sense of Definition 6.2.5) and that every $\mathbb{T}$-model homomorphism between two models in $\mathcal{K}$ is induced by a $\mathbb{T}$-provably functional formula between formulae which (strongly) present them. Then, in light of the discussion preceding Definition 6.2.5, we can express the conditions for the functor H_M to be flat (equivalently, filtering) in terms of the satisfaction in M of certain geometric sequents involving these formulae. Specifically, we have the following result:

Theorem 6.3.6 *Let $\mathbb{T}$ be a geometric theory over a signature Σ, $\mathcal{K}$ a full subcategory of f.p.$\mathbb{T}$-mod(**Set**) and $\mathcal{P}$ a set of geometric formulae over Σ such that every $\mathbb{T}$-model in $\mathcal{K}$ is strongly presented by a formula in $\mathcal{P}$ and for any two formulae $\phi(\vec{x})$ and $\psi(\vec{y})$ in $\mathcal{P}$ presenting respectively models c and d in $\mathcal{K}$, any $\mathbb{T}$-model homomorphism $d \to c$ is induced by a $\mathbb{T}$-provably functional formula from $\phi(\vec{x})$ to $\psi(\vec{y})$. Then $\mathbb{T}$ satisfies condition (i) of Theorem 6.3.1 with respect* to the category $\mathcal{K}$ if and only if the following conditions are satisfied for every $\mathbb{T}$-model M in a Grothendieck topos $\mathcal{E}$:

(i) The sequent
$$\left(\top \vdash_{[]} \bigvee_{\phi(\vec{x}) \in \mathcal{P}} (\exists \vec{x}) \phi(\vec{x})\right)$$
is valid in M.

(ii) For any formulae $\phi(\vec{x})$ and $\psi(\vec{y})$ in $\mathcal{P}$, the sequent
$$\left(\phi(\vec{x}) \wedge \psi(\vec{y}) \vdash_{\vec{x},\vec{y}} \bigvee_{\substack{\chi(\vec{z}) \in \mathcal{P}, \\ \{\vec{x}.\phi\} \xleftarrow{[\theta_1]} \{\vec{z}.\chi\} \xrightarrow{[\theta_2]} \{\vec{y}.\psi\} \text{ in } \mathcal{C}_\mathbb{T}}} (\exists \vec{z})(\theta_1(\vec{z},\vec{x}) \wedge \theta_2(\vec{z},\vec{y}))\right)$$
is valid in M.

(iii) For any $\mathbb{T}$-provably functional formulae $\theta_1, \theta_2 : \phi(\vec{x}) \to \psi(\vec{y})$ between two formulae $\phi(\vec{x})$ and $\psi(\vec{y})$ in $\mathcal{P}$, the sequent
$$\left(\theta_1(\vec{x},\vec{y}) \wedge \theta_2(\vec{x},\vec{y}) \vdash_{\vec{x},\vec{y}} \bigvee_{\substack{\chi(\vec{z}) \in \mathcal{P}, \\ \{\vec{z}.\chi\} \xrightarrow{[\tau]} \{\vec{x}.\phi\} \text{ in } \mathcal{C}_\mathbb{T} \mid \\ \tau \wedge \theta_1 \dashv\vdash_\mathbb{T} \tau \wedge \theta_2}} (\exists \vec{z}) \tau(\vec{z},\vec{x})\right)$$
is valid in M.

Proof The fact that any model c in $\mathcal{K}$ is strongly finitely presented by a formula $\phi(\vec{x})$ in $\mathcal{P}$ ensures that for any object E of $\mathcal{E}$ and any $\mathbb{T}$-model M in $\mathcal{E}$, the Σ-homomorphisms $c \to \text{Hom}_\mathcal{E}(E, M)$ are in bijective correspondence with the generalized elements $E \to [[\vec{x}.\phi]]_M$ in $\mathcal{E}$. Our thesis thus follows from the Kripke-Joyal semantics for the topos $\mathcal{E}$, noticing that, by our hypotheses, for any two formulae $\phi(\vec{x})$ and $\psi(\vec{y})$ in $\mathcal{P}$ presenting respectively models c and d in $\mathcal{K}$, any $\mathbb{T}$-model homomorphism $d \to c$ is induced by a $\mathbb{T}$-provably functional formula from $\phi(\vec{x})$ to $\psi(\vec{y})$. $\square$

Remarks 6.3.7 (a) Under the hypotheses of Theorem 6.3.6, all the sequents in the statement of the theorem are satisfied by every model in $\mathcal{K}$. Therefore, adding them to $\mathbb{T}$ yields a quotient of $\mathbb{T}$ satisfying condition (i) of Theorem 6.3.1 with respect to the category $\mathcal{K}$.

(b) Under the alternative hypothesis that every model of $\mathcal{K}$ is both strongly finitely presentable and finitely generated (with respect to the same generators), for any two formulae $\phi(\vec{x})$ and $\psi(\vec{y})$ in $\mathcal{P}$ presenting respectively models c and d in $\mathcal{K}$, the $\mathbb{T}$-model homomorphisms $c \to d$ are in bijection (via the evaluation of such homomorphisms at the generators of c) with the tuples of elements of d which satisfy ϕ, each of which has the form $(t_1(\vec{y}), \ldots, t_n(\vec{y}))$ (where n is the length of the context $\vec{x}$) for some terms $t_1, \ldots, t_n$ in the context $\vec{y}$ over the signature Σ. Conditions (ii) and (iii) of Theorem 6.3.6 can thus be reformulated more explicitly as follows (using the notation of section 6.1.4):

(ii') For any formulae $\phi(\vec{x})$ and $\psi(\vec{y})$ in $\mathcal{P}$, where $\vec{x} = (x_1^{A_1}, \ldots, x_n^{A_n})$ and $\vec{y} = (y_1^{B_1}, \ldots, y_m^{B_m})$, the sequent

$$\left(\phi(\vec{x}) \wedge \psi(\vec{y}) \vdash_{\vec{x},\vec{y}} \bigvee_{\substack{\chi(\vec{z})\in\mathcal{P},\, t_1^{A_1}(\vec{z}),\ldots,t_n^{A_n}(\vec{z}) \\ s_1^{B_1}(\vec{z}),\ldots,s_m^{B_m}(\vec{z})}} (\exists \vec{z})\left(\chi(\vec{z}) \wedge \bigwedge_{\substack{i\in\{1,\ldots,n\},\\ j\in\{1,\ldots,m\}}} (x_i = t_i(\vec{z}) \wedge y_j = s_j(\vec{z}))\right)\right),$$

where the disjunction is taken over all the formulae $\chi(\vec{z})$ in $\mathcal{P}$ and all the tuples of terms $t_1^{A_1}(\vec{z}), \ldots, t_n^{A_n}(\vec{z})$ and $s_1^{B_1}(\vec{z}), \ldots, s_m^{B_m}(\vec{z})$ such that

$$(t_1^{A_1}(\vec{\xi_\chi}), \ldots, t_n^{A_n}(\vec{\xi_\chi})) \in [[\vec{x}.\phi]]_{M_{\{\vec{z}.\chi\}}}$$

and

$$(s_1^{B_1}(\vec{\xi_\chi}), \ldots, s_m^{B_m}(\vec{\xi_\chi})) \in [[\vec{y}.\psi]]_{M_{\{\vec{z}.\chi\}}},$$

is valid in M.

(iii') For any formulae $\phi(\vec{x})$ and $\psi(\vec{y})$ in $\mathcal{P}$, where $\vec{x} = (x_1^{A_1}, \ldots, x_n^{A_n})$ and $\vec{y} = (y_1^{B_1}, \ldots, y_m^{B_m})$, and any terms $t_1^{A_1}(\vec{y}), s_1^{A_1}(\vec{y}), \ldots, t_n^{A_n}(\vec{y}), s_n^{A_n}(\vec{y})$ such that

$$(t_1(\vec{\xi_\phi}), \ldots, t_n(\vec{\xi_\phi})) \in [[\vec{y}.\phi]]_{M_{\{\vec{y}.\psi\}}}$$

and

$$(s_1(\vec{\xi_\phi}), \ldots, s_n(\vec{\xi_\phi})) \in [[\vec{y}.\phi]]_{M_{\{\vec{y}.\psi\}}},$$

the sequent

$$\left(\bigwedge_{i\in\{1,\ldots,n\}} (t_i(\vec{y}) = s_i(\vec{y})) \wedge \psi(\vec{y}) \vdash_{\vec{y}} \bigvee_{\chi(\vec{z})\in\mathcal{P},\, u_1^{B_1}(\vec{z}),\ldots,u_m^{B_m}(\vec{z})} (\exists\vec{z})\left(\chi(\vec{z}) \wedge \bigwedge_{j\in\{1,\ldots,m\}} (y_j = u_j(\vec{z}))\right)\right),$$

where the disjunction is taken over all the formulae $\chi(\vec{z})$ in $\mathcal{P}$ and all the tuples of terms $u_1^{B_1}(\vec{z}), \ldots, u_m^{B_m}(\vec{z})$ such that

$$(u_1(\vec{\xi}_\chi), \ldots, u_m(\vec{\xi}_\chi)) \in [[\vec{y} \cdot \psi]]_{M_{\{\vec{z} \cdot \chi\}}}$$

and

$$t_i(u_1(\vec{\xi}_\chi), \ldots, u_m(\vec{\xi}_\chi)) = s_i(u_1(\vec{\xi}_\chi), \ldots, u_m(\vec{\xi}_\chi)) \text{ in } M_{\{\vec{z} \cdot \chi\}}$$

for all $i \in \{1, \ldots, n\}$, is valid in M.

Condition (ii)

In this section we shall give more concrete reformulations of condition (ii) of Theorem 6.3.1.

Let $\mathcal{E}$ be a Grothendieck topos with a separating family of objects $\mathcal{S}$ and M a $\mathbb{T}$-model in $\mathcal{E}$ satisfying condition (i) of Theorem 6.3.1, i.e. such that the functor H_M is flat. The canonical morphism $\widetilde{H_M}(M_\mathbb{T}) \to M$ considered in condition (ii) is an isomorphism if and only if, for every sort A of Σ, the induced arrow $\widetilde{H_M}(\{x^A \cdot \top\}) \to MA$ is an isomorphism in $\mathcal{E}$.

By Theorem 5.2.5, we have that $\widetilde{H_M}(\{x^A \cdot \top\}) \cong \text{colim}(H_M \circ \pi^\mathcal{K}_{\{x^A \cdot \top\}})$, where $\pi^\mathcal{K}_{\{x^A \cdot \top\}}$ is the projection functor $\mathcal{A}^\mathcal{K}_{\{x^A \cdot \top\}} \to \mathcal{K}^{\text{op}}$ from the category $\mathcal{A}^\mathcal{K}_{\{x^A \cdot \top\}}$ having as objects the pairs (c, z), where c is a $\mathbb{T}$-model in $\mathcal{K}$ and z is an element of the set cA, and as arrows $(c, z) \to (d, w)$ the $\mathbb{T}$-model homomorphisms $g : d \to c$ in $\mathcal{K}$ such that $g_A(w) = z$. Proposition 5.2.8 provides the following explicit representation of $\text{colim}(H_M \circ \pi^\mathcal{K}_{\{x^A \cdot \top\}})$ as the quotient

$$\left(\coprod_{(c,z) \in \mathcal{A}^\mathcal{K}_{\{x^A \cdot \top\}}} H_M(c) \right) / R_{\{x^A \cdot \top\}},$$

where $R_{\{x^A \cdot \top\}}$ is the equivalence relation defined by the condition that, for any objects $(c, z), (d, w)$ of $\mathcal{A}^\mathcal{K}_{\{x^A \cdot \top\}}$ and any generalized elements $x : E \to F(c)$ and $x' : E \to F(d)$, $(x, x') \in R_{\{x^A \cdot \top\}}$ if and only if there exist an epimorphic family $\{e_i : E_i \to E \mid i \in I\}$ in $\mathcal{E}$ and for each $i \in I$ a $\mathbb{T}$-model a_i in $\mathcal{K}$, a generalized element $h_i : E_i \to F(b_i)$ and two $\mathbb{T}$-model homomorphisms $f_i : c \to b_i$ and $f'_i : d \to b_i$ in $\mathcal{K}$ such that $f_{iA}(z) = f'_{iA}(w)$ and $< F(f_i), F(f'_i) > \circ h_i = < x, x' > \circ e_i$. From this representation it follows that the canonical arrow $\widetilde{H_M}(\{x^A \cdot \top\}) \to MA$ is an isomorphism if and only if (using the notation of Proposition 5.2.8)

(1) the canonical arrows $\kappa_{(c,z)} : H_M(c) \to MA$ for $(c, z) \in \mathcal{A}_{\{x^A \cdot \top\}}$ are jointly epimorphic and

(2) for any two objects (c, z) and (d, w) of the category $\mathcal{A}_{\{x^A \cdot \top\}}$ and any generalized elements $x : E \to H_M(c)$ and $x' : E \to H_M(d)$ (where $E \in \mathcal{S}$), $\kappa_{(c,z)} \circ x = \kappa_{(d,w)} \circ x'$ if and only if $(x, x') \in R_{\{x^A \cdot \top\}}$.

Thanks to the identification between the generalized elements $x : E \to H_M(c)$ and the Σ-structure homomorphisms $f_x : c \to \text{Hom}_\mathcal{E}(E, M)$ provided by Proposition 6.2.3, we can rewrite conditions (1) and (2) more explicitly, as follows.

We preliminarily notice that, for any object (c, z) of the category $\mathcal{A}_{\{x^A . \mathbb{T}\}}$, the canonical arrow $\kappa_{(c,z)} : H_M(c) = \mathrm{Hom}^{\mathcal{E}}_{\mathbb{T}\text{-mod}(\mathcal{E})}(\gamma^*_{\mathcal{E}}(c), M) \to MA$ can be described in terms of generalized elements as the arrow sending any generalized element $x : E \to H_M(c)$, corresponding via the identification of Proposition 6.2.3 to a Σ-structure homomorphism $f_x : c \to \mathrm{Hom}_{\mathcal{E}}(E, M)$, to the generalized element $E \to MA$ of MA given by $(f_x)_A(z)$. Therefore, for any generalized elements $x : E \to H_M(c)$ and $x' : E \to H_M(d)$ corresponding respectively to Σ-structure homomorphisms $f_x : c \to \mathrm{Hom}_{\mathcal{E}}(E, M)$ and $f_{x'} : d \to \mathrm{Hom}_{\mathcal{E}}(E, M)$, we have that $\kappa_{(c,z)} \circ x = \kappa_{(d,w)} \circ x'$ if and only if $(f_x)_A(z) = (f_{x'})_A(w)$.

Now, condition (1) can be formulated by saying that for any generalized element $x : E \to MA$ (where $E \in \mathcal{S}$) there exist an epimorphic family $\{e_i : E_i \to E \mid i \in I, E_i \in \mathcal{S}\}$ in $\mathcal{E}$, a family $\{(c_i, z_i) \mid i \in I\}$ of objects of the category $\mathcal{A}_{\{x^A . \mathbb{T}\}}$ and generalized elements $y_i : E_i \to H_M(c_i)$ (for each $i \in I$) such that $x \circ e_i = \kappa_{(c_i, z_i)} \circ y_i$. By Proposition 6.2.3, it thus rewrites as follows: for any generalized element $x : E \to MA$ (where $E \in \mathcal{S}$), there exist an epimorphic family $\{e_i : E_i \to E \mid i \in I, E_i \in \mathcal{S}\}$, a family $\{(c_i, z_i) \mid i \in I\}$ of objects of the category $\mathcal{A}_{\{x^A . \mathbb{T}\}}$ and Σ-structure homomorphisms $f_i : c_i \to \mathrm{Hom}_{\mathcal{E}}(E_i, M)$ (for each $i \in I$) such that $(f_i)_A(z_i) = x \circ e_i$.

Concerning condition (2), we observe that, for any objects (c, z) and (d, w) of $\mathcal{A}_{\{x^A . \mathbb{T}\}}$ and any Σ-structure homomorphisms $f_x : c \to \mathrm{Hom}_{\mathcal{E}}(E, M)$ and $f_{x'} : d \to \mathrm{Hom}_{\mathcal{E}}(E, M)$ (where $E \in \mathcal{S}$) corresponding respectively to generalized elements $x : E \to H_M(c)$ and $x' : E \to H_M(d)$ via the identification of Proposition 6.2.3, we have that $(x, x') \in R_{\{x^A . \mathbb{T}\}}$ if and only if there exist an epimorphic family $\{e_i : E_i \to E \mid i \in I, E_i \in \mathcal{S}\}$ in $\mathcal{E}$ and for each $i \in I$ a $\mathbb{T}$-model b_i in $\mathcal{K}$, a Σ-structure homomorphism $h_i : b_i \to \mathrm{Hom}_{\mathcal{E}}(E_i, M)$ and two $\mathbb{T}$-model homomorphisms $f_i : c \to b_i$ and $f'_i : d \to b_i$ in $\mathcal{K}$ such that $(f_i)_A(z) = (f'_i)_A(w)$ and $h_i \circ f_i = \mathrm{Hom}_{\mathcal{E}}(e_i, M) \circ f_x$ and $h_i \circ f'_i = \mathrm{Hom}_{\mathcal{E}}(e_i, M) \circ f_{x'}$.

Summarizing, we have the following result:

Theorem 6.3.8 *Let $\mathbb{T}$ be a geometric theory over a signature Σ, $\mathcal{K}$ a full subcategory of f.p.$\mathbb{T}$-mod(**Set**), $\mathcal{E}$ a Grothendieck topos with a separating family of objects $\mathcal{S}$ and M a $\mathbb{T}$-model in $\mathcal{E}$. Then, under the assumption that $\mathbb{T}$ satisfies condition (i) of Theorem 6.3.1 with respect to $\mathcal{K}$, the following conditions are equivalent:*

(i) *The extension $\widetilde{H_M} : \mathcal{C}_{\mathbb{T}} \to \mathcal{E}$ of the functor $H_M : \mathcal{K}^{\mathrm{op}} \to \mathcal{E}$ to the syntactic category $\mathcal{C}_{\mathbb{T}}$ (in the sense of section 5.2.3) satisfies the property that the canonical morphism $\widetilde{H_M}(M_{\mathbb{T}}) \to M$ is an isomorphism.*

(ii) *For every sort A of Σ, the following conditions are satisfied:*

(a) *For any generalized element $x : E \to MA$ (where $E \in \mathcal{S}$), there exist an epimorphic family $\{e_i : E_i \to E \mid i \in I, E_i \in \mathcal{S}\}$ in $\mathcal{E}$ and for each $i \in I$ a $\mathbb{T}$-model c_i in $\mathcal{K}$, an element z_i of $c_i A$ and a Σ-structure homomorphism $f_i : c_i \to \mathrm{Hom}_{\mathcal{E}}(E_i, M)$ such that $(f_i)_A(z_i) = x \circ e_i$.*

(b) *For any two pairs (c, z) and (d, w), where c and d are $\mathbb{T}$-models in $\mathcal{K}$ and $z \in cA, w \in dA$, and any Σ-structure homomorphisms $f : c \to \mathrm{Hom}_{\mathcal{E}}(E, M)$ and $f' : d \to \mathrm{Hom}_{\mathcal{E}}(E, M)$ (where $E \in \mathcal{S}$), we have that $f_A(z) = f'_A(w)$ if and only if there exist an epimorphic family $\{e_j : E_j \to E \mid j \in J, E_j \in \mathcal{S}\}$ in $\mathcal{E}$ and for each $j \in J$ a $\mathbb{T}$-model b_j in $\mathcal{K}$, a Σ-structure homomorphism $h_j : b_j \to \mathrm{Hom}_{\mathcal{E}}(E_j, M)$ and two $\mathbb{T}$-model homomorphisms $f_j : c \to b_j$ and $f'_j : d \to b_j$ in $\mathcal{K}$ such that $(f_j)_A(z) = (f_j)_{A'}(w)$, $h_j \circ f_j = \mathrm{Hom}_{\mathcal{E}}(e_j, M) \circ f$ and $h_j \circ f'_j = \mathrm{Hom}_{\mathcal{E}}(e_j, M) \circ f'$.*

Remarks 6.3.9 (a) If all the $\mathbb{T}$-model homomorphisms in any Grothendieck topos are monic and $\mathbb{T}$ satisfies condition (ii)(b) of Theorem 6.3.4 then condition (ii)(b) of Theorem 6.3.8 is automatically satisfied.

(b) A sufficient condition for condition (ii)(a) of Theorem 6.3.8 to hold is that the sequent $\left(\top \vdash_x \bigvee_{\phi(x) \in \mathcal{I}_{\mathcal{K}}^x} \phi(x)\right)$ be provable in $\mathbb{T}$, where $\mathcal{I}_{\mathcal{K}}^x$ is the set of (representatives of $\mathbb{T}$-provable-equivalence classes of) geometric formulae in one variable which strongly finitely present a $\mathbb{T}$-model in $\mathcal{K}$ (in the sense of Definition 6.2.5).

The following result provides an explicit reformulation of condition (ii) of Theorem 6.3.1 holding for theories $\mathbb{T}$ with respect to full subcategories $\mathcal{K} \hookrightarrow$ f.p.$\mathbb{T}$-mod(**Set**) such that every model of $\mathcal{K}$ is both strongly finitely presentable and finitely generated (with respect to the same generators).

Theorem 6.3.10 *Let $\mathbb{T}$ be a geometric theory over a signature Σ and $\mathcal{K}$ a full subcategory of f.p.$\mathbb{T}$-mod(**Set**). If every model in $\mathcal{K}$ is both strongly finitely presentable and finitely generated (with respect to the same generators) then $\mathbb{T}$ satisfies condition (ii) of Theorem 6.3.1 with respect to the category $\mathcal{K}$ if and only if for every model M of $\mathbb{T}$ in a Grothendieck topos satisfying condition (i) of Theorem 6.3.1 (i.e. such that the functor H_M is flat), the following sequents are satisfied in M (where $\mathcal{P}$ denotes the set of (representatives of) formulae over Σ which present a model in $\mathcal{K}$ and the notation is that of section 6.1.4):*

(i) *for any sort A of Σ, the sequent*

$$\left(\top \vdash_{x^A} \bigvee_{\chi(\vec{z}) \in \mathcal{P}, \, t^A(\vec{z})} (\exists \vec{z})(\chi(\vec{z}) \wedge x = t(\vec{z}))\right),$$

where the disjunction is taken over all the formulae $\chi(\vec{z})$ in $\mathcal{P}$ and all the terms $t^A(\vec{z})$;

(ii) *for any sort A of Σ, any formulae $\phi(\vec{x})$ and $\psi(\vec{y})$ in $\mathcal{P}$, where $\vec{x} = (x_1^{A_1}, \ldots, x_n^{A_n})$ and $\vec{y} = (y_1^{B_1}, \ldots, y_m^{B_m})$, and any terms $t^A(\vec{x})$ and $s^A(\vec{y})$, the sequent*

$$\left(\phi(\vec{x}) \wedge \psi(\vec{y}) \wedge t(\vec{x}) = s(\vec{y}) \vdash_{\vec{x},\vec{y}} \right.$$

$$\left. \bigvee_{\substack{\chi(\vec{z}) \in \mathcal{P},\, p_1^{A_1}(\vec{z}),\ldots,p_n^{A_n}(\vec{z}) \\ q_1^{B_1}(\vec{z}),\ldots,q_m^{B_m}(\vec{z})}} (\exists \vec{z}) \left(\chi(\vec{z}) \wedge \bigwedge_{\substack{i \in \{1,\ldots,n\}, \\ j \in \{1,\ldots,m\}}} (x_i = p_i(\vec{z}) \wedge y_j = q_j(\vec{z})) \right) \right),$$

where the disjunction is taken over all the formulae $\chi(\vec{z})$ in $\mathcal{P}$ and all the tuples of terms $p_1^{A_1}(\vec{z}),\ldots,p_n^{A_n}(\vec{z})$ and $q_1^{B_1}(\vec{z}),\ldots,q_m^{B_m}(\vec{z})$ such that $(p_1(\vec{\xi_\chi}),\ldots,p_n(\vec{\xi_\chi})) \in [[\vec{x}.\phi]]_{M_{\{\vec{z}.\chi\}}}$, $(q_1(\vec{\xi}),\ldots,q_m(\vec{\xi_\chi})) \in [[\vec{y}.\psi]]_{M_{\{\vec{z}.\chi\}}}$ and $t(p_1(\vec{\xi_\chi}),\ldots,p_n(\vec{\xi_\chi})) = s(q_1(\vec{\xi_\chi}),\ldots,q_m(\vec{\xi_\chi}))$ in $M_{\{\vec{z}.\chi\}}$.

Proof The thesis follows from Theorem 6.3.8 in light of the Kripke-Joyal semantics and the fact that for every model c in $\mathcal{K}$ strongly finitely presented by a formula $\phi(\vec{x})$ in $\mathcal{P}$, any object E of $\mathcal{E}$ and any $\mathbb{T}$-model M in $\mathcal{E}$, the Σ-homomorphisms $c \to \text{Hom}_{\mathcal{E}}(E,M)$ are in bijective correspondence with the generalized elements $E \to [[\vec{x}.\phi]]_M$ in $\mathcal{E}$. □

Remark 6.3.11 All the sequents in the statement of Theorem 6.3.10 are satisfied by every $\mathbb{T}$-model in $\mathcal{K}$ (cf. Remark 6.3.2(d)); therefore, adding them to any theory $\mathbb{T}$ satisfying the hypotheses of the theorem and condition (i) of Theorem 6.3.1 yields a quotient of $\mathbb{T}$ satisfying condition (ii) of Theorem 6.3.1 with respect to the category $\mathcal{K}$.

Condition (iii)

In this section, we shall give 'concrete' reformulations of conditions (iii)(a)-(b)-(c) of Theorem 6.3.1.

The following proposition shows that, under some natural assumptions which are often verified in practice, condition (iii)(a) of Theorem 6.3.1 is satisfied.

Proposition 6.3.12 *Let $\mathbb{T}$ be a geometric theory over a signature Σ and $\mathcal{K}$ a full subcategory of* f.p.$\mathbb{T}$-mod(**Set**).

(i) *If $\mathbb{T}$ is a quotient of a theory $\mathbb{S}$ satisfying condition (iii)(a) of Theorem 6.3.1 with respect to a full subcategory $\mathcal{H}$ of* f.p.$\mathbb{S}$-mod(**Set**) *and $\mathcal{K}$ is a subcategory of $\mathcal{H}$ then $\mathbb{T}$ satisfies condition (iii)(a) of Theorem 6.3.1 with respect to the category $\mathcal{K}$.*

(ii) *If every $\mathbb{T}$-model in $\mathcal{K}$ is strongly finitely presented (in the sense of section 6.2.2) then $\mathbb{T}$ satisfies condition (iii)(a) of Theorem 6.3.1 with respect to the category $\mathcal{K}$.*

(iii) *If for every $\mathbb{T}$-model c in $\mathcal{K}$ the object $\text{Hom}^{\mathcal{E}}_{\mathbb{T}\text{-mod}(\mathcal{E})}(\gamma^*_{\mathcal{E}}(c), M)$ can be built from c and M by only using geometric constructions (i.e. constructions only involving finite limits and small colimits) then $\mathbb{T}$ satisfies condition (iii)(a) of Theorem 6.3.1 with respect to the category $\mathcal{K}$.*

Proof

(i) If $\mathbb{T}$ is a quotient of $\mathbb{S}$ then, for every Grothendieck topos $\mathcal{E}$, the category $\mathbb{T}\text{-mod}(\mathcal{E})$ is a full subcategory of the category $\mathbb{S}\text{-mod}(\mathcal{E})$. This clearly implies that for any $\mathbb{T}$-models M and N in a Grothendieck topos $\mathcal{E}$, we have

$$\mathrm{Hom}^{\mathcal{E}}_{\mathbb{T}\text{-mod}(\mathcal{E})}(M, N) \cong \mathrm{Hom}^{\mathcal{E}}_{\mathbb{S}\text{-mod}(\mathcal{E})}(M, N);$$

in particular, for any $\mathbb{T}$-model c in $\mathcal{K}$ and any $\mathbb{T}$-model M in a Grothendieck topos $\mathcal{E}$, $\mathrm{Hom}^{\mathcal{E}}_{\mathbb{T}\text{-mod}(\mathcal{E})}(\gamma_{\mathcal{E}}^*(c), M) \cong \mathrm{Hom}^{\mathcal{E}}_{\mathbb{S}\text{-mod}(\mathcal{E})}(\gamma_{\mathcal{E}}^*(c), M)$. The fact that $\mathbb{S}$ satisfies condition (iii)(a) of Theorem 6.3.1 with respect to the category $\mathcal{H}$ thus implies that $\mathbb{T}$ does so with respect to the category $\mathcal{K}$, as required.

(ii) If every $\mathbb{T}$-model c in $\mathcal{K}$ is strongly finitely presented by a formula $\phi(\vec{x})$ over the signature of $\mathbb{T}$ then, for any $\mathbb{T}$-model M in a Grothendieck topos $\mathcal{E}$, $\mathrm{Hom}^{\mathcal{E}}_{\mathbb{T}\text{-mod}(\mathcal{E})}(\gamma_{\mathcal{E}}^*(c), M) \cong [[\vec{x}\,.\,\phi]]_M$ (cf. section 6.2.2). Therefore, as the interpretation of geometric formulae is preserved by inverse image functors of geometric morphisms, condition (iii)(a) of Theorem 6.3.1 is satisfied by the theory $\mathbb{T}$ with respect to the category $\mathcal{K}$.

(iii) Condition (iii) of Proposition 6.3.12 implies condition (iii)(a) of Theorem 6.3.1 since geometric constructions are preserved by inverse image functors of geometric morphisms.

□

We shall now proceed to give concrete reformulations of conditions (iii)(b)-(1) and (iii)(b)-(2) of Remark 6.3.2(f), in order to make them more easily verifiable in practice.

First, let us explicitly describe, for any object c of $\mathcal{K}$, the arrow

$$\eta^F(c) : F(c) \to \mathrm{Hom}^{\mathcal{E}}_{\mathbb{T}\text{-mod}(\mathcal{E})}(\gamma_{\mathcal{E}}^*(c), \tilde{F}(M_{\mathbb{T}}))$$

of condition (iii)(b) of Theorem 6.3.1 in terms of generalized elements.

For any generalized element $x : E \to F(c)$, $\eta^F(c) \circ x$ corresponds under the identification of Proposition 6.2.3 to the Σ-structure homomorphism $z_x : c \to \mathrm{Hom}_{\mathcal{E}}(E, \tilde{F}(M_{\mathbb{T}}))$ defined at each sort A of Σ as the function $cA \to \mathrm{Hom}_{\mathcal{E}}(E, \tilde{F}(\{x^A\,.\,\top\}))$ sending any element $y \in cA$ to the generalized element $E \to \tilde{F}(\{x^A\,.\,\top\})$ obtained by composing the canonical colimit arrow $\kappa^F_{(c,y)} : F(c) \to \tilde{F}(\{x^A\,.\,\top\})$ with the generalized element $x : E \to F(c)$. Given a separating family of objects $\mathcal{S}$ for $\mathcal{E}$, it follows that $\eta^F(c)$ is a monomorphism if and only if for any generalized elements $x, x' : E \to F(c)$ (where $E \in \mathcal{S}$), $\kappa^F_{(c,y)} \circ x = \kappa^F_{(c,y)} \circ x'$ for every sort A of Σ and every element $y \in cA$ implies $x = x'$. By Proposition 5.2.8 we have that $\kappa^F_{(c,y)} \circ x = \kappa^F_{(c,y)} \circ x'$ if and only if there exist an epimorphic family $\{e_i : E_i \to E \mid i \in I, E_i \in \mathcal{S}\}$ in $\mathcal{E}$ and for each $i \in I$ a $\mathbb{T}$-model a_i in $\mathcal{K}$, a generalized element $h_i : E_i \to F(a_i)$ and two

$\mathbb{T}$-model homomorphisms $f_i, f'_i : c \to a_i$ in $\mathcal{K}$ such that $(f_i)_A(y) = (f'_i)_A(y)$ and $< F(f_i), F(f'_i) > \circ h_i = < x, x' > \circ e_i$.

To obtain an explicit characterization of the condition for $\eta^F(c)$ to be an epimorphism, we notice that an arrow $f : A \to B$ in a Grothendieck topos $\mathcal{E}$ with a separating family of objects $\mathcal{S}$ is an epimorphism if and only if for every generalized element $x : E \to B$ of B (where $E \in \mathcal{S}$) there exist an epimorphic family $\{e_i : E_i \to E \mid i \in I, E_i \in \mathcal{S}\}$ in $\mathcal{E}$ and for each $i \in I$ a generalized element $y_i : E_i \to A$ such that $f \circ y_i = x \circ e_i$. Applying this characterization to the arrow $\eta^F(c)$ in light of Proposition 6.2.3, we obtain the following criterion: $\eta^F(c)$ is an epimorphism if and only if for every object E of $\mathcal{E}$ and any Σ-structure homomorphism $v : c \to \mathrm{Hom}_{\mathcal{E}}(E, \tilde{F}(M_{\mathbb{T}}))$, there exist an epimorphic family $\{e_i : E_i \to E \mid i \in I, E_i \in \mathcal{S}\}$ in $\mathcal{E}$ and for each $i \in I$ a generalized element $x_i : E_i \to F(c)$ such that $z_{x_i} = \mathrm{Hom}_{\mathcal{E}}(e_i, \tilde{F}(M_{\mathbb{T}})) \circ v$ for all i (where $z_{x_i} = \xi_{\eta^F(c)} \circ x_i$ using the notation of Proposition 6.2.3). The condition '$z_{x_i} = \mathrm{Hom}_{\mathcal{E}}(e_i, \tilde{F}(M_{\mathbb{T}})) \circ v$' can be explicitly reformulated as the requirement that, for every sort A of Σ and every element $y \in cA$, $\kappa^F_{(c,y)} \circ x_i = v_A(y) \circ e_i$.

Summarizing, we have the following result:

Theorem 6.3.13 *Let $\mathbb{T}$ be a geometric theory over a signature Σ, $\mathcal{K}$ a full subcategory of f.p.$\mathbb{T}$-mod(**Set**), $\mathcal{E}$ a Grothendieck topos with a separating family of objects $\mathcal{S}$ and $F : \mathcal{K}^{\mathrm{op}} \to \mathcal{E}$ a flat functor. Then*

(i) *F satisfies condition (iii)(b)-(1) of Remark 6.3.2(f) if and only if for any $\mathbb{T}$-model c in $\mathcal{K}$ and any generalized elements $x, x' : E \to F(c)$ (where $E \in \mathcal{S}$), if for every sort A of Σ and every element $y \in cA$ there exist an epimorphic family $\{e_i : E_i \to E \mid i \in I, E_i \in \mathcal{S}\}$ in $\mathcal{E}$ and for each $i \in I$ a $\mathbb{T}$-model a_i in $\mathcal{K}$, a generalized element $h_i : E_i \to F(a_i)$ and two $\mathbb{T}$-model homomorphisms $f_i, f'_i : c \to a_i$ such that $(f_i)_A(y) = (f'_i)_A(y)$ and $< F(f_i), F(f'_i) > \circ h_i = < x, x' > \circ e_i$ then $x = x'$.*

(ii) *F satisfies condition (iii)(b)-(2) of Remark 6.3.2(f) if and only if for any $\mathbb{T}$-model c in $\mathcal{K}$, any object E of $\mathcal{S}$ and any Σ-structure homomorphism $v : c \to \mathrm{Hom}_{\mathcal{E}}(E, \tilde{F}(M_{\mathbb{T}}))$, there exist an epimorphic family $\{e_i : E_i \to E \mid i \in I, E_i \in \mathcal{S}\}$ in $\mathcal{E}$ and for each $i \in I$ a generalized element $x_i : E_i \to F(c)$ such that, for every sort A of Σ and every element $y \in cA$, $\kappa^F_{(c,y)} \circ x_i = v_A(y) \circ e_i$.* □

Under the hypothesis that for every sort A of Σ the formula $\{x^A \, . \, \top\}$ presents a $\mathbb{T}$-model F_A in $\mathcal{K}$, we have that $\tilde{F}(M_{\mathbb{T}})A = \tilde{F}(\{x^A \, . \, \top\}) \cong F(F_A)$, and the homomorphism $z_x : c \to \mathrm{Hom}_{\mathcal{E}}(E, \tilde{F}(M_{\mathbb{T}}))$ corresponding to a generalized element $x : E \to F(c)$ can be described as follows: for any sort A of Σ, $(z_x)_A : cA \to \mathrm{Hom}_{\mathcal{E}}(E, F(F_A))$ assigns to any element $y \in cA$, corresponding to a $\mathbb{T}$-model homomorphism $s_y : F_A \to c$ via the universal property of F_A, the generalized element $F(s_y) \circ x$. Therefore, for any generalized elements $x, x' : E \to F(c)$ and any element $y \in cA$, $(z_x)_A(y) = (z_{x'})_A(y)$ if and only if $F(s_y) \circ x = F(s_y) \circ x'$.

This gives the following result:

Proposition 6.3.14 *Let $\mathbb{T}$ be a geometric theory over a signature Σ, $\mathcal{K}$ a full subcategory of f.p.$\mathbb{T}$-mod(**Set**) and $F : \mathcal{K}^{op} \to \mathcal{E}$ a flat functor with values in a Grothendieck topos $\mathcal{E}$. Suppose that for every sort A of Σ the formula $\{x^A . \top\}$ presents a $\mathbb{T}$-model F_A and that all the models F_A lie in $\mathcal{K}$. Then F satisfies condition (iii)(b)-(1) of Remark 6.3.2(f) with respect to the category $\mathcal{K}$ if and only if for every $\mathbb{T}$-model c in $\mathcal{K}$ the family of arrows $\{F(s_y) : F(c) \to F(F_A) \mid A \in \Sigma, y \in cA\}$ is jointly monic.* □

The characterization of $\tilde{F}(M_\mathbb{T})$ as an $\mathcal{E}$-filtered $\mathcal{E}$-indexed colimit established in section 5.2.3 allows us to obtain a different reformulation, in terms of the Σ-structure homomorphisms $\xi_{(c,x)} : F(c) \to \tilde{F}(M_\mathbb{T})$ defined in that context, of Theorem 6.3.13, as follows.

Theorem 6.3.15 *Let $\mathbb{T}$ be a geometric theory over a signature Σ, $\mathcal{K}$ a full subcategory of f.p.$\mathbb{T}$-mod(**Set**) and $F : \mathcal{K}^{op} \to \mathcal{E}$ a flat functor with values in a Grothendieck topos $\mathcal{E}$. Then*

(i) *F satisfies condition (iii)(b)-(1) of Remark 6.3.2(f) if and only if for any $\mathbb{T}$-model c in $\mathcal{K}$ and any generalized elements $x, x' : E \to F(c)$, the Σ-structure homomorphisms $\xi_{(c,x)}$ and $\xi_{(c,x')}$ are equal if and only if $x = x'$.*

(ii) *F satisfies condition (iii)(b)-(2) of Remark 6.3.2(f) if and only if for any $\mathbb{T}$-model c in $\mathcal{K}$, any object E of $\mathcal{E}$ and any Σ-structure homomorphism $v : c \to \text{Hom}_\mathcal{E}(E, \tilde{F}(M_\mathbb{T}))$, there exist an epimorphic family $\{e_i : E_i \to E \mid i \in I\}$ in $\mathcal{E}$ and for each $i \in I$ a generalized element $x_i : E_i \to F(c)$ such that $\text{Hom}_\mathcal{E}(e_i, \tilde{F}(M_\mathbb{T})) \circ v = \xi_{(c,x_i)}$.*

Proof

(i) The equality $\xi_{(c,x)} = \xi_{(c,x')}$ holds if and only if, for every sort A of Σ and every $y \in cA$, $\xi_{(c,x)\,A}(y) = \xi_{(c,x')\,A}(y)$. But by Proposition 5.2.11 this latter condition is satisfied if and only if there exist an epimorphic family $\{e_i : E_i \to E \mid i \in I\}$ in $\mathcal{E}$ and for each $i \in I$ a $\mathbb{T}$-model a_i in $\mathcal{K}$, a generalized element $h_i : E_i \to F(a_i)$ and two $\mathbb{T}$-model homomorphisms $f_i, f'_i : c \to a_i$ such that $(f_i)_A(y) = (f'_i)_A(y)$ and $< F(f_i), F(f'_i) > \circ h_i = < x, x' > \circ e_i$. Our thesis thus follows from Theorem 6.3.13(i).

(ii) By Theorem 6.3.13(ii), it suffices to notice that, for every sort A of Σ and every element $y \in cA$, the condition '$\kappa^F_{(c,y)} \circ x_i = v_A(y) \circ e_i$' can be reformulated as the requirement that $\text{Hom}_\mathcal{E}(e_i, \tilde{F}(M_\mathbb{T})) \circ v = \xi_{(c,x_i)}$ (for any $i \in I$). □

Thanks to this different reformulation of condition (iii)(b)-(1) of Remark 6.3.2(f), we can prove that if all the arrows in $\mathcal{K}$ are sortwise injective then $\mathbb{T}$ satisfies this condition.

First, we need a lemma.

Lemma 6.3.16 *Under the hypotheses of Theorem 6.3.15, if all the homomorphisms in $\mathcal{K}$ are sortwise injective then, for any pair (c, x) consisting of an object c of $\mathcal{K}$ and a generalized element $x : E \to F(c)$ such that $E \not\cong 0_{\mathcal{E}}$, the Σ-structure homomorphism $\xi_{(c,x)} : c \to \mathrm{Hom}_{\mathcal{E}}(E, \tilde{F}(M_{\mathbb{T}}))$ is sortwise injective.*

Proof For any sort A of Σ, the function $\xi_{(c,x)\,A} : cA \to \mathrm{Hom}_{\mathcal{E}}(E, MA)$ sends any element $y \in cA$ to the generalized element $\kappa_{(c,y)} \circ x$. Now, for any $y_1, y_2 \in cA$, we have, by Proposition 5.2.8, that $\kappa_{(c,y_1)} \circ x = \kappa_{(c,y_2)} \circ x$ if and only if there exist an epimorphic family $\{e_i : E_i \to E \mid i \in I\}$ in $\mathcal{E}$ and for each $i \in I$ a $\mathbb{T}$-model a_i in $\mathcal{K}$, a generalized element $h_i : E_i \to F(a_i)$ and a $\mathbb{T}$-model homomorphism $f_i : c \to a_i$ in $\mathcal{K}$ such that $(f_i)_A(y_1) = (f_i)_A(y_2)$ and $F(f_i) \circ h_i = x \circ e_i$. If $E \not\cong 0_{\mathcal{E}}$ then the set I is non-empty; given $i \in I$, we thus have that $(f_i)_A(y_1) = (f_i)_A(y_2)$, which entails $y_1 = y_2$ as f_i is sortwise injective by our hypothesis. $\square$

Corollary 6.3.17 *Let $\mathbb{T}$ be a geometric theory over a signature Σ and $\mathcal{K}$ a full subcategory of f.p.$\mathbb{T}$-mod(**Set**) whose arrows are all sortwise injective. Then $\mathbb{T}$ satisfies condition (iii)(b)-(1) of Remark 6.3.2(f) with respect to the category $\mathcal{K}$.*

Proof By Proposition 5.2.11, for any flat functor $F : \mathcal{K}^{\mathrm{op}} \to \mathcal{E}$ and any generalized elements $x : E \to F(c)$ and $x' : E \to F(c)$, the following 'joint embedding property' holds: there exist an epimorphic family $\{e_i : E_i \to E \mid i \in I\}$ in $\mathcal{E}$, where we suppose without loss of generality all the objects E_i to be non-zero, and for each $i \in I$ a $\mathbb{T}$-model c_i in $\mathcal{K}$, $\mathbb{T}$-model homomorphisms $f_i, g_i : c \to c_i$ and a generalized element $x_i : E_i \to F(c_i)$ such that $<x, x'> \circ e_i = <F(f_i), F(g_i)> \circ x_i$, $\mathrm{Hom}_{\mathcal{E}}(e_i, M) \circ \xi_{(c,x)} = \xi_{(c_i, x_i)} \circ f_i$ and $\mathrm{Hom}_{\mathcal{E}}(e_i, M) \circ \xi_{(c,x')} = \xi_{(c_i, x_i)} \circ g_i$ (for all $i \in I$).

Now, since all the arrows in $\mathcal{K}$ are sortwise injective homomorphisms, by Lemma 6.3.16 the homomorphism $\xi_{(c_i, x_i)}$ is sortwise injective (for each $i \in I$) and hence $\xi_{(c,x)} = \xi_{(c,x')}$ implies $f_i = g_i$. But then $x \circ e_i = F(f_i) \circ x_i = F(g_i) \circ x_i = x' \circ e_i$ for all $i \in I$, whence $x = x'$, as required. $\square$

We shall now proceed to identify some natural sufficient conditions for a theory $\mathbb{T}$ to satisfy condition (iii)(b)-(2) of Remark 6.3.2(f). Before doing so, we need a number of preliminary results.

Lemma 6.3.18 *Let Σ be a signature without relation symbols and $\mathbb{T}$ a geometric theory over a signature Σ' obtained from Σ by solely adding relation symbols whose interpretation in any $\mathbb{T}$-model in a Grothendieck topos is the complement of the interpretation of a geometric formula over Σ (for instance, $\mathbb{T}$ can be the injectivization of a geometric theory over Σ in the sense of Definition 7.2.10). Let $f : M \to N$ and $g : P \to N$ be homomorphisms of $\mathbb{T}$-models in a Grothendieck topos $\mathcal{E}$, and let k be an assignment to any sort A of Σ of an arrow $k_A : MA \to PA$ such that $g_A \circ k_A = f_A$. Then, if g is sortwise monic, k defines a $\mathbb{T}$-model homomorphism $M \to P$ such that $g \circ k = f$.*

Proof The fact that k preserves the interpretation of function symbols of Σ' follows from the fact that f and g do, as g is sortwise monic. It remains to prove that, for any relation symbol $R \rightarrowtail A_1 \cdots A_n$ of Σ' and any generalized element $x : E \to MA_1 \times \cdots \times MA_n$ in $\mathcal{E}$, if x factors through $MR \rightarrowtail MA_1 \times \cdots \times MA_n$ then $(k_{A_1} \times \cdots \times k_{A_n}) \circ x$ factors through $PR \rightarrowtail PA_1 \times \cdots \times PA_n$. Now, if PR is the complement of the interpretation $i_{\{\vec{x}.\phi\}} : [[\vec{x}.\phi]]_P \rightarrowtail PA_1 \times \cdots \times PA_n$ of a geometric formula $\phi(\vec{x}) = \phi(x^{A_1}, \ldots, x^{A_n})$ in the model P, this latter condition is equivalent to the requirement that the equalizer e of $(k_{A_1} \times \cdots \times k_{A_n}) \circ x$ and $i_{\{\vec{x}.\phi\}}$ be zero; but, since g is a Σ'-structure homomorphism, the subobject e is contained in the equalizer of the arrows $(g_{A_1} \times \cdots \times g_{A_n}) \circ (k_{A_1} \times \cdots \times k_{A_n}) \circ x = (f_{A_1} \times \cdots \times f_{A_n}) \circ x$ and $[[\vec{x}.\phi]]_N \rightarrowtail NA_1 \times \cdots \times NA_n$, which is zero since f is a Σ'-structure homomorphism and NR is the complement of $[[\vec{x}.\phi]]_N$. Therefore e is zero, as required. □

Lemma 6.3.19 *Let $\mathcal{E}$ be a Grothendieck topos. Then the inverse image functor $\gamma_{\mathcal{E}}^* : \mathbf{Set} \to \mathcal{E}$ of the unique geometric morphism $\gamma_{\mathcal{E}} : \mathcal{E} \to \mathbf{Set}$ is faithful if and only if $\mathcal{E}$ is non-trivial (i.e. $1_{\mathcal{E}} \not\cong 0_{\mathcal{E}}$).*

Proof It is clear that if $\mathcal{E}$ is trivial then $\gamma_{\mathcal{E}}^* : \mathbf{Set} \to \mathcal{E}$ is not faithful, it being the constant functor with value $0_{\mathcal{E}}$.

In the converse direction, suppose that $\mathcal{E}$ is non-trivial. Given two functions $f, g : A \to B$ in $\mathbf{Set}$, the arrows $\gamma_{\mathcal{E}}^*(f), \gamma_{\mathcal{E}}^*(g) : \coprod_{a \in A} 1_{\mathcal{E}} \to \coprod_{b \in B} 1_{\mathcal{E}}$ in $\mathcal{E}$ are characterized by the following identities: for any $a \in A$, $\gamma_{\mathcal{E}}^*(f) \circ s_a = t_{f(a)}$ and $\gamma_{\mathcal{E}}^*(g) \circ s_a = t_{g(a)}$, where $s_a : 1_{\mathcal{E}} \to \coprod_{a \in A} 1_{\mathcal{E}}$ and $t_b : 1_{\mathcal{E}} \to \coprod_{b \in B} 1_{\mathcal{E}}$ are respectively the a-th and b-th coproduct arrows (for any $a \in A$ and $b \in B$); so $\gamma_{\mathcal{E}}^*(f) = \gamma_{\mathcal{E}}^*(g)$ if and only if $t_{f(a)} = t_{g(a)}$ for every $a \in A$. Since distinct coproduct arrows are disjoint from each other and $\mathcal{E}$ is non-trivial, it follows that $f(a) = g(a)$ for every $a \in A$. Therefore $f = g$, as required. □

Corollary 6.3.20 *Let $\mathbb{T}$ be a geometric theory over a signature Σ, $\mathcal{E}$ a non-trivial Grothendieck topos (i.e. $1_{\mathcal{E}} \not\cong 0_{\mathcal{E}}$), M and N two $\mathbb{T}$-models in $\mathbf{Set}$ and f a function sending each sort A of Σ to a map $f_A : MA \to NA$ in such a way that the assignment $A \to \gamma_{\mathcal{E}}^*(f_A) : \gamma_{\mathcal{E}}^*(MA) \to \gamma_{\mathcal{E}}^*(NA)$ is a $\mathbb{T}$-model homomorphism in $\mathcal{E}$. Then the assignment $f : A \to f_A$ defines a $\mathbb{T}$-model homomorphism $M \to N$ in $\mathbf{Set}$.*

Proof We have to prove that f preserves the interpretation of function symbols of Σ and the satisfaction of atomic formulae over Σ. The preservation by f of the interpretation of function symbols of Σ can be expressed as the commutativity of certain squares involving finite products of sets of the form MA and NA and hence follows from that of $\gamma_{\mathcal{E}}^*(f)$ by virtue of Lemma 6.3.19. Let $R \rightarrowtail A_1 \cdots A_n$ be a relation symbol of Σ. From the fact that $\gamma_{\mathcal{E}}^*(f) : \gamma_{\mathcal{E}}^*(M) \to \gamma_{\mathcal{E}}^*(N)$ preserves the satisfaction by R it follows that, for any n-tuple $\vec{a} = (a_1, \ldots, a_n) \in MR$, the coproduct arrow: $1_{\mathcal{E}} \to \coprod_{(b_1, \ldots, b_n) \in NA_1 \times \cdots \times NA_n} 1_{\mathcal{E}}$ corresponding to the n-tuple $(f_{A_1}(a_1), \ldots, f_{A_n}(a_n))$ factors through the subobject

$$\coprod_{(b_1,\ldots,b_n)\in NR} 1_{\mathcal{E}} \rightarrowtail \coprod_{(b_1,\ldots,b_n)\in NA_1\times\cdots\times NA_n} 1_{\mathcal{E}}.$$ But this implies, since distinct coproduct arrows are disjoint from each other and the topos $\mathcal{E}$ is non-trivial, that $f(\vec{a}) \in NR$, as required. □

Corollary 6.3.21 *Let Σ be a signature without relation symbols and $\mathbb{T}$ a geometric theory over a signature Σ' obtained from Σ by solely adding relation symbols whose interpretation in any $\mathbb{T}$-model in a Grothendieck topos is the complement of the interpretation of a geometric formula over Σ. Let $\mathcal{K}$ be a full subcategory of f.p.$\mathbb{T}$-mod(Set), a and b be $\mathbb{T}$-models in $\mathcal{K}$, M be a $\mathbb{T}$-model in a Grothendieck topos $\mathcal{E}$ and $f : a \to \mathrm{Hom}_{\mathcal{E}}(E,M)$, $g : b \to \mathrm{Hom}_{\mathcal{E}}(E,M)$ be Σ'-structure homomorphisms, where E is an object of $\mathcal{E}$. Let k be an assignment sending any sort A of Σ to a function $k_A : aA \to bA$ such that $g_A \circ k_A = f_A$. Suppose that either of the following conditions is satisfied:*

(i) *There are no relation symbols in Σ' except possibly for binary relation symbols which are $\mathbb{T}$-provable complements of the equality relation on some sort of Σ, and the Σ'-structure homomorphisms f and g are sortwise monic.*

(ii) *$E \not\cong 0_{\mathcal{E}}$ and the $\mathbb{T}$-model homomorphism $\tilde{g} : \gamma^*_{\mathcal{E}/E}(b) \to M$ corresponding to g via the identification of Proposition 6.2.3 is sortwise monic.*

Then the assignment $k : A \to k_A$ defines a $\mathbb{T}$-model homomorphism $a \to b$ in Set.

Proof The assignment k preserves the interpretation of any binary relation which is $\mathbb{T}$-provably complemented to the equality relation on some sort of Σ if and only if it is monic at this sort (cf. Lemma 7.2.11 below). This holds under hypothesis (i) since f is sortwise monic and, for every sort A of Σ, $g_A \circ k_A = f_A$. On the other hand, under hypothesis (i), the fact that k preserves the interpretation of function symbols of Σ' follows from the fact that g and f do, since g is sortwise monic and $g_A \circ k_A = f_A$ for every sort A of Σ. This shows that our thesis is satisfied under hypothesis (i).

Let us now suppose that hypothesis (ii) holds. Consider the $\mathbb{T}$-model homomorphisms $\tilde{f} : \gamma^*_{\mathcal{E}/E}(a) \to M$ and $\tilde{g} : \gamma^*_{\mathcal{E}/E}(b) \to M$ corresponding to f and g via the identification of Proposition 6.2.3. Clearly, the assignment $A \to \tilde{k}_A = \gamma^*_{\mathcal{E}/E}(k_A)$ satisfies $\tilde{g}_A \circ \tilde{k}_A = \tilde{f}_A$ for all sorts A of Σ. It thus follows from Lemma 6.3.18 that $A \to \tilde{k}_A$ is a $\mathbb{T}$-model homomorphism $\gamma^*_{\mathcal{E}/E}(a) \to \gamma^*_{\mathcal{E}/E}(b)$; but this in turn implies, by Corollary 6.3.20 (notice that $E \not\cong 0_{\mathcal{E}}$ if and only if the topos $\mathcal{E}/E$ is non-trivial), that the assignment $A \to k_A$ defines a $\mathbb{T}$-model homomorphism $a \to b$, as required. □

Theorem 6.3.22 *Let Σ be a signature without relation symbols and $\mathbb{T}$ a geometric theory over a signature Σ' obtained from Σ by solely adding relation symbols whose interpretation in any $\mathbb{T}$-model in a Grothendieck topos is the complement of the interpretation of some geometric formula over Σ (for instance, $\mathbb{T}$ can be the injectivization of a geometric theory over Σ in the sense of Definition 7.2.10). Let*

Semantic criteria for a theory to be of presheaf type 237

$\mathcal{K}$ be a full subcategory of f.p.$\mathbb{T}$-mod(**Set**) whose objects are all finitely generated $\mathbb{T}$-models. Suppose that either of the following conditions is satisfied:

(i) All the arrows in $\mathcal{K}$ are sortwise injective homomorphisms and there are no relation symbols except possibly for binary relation symbols which are $\mathbb{T}$-provable complements of the equality relation on some sort of Σ.

(ii) All the $\mathbb{T}$-model homomorphisms in any Grothendieck topos are sortwise monic.

Then $\mathbb{T}$ satisfies condition (iii)(b)-(2) of Remark 6.3.2(f) with respect to the category $\mathcal{K}$.

Proof By Theorem 6.3.15, we have to verify that for any flat functor $F : \mathcal{K}^{\mathrm{op}} \to \mathcal{E}$, any $\mathbb{T}$-model c in $\mathcal{K}$, any object E of $\mathcal{E}$ and any Σ-structure homomorphism $z : c \to \mathrm{Hom}_{\mathcal{E}}(E, \tilde{F}(M_{\mathbb{T}}))$, there exist an epimorphic family $\{e_i : E_i \to E \mid i \in I\}$ in $\mathcal{E}$ and for each $i \in I$ a generalized element $x_i : E_i \to F(c)$ such that $\mathrm{Hom}_{\mathcal{E}}(e_i, \tilde{F}(M_{\mathbb{T}})) \circ z = \xi_{(c,x_i)}$.

If $E \cong 0_{\mathcal{E}}$ then the condition is trivially satisfied; indeed, one can take $I = \emptyset$. We shall therefore suppose $E \not\cong 0_{\mathcal{E}}$ or, equivalently, that the topos $\mathcal{E}/E$ is non-trivial.

In this proof, we shall suppose for simplicity that Σ is one-sorted, but all our arguments can be readily extended to the general case.

Let $\{r_1, \ldots, r_n\}$ be a set of generators for the model c. Consider their images $z(r_1), \ldots, z(r_n) : E \to \tilde{F}(M_{\mathbb{T}})$ under the homomorphism z. By Proposition 5.2.11, for any $i \in \{1, \ldots, n\}$ there exist an epimorphic family $\{e^i_j : E^i_j \to E \mid j \in I_i\}$ in $\mathcal{E}$ and for each $j \in J_i$ a $\mathbb{T}$-model a^i_j in $\mathcal{K}$, a generalized element $x^i_j : E^i_j \to F(a^i_j)$ and an element $y^i_j \in a^i_j$ such that $\xi_{(a^i_j, x^i_j)}(y^i_j) = z(r_i) \circ e^i_j$.

For any tuple $\vec{k} = (k_1, \ldots, k_n) \in J_1 \times \cdots \times J_n$, consider the iterated pullback $e_{\vec{k}} : E_{\vec{k}} := E^1_{k_1} \times_E \cdots \times_E E^n_{k_n} \to E$. The family of arrows $\{e_{\vec{k}} : E_{\vec{k}} \to E \mid \vec{k} \in J_1 \times \cdots \times J_n\}$ is clearly epimorphic. For any $i \in \{1, \ldots, n\}$ and $k_i \in J_i$, set $x^i_{\vec{k}} : E_{\vec{k}} \to F(a^i_{k_i})$ equal to the composite of the generalized element $x^i_{k_i} : E^i_{k_i} \to F(a^i_{k_i})$ with the canonical pullback arrow $p^i_{k_i} : E_{\vec{k}} \to E^i_{k_i}$. For any fixed $\vec{k} \in J_1 \times \cdots \times J_n$, by inductively applying the 'joint embedding property' of Proposition 5.2.11 we can find an epimorphic family $\{u^{\vec{k}}_l : U^{\vec{k}}_l \to E_{\vec{k}} \mid l \in L_{\vec{k}}\}$ and for each $l \in L_{\vec{k}}$ a $\mathbb{T}$-model $d^{\vec{k}}_l$ in $\mathcal{K}$, a generalized element $w^{\vec{k}}_l : U^{\vec{k}}_l \to F(d^{\vec{k}}_l)$ and arrows $f^{\vec{k}}_{i,l} : a^i_{k_i} \to d^{\vec{k}}_l$ in $\mathcal{K}$ such that $x^i_{\vec{k}} \circ u^{\vec{k}}_l = F(f^{\vec{k}}_{i,l}) \circ w^{\vec{k}}_l$. Let us prove that $z(r_i) \circ e_{\vec{k}} \circ u^{\vec{k}}_l = \xi_{(d^{\vec{k}}_l, w^{\vec{k}}_l)}(f^{\vec{k}}_{i,l}(y^i_{k_i}))$ (for any $i, \vec{k}$ and l). We have already observed that for any $i \in \{1, \ldots, n\}$, we have $z(r_i) \circ e^i_{k_i} = \xi_{(a^i_{k_i}, x^i_{k_i})}(y^i_{k_i})$. Composing both terms of this equation with $p^i_{k_i}$ and applying Lemma 5.2.10(ii) yields the equality $z(r_i) \circ e_{\vec{k}} = \xi_{(a^i_{k_i}, x^i_{\vec{k}})}(y^i_{k_i})$. On the other hand, by Lemma 5.2.10(i) we have that $\xi_{(a^i_{k_i}, x^i_{\vec{k}})}(y^i_{k_i}) \circ u^{\vec{k}}_l = \xi_{(d^{\vec{k}}_l, w^{\vec{k}}_l)}(f^{\vec{k}}_{i,l}(y^i_{k_i}))$. Our claim thus follows by combining the former identity with the one obtained by composing both sides of the latter identity with $u_l^{\vec{k}}$.

Let us now consider the Σ'-structure homomorphisms

$$\text{Hom}_{\mathcal{E}}(e_{\vec{k}} \circ u_l^{\vec{k}}, \tilde{F}(M_{\mathbb{T}})) \circ z : c \to \text{Hom}_{\mathcal{E}}(U_l^{\vec{k}}, \tilde{F}(M_{\mathbb{T}}))$$

and

$$\xi_{(d_l^{\vec{k}}, w_l^{\vec{k}})} : d_l^{\vec{k}} \to \text{Hom}_{\mathcal{E}}(U_l^{\vec{k}}, \tilde{F}(M_{\mathbb{T}})).$$

Since, under either hypothesis (i) or (ii), the arrows of the category $\mathcal{K}$ are sortwise monic, by Proposition 6.3.16 the homomorphism $\xi_{(d_l^{\vec{k}}, w_l^{\vec{k}})}$ is injective. Therefore, since the image of all the generators $r_1, \ldots, r_n$ of c under the homomorphism $\text{Hom}_{\mathcal{E}}(e_{\vec{k}} \circ u_l^{\vec{k}}, \tilde{F}(M_{\mathbb{T}})) \circ z$ is contained in the image of the homomorphism $\xi_{(d_l^{\vec{k}}, w_l^{\vec{k}})}$, an easy induction on the structure $t(r_1, \ldots, r_n)$ of the elements of c shows that the image of every element of c belongs to the image of $\xi_{(d_l^{\vec{k}}, w_l^{\vec{k}})}$; in other words, there exists a factorization $n_l^{\vec{k}} : c \to d_l^{\vec{k}}$ of $\text{Hom}_{\mathcal{E}}(e_{\vec{k}} \circ u_l^{\vec{k}}, \tilde{F}(M_{\mathbb{T}})) \circ z$ across $\xi_{(d_l^{\vec{k}}, w_l^{\vec{k}})}$. By Corollary 6.3.21, this factorization is a $\mathbb{T}$-model homomorphism. We have thus found an epimorphic family on E, namely $\{e_{\vec{k}} \circ u_l^{\vec{k}} \mid \vec{k} \in J_1 \times \cdots \times J_n, \, l \in L_{\vec{k}}\}$, and for every $\vec{k}$ and l a $\mathbb{T}$-model homomorphism $n_l^{\vec{k}} : c \to d_l^{\vec{k}}$ in $\mathcal{K}$ such that

$$\text{Hom}_{\mathcal{E}}(e_{\vec{k}} \circ u_l^{\vec{k}}, \tilde{F}(M_{\mathbb{T}})) \circ z = \xi_{(d_l^{\vec{k}}, w_l^{\vec{k}})} \circ n_l^{\vec{k}}.$$

The composites $v_l^{\vec{k}} : U_l^{\vec{k}} \to F(c)$ of the generalized elements $w_l^{\vec{k}} : U_l^{\vec{k}} \to F(d_l^{\vec{k}})$ with the arrows $F(n_l^{\vec{k}})$ thus yield generalized elements such that $\text{Hom}_{\mathcal{E}}(e_{\vec{k}} \circ u_l^{\vec{k}}, \tilde{F}(M_{\mathbb{T}})) \circ z = \xi_{(d_l^{\vec{k}}, w_l^{\vec{k}})} \circ n_l^{\vec{k}} = \xi_{(c, v_l^{\vec{k}})}$, where the last equality follows by Lemma 5.2.10(i). This completes our proof. □

Remark 6.3.23 The assumption in condition (i) of the theorem that all the arrows in $\mathcal{K}$ be sortwise monic is weaker than the requirement of condition (ii) that all the $\mathbb{T}$-model homomorphisms in any Grothendieck topos be sortwise monic. Anyway, condition (ii) is necessary for $\mathbb{T}$ to be classified by the topos $[\mathcal{K}, \mathbf{Set}]$ if all the homomorphisms in $\mathcal{K}$ are sortwise injective (cf. Corollary 7.2.9 below).

Corollary 6.3.24 *Let $\mathbb{T}$ be a geometric theory whose model homomorphisms in any Grothendieck topos are monic and which satisfies the hypotheses of Theorem 6.3.22 with respect to a full subcategory $\mathcal{K}$ of f.p.$\mathbb{T}$-mod$(\mathbf{Set})$. Then $\mathbb{T}$ is of presheaf type classified by the topos $[\mathcal{K}, \mathbf{Set}]$ if and only if it satisfies condition (i) of Theorem 6.3.1 and condition (ii)(a) of Theorem 6.3.8 with respect to the category $\mathcal{K}$.*

Proof The theory $\mathbb{T}$ satisfies condition (iii) of Theorem 6.3.1 with respect to the category $\mathcal{K}$ since condition (iii)(b)-(1) of Remark 6.3.2(f) holds by Corollary

6.3.17 and condition (iii)(b)-(2) of Remark 6.3.2(f) holds by Theorem 6.3.22. By Remark 6.3.9(a) and Theorem 6.3.4, if $\mathbb{T}$ satisfies condition (i) of Theorem 6.3.1 with respect to $\mathcal{K}$ then it satisfies condition (ii)(b) of Theorem 6.3.8 whence by Theorem 6.3.8 it satisfies condition (ii) of Theorem 6.3.1 if and only if it satisfies condition (ii)(a) of Theorem 6.3.8. □

6.3.3 Abstract reformulation

Thanks to the general theory of indexed filtered colimits developed in section 5.1, we can reformulate the conditions of Theorem 6.3.1 in more abstract, though less explicit terms, as follows.

Let $\mathbb{T}$ be a geometric theory over a signature Σ and $\mathcal{K}$ a full subcategory of f.p.$\mathbb{T}$-mod(**Set**).

Condition (i) of Theorem 6.3.1 for $\mathbb{T}$ with respect to $\mathcal{K}$ can be reformulated as the requirement that for every $\mathbb{T}$-model M in a Grothendieck topos $\mathcal{E}$, the $\mathcal{E}$-indexed category of elements $\int \overline{H_M}^{\text{f}}_{\mathcal{E}}$ of the functor H_M of Theorem 6.3.1 (or equivalently its $\mathcal{E}$-final subcategory $\int H_{M_{\mathcal{E}}}$, cf. Theorem 5.1.12) be $\mathcal{E}$-filtered.

By the discussion at the end of section 5.2.3, condition (ii) of Theorem 6.3.1 can be reformulated as the requirement that for any $\mathbb{T}$-model M in a Grothendieck topos $\mathcal{E}$ such that the functor H_M is flat, the canonical $\mathcal{E}$-indexed cocone with vertex M over the $\mathcal{E}$-indexed functor $P \circ \pi^{\text{f}}_{\overline{H_M}}$ (where P is the $\mathcal{E}$-indexed functor defined at the end of section 5.2.3), or equivalently over its restriction to the $\mathcal{E}$-strictly final subcategory $\int H_{M_{\mathcal{E}}}$ of $\int \overline{H_M}^{\text{f}}_{\mathcal{E}}$ (cf. Theorem 5.1.12), be colimiting.

Conditions (i) and (ii) of Theorem 6.3.1 can thus be interpreted by saying that every $\mathbb{T}$-model in a Grothendieck topos $\mathcal{E}$ is canonically an $\mathcal{E}$-filtered colimit of 'constant' $\mathbb{T}$-models coming from $\mathcal{K}$.

Under the assumption that conditions (i) and (ii) are satisfied, condition (iii) of Theorem 6.3.1 is equivalent to the requirement that for any $\mathbb{T}$-model c in $\mathcal{K}$ and any Grothendieck topos $\mathcal{E}$, the $\mathbb{T}$-model $\gamma^*_{\mathcal{E}}(c)$ be $\mathcal{E}$-finitely presentable (in the sense of section 6.2.3). Indeed, the fact that this condition is necessary for $\mathbb{T}$ to be of presheaf type follows from Proposition 6.2.13, while the fact that it is sufficient, together with conditions (i) and (ii) of Theorem 6.3.1, for $\mathbb{T}$ to be classified by the topos $[\mathcal{K}, \mathbf{Set}]$ can be proved by showing that it implies condition (iii)(a) of Theorem 6.3.1. In fact, this immediately follows from the fact that inverse image functors of geometric morphisms preserve internal colimits in view of the representation of every $\mathbb{T}$-model M in a Grothendieck topos $\mathcal{E}$ as an $\mathcal{E}$-filtered colimit of 'constant' $\mathbb{T}$-models coming from $\mathcal{K}$ provided by condition (ii).

We can thus conclude that $\mathbb{T}$ is of presheaf type classified by the topos $[\mathcal{K}, \mathbf{Set}]$ if and only if every $\mathbb{T}$-model in any Grothendieck topos $\mathcal{E}$ is an $\mathcal{E}$-filtered colimit of its canonical diagram made of 'constant' models coming from $\mathcal{K}$ and these models are $\mathcal{E}$-finitely presentable.

7

EXPANSIONS AND FAITHFUL INTERPRETATIONS

In this chapter we introduce the concepts of expansion and faithful interpretation of a geometric theory, establish a number of general results about them and apply them in the context of theories of presheaf type.

In section 7.1, we investigate expansions of geometric theories from the point of view of the geometric morphisms that they induce between the respective classifying toposes. In particular, we introduce the notion of localic (resp. hyperconnected) expansion, and show that it naturally corresponds to the notion of localic (resp. hyperconnected) geometric morphism at the level of classifying toposes; as a result, we obtain a syntactic description of the hyperconnected-localic factorization of a geometric morphism. Next, we address the problem of expanding a given geometric theory $\mathbb{T}$ to a theory classified by the topos $[\text{f.p.}\mathbb{T}\text{-mod}(\mathbf{Set}), \mathbf{Set}]$, and describe a general method for constructing such expansions (which we call *presheaf completions*). Along the way, we establish a new criterion for a geometric theory to be of presheaf type.

In section 7.2, we investigate to what extent the satisfaction of the conditions of Theorem 6.3.1 is preserved by faithful interpretations of geometric theories. Notable examples of faithful interpretations are given by quotients and injectivizations (the latter being theories obtained from a given geometric theory by adding, for each sort of its signature, a binary predicate which is provably a complement of the equality relation on that sort). We focus on the case of injectivizations – the case of quotients being treated in section 8.2 – providing sufficient conditions for the injectivization $\mathbb{T}_m$ of a geometric theory $\mathbb{T}$ to be of presheaf type if $\mathbb{T}$ is so. To this end, we carry out a general analysis of the relationship between finitely presentable and finitely generated models of a given geometric theory.

7.1 Expansions of geometric theories

7.1.1 *General theory*

In this section we introduce the syntactic notion of expansion of a geometric theory, and show that it corresponds in a natural way to that of geometric morphism between the respective classifying toposes. Further, we identify classes of expansions of theories whose corresponding geometric morphisms are localic (resp. hyperconnected). This development has been inspired by section 6.2 of [41], which contains a brief informal discussion of this topic.

Definition 7.1.1 Let $\mathbb{T}$ be a geometric theory over a signature Σ.

(a) A geometric *expansion* of $\mathbb{T}$ is a geometric theory obtained from $\mathbb{T}$ by adding sorts, relation or function symbols to the signature Σ and geometric axioms over the resulting extended signature; equivalently, a geometric expansion of $\mathbb{T}$ is a geometric theory $\mathbb{T}'$ over a signature containing Σ such that every axiom of $\mathbb{T}$, regarded as a geometric sequent over the signature of $\mathbb{T}'$, is provable in $\mathbb{T}'$.

(b) A geometric expansion $\mathbb{T}'$ of $\mathbb{T}$ is said to be *localic* if no new sorts are added to Σ to obtain the signature of $\mathbb{T}'$.

(c) A geometric expansion $\mathbb{T}'$ of $\mathbb{T}$ is said to be *hyperconnected* if every geometric formula over Σ' in a context entirely consisting of sorts of Σ is provably equivalent in $\mathbb{T}'$ to a geometric formula in the same context over Σ, and for any geometric sequent σ over Σ, σ is provable in $\mathbb{T}'$ if and only if it is provable in $\mathbb{T}$.

If $\mathbb{T}'$ is an expansion of a geometric theory $\mathbb{T}$ then there is a canonical morphism of sites $(\mathcal{C}_\mathbb{T}, J_\mathbb{T}) \to (\mathcal{C}_{\mathbb{T}'}, J_{\mathbb{T}'})$ inducing a geometric morphism $p_\mathbb{T}^{\mathbb{T}'} : \mathbf{Set}[\mathbb{T}'] \to \mathbf{Set}[\mathbb{T}]$.

We know from Theorem 4.2.13 that the inclusion part of the surjection-inclusion factorization of the geometric morphism $p_\mathbb{T}^{\mathbb{T}'} : \mathbf{Set}[\mathbb{T}'] \to \mathbf{Set}[\mathbb{T}]$ can be identified with the classifying topos of the quotient of $\mathbb{T}$ consisting of all the geometric sequents over Σ which are provable in the theory $\mathbb{T}'$.

Recall (cf. for instance section A4.6 of [55]) that a geometric morphism $f : \mathcal{F} \to \mathcal{E}$ is *localic* if every object of $\mathcal{F}$ is a subquotient (i.e. a quotient of a subobject) of an object of the form $f^*(a)$, where a is an object of $\mathcal{E}$; the morphism f is *hyperconnected* if f^* is full and faithful and its image is closed under subobjects in $\mathcal{F}$.

The following technical lemma will be useful in the sequel.

Lemma 7.1.2 *Let $f : \mathcal{F} \to \mathcal{E}$ be a geometric morphism between Grothendieck toposes and $\mathcal{C}$ a small, full subcategory of $\mathcal{E}$ which is separating for $\mathcal{E}$. Suppose that the following conditions are satisfied:*

(i) *$\mathcal{C}$ is closed in $\mathcal{E}$ under finite products.*

(ii) *f satisfies the property that the restriction $f^*|_\mathcal{C} : \mathcal{C} \to \mathcal{F}$ of f^* to $\mathcal{C}$ is full and faithful.*

(iii) *For any family of arrows $\mathcal{T}$ in $\mathcal{C}$ with common codomain, if the image of $\mathcal{T}$ under f^* is epimorphic in $\mathcal{F}$ then $\mathcal{T}$ is epimorphic in $\mathcal{E}$.*

(iv) *Every subobject in $\mathcal{F}$ of an object of the form $f^*(c)$, where c is an object of $\mathcal{C}$, is, up to isomorphism, of the form $f^*(m)$ (where m is a subobject of c in $\mathcal{E}$).*

Then f is hyperconnected.

Proof We have to prove that, under the specified hypotheses, $f^* : \mathcal{E} \to \mathcal{F}$ is full and faithful and its image is closed under subobjects in $\mathcal{F}$.

To prove that f^* is faithful, we have to verify that for any arrows $h, k : u \to v$ in $\mathcal{E}$, $f^*(h) = f^*(k)$ implies $h = k$. Since $\mathcal{C}$ is separating for $\mathcal{E}$, we can suppose without loss of generality that u lies in $\mathcal{C}$. Now, $h = k$ if and only if the equalizer $z : w \rightarrowtail u$ of h and k is an isomorphism. The family of arrows from objects of $\mathcal{C}$ to w is epimorphic, so, as $f^*(z)$ is an isomorphism (since $f^*(h) = f^*(k)$), the family formed by the composition of these arrows with z is epimorphic on u; indeed, the members of this latter family all lie in $\mathcal{C}$ and the image of this family under f^* is epimorphic. It follows that z is an epimorphism, equivalently an isomorphism, that is, $h = k$, as required.

To prove the fullness of f^*, we have to verify that for any objects a and b of $\mathcal{E}$ and any arrow $s : f^*(a) \to f^*(b)$ in $\mathcal{F}$, there exists a (unique) arrow $t : a \to b$ in $\mathcal{E}$ such that $f^*(t) = s$. As $\mathcal{C}$ is separating for $\mathcal{E}$, there exist epimorphic families $\{f_i : c_i \to b \mid i \in I\}$ and $\{g_j : d_j \to a \mid j \in J\}$ in $\mathcal{E}$ consisting of arrows whose domains lie in $\mathcal{C}$.

For any $i \in I$ and $j \in J$, consider the following pullback square:

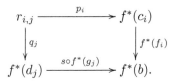

By the universal property of pullbacks, the arrow $<p_i, q_j> : r_{i,j} \to f^*(c_i) \times f^*(d_j) \cong f^*(c_i \times d_j)$ is a monomorphism. Since $c_i \times d_j$ lies in $\mathcal{C}$ by our hypothesis, $r_{i,j}$ lies in the image of f^* and hence can be covered by arrows $h_k^{i,j} : f^*(e_k^{i,j}) \to r_{i,j}$ (for $k \in K_{i,j}$) where each object $e_k^{i,j}$ lies in $\mathcal{C}$. Consider the arrows $p_i \circ h_k^{i,j} : f^*(e_k^{i,j}) \to f^*(c_i)$ and $q_j \circ h_k^{i,j} : f^*(e_k^{i,j}) \to f^*(d_j)$. By our hypotheses, there exist arrows $l_k^{i,j} : e_k^{i,j} \to c_i$ and $m_k^{i,j} : e_k^{i,j} \to d_j$ in $\mathcal{C}$ such that $f^*(l_k^{i,j}) = p_i \circ h_k^{i,j}$ and $f^*(m_k^{i,j}) = q_j \circ h_k^{i,j}$. Since the family $\{f^*(g_j \circ m_k^{i,j}) \mid i \in I, j \in J, k \in K_{i,j}\}$ is epimorphic in $\mathcal{F}$, the faithfulness of f^* ensures that the family $\{g_j \circ m_k^{i,j} : e_k^{i,j} \to a \mid i \in I, j \in J, k \in K_{i,j}\}$ is epimorphic in $\mathcal{E}$. Defining an arrow $t : a \to b$ in $\mathcal{E}$ is thus equivalent to specifying, for each $i \in I$, $j \in J$ and $k \in K_{i,j}$, an arrow $v_k^{i,j} : e_k^{i,j} \to b$ in such a way that the compatibility conditions of Corollary 5.1.7 are satisfied. Take $v_k^{i,j}$ equal to $f_i \circ l_k^{i,j}$. The compatibility conditions hold since the images of them under the functor f^* satisfy them and, as we have proved above, f^* is faithful. Thus we have a unique arrow $t : a \to b$ in $\mathcal{E}$ such that $t \circ g_j \circ m_k^{i,j} = f_i \circ l_k^{i,j}$ (for all $i \in I$, $j \in J$, $k \in K_{i,j}$). The fact that $f^*(t) = s$ follows at once.

To conclude the proof of the lemma, it remains to show that the image of f^* is closed under subobjects in $\mathcal{F}$. Let a be an object of $\mathcal{E}$ and $k : z \rightarrowtail f^*(a)$ a subobject in $\mathcal{F}$. Since $\mathcal{C}$ is separating for $\mathcal{E}$, the canonical arrow from the coproduct of all the objects in $\mathcal{C}$ to a is an epimorphism, call it y. Pulling back $f^*(y)$ along the monomorphism k, we obtain a monomorphism $u : z' \to \text{dom}(f^*(y))$ and an epimorphism $e : z' \to z$. Now, by our hypotheses, the pullbacks of u along the coproduct arrows belong to the image of f^*; from the fact that f^* preserves coproducts it thus follows that u itself belongs to the image of f^*, it being of the form $f^*(m)$ where m is a coproduct of subobjects in $\mathcal{E}$ of objects of $\mathcal{C}$. Now, since e is an epimorphism, z is isomorphic to the coequalizer of its kernel pair $r : w \rightarrowtail f^*(b) \times f^*(b) \cong f^*(b \times b)$ where $b = \mathrm{dom}(m)$. Using the coproduct representation of m and the fact that $\mathcal{C}$ is closed under finite products in $\mathcal{E}$, one can prove, by considering the pullbacks of r along the images under f^* of the induced coproduct arrows to $b \times b$, that r lies, up to isomorphism, in the image of f^*; therefore, as f^* preserves coequalizers, k is isomorphic to a subobject in the image of f^*, as required. $\square$

Theorem 7.1.3 *Let $\mathbb{T}$ be a geometric theory over a signature Σ and $\mathbb{T}'$ a geometric expansion of $\mathbb{T}$ over a signature Σ'. If $\mathbb{T}'$ is a hyperconnected (resp. localic) expansion of $\mathbb{T}$ then $p_{\mathbb{T}}^{\mathbb{T}'} : \mathbf{Set}[\mathbb{T}'] \to \mathbf{Set}[\mathbb{T}]$ is a hyperconnected (resp. localic) geometric morphism. In particular, if we identify $\mathbb{T}'$ with its (Morita-equivalent) expansion obtained by adding, for each finite string of sorts $A_1 \cdots A_n$ of Σ and each geometric formula $\phi(\vec{x}) = \phi(x_1^{A_1}, \ldots, x_n^{A_n})$ over Σ' (considered up to provable equivalence), a relation symbol $R_{\phi(\vec{x})} \rightarrowtail A_1 \cdots A_n$ and the axiom $(R_{\phi(\vec{x})} \dashv\vdash_{\vec{x}} \phi)$, the hyperconnected-localic factorization of the geometric morphism $p_{\mathbb{T}}^{\mathbb{T}'} : \mathbf{Set}[\mathbb{T}'] \to \mathbf{Set}[\mathbb{T}]$ is given by $p_{\mathbb{T}}^{\mathbb{T}''} \circ p_{\mathbb{T}''}^{\mathbb{T}'}$, where $\mathbb{T}''$ is the intermediate expansion of $\mathbb{T}$ obtained by adding to the signature Σ of $\mathbb{T}$ no new sorts and the relation symbols $R_{\phi(\vec{x})}$ and taking as axioms all the sequents over this extended signature which are provable in $\mathbb{T}'$ when each occurrence of $R_{\phi(\vec{x})}$ is replaced by the formula $\phi(\vec{x})$ which defines it.*

Proof Suppose that $\mathbb{T}'$ is a localic expansion of $\mathbb{T}$. We have to prove that every object of $\mathbf{Set}[\mathbb{T}']$ is a quotient of a subobject of an object of the form $(p_{\mathbb{T}}^{\mathbb{T}'})^*(H)$ where H is an object of $\mathbf{Set}[\mathbb{T}]$. Since the signature of $\mathbb{T}'$ does not contain any sorts not already present in the signature of $\mathbb{T}$ and the images of the objects of the category $\mathcal{C}_{\mathbb{T}'}$ under the Yoneda embedding $y_{\mathbb{T}'} : \mathcal{C}_{\mathbb{T}'} \to \mathbf{Sh}(\mathcal{C}_{\mathbb{T}'}, J_{\mathbb{T}'})$ form a separating (essentially small) family of objects for the topos $\mathbf{Set}[\mathbb{T}']$, we can conclude that the subobjects of the objects of the form $(p_{\mathbb{T}}^{\mathbb{T}'})^*(y_{\mathbb{T}}(\{\vec{x} \,.\, \top\})) \cong y_{\mathbb{T}'}(\{\vec{x} \,.\, \top\})$, where $y_{\mathbb{T}}$ is the Yoneda embedding $\mathcal{C}_{\mathbb{T}} \to \mathbf{Sh}(\mathcal{C}_{\mathbb{T}}, J_{\mathbb{T}})$, form a separating, essentially small family of objects for the topos $\mathbf{Set}[\mathbb{T}']$. Therefore any object of $\mathbf{Set}[\mathbb{T}']$ is a quotient of a coproduct of subobjects of objects of the form $(p_{\mathbb{T}}^{\mathbb{T}'})^*(y_{\mathbb{T}}(\{\vec{x} \,.\, \top\}))$; but the fact that $(p_{\mathbb{T}}^{\mathbb{T}'})^*$ preserves coproducts implies that any such coproduct is a subobject of an object in the image of $(p_{\mathbb{T}}^{\mathbb{T}'})^*$.

Suppose instead that $\mathbb{T}'$ is a hyperconnected expansion of $\mathbb{T}$. We have to prove that $(p_{\mathbb{T}}^{\mathbb{T}'})^*$ is hyperconnected. It suffices to apply Lemma 7.1.2 by taking $\mathcal{E}$ equal to $\mathbf{Set}[\mathbb{T}]$ and $\mathcal{C}$ equal to (a skeleton of) the syntactic category $\mathcal{C}_{\mathbb{T}}$; the fact that the hypotheses of the lemma are satisfied follows immediately from the fact that $\mathbb{T}'$ is hyperconnected over $\mathbb{T}$. $\square$

Remark 7.1.4 Let $\mathbb{T}'$ be a geometric expansion of a geometric theory $\mathbb{T}$. It is immediate to see that the property of this expansion to be hyperconnected is not only sufficient for the geometric morphism $p_{\mathbb{T}}^{\mathbb{T}'}$ to be hyperconnected (by Theorem 7.1.3), but also necessary. On the other hand, it is not necessary for $p_{\mathbb{T}}^{\mathbb{T}'}$ to be localic that $\mathbb{T}'$ be a localic expansion of $\mathbb{T}$; indeed, there are non-propositional theories classified by localic toposes (see, for instance, sections 3.4 and 3.5 of [74]).

The following theorem provides a converse to Theorem 7.1.3.

Theorem 7.1.5 *Let $p : \mathcal{E} \to \mathbf{Set}[\mathbb{T}]$ be a geometric morphism to the classifying topos of a geometric theory $\mathbb{T}$. Then p is, up to isomorphism, of the form $p_{\mathbb{T}}^{\mathbb{T}'}$ for some geometric expansion $\mathbb{T}'$ of $\mathbb{T}$. If p is hyperconnected (resp. localic) then we can take $\mathbb{T}'$ to be a hyperconnected (resp. localic) expansion of $\mathbb{T}$.*

Proof Choose a triple $\mathcal{T} = (\mathcal{C}_{\mathrm{ob}}, \mathcal{C}_{\mathrm{arr}}, \mathcal{C}_{\mathrm{rel}})$ consisting of a set $\mathcal{C}_{\mathrm{ob}}$ of objects of $\mathcal{E}$, a set $\mathcal{C}_{\mathrm{arr}}$ of arrows in $\mathcal{E}$ from finite products of objects of $\mathcal{C}_{\mathrm{ob}}$ to objects of $\mathcal{C}_{\mathrm{ob}}$ and a set $\mathcal{C}_{\mathrm{rel}}$ of subobjects of finite products of objects of $\mathcal{C}_{\mathrm{ob}}$ with the property that the family of objects of $\mathcal{E}$ which can be built up from objects in $\mathcal{C}_{\mathrm{ob}}$, arrows in $\mathcal{C}_{\mathrm{arr}}$ and subobjects in $\mathcal{C}_{\mathrm{rel}}$ by using geometric logic constructions is separating for $\mathcal{E}$. By definition of a Grothendieck topos, such a triple always exists. Let us define an expansion $\mathbb{T}_{\mathcal{T}}$ of $\mathbb{T}$ as follows. The signature $\Sigma_{\mathbb{T}_{\mathcal{T}}}$ of $\mathbb{T}_{\mathcal{T}}$ is obtained by adding to the signature of $\mathbb{T}$ one sort $\ulcorner c \urcorner$ for each object c in $\mathcal{C}_{\mathrm{ob}}$ (or, more economically, for each c which is not, up to isomorphism, of the form $f^*(H)$ for some object H of $\mathcal{C}_{\mathbb{T}} \hookrightarrow \mathbf{Set}[\mathbb{T}]$), one function symbol $\ulcorner f \urcorner$ for each arrow f in $\mathcal{C}_{\mathrm{arr}}$ (or, more economically, for each f whose domain or codomain is not, up to isomorphism, of the form $f^*(H)$ for some object H of $\mathcal{C}_{\mathbb{T}} \hookrightarrow \mathbf{Set}[\mathbb{T}]$), whose sorts are the obvious ones, one relation symbol for each subobject in $\mathcal{C}_{\mathrm{rel}}$ (whose sorts are the obvious ones, those corresponding to an object of the form $f^*(\{\vec{x} . \phi\})$ being the sorts of the variables in $\vec{x}$) and an additional relation symbol $\ulcorner R \urcorner$ for any subobject $R \rightarrowtail c_1 \times \cdots \times c_n$ in $\mathcal{E}$ where $c_1, \ldots, c_n$ are objects in $\mathcal{C}_{\mathrm{ob}}$ (or, more economically, for each R which cannot be obtained, up to isomorphism, from the data in $\mathcal{T}$ by means of geometric logic constructions), whose sorts are the obvious ones.

Consider the tautological $\Sigma_{\mathbb{T}_{\mathcal{T}}}$-structure M in $\mathcal{E}$ obtained by interpreting each sort $\ulcorner c \urcorner$ by the corresponding object c (and similarly for the function and relation symbols added to the signature of $\mathbb{T}$), and each sort A of the signature of $\mathbb{T}$ by the object $f^*(\{x^A . \top\})$ (and similarly for the function and relation symbols of the signature of $\mathbb{T}$). Define $\mathbb{T}_{\mathcal{T}}$ to be the theory $\mathrm{Th}(M)$ of M over the signature $\Sigma_{\mathbb{T}_{\mathcal{T}}}$ (in the sense of Definition 4.2.12). The theory $\mathbb{T}' := \mathbb{T}_{\mathcal{T}}$ satisfies the conditions of Theorem 2.1.29; the validity of the first two conditions is obvious, while the validity of the third follows from the fact that any subobject of a product of objects in $\mathcal{C}_{\mathrm{ob}}$ is definable in $\mathbb{T}'$ and the model M is conservative for $\mathbb{T}'$, whence the formula defining the graph of the given arrow is $\mathbb{T}'$-provably functional from the formula defining the domain to the formula defining the codomain. Therefore $\mathbb{T}'$ is classified by the topos $\mathcal{E}$ with universal model M. Since $\mathbb{T}'$ is an expansion of $\mathbb{T}$, this proves the first part of the theorem.

Suppose now that f is localic. We can define a triple $\mathcal{T} = (\mathcal{C}_{\text{ob}}, \mathcal{C}_{\text{arr}}, \mathcal{C}_{\text{rel}})$ satisfying the conditions specified above as follows: we set $\mathcal{C}_{\text{ob}}$ equal to the set of objects of the form $f^*(H)$ where H is an object of $\mathcal{C}_\mathbb{T} \hookrightarrow \mathbf{Set}[\mathbb{T}]$, $\mathcal{C}_{\text{arr}}$ equal to the empty set and $\mathcal{C}_{\text{rel}}$ equal to the set of all subobjects of (finite products of) objects in $\mathcal{C}_{\text{ob}}$. Alternatively, one could take $\mathcal{C}_{\text{ob}}$ to be the set of objects of the form $f^*(\{x^A . \top\})$ (where A is any sort of the signature of $\mathbb{T}$), $\mathcal{C}_{\text{arr}}$ to be the set of arrows of the form $f^*([\theta])$, where $[\theta]$ is an arrow in $\mathcal{C}_\mathbb{T}$, and $\mathcal{C}_{\text{rel}}$ to be the set of all subobjects of (finite products of) objects in $\mathcal{C}_{\text{ob}}$. In either case, the most economical version of the signature $\Sigma_{\mathbb{T}_\mathcal{T}}$ of $\mathbb{T}_\mathcal{T}$ will contain no new sorts with respect to the signature of $\mathbb{T}$ and hence $\mathbb{T}_\mathcal{T}$ will be a localic expansion of $\mathbb{T}$.

Suppose instead that f is hyperconnected. Given any set $\mathcal{K}$ of objects of $\mathcal{E}$ which, together with the objects of the form $f^*(H)$ (where H is an object of $\mathcal{C}_\mathbb{T} \hookrightarrow \mathbf{Set}[\mathbb{T}]$), is separating for $\mathcal{E}$, we can define a triple $\mathcal{T} = (\mathcal{C}_{\text{ob}}, \mathcal{C}_{\text{arr}}, \mathcal{C}_{\text{rel}})$ satisfying the required conditions by setting $\mathcal{C}_{\text{ob}} = \mathcal{K}$, $\mathcal{C}_{\text{arr}} = \emptyset$ and $\mathcal{C}_{\text{rel}} = \emptyset$. Since f is hyperconnected, the image of f^* is closed under subobjects; hence every geometric formula over the signature of $\mathbb{T}'$ in a context entirely consisting of sorts of Σ is provably equivalent in $\mathbb{T}'$ to a geometric formula in the same context over the signature of $\mathbb{T}$. On the other hand, any geometric sequent over the signature of $\mathbb{T}$ is provable in $\mathbb{T}'$ if and only if it is provable in $f^*(U_\mathbb{T})$, where $U_\mathbb{T}$ is 'the' universal model of $\mathbb{T}$ lying in $\mathbf{Set}[\mathbb{T}]$ (since f^* is full and faithful), i.e. if and only if it is provable in $\mathbb{T}$. Hence $\mathbb{T}'$ is a hyperconnected expansion of $\mathbb{T}$, as required. □

7.1.2 Another criterion for a theory to be of presheaf type

The following criterion for a geometric theory to be of presheaf type will be useful in connection with the problem of expanding a given geometric theory to a theory of presheaf type addressed in section 7.1.3.

Theorem 7.1.6 *Let $\mathbb{T}$ be a geometric theory over a signature Σ. Then $\mathbb{T}$ is of presheaf type if and only if the following conditions are satisfied:*

(i) *Every finitely presentable model is presented by a geometric formula over Σ.*

(ii) *Every property of finite tuples of elements of a finitely presentable $\mathbb{T}$-model which is preserved by $\mathbb{T}$-model homomorphisms is definable (in finitely presentable $\mathbb{T}$-models) by a geometric formula over Σ.*

(iii) *The finitely presentable $\mathbb{T}$-models are jointly conservative for $\mathbb{T}$.*

Proof The fact that every theory of presheaf type satisfies the given conditions has been established in section 6.1 (cf. Theorems 6.1.4 and 6.1.13 and Remark 6.1.2(c)). It thus remains to prove the 'if' part of the theorem.

Consider the Σ-structure U in the topos $[\text{f.p.}\mathbb{T}\text{-mod}(\mathbf{Set}), \mathbf{Set}]$ given by the forgetful functors at each sort. Clearly, U is a $\mathbb{T}$-model. To deduce our thesis, we shall verify that $\mathbb{T}$ satisfies the three conditions of Theorem 2.1.29 with respect to the model U.

Since every finitely presentable $\mathbb{T}$-model is presented by a geometric formula over Σ, the first condition of the theorem is satisfied; indeed, every representable functor $\mathrm{Hom}_{\mathrm{f.p.}\mathbb{T}\text{-mod}(\mathbf{Set})}(c, -)$ is isomorphic to the interpretation of a geometric formula $\phi(\vec{x})$ in the model U (take as $\phi(\vec{x})$ any formula presenting c). The second condition of the theorem follows immediately from the fact that the finitely presentable $\mathbb{T}$-models are jointly conservative for $\mathbb{T}$. It remains to show that the third condition of Theorem 2.1.29 is satisfied. To this end, we observe that for any geometric formulae $\phi(\vec{x})$ and $\psi(\vec{y})$ over the signature of $\mathbb{T}$, the graph of any arrow $[[\vec{x}.\phi]]_U \to [[\vec{y}.\psi]]_U$ in the topos $[\mathrm{f.p.}\mathbb{T}\text{-mod}(\mathbf{Set}), \mathbf{Set}]$ is a subobject of the product $[[\vec{x}.\phi]]_U \times [[\vec{y}.\psi]]_U$ in $[\mathrm{f.p.}\mathbb{T}\text{-mod}(\mathbf{Set}), \mathbf{Set}]$ and hence it defines a property of tuples of elements of finitely presentable models of $\mathbb{T}$ which is preserved by $\mathbb{T}$-model homomorphisms; so, by our assumptions, there exists a formula $\theta(\vec{x}, \vec{y})$ over Σ whose interpretation in U coincides with this subobject. Since U is conservative and this subobject is the graph of an arrow in the topos $[\mathrm{f.p.}\mathbb{T}\text{-mod}(\mathbf{Set}), \mathbf{Set}]$, it follows that the formula $\theta(\vec{x}, \vec{y})$ is $\mathbb{T}$-provably functional from $\phi(\vec{x})$ to $\psi(\vec{y})$, as required. $\square$

7.1.3 Expanding a geometric theory to a theory of presheaf type

In this section we discuss the problem of expanding a geometric theory $\mathbb{T}$ to a theory of presheaf type classified by the topos $[\mathrm{f.p.}\mathbb{T}\text{-mod}(\mathbf{Set}), \mathbf{Set}]$. We shall say that a geometric theory is a *presheaf completion* of a geometric theory $\mathbb{T}$ if it is an expansion of $\mathbb{T}$ such that the geometric morphism $p_{\mathbb{T}}^{\mathbb{T}'} : \mathbf{Set}[\mathbb{T}'] \to \mathbf{Set}[\mathbb{T}]$ is isomorphic to the canonical geometric morphism

$$p_{\mathrm{f.p.}\mathbb{T}\text{-mod}(\mathbf{Set})} : [\mathrm{f.p.}\mathbb{T}\text{-mod}(\mathbf{Set}), \mathbf{Set}] \to \mathbf{Set}[\mathbb{T}]$$

(cf. section 5.2.3).

Theorem 7.1.6 shows, in light of the results of section 7.1.1, that in order to obtain a presheaf completion of a given geometric theory $\mathbb{T}$ we should add a new sort $\ulcorner c \urcorner$ for each finitely presentable $\mathbb{T}$-model c which is not presented by a geometric formula over the signature of $\mathbb{T}$, and a relation symbol for each subobject of a finite product of objects of $[\mathrm{f.p.}\mathbb{T}\text{-mod}(\mathbf{Set}), \mathbf{Set}]$ which are either of the form $\mathrm{Hom}_{\mathrm{f.p.}\mathbb{T}\text{-mod}(\mathbf{Set})}(c, -)$ (where c is not presented by any geometric formula over the signature of $\mathbb{T}$), or of the form U_A for a sort A over the signature of $\mathbb{T}$ (where U_A is the forgetful functor $\mathrm{f.p.}\mathbb{T}\text{-mod}(\mathbf{Set}) \to \mathbf{Set}$ at the sort A), the sort corresponding to such a representable $\mathrm{Hom}_{\mathcal{C}}(c, -)$ being $\ulcorner c \urcorner$ and to such a functor U_A being A. Indeed, by Theorem 2.1.29, the theory of the tautological structure over this extended signature will be an expansion of $\mathbb{T}$ classified by the topos $[\mathrm{f.p.}\mathbb{T}\text{-mod}(\mathbf{Set}), \mathbf{Set}]$. Notice that if $\mathbb{T}$ satisfies the property that every finitely presentable model of $\mathbb{T}$ is presented by a geometric formula over its signature then this expansion of $\mathbb{T}$ will be localic over $\mathbb{T}$.

Of course, as is clear from the results of section 7.1.1, there are in general many ways of 'completing' a given geometric theory to a theory of presheaf type; the procedure described above represents just a particular choice which is by no means canonical. In fact, what is most interesting in practice is to obtain explicit axiomatizations of presheaf-type completions of a given theory $\mathbb{T}$ directly from the axioms of $\mathbb{T}$. Theorem 7.1.6 and the above remarks provide a useful guide in seeking such axiomatizations, as they indicate that, in order to complete a geometric theory $\mathbb{T}$ to a theory classified by the topos $[\text{f.p.}\mathbb{T}\text{-mod}(\mathbf{Set}), \mathbf{Set}]$, one should expand the language of $\mathbb{T}$ so as to make each finitely presentable $\mathbb{T}$-model presented by a formula over the extended signature and every property of finite tuples of elements of finitely presentable models of $\mathbb{T}$ which is preserved by $\mathbb{T}$-model homomorphisms definable by a geometric formula in the extended signature; we shall see concrete applications of this general method in Chapter 9. We note moreover that, if one is able to prove that every finitely presentable $\mathbb{T}$-model is strongly finitely presented as a model of the theory $\mathbb{T}$ when the latter is considered over a signature Σ' expanding that of $\mathbb{T}$, condition (iii) of Theorem 6.3.1 will automatically be satisfied by this latter theory, while conditions (i) and (ii) can be forced to hold by adding further axioms to $\mathbb{T}$ over the signature Σ' (cf. Theorem 8.2.10 below).

The following lemma shows that, under appropriate conditions, models which are finitely presented over a given signature remain finitely presented over a larger signature obtained from the former by adding relation symbols characterized by disjunctive sequents of a certain form.

Lemma 7.1.7 *Let $\mathbb{T}$ be a geometric theory over a signature Σ, Σ' a signature obtained from Σ by only adding relation symbols R and $\mathbb{S}$ a geometric theory over Σ' obtained from $\mathbb{T}$ by adding, for each such R, a pair of sequents of the form $(\top \vdash_{\vec{z}} R(\vec{z}) \vee \bigvee_{i \in I} \phi_i^R(\vec{z}))$ and $(R(\vec{z}) \wedge \bigvee_{i \in I} \phi_i^R(\vec{z}) \vdash_{\vec{z}} \bot)$, where, for each $i \in I$, $\phi_i^R(\vec{z})$ is a geometric formula-in-context over Σ such that there exists a geometric formula $\psi_i^R(\vec{z})$ over Σ with the property that the sequents $(\phi_i^R \wedge \psi_i^R \vdash_{\vec{z}} \bot)$ and $(\top \vdash_{\vec{z}} \phi_i^R \vee \psi_i^R)$ are provable in $\mathbb{S}$.*

Let $\mathbb{R}$ be the cartesianization of $\mathbb{S}$ (in the sense of Remark 8.2.2(c)) and $\phi(\vec{x}) = \phi(x_1, \ldots, x_n)$ an $\mathbb{R}$-cartesian formula over Σ' with the property that there exists an $\mathbb{R}$-model $M_{\{\vec{x}.\phi\}}$ with n generators $\vec{\xi} = (\xi_1, \ldots, \xi_n) \in [[\vec{x}.\phi]]_{M_{\{\vec{x}.\phi\}}}$ such that for any $\mathbb{R}$-model N the elements of $[[\vec{x}.\phi]]_N$ are in natural bijective correspondence with the Σ-structure homomorphisms $f : M_{\{\vec{x}.\phi\}} \to N$ such that $f(\vec{\xi}) \in [[\vec{x}.\phi]]_N$ through the assignment $f \to f(\vec{\xi})$. Then $M_{\{\vec{x}.\phi\}}$ is finitely presented by the formula $\{\vec{x}.\phi\}$ as an $\mathbb{R}$-model.

Proof We have to prove that, for any $\mathbb{R}$-model N and any tuple $\vec{a} \in [[\vec{x}.\phi]]_N$, the unique Σ-structure homomorphism $f : M_{\{\vec{x}.\phi\}} \to N$ such that $f(\vec{\xi}) = \vec{a}$ preserves the satisfaction of all the relation symbols in Σ', i.e. that for any such symbol R of arity m and any m-tuple $(y_1, \ldots, y_m)$ of elements of $M_{\{\vec{x}.\phi\}}$ belonging to $M_{\{\vec{x}.\phi\}}R$, the m-tuple $(f(y_1), \ldots, f(y_m))$ belongs to NR.

As $M_{\{\vec{x}.\phi\}}$ is by our hypothesis generated by the elements $\xi_1, \ldots, \xi_n$, each y_i is equal to the interpretation in $M_{\{\vec{x}.\phi\}}$ of a term $t_i(\xi_1, \ldots, \xi_n)$ evaluated at the tuple $\vec{\xi} = (\xi_1, \ldots, \xi_n)$. For each $i \in I$, consider the formula $\psi_i'^R(x_1, \ldots, x_n)$ obtained by making the substitution $z_i \to t_i(x_1, \ldots, x_n)$ in the formula $\psi_i^R(\vec{z})$

(for each $i = 1,\ldots, m$). Let us show that the sequent $(\phi \vdash_{\vec{x}} \psi_i'^R)$ is valid in every $\mathbb{R}$-model. This amounts to showing that for any $\mathbb{R}$-model P and any tuple $\vec{b} \in [[\vec{x}.\phi]]_P$, $\vec{b}$ satisfies the formula $\psi_i'^R$. To prove this, we observe that, since $M_{\{\vec{x}.\phi\}}$ is by our hypotheses a $\mathbb{S}$-model, the m-tuple $(y_1,\ldots, y_m)$ satisfies the formula ψ_i^R (for each $i \in I$). Now, $(y_1,\ldots, y_m) \in [[\vec{y}.\psi_i^R]]_{M_{\{\vec{x}.\phi\}}}$ means that $\vec{\xi} \in [[\vec{x}.\psi_i'^R]]_{M_{\{\vec{x}.\phi\}}}$, which implies, since by our hypotheses there exists a Σ-structure homomorphism $M_{\{\vec{x}.\phi\}} \to P$ sending $\vec{\xi}$ to $\vec{b}$, that $\vec{b} \in [[\vec{x}.\psi_i'^R]]_N$, as required. Since $\mathbb{R}$ has enough set-based models as it is cartesian, we can conclude that the sequent $(\phi \vdash_{\vec{x}} \psi_i'^R)$ is provable in $\mathbb{R}$ and hence in $\mathbb{S}$ (for each $i \in I$). It follows that the cartesian sequent $(\phi \vdash_{\vec{x}} R(t_1/z_1,\ldots, t_m/z_m))$ is provable in $\mathbb{R}$. By evaluating this sequent in the model N at the tuple $f(\vec{\xi}) \in [[\vec{x}.\phi]]_N$ and using the fact that Σ-structure homomorphisms commute with the interpretation of Σ-terms, we obtain that $(f(y_1),\ldots, f(y_m)) = ([[\vec{x}.t_1]]_N(f(\vec{\xi})),\ldots, [[\vec{x}.t_m]]_N(f(\vec{\xi}))) \in NR$, as required. □

7.1.4 Presheaf-type expansions

In this section we shall consider presheaf-type expansions of presheaf-type theories and prove, under some natural hypotheses, a theorem concerning the formulae which present the models of such theories.

Let $\mathbb{T}$ be a geometric theory over a signature Σ and $\mathbb{T}'$ a geometric expansion of $\mathbb{T}$ over a signature Σ'.

Suppose that $\mathcal{C}$ is a subcategory of the category of finitely presentable $\mathbb{T}'$-models such that the canonical functor

$$(p_{\mathbb{T}}^{\mathbb{T}'})_{\mathbf{Set}} : \mathbb{T}'\text{-mod}(\mathbf{Set}) \to \mathbb{T}\text{-mod}(\mathbf{Set})$$

induced by the geometric morphism $p_{\mathbb{T}}^{\mathbb{T}'} : \mathbf{Set}[\mathbb{T}'] \to \mathbf{Set}[\mathbb{T}]$ restricts to a functor

$$j : \mathcal{C} \to \text{f.p.}\mathbb{T}\text{-mod}(\mathbf{Set}).$$

Then we have a commutative diagram

$$\begin{array}{ccc} \mathbf{Sh}(\mathcal{C}_{\mathbb{T}'}, J_{\mathbb{T}'}) & \xrightarrow{p_{\mathbb{T}}^{\mathbb{T}'}} & \mathbf{Sh}(\mathcal{C}_{\mathbb{T}}, J_{\mathbb{T}}) \\ {\scriptstyle s_{\mathcal{C}}^{\mathbb{T}'}}\uparrow & & \uparrow {\scriptstyle t^{\mathbb{T}}} \\ [\mathcal{C}, \mathbf{Set}] & \xrightarrow{[j, \mathbf{Set}]} & [\text{f.p.}\mathbb{T}\text{-mod}(\mathbf{Set}), \mathbf{Set}], \end{array}$$

where the geometric morphisms

$$s_{\mathcal{C}}^{\mathbb{T}'} : [\mathcal{C}, \mathbf{Set}] \to \mathbf{Sh}(\mathcal{C}_{\mathbb{T}'}, J_{\mathbb{T}'})$$

and

$$t^{\mathbb{T}} : [\text{f.p.}\mathbb{T}\text{-mod}(\mathbf{Set}), \mathbf{Set}] \to \mathbf{Sh}(\mathcal{C}_{\mathbb{T}}, J_{\mathbb{T}})$$

are the canonical ones and $[j, \mathbf{Set}]$ is the geometric morphism canonically induced by the functor j.

Theorem 7.1.8 *Let $\mathbb{T}$ be a theory of presheaf type over a signature Σ and $\mathbb{T}'$ a geometric expansion of $\mathbb{T}$ over a signature Σ'. Suppose that $\mathbb{T}'$ is classified by a topos $[\mathcal{C}, \mathbf{Set}]$, where $\mathcal{C}$ is a full subcategory of f.p.$\mathbb{T}'$-mod($\mathbf{Set}$) such that the functor $(p_{\mathbb{T}}^{\mathbb{T}'})_{\mathbf{Set}} : \mathbb{T}'\text{-mod}(\mathbf{Set}) \to \mathbb{T}\text{-mod}(\mathbf{Set})$ restricts to a faithful functor $j : \mathcal{C} \to$ f.p.$\mathbb{T}$-mod($\mathbf{Set}$). Then, for any model c of $\mathbb{T}'$ in $\mathcal{C}$ whose associated $\mathbb{T}$-model $j(c)$ is presented by a geometric formula $\phi(\vec{x})$ over Σ, there exists a geometric formula $\psi(\vec{x})$ over Σ' in the context $\vec{x}$ which presents the $\mathbb{T}'$-model c and such that the sequent $(\psi \vdash_{\vec{x}} \phi)$ is provable in $\mathbb{T}'$.*

Before giving the proof of the theorem, we need to recall the following well-known lemma (of which we give a proof for the reader's convenience):

Lemma 7.1.9 *Let $R : \mathcal{A} \to \mathcal{B}$ and $L : \mathcal{B} \to \mathcal{A}$ be a pair of adjoint functors, where R is the right adjoint and L is the left adjoint. Let $\eta : 1_{\mathcal{B}} \to R \circ L$ be the unit of the adjunction and b an object of $\mathcal{B}$. Then $\eta(b)$ is monic if and only if for any arrows f, g in $\mathcal{B}$ with codomain b, $Lf = Lg$ implies $f = g$. In particular, η is pointwise monic if and only if L is faithful.*

Proof Let us denote by $\tau_{a,b}$ the bijection between the sets $\mathrm{Hom}_{\mathcal{A}}(Lb, a)$ and $\mathrm{Hom}_{\mathcal{B}}(b, Ra)$ given by the adjunction. By the naturality in b of $\tau_{a,b}$, for any arrow $h : b' \to b$ in $\mathcal{B}$ the arrow $\eta_b \circ h$ corresponds under $\tau_{Lb, b'}$ to the arrow Lh. Therefore, as $\tau_{Lb, b'}$ is a bijection, $\eta_b \circ f = \eta_b \circ g$ if and only if $Lg = Lf$. From this our thesis follows at once. $\square$

Proof (of Theorem 7.1.8) The geometric morphism

$$[j, \mathbf{Set}] : [\mathcal{C}, \mathbf{Set}] \to [\text{f.p.}\mathbb{T}\text{-mod}(\mathbf{Set}), \mathbf{Set}]$$

is essential (cf. Example 1.1.15(d)), that is, its inverse image $[j, \mathbf{Set}]^*$ admits a left adjoint $[j, \mathbf{Set}]_!$, namely the left Kan extension along the functor j, and the following diagram (where $y_{\mathcal{C}}$ and y are the Yoneda embeddings) commutes:

$$\begin{array}{ccc} [\mathcal{C}, \mathbf{Set}] & \xrightarrow{[j, \mathbf{Set}]_!} & [\text{f.p.}\mathbb{T}\text{-mod}(\mathbf{Set}), \mathbf{Set}] \\ y_{\mathcal{C}} \uparrow & & \uparrow y \\ \mathcal{C}^{\mathrm{op}} & \xrightarrow{j^{\mathrm{op}}} & \text{f.p.}\mathbb{T}\text{-mod}(\mathbf{Set})^{\mathrm{op}}. \end{array}$$

The functor

$$[j, \mathbf{Set}]_! : [\mathcal{C}, \mathbf{Set}] \to [\text{f.p.}\mathbb{T}\text{-mod}(\mathbf{Set}), \mathbf{Set}]$$

satisfies the property that for any object c of $\mathcal{C}$ and any arrows $\alpha, \beta : P \to y_{\mathcal{C}}(c)$, where P is an object of $[\mathcal{C}, \mathbf{Set}]$, $[j, \mathbf{Set}]_!(\alpha) = [j, \mathbf{Set}]_!(\beta)$ implies $\alpha = \beta$. Indeed, this is clearly true for P equal to a representable by the commutativity of the above diagram, the fullness and faithfulness of the Yoneda embeddings $y_{\mathcal{C}}$ and y' and the fact that the functor j is faithful by our hypotheses, and one can always reduce to this case by considering a covering of P in $[\mathcal{C}, \mathbf{Set}]$ by representables.

Let $y_\mathbb{T} : \mathcal{C}_\mathbb{T} \to \mathbf{Sh}(\mathcal{C}_\mathbb{T}, J_\mathbb{T})$ and $y_{\mathbb{T}'} : \mathcal{C}_{\mathbb{T}'} \to \mathbf{Sh}(\mathcal{C}_{\mathbb{T}'}, J_{\mathbb{T}'})$ be the Yoneda embeddings. By our hypotheses the geometric morphisms $s_\mathcal{C}^{\mathbb{T}'}$ and $t^\mathbb{T}$ defined above are equivalences. Therefore the geometric morphism $p_\mathbb{T}^{\mathbb{T}'}$ is essential and by Lemma 7.1.9 the unit of the adjunction between $(p_\mathbb{T}^{\mathbb{T}'})^*$ (right adjoint) and $(p_\mathbb{T}^{\mathbb{T}'})_!$ (left adjoint) is monic when evaluated at any object of the form $y_{\mathbb{T}'}(\{\vec{y}.\chi\})$, where $\chi(\vec{y})$ is a formula presenting a $\mathbb{T}'$-model in $\mathcal{C}$. Let c be a $\mathbb{T}'$-model in $\mathcal{C}$. Since $\mathbb{T}'$ is classified by the topos $[\mathcal{C}, \mathbf{Set}]$, c is finitely presented by a formula $\{\vec{y}.\chi\}$ over the signature Σ' of $\mathbb{T}'$. The commutativity of the above square and of the diagram preceding the statement of Theorem 7.1.8 thus implies that $(p_\mathbb{T}^{\mathbb{T}'})_!(y_{\mathbb{T}'}(\{\vec{y}.\chi\})) \cong y_\mathbb{T}(\{\vec{x}.\phi\})$, where $\phi(\vec{x})$ is a formula over the signature Σ which presents the model $j(c)$. By definition of the geometric morphism $p_\mathbb{T}^{\mathbb{T}'}$, we have that $(p_\mathbb{T}^{\mathbb{T}'})^*(y_\mathbb{T}(\{\vec{x}.\phi\})) = y_{\mathbb{T}'}(\{\vec{x}.\phi\})$. The unit of the adjunction between $(p_\mathbb{T}^{\mathbb{T}'})^*$ and $(p_\mathbb{T}^{\mathbb{T}'})_!$ thus yields a monic arrow

$$y_{\mathbb{T}'}(\{\vec{y}.\chi\}) \rightarrowtail p_\mathbb{T}^{\mathbb{T}'^*}(p_\mathbb{T}^{\mathbb{T}'}{}_!(y_{\mathbb{T}'}(\{\vec{y}.\chi\}))) \cong y_{\mathbb{T}'}(\{\vec{x}.\phi\})$$

in the topos $\mathbf{Sh}(\mathcal{C}_{\mathbb{T}'}, J_{\mathbb{T}'})$, in other words a monic arrow $\{\vec{y}.\chi\} \rightarrowtail \{\vec{x}.\phi\}$ in the geometric syntactic category $\mathcal{C}_{\mathbb{T}'}$. Therefore $\{\vec{y}.\chi\}$ is isomorphic, as an object of $\mathcal{C}_{\mathbb{T}'}$, to an object $\{\vec{x}.\phi'\}$ such that the sequent $(\phi' \vdash_{\vec{x}} \phi)$ is provable in $\mathbb{T}'$ (cf. Lemma 1.4.2); hence $\phi'(\vec{x})$ presents c as a $\mathbb{T}'$-model, as required. □

7.2 Faithful interpretations of theories of presheaf type

In this section we introduce the notion of faithful interpretation of geometric theories and establish sufficient criteria for theories admitting faithful interpretations of theories of presheaf type (in the sense of the definition below) to be again of presheaf type.

7.2.1 General results

Definition 7.2.1 A geometric morphism $a : \mathbf{Set}[\mathbb{T}] \to \mathbf{Set}[\mathbb{T}']$ between the classifying toposes of two geometric theories $\mathbb{T}'$ and $\mathbb{T}$ is said to be a *faithful interpretation* of $\mathbb{T}'$ into $\mathbb{T}$ if for every Grothendieck topos $\mathcal{E}$ the induced functor

$$a_\mathcal{E} : \mathbb{T}\text{-mod}(\mathcal{E}) \to \mathbb{T}'\text{-mod}(\mathcal{E})$$

between the categories of models is faithful and full on isomorphisms.

For any faithful interpretation of $\mathbb{T}'$ into $\mathbb{T}$ and any Grothendieck topos $\mathcal{E}$, we have a subcategory $\mathrm{Im}(a_\mathcal{E})$ of $\mathbb{T}'$-$\mathrm{mod}(\mathcal{E})$ whose objects (resp. arrows) are exactly the models (resp. the model homomorphisms) in the image of the functor $a_\mathcal{E}$. Notice that for any functor $F : \mathcal{A} \to \mathbb{T}\text{-mod}(\mathcal{E})$, F is full and faithful if and only if its composite $a_\mathcal{E} \circ F$ with $a_\mathcal{E}$ is full and faithful as a functor $\mathcal{A} \to \mathrm{Im}(a_\mathcal{E})$.

Remarks 7.2.2 (a) Any quotient $\mathbb{S}$ of a geometric theory $\mathbb{T}$ defines a faithful interpretation of $\mathbb{T}$ into $\mathbb{S}$.

(b) The *injectivization* $\mathbb{T}_m$ (in the sense of Definition 7.2.10) of a geometric theory $\mathbb{T}$ defines a faithful interpretation of $\mathbb{T}$ into $\mathbb{T}_m$.

Given a faithful interpretation of $\mathbb{T}'$ into $\mathbb{T}$ as above, we can reformulate the conditions of Theorem 6.3.1 for the theory $\mathbb{T}$ with respect to a full subcategory $\mathcal{K}$ of f.p.$\mathbb{T}$-mod(**Set**) in alternative ways, as follows.

Condition (iii)(c) of Theorem 6.3.1 for $\mathbb{T}$ with respect to $\mathcal{K}$ can be reformulated as follows: the composite functor

$$q_{\mathcal{E}} := a_{\mathcal{E}} \circ u_{\mathcal{K}} : \mathbf{Flat}(\mathcal{K}^{\mathrm{op}}, \mathcal{E}) \to \mathbb{T}'\text{-mod}(\mathcal{E})$$

is full and faithful into the image $\mathrm{Im}(a_{\mathcal{E}})$ of $a_{\mathcal{E}}$, where $u_{\mathcal{K}}$ is the functor

$$\mathbf{Flat}(\mathcal{K}^{\mathrm{op}}, \mathcal{E}) \to \mathbb{T}\text{-mod}(\mathcal{E})$$

induced by the canonical geometric morphism $p_{\mathcal{K}} : [\mathcal{K}, \mathbf{Set}] \to \mathbf{Set}[\mathbb{T}]$.

Condition (ii) of Theorem 6.3.1 for $\mathbb{T}$ with respect to $\mathcal{K}$ can be reformulated by saying that for any $\mathbb{T}$-model M in a Grothendieck topos $\mathcal{E}$, the canonical morphism

$$q_{\mathcal{E}}(H_M) \to a_{\mathcal{E}}(M)$$

is an isomorphism. Indeed, this morphism is the image under $a_{\mathcal{E}}$ of the canonical morphism $u_{\mathcal{K}}(H_M) \to M$ considered in Remark 6.3.2(c).

Suppose moreover that the functor

$$a_{\mathbf{Set}} : \mathbb{T}\text{-mod}(\mathbf{Set}) \to \mathbb{T}'\text{-mod}(\mathbf{Set})$$

restricts to a functor

$$f : \mathcal{K} \to \mathcal{H},$$

where $\mathcal{H}$ is a full subcategory of f.p.$\mathbb{T}'$-mod(**Set**). We have a commutative diagram

$$\begin{array}{ccc} [\mathcal{K}, \mathbf{Set}] & \xrightarrow{[f, \mathbf{Set}]} & [\mathcal{H}, \mathbf{Set}] \\ \downarrow{p_{\mathcal{K}}} & & \downarrow{p_{\mathcal{H}}} \\ \mathbf{Set}[\mathbb{T}] & \xrightarrow{a} & \mathbf{Set}[\mathbb{T}'], \end{array}$$

where $p_{\mathcal{K}}$ (resp. $p_{\mathcal{H}}$) is the canonical geometric morphism determined by the universal property of the classifying topos of $\mathbb{T}$ (resp. of $\mathbb{T}'$).

Also, for any $\mathbb{T}'$-model M in a Grothendieck topos $\mathcal{E}$, the functors $a_{\mathcal{E}/E}$ (for $E \in \mathcal{E}$) induce an $\mathcal{E}$-indexed functor

$$\int a : \int H_M^{\mathbb{T}} \underset{\mathcal{E}}{\longrightarrow} \int H_{a_{\mathcal{E}}(M)}^{\mathbb{T}'} \underset{\mathcal{E}}{}$$

between the $\mathcal{E}$-indexed categories of elements $\int H_M^{\mathbb{T}}{}_{\mathcal{E}}$ and $\int H_{a_{\mathcal{E}}(M)}^{\mathbb{T}'}{}_{\mathcal{E}}$ of the functors

$$H_M^{\mathbb{T}} := \mathrm{Hom}_{\mathbb{T}\text{-mod}(\mathcal{E})}^{\mathcal{E}}(\gamma_{\mathcal{E}}^*(-), M) : \mathcal{K}^{\mathrm{op}} \to \mathcal{E}$$

and

$$H_{a_{\mathcal{E}}(M)}^{\mathbb{T}'} := \mathrm{Hom}_{\mathbb{T}'\text{-mod}(\mathcal{E})}^{\mathcal{E}}(\gamma_{\mathcal{E}}^*(-), a_{\mathcal{E}}(M)) : \mathcal{H}^{\mathrm{op}} \to \mathcal{E}.$$

Since a is by our hypothesis a faithful interpretation of theories, $\int a$ is faithful at each fibre. Suppose now that $\int a$ is moreover $\mathcal{E}$-full (in the sense of Definition 5.1.8). Since for any functor F with values in a Grothendieck topos, F is flat if and only if its $\mathcal{E}$-indexed category of elements is $\mathcal{E}$-filtered, we conclude by Proposition 5.1.10 that if $\int a$ is $\mathcal{E}$-final then the theory $\mathbb{T}$ satisfies condition (i) of Theorem 6.3.1 with respect to the category $\mathcal{K}$ if $\mathbb{T}'$ does so with respect to the category $\mathcal{H}$. Moreover, Proposition 5.2.14 and Theorem 5.1.11 ensure that if $\mathbb{T}'$ satisfies condition (ii) of Theorem 6.3.1 with respect to the category $\mathcal{H}$ then $\mathbb{T}$ does so with respect to the category $\mathcal{K}$.

Summarizing, we have the following result:

Theorem 7.2.3 *Let* $a : \mathbf{Set}[\mathbb{T}] \to \mathbf{Set}[\mathbb{T}']$ *be a faithful interpretation of geometric theories and let $\mathcal{K}$ and $\mathcal{H}$ be full subcategories respectively of* f.p.$\mathbb{T}$-mod($\mathbf{Set}$) *and of* f.p.$\mathbb{T}'$-mod($\mathbf{Set}$) *such that the functor*

$$a_{\mathbf{Set}} : \mathbb{T}\text{-mod}(\mathbf{Set}) \to \mathbb{T}'\text{-mod}(\mathbf{Set})$$

restricts to a functor $\mathcal{K} \to \mathcal{H}$ *and for any* $\mathbb{T}$*-model M in a Grothendieck topos $\mathcal{E}$, the $\mathcal{E}$-indexed functor*

$$\int a : \int_{\mathcal{E}} H_M^{\mathbb{T}} \to \int_{\mathcal{E}} H_{a_{\mathcal{E}}(M)}^{\mathbb{T}'}$$

is $\mathcal{E}$-full and $\mathcal{E}$-final.

Then the theory $\mathbb{T}$ satisfies condition (i) (resp. condition (ii)) of Theorem 6.3.1 with respect to the category $\mathcal{K}$ if $\mathbb{T}'$ satisfies condition (i) (resp. condition (ii)) of Theorem 6.3.1 with respect to the category $\mathcal{H}$. □

Remark 7.2.4 If $\mathbb{T}$ is the injectivization of $\mathbb{T}'$ (in the sense of Definition 7.2.10) or if $\mathbb{T}$ is a quotient of $\mathbb{T}'$ then the $\mathcal{E}$-indexed functor $\int a$ in the statement of Theorem 7.2.3 is $\mathcal{E}$-full. Indeed, in the first case this follows from the fact that if the composite of two arrows is monic then the first one is monic, while in the second it follows from the fact that the category of models of $\mathbb{T}$ in any Grothendieck topos is a full subcategory of the category of models of $\mathbb{T}'$ in it.

The following proposition provides an explicit rephrasing of the finality condition for the $\mathcal{E}$-indexed functor $\int a$ of Theorem 7.2.3.

Proposition 7.2.5 *Let* $a : \mathbf{Set}[\mathbb{T}] \to \mathbf{Set}[\mathbb{T}']$ *be a faithful interpretation of geometric theories and let M a $\mathbb{T}$-model in a Grothendieck topos $\mathcal{E}$ such that the functor $H_{a_{\mathcal{E}}(M)}^{\mathbb{T}'}$ is flat. Let Σ' be the signature of $\mathbb{T}'$. Then the following conditions are equivalent:*

(i) The functor $\int a$ is $\mathcal{E}$-final.

(ii) For any object E of $\mathcal{E}$, $\mathbb{T}'$-model c in $\mathcal{H}$, $\mathbb{T}$-model M in $\mathcal{E}$ and $\mathbb{T}'$-model homomorphism $x : \gamma^*_{\mathcal{E}/E}(c) \to !^*_E(a_\mathcal{E}(M))$, there exist an epimorphic family $\{e_i : E_i \to E \mid i \in I\}$ in $\mathcal{E}$ and for each $i \in I$ a $\mathbb{T}'$-model homomorphism $f_i : c \to a_{\mathbf{Set}}(c_i)$, where c_i is a $\mathbb{T}$-model in $\mathcal{K}$, and a $\mathbb{T}'$-model homomorphism $x_i : \gamma^*_{\mathcal{E}/E_i}(a_{\mathbf{Set}}(c_i)) \to !^*_{E_i}(a_\mathcal{E}(M))$ such that $x_i \circ \gamma^*_{\mathcal{E}/E_i}(f_i) = e_i^*(x)$.

(iii) For any object E of $\mathcal{E}$ and any Σ'-structure homomorphism $x : c \to \mathrm{Hom}_\mathcal{E}(E, a_\mathcal{E}(M))$, where c is a $\mathbb{T}'$-model in $\mathcal{H}$, there exist an epimorphic family $\{e_i : E_i \to E \mid i \in I\}$ in $\mathcal{E}$ and for each $i \in I$ a $\mathbb{T}'$-model homomorphism $f_i : c \to a_{\mathbf{Set}}(c_i)$, where c_i is a $\mathbb{T}$-model in $\mathcal{K}$, and a Σ'-structure homomorphism $x_i : a_{\mathbf{Set}}(c_i) \to \mathrm{Hom}_\mathcal{E}(E_i, a_\mathcal{E}(M))$ such that $x_i \circ f_i = \mathrm{Hom}_\mathcal{E}(e_i, a_\mathcal{E}(M)) \circ x$.

Proof The equivalence between the first two conditions immediately follows from the fact that, H_M being flat, the $\mathcal{E}$-indexed category $\int H_{M_\mathcal{E}}$ is $\mathcal{E}$-filtered and hence Remark 5.1.9(c) applies, while the equivalence between the second and the third conditions follows from Proposition 6.2.3. $\square$

Remark 7.2.6 If $\mathcal{E} = \mathbf{Set}$ and $\mathbb{T}$ is a quotient of $\mathbb{T}'$ then condition (iii) of Proposition 7.2.5 rewrites as follows: for any $\mathbb{T}$-model M in $\mathbf{Set}$, any $\mathbb{T}'$-model c in $\mathcal{H}$ and any $\mathbb{T}'$-model homomorphism $f : c \to M$, there exist a $\mathbb{T}$-model d in $\mathcal{K}$ and $\mathbb{T}'$-model homomorphisms $g : c \to d$ and $h : d \to M$ such that $h \circ g = f$. In the particular case where $\mathbb{T}'$ is the empty theory over a signature Σ, $\mathbb{T}$ is an $\mathcal{F}$-finitary geometric theory over Σ (in the sense of Definition 3.1 [73]), $\mathcal{H}$ is the category of finite Σ-structures and $\mathcal{K}$ is the category of finite $\mathbb{T}$-models, the condition, required for every $\mathbb{T}$-model M in $\mathbf{Set}$, specializes precisely to the 'finite structure condition' of [73].

The following result shows a relation between the action on points of a morphism of classifying toposes and the related induced action on flat functors:

Proposition 7.2.7 *Let $a : \mathbf{Set}[\mathbb{T}] \to \mathbf{Set}[\mathbb{T}']$ be a faithful interpretation of geometric theories and $\mathcal{K}$ and $\mathcal{H}$ full subcategories respectively of f.p.$\mathbb{T}$-mod($\mathbf{Set}$) and of f.p.$\mathbb{T}'$-mod($\mathbf{Set}$) such that the functor*

$$a_{\mathbf{Set}} : \mathbb{T}\text{-mod}(\mathbf{Set}) \to \mathbb{T}'\text{-mod}(\mathbf{Set})$$

restricts to a functor $f : \mathcal{K} \to \mathcal{H}$. Then, for every Grothendieck topos $\mathcal{E}$, the extension functor

$$\mathbf{Flat}(\mathcal{K}^{\mathrm{op}}, \mathcal{E}) \to \mathbf{Flat}(\mathcal{H}^{\mathrm{op}}, \mathcal{E}) \to \mathbb{T}'\text{-mod}(\mathcal{E})$$

along the geometric morphism $[f, \mathbf{Set}] : [\mathcal{K}, \mathbf{Set}] \to [\mathcal{H}, \mathbf{Set}]$ (as in section 5.2.2) takes values in the subcategory $\mathrm{Im}(a_\mathcal{E})$ of $\mathbb{T}'$-mod($\mathcal{E}$).

Proof The diagram

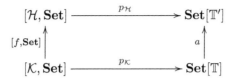

clearly commutes by definition of the functor f. Therefore, by Diaconescu's equivalence, for any Grothendieck topos $\mathcal{E}$ we have a commutative diagram

$$\begin{array}{ccc} \mathbf{Flat}(\mathcal{H}^{\mathrm{op}}, \mathcal{E}) & \xrightarrow{u^{\mathbb{T}'}_{(\mathcal{H},\mathcal{E})}} & \mathbb{T}'\text{-mod}(\mathcal{E}) \\ {\scriptstyle \xi_{\mathcal{E}}} \uparrow & & \uparrow {\scriptstyle a_{\mathcal{E}}} \\ \mathbf{Flat}(\mathcal{K}^{\mathrm{op}}, \mathcal{E}) & \xrightarrow{u^{\mathbb{T}}_{(\mathcal{K},\mathcal{E})}} & \mathbb{T}\text{-mod}(\mathcal{E}), \end{array}$$

where $u^{\mathbb{T}}_{(\mathcal{K},\mathcal{E})}$ and $u^{\mathbb{T}'}_{(\mathcal{H},\mathcal{E})}$ are the functors of section 5.2.3 and $\xi_{\mathcal{E}}$ is the extension functor induced by the geometric morphism $[f, \mathbf{Set}]$ as in section 5.2.2. From this our thesis immediately follows. □

The following result is a particular case of Proposition 7.2.7, obtained by taking $\mathbb{T}'$ to be the geometric theory of flat functors on $\mathcal{C}^{\mathrm{op}}$, $\mathbb{T}$ its injectivization (in the sense of Definition 7.2.10), $\mathcal{H}$ equal to $\mathcal{C}$ and $\mathcal{K}$ equal to $\mathcal{D}$.

Recall that an object a in a topos $\mathcal{E}$ is said to be *decidable* if the diagonal subobject $a \rightarrowtail a \times a$ has a complement in $\mathrm{Sub}_{\mathcal{E}}(a \times a)$.

Corollary 7.2.8 *Let $\mathcal{D}$ be a subcategory of a small category $\mathcal{C}$ and $\mathcal{E}$ a Grothendieck topos. If all the arrows in $\mathcal{D}$ are monic in $\mathcal{C}$ then*

(i) *For any flat functor $F : \mathcal{D}^{\mathrm{op}} \to \mathcal{E}$ and any object $d \in \mathcal{D}$, $\tilde{F}(d)$ is a decidable object of $\mathcal{E}$.*

(ii) *For any natural transformation $\alpha : F \to G$ between two flat functors $F, G : \mathcal{D}^{\mathrm{op}} \to \mathcal{E}$, $\tilde{\alpha}(c) : \tilde{F}(c) \to \tilde{G}(c)$ is monic in $\mathcal{E}$ for every $c \in \mathcal{C}$; in particular, $\alpha(d) : F(d) \to G(d)$ is monic in $\mathcal{E}$ for every $d \in \mathcal{D}$.*

□

The following corollary is obtained by applying Proposition 7.2.7 in the case when $\mathbb{T}$ is equal to the injectivization of $\mathbb{T}'$ (in the sense of Definition 7.2.10), $\mathcal{H}$ is a full subcategory of f.p.$\mathbb{T}'$-mod($\mathbf{Set}$) and $\mathcal{K} = \mathcal{H}$.

Corollary 7.2.9 *Let $\mathbb{T}$ be a theory of presheaf type classified by a topos $[\mathcal{K}, \mathbf{Set}]$, where $\mathcal{K}$ is a full subcategory of f.p.$\mathbb{T}$-mod($\mathbf{Set}$). If all the $\mathbb{T}$-model homomorphisms in $\mathcal{K}$ are sortwise injective then every $\mathbb{T}$-model homomorphism in a Grothendieck topos is sortwise monic.*

□

7.2.2 Injectivizations of theories

For any geometric theory $\mathbb{T}$, we can slightly modify its syntax so as to obtain a geometric theory whose models in **Set** are the same as those of $\mathbb{T}$ and whose model homomorphisms are precisely the sortwise injective $\mathbb{T}$-model homomorphisms. This construction is useful in many contexts. For instance, in section 8.1.1 we show that if the category f.p.$\mathbb{T}$-mod(**Set**) of finitely presentable models of a theory of presheaf type $\mathbb{T}$ satisfies the amalgamation property (i.e. the dual of the right Ore condition, cf. Definition 10.4.1) then the quotient of $\mathbb{T}$ corresponding to the subtopos **Sh**(f.p.$\mathbb{T}$-mod(**Set**)^{op}, J_{at}) of [f.p.$\mathbb{T}$-mod(**Set**), **Set**] (where J_{at} is the atomic topology) via the duality of Theorem 3.2.5 axiomatizes the homogeneous $\mathbb{T}$-models (in the sense of section 8.1.1) in any Grothendieck topos. Now, the notion of homogeneous $\mathbb{T}$-model, which is strictly related to the notion of weakly homogeneous model in classical Model Theory (cf. [48]), is mostly interesting when the arrows of the category $\mathbb{T}$-mod(**Set**) are all monic; indeed, as shown in section 10.5, a necessary condition for $\mathbb{T}$ to admit an associated 'concrete' Galois theory (i.e. a Galois equivalence holding at the level of sites) is that all the arrows in f.p.$\mathbb{T}$-mod(**Set**) be strict monomorphisms.

This motivates the following definition:

Definition 7.2.10 Let $\mathbb{T}$ be a geometric theory over a signature Σ. The *injectivization* $\mathbb{T}_m$ of $\mathbb{T}$ is the geometric theory obtained from $\mathbb{T}$ by adding a binary predicate $D_A \rightarrowtail A, A$ for each sort A of Σ and the coherent sequents

$$(D_A(x^A, y^A) \wedge x^A = y^A) \vdash_{x^A, y^A} \bot)$$

and

$$(\top \vdash_{x^A, y^A} D_A(x^A, y^A) \vee x^A = y^A).$$

The models of $\mathbb{T}_m$ in an arbitrary topos $\mathcal{E}$ coincide with the models M of $\mathbb{T}$ in $\mathcal{E}$ which are sortwise decidable, in the sense that for every sort A of the signature of $\mathbb{T}$ the object MA of $\mathcal{E}$ is *decidable*, i.e. the diagonal subobject of MA is complemented (by the interpretation of D_A in M).

As shown by the following lemma, the arrows $M \to N$ in the category $\mathbb{T}_m$-mod(**Set**) are precisely the $\mathbb{T}$-model homomorphisms $f : M \to N$ such that for every sort A over the signature of $\mathbb{T}$, the arrow $f_A : MA \to NA$ is a monomorphism in $\mathcal{E}$:

Lemma 7.2.11 *Let A and B be decidable objects in a topos $\mathcal{E}$ and $f : A \to B$ an arrow in $\mathcal{E}$. Let $D_A \rightarrowtail A \times A$ and $D_B \rightarrowtail B \times B$ be respectively the complements of the diagonal subobjects $\delta_A : A \rightarrowtail A \times A$ and $\delta_B : B \rightarrowtail B \times B$. Then f is a monomorphism if and only if the arrow $f \times f : A \times A \to B \times B$ restricts to an arrow $D_A \to D_B$.*

Proof It is immediate to see that $f : A \to B$ is a monomorphism if and only if the diagram

$$\begin{array}{ccc} A & \xrightarrow{f} & B \\ \downarrow{\delta_A} & & \downarrow{\delta_B} \\ A \times A & \xrightarrow{f \times f} & B \times B \end{array}$$

is a pullback square.

Since pullback functors preserve arbitrary unions and intersections of subobjects in a topos (as they have both a left and a right adjoint), we have that $(f \times f)^*(D_B) \cong \neg(f \times f)^*(\delta_B)$. Now, $f \times f : A \times A \to B \times B$ restricts to an arrow $D_A \to D_B$ if and only if $D_A \leq (f \times f)^*(D_B)$. But this condition holds if and only if $D_A \cap (f \times f)^*(\delta_B) \cong 0$, i.e. if and only if $(f \times f)^*(\delta_B) \leq \delta_A$, which is equivalent to the condition $(f \times f)^*(\delta_B) \cong \delta_A$ (as $\delta_A \leq (f \times f)^*(\delta_B)$). □

Several injectivizations of theories of presheaf type have been considered in the literature, e.g. in [63] and in [55]; see also sections 10.4 and 10.5 for other examples arising in the context of topos-theoretic Galois-type equivalences.

As we shall see in section 7.2.5, under certain conditions, the injectivization of a theory of presheaf type is again of presheaf type.

Lemma 7.2.12 *Let $\{f_i : A_i \to B \mid i \in I\}$ be a family of arrows in a Grothendieck topos $\mathcal{E}$. Then the arrow $\coprod_{i \in I} f_i : \coprod_{i \in I} A_i \to B$ is monic if and only if, for every $i \in I$, f_i is monic and, for every $i, i' \in I$, either $i = i'$ or the subobjects $f_i : A_i \rightarrowtail B$ and $f'_i : A_{i'} \rightarrowtail B$ are disjoint.*

Proof The 'if' direction follows from Proposition IV 7.6 [63], while the 'only if' one follows from the fact that coproducts in a topos are always disjoint (cf. for instance Corollary IV 10.5 [63]). □

Remark 7.2.13 If in the statement of Lemma 7.2.12 the objects A_i are all equal to the terminal object $1_\mathcal{E}$ of $\mathcal{E}$ then the arrows f_i are automatically monic and any two of them are disjoint if and only if their equalizer is zero.

The following proposition is a corollary of Theorem 6.2.1:

Proposition 7.2.14 *Let $\mathbb{T}$ be a geometric theory. Then, for any $\mathbb{T}$-models M and N in a Grothendieck topos $\mathcal{E}$ which are sortwise decidable, there exists an object $\underline{\mathrm{Hom}}^{\mathcal{E}}_{\mathbb{T}_m\text{-mod}(\mathcal{E})}(M, N)$ of $\mathcal{E}$ satisfying the following universal property: for any object E of $\mathcal{E}$ we have an equivalence*

$$\mathrm{Hom}_{\mathcal{E}}(E, \underline{\mathrm{Hom}}^{\mathcal{E}}_{\mathbb{T}_m\text{-mod}(\mathcal{E})}(M, N)) \cong \mathrm{Hom}_{\mathbb{T}_m\text{-mod}(\mathcal{E}/E)}(!^*_E(M), !^*_E(N))$$

natural in $E \in \mathcal{E}$.

Remarks 7.2.15 (a) For any $\mathcal{E}$, M and N, $\underline{\mathrm{Hom}}^{\mathcal{E}}_{\mathbb{T}_m\text{-mod}(\mathcal{E})}(M, N)$ embeds canonically as a subobject of $\underline{\mathrm{Hom}}^{\mathcal{E}}_{\mathbb{T}\text{-mod}(\mathcal{E})}(M, N)$.

(b) Let $\mathbb{T}$ be a geometric theory over a signature Σ, c a finitely presentable $\mathbb{T}$-model and M a sortwise decidable model of $\mathbb{T}$ in a Grothendieck topos $\mathcal{E}$. Then the subobject

$$\mathrm{Hom}^{\mathcal{E}}_{\mathbb{T}_m\text{-mod}(\mathcal{E})}(\gamma_{\mathcal{E}}^*(c), M) \hookrightarrow \mathrm{Hom}^{\mathcal{E}}_{\mathbb{T}\text{-mod}(\mathcal{E})}(\gamma_{\mathcal{E}}^*(c), M)$$

can be identified with the interpretation of the formula

$$\chi_c(f) := \bigwedge_{\substack{A \text{ sort of } \Sigma, \\ x,y \in cA,\ x \neq y}} (\pi_A(f)(\gamma_{\mathcal{E}}^*(\overline{x})), \pi_A(f)(\gamma_{\mathcal{E}}^*(\overline{y}))) \in D_{MA},$$

written in the internal language of the topos $\mathcal{E}$, where

$$\pi_A : \mathrm{Hom}^{\mathcal{E}}_{\mathbb{T}\text{-mod}(\mathcal{E})}(M, N) \to NA^{MA}$$

is the arrow defined in Remark 6.2.2(b) and $\overline{x}, \overline{y} : 1 \to cA$ are the arrows in **Set** corresponding respectively to the elements x and y of cA. Indeed, a $\mathbb{T}$-model homomorphism $f : \gamma_{\mathcal{E}}^*(c) \to M$ is sortwise monic if and only if, for every sort A of Σ and any distinct elements $x, y \in cA$, $f_A(\gamma_{\mathcal{E}}^*(\overline{x}))$ and $f_A(\gamma_{\mathcal{E}}^*(\overline{y}))$ are disjoint, i.e. they satisfy the relation D_{MA}. It follows that if for every finitely presentable $\mathbb{T}$-model c, the formula $\chi_c(f)$ is equivalent to a geometric formula over the signature of $\mathbb{T}_m$ (e.g. when Σ only contains a finite number of sorts and for every sort A and finitely presentable $\mathbb{T}$-model c the set cA is finite), then the theory $\mathbb{T}_m$ satisfies condition (iii)(a) of Theorem 6.3.1 with respect to every finitely presentable $\mathbb{T}$-model if $\mathbb{T}$ does.

Lemma 7.2.16 *Let $\mathbb{T}$ be a geometric theory over a signature Σ, c a set-based $\mathbb{T}$-model and M a $\mathbb{T}$-model in a Grothendieck topos $\mathcal{E}$. Then any generalized element $x : E \to \mathrm{Hom}^{\mathcal{E}}_{\mathbb{T}_m\text{-mod}(\mathcal{E})}(\gamma_{\mathcal{E}}^*(c), M)$ can be identified with a Σ-structure homomorphism $\xi_x : c \to \mathrm{Hom}_{\mathcal{E}}(E, M)$ which is sortwise disjunctive in the sense that, for every sort A of Σ, the function $(\xi_x)_A : cA \to \mathrm{Hom}_{\mathcal{E}}(E, MA)$ has the property that for any distinct elements $z, w \in cA$, the arrows $(\xi_x)_A(z), (\xi_x)_A(w) : E \to MA$ have equalizer zero in $\mathcal{E}$.*

Proof The thesis follows from Proposition 6.2.3 by observing that, by Lemma 7.2.12 and Remark 7.2.13, an arrow $\tau_A : \gamma_{\mathcal{E}/E}^*(cA) \to !_E^*(MA)$ is monic in $\mathcal{E}/E$ if and only if the corresponding function $\xi_A : cA \to \mathrm{Hom}_{\mathcal{E}}(E, MA)$ satisfies the property that for any distinct elements $z, w \in cA$, the arrows $\xi_A(z) : E \to MA$ and $\xi_A(w) : E \to MA$ have equalizer zero in $\mathcal{E}$ (notice that two arrows $s, t : (a : A \to E) \to (b : B \to E)$ in $\mathcal{E}/E$ have equalizer zero in $\mathcal{E}/E$ if and only if the arrows $s : A \to B$ and $t : A \to B$ have equalizer zero in $\mathcal{E}$). $\square$

7.2.3 Finitely presentable and finitely generated models

In this section we discuss, for the purpose of establishing our main criteria for the injectivization of a theory of presheaf type to be again of presheaf type, the relationship between finitely presentable and finitely generated models of a given geometric theory.

Throughout this section, we shall assume for simplicity all our first-order signatures to be one-sorted, if not otherwise stated; it is certainly possible to extend our definitions and results to the general situation, but we shall not embark on the routine task of making this explicit.

Let Σ be a (one-sorted) first-order signature. Recall that, given a Σ-structure M and a subset $A \subseteq M$, the Σ-structure $<A>$ generated by A is the smallest Σ-substructure of M containing A; it can be concretely represented as the union $\bigcup_{n \in \mathbb{N}} A_n$, where the subsets A_n are defined inductively as follows:

$$A_0 = A \cup \{Mc \mid c \text{ constant of } \Sigma\},$$

and

$$A_{n+1} = A_n \cup \{(Mf)(\vec{a}) \mid f \text{ is an } m\text{-ary function symbol of } \Sigma \text{ for some } \\ m > 0 \text{ and } \vec{a} \text{ is an } m\text{-tuple of elements of } A_n\}.$$

A Σ-structure M is said to be *finitely generated* if there exists a finite subset $A \subseteq M$ such that $M =<A>$.

We shall say that a set-based model N of a geometric theory $\mathbb{T}$ is a *quotient* of a set-based $\mathbb{T}$-model M if there exists a surjective $\mathbb{T}$-model homomorphism $M \to N$.

Proposition 7.2.17 *Let $\mathbb{T}$ be a geometric theory over a signature Σ. Then every quotient in $\mathbb{T}$-mod(**Set**) of a finitely generated $\mathbb{T}$-model is finitely generated.*

Proof Let $q : M \to N$ be a surjective homomorphism of $\mathbb{T}$-models. Suppose that M is finitely generated, by a finite subset $A \subseteq M$. Set $B = q(A)$. This set is clearly finite, and N is equal to $<B>$. □

Proposition 7.2.18 *Let $\mathbb{T}$ be a geometric theory over a signature Σ whose axioms are all of the form $(\phi \vdash_{\vec{x}} \psi)$, where ψ is a quantifier-free geometric formula. Then every finitely presentable $\mathbb{T}$-model is finitely generated.*

Proof First, we notice that any geometric theory $\mathbb{T}$ over a signature Σ such that all its axioms are of the form $(\phi \vdash_{\vec{x}} \psi)$, where ψ is a quantifier-free geometric formula, satisfies the property that every substructure of a model of $\mathbb{T}$ is a model of $\mathbb{T}$.

Let M be a finitely presentable $\mathbb{T}$-model. We can clearly represent M as the directed union of its finitely generated substructures (equivalently, submodels). Since M is finitely presentable, the identity on M factors through one of the embeddings of a finitely generated $\mathbb{T}$-submodel of M into M; such an embedding is thus necessarily an isomorphism, whence M is finitely generated, as required. □

Recall that a category $\mathcal{A}$ is said to be *finitely accessible* if it has filtered colimits and a set P of finitely presentable objects such that every object of $\mathcal{A}$ is a filtered colimit of objects in P, equivalently if it is of the form Ind-$\mathcal{C}$ for some small category $\mathcal{C}$.

Proposition 7.2.19 *Let $\mathbb{T}$ be a geometric theory over a signature Σ whose category of set-based models is finitely accessible (for instance, a theory of presheaf type). Then every finitely generated $\mathbb{T}$-model is a quotient of a finitely presentable $\mathbb{T}$-model.*

*If moreover all the axioms of $\mathbb{T}$ are of the form $(\phi \vdash_{\vec{x}} \psi)$, where ψ is a quantifier-free geometric formula, then the finitely generated $\mathbb{T}$-models are precisely the quotients of the finitely presentable $\mathbb{T}$-models in the category $\mathbb{T}$-mod(**Set**).*

Proof As the category $\mathbb{T}$-mod(**Set**) is finitely accessible, M can be expressed as a filtered colimit $(M = \text{colim}(N_i), \{J_i : N_i \to M \mid i \in \mathcal{I}\})$ of finitely presentable $\mathbb{T}$-models N_i. Let $a_1, \ldots, a_n$ be a finite set of generators for M. As the family of arrows $\{J_i : N_i \to M \mid i \in \mathcal{I}\}$ is jointly surjective (since filtered colimits in $\mathbb{T}$-mod(**Set**) are computed sortwise as in **Set**), for any generator a_j there exist an index $k_j \in \mathcal{I}$ and an element d_j of N_{k_j} such that $J_{k_j}(d_j) = a_j$. Now, the set of generators a_j being finite and the indexing category $\mathcal{I}$ being filtered, we can suppose without loss of generality that all the k_j are equal. We thus have an index $k \in \mathcal{I}$ and a string of elements of N_k which are sent by J_k to the generators of M, whence the homomorphism $J_k : N_k \to M$ is surjective.

The second part of the proposition follows from Propositions 7.2.17 and 7.2.18, which ensure that every quotient of a finitely presentable $\mathbb{T}$-model is finitely generated. □

Proposition 7.2.20 *Let Σ be a signature with no relation symbols and $\mathbb{T}$ a geometric theory over a larger signature Σ' obtained from Σ by only adding relation symbols whose interpretation in any set-based $\mathbb{T}$-model coincides with the complement of that of a geometric formula over Σ (e.g. $\mathbb{T}$ can be the injectivization of a geometric theory over Σ). Suppose that all the $\mathbb{T}$-model homomorphisms between set-based $\mathbb{T}$-models are injective. Then every finitely generated $\mathbb{T}$-model is finitely presentable.*

Proof We observed in Lemma 6.1.12 that, given a category $\mathcal{D}$ with filtered colimits, an object M of $\mathcal{D}$ is finitely presentable in $\mathcal{D}$ if and only if, for any diagram $D : \mathcal{I} \to \mathcal{D}$ defined on a small filtered category $\mathcal{I}$, every arrow $M \to \text{colim}(D)$ factors through one of the canonical colimit arrows $J_i : D(i) \to \text{colim}(D)$ and, for any arrows $f : M \to D(i)$ and $g : M \to D(j)$ in $\mathcal{D}$ such that $J_i \circ f = J_j \circ g$, there exist $k \in \mathcal{I}$ and two arrows $s : i \to k$ and $t : j \to k$ in $\mathcal{I}$ such that $D(s) \circ f = D(t) \circ g$. Now, the latter condition is automatically satisfied if all the arrows of $\mathcal{D}$ are monic (since the category $\mathcal{I}$ is filtered); in particular, applying this criterion in the case of our theory $\mathbb{T}$ yields the following characterization of its finitely presentable models: a set-based $\mathbb{T}$-model M is finitely presentable if and only if for any filtered colimit $\text{colim}(N_i)$ of set-based $\mathbb{T}$-models with colimit arrows $J_i : N_i \to \operatorname{colim}(N_i)$, every $\mathbb{T}$-model homomorphism $f : M \to \operatorname{colim}(N_i)$ factors through at least one arrow J_i in the category $\mathbb{T}\text{-mod}(\mathbf{Set})$.

Now, if M is finitely generated by elements $a_1, \ldots, a_n$, for any $i \in \{1, \ldots, n\}$ there exist an object k_i of $\mathcal{I}$ and an element $b_i \in N_{k_i}$ such that $f(a_i) = J_i(b_i)$. As the category $\mathcal{I}$ is filtered, we can suppose without loss of generality that all the k_i are equal to each other. Therefore there exist an object $k \in \mathcal{I}$ and elements $b_1, \ldots, b_n$ of N_k such that $f(a_i) = J_k(b_i)$ for all i. The image of M under f is thus entirely contained in N_k and hence we have a function $g : M \to N_k$ such that $J_k \circ g = f$. Let us prove that g is a Σ'-structure homomorphism. The fact that g is a Σ-structure homomorphism, i.e. that it commutes with the interpretation of the function symbols of Σ, follows from the fact that f is so, by virtue of the injectivity of J_k. The fact that g preserves the satisfaction of atomic relations over Σ' follows from the fact that f is a Σ'-structure homomorphism, since the interpretation of any of these relations in a set-based $\mathbb{T}$-model is the complement of that of a geometric formula over Σ. □

A useful sufficient condition for condition (iii)(b)-(1) of Remark 6.3.2(f) to hold, which is applicable to categories $\mathcal{K}$ whose arrows are not necessarily monomorphisms, is given by the following result:

Theorem 7.2.21 *Let $\mathbb{T}$ be a geometric theory over a signature Σ and $\mathcal{K}$ a full subcategory of f.p.$\mathbb{T}$-mod(**Set**) such that every model in $\mathcal{K}$ is a quotient of a $\mathbb{T}$-model which is finitely presented by a geometric formula over Σ. Then $\mathbb{T}$ satisfies condition (iii)(b)-(1) of Remark 6.3.2(f) with respect to the category $\mathcal{K}$.*

Proof We have to show that, for any Grothendieck topos $\mathcal{E}$, the functor

$$u^{\mathbb{T}}_{(\mathcal{K},\mathcal{E})} : \mathbf{Flat}(\mathcal{K}^{\mathrm{op}}, \mathcal{E}) \to \mathbf{Flat}_{J_{\mathbb{T}}}(\mathcal{C}_{\mathbb{T}}, \mathcal{E})$$

of section 5.2.3 is faithful.

By Theorem 5.2.5, for any formula $\{\vec{x}.\phi\}$ presenting a $\mathbb{T}$-model $M_{\{\vec{x}.\phi\}}$ and any flat functor $F : \mathcal{K}^{\mathrm{op}} \to \mathcal{E}$,

$$u^{\mathbb{T}}_{(\mathcal{K},\mathcal{E})}(F)(\{\vec{x}.\phi\}) \cong F(M_{\{\vec{x}.\phi\}}).$$

Our thesis thus follows at once from Lemma 7.2.24. □

7.2.4 Further reformulations of condition (iii) of Theorem 6.3.1

In this section we present other reformulations of condition (iii) of Theorem 6.3.1 which are applicable in a variety of different contexts.

Let us work in the context of a faithful interpretation of theories as in section 7.2. Consider the functors

$$u^{\mathbb{T}}_{(\mathcal{K},\mathcal{E})} : \mathbf{Flat}(\mathcal{K}^{\mathrm{op}}, \mathcal{E}) \to \mathbb{T}\text{-mod}(\mathcal{E})$$

and

$$u^{\mathbb{T}'}_{(\mathcal{H},\mathcal{E})} : \mathbf{Flat}(\mathcal{H}^{\mathrm{op}}, \mathcal{E}) \to \mathbb{T}'\text{-mod}(\mathcal{E})$$

of section 5.2.3.

Suppose that $\mathbb{T}'$ is of presheaf type and that $\mathcal{H}$ contains a category $\mathcal{P}$ whose Cauchy completion is the category of finitely presentable $\mathbb{T}'$-models (so that $\mathbb{T}'$ is classified by the topos $[\mathcal{P}, \mathbf{Set}]$). Notice that, since the functor $f : \mathcal{K} \to \mathcal{H}$ is a restriction of the functor $a_{\mathbf{Set}}$, $\mathcal{K}$ gets identified under it with a subcategory of $\mathcal{H}$ and hence, by Remark 5.2.2(c), the extension functor

$$\xi_{\mathcal{E}} : \mathbf{Flat}(\mathcal{K}^{\mathrm{op}}, \mathcal{E}) \to \mathbf{Flat}(\mathcal{H}^{\mathrm{op}}, \mathcal{E})$$

along the geometric morphism $[f, \mathbf{Set}]$ is faithful.

From the fact that $\mathbb{T}'$ faithfully interprets in $\mathbb{T}$ (in the sense of Definition 7.2.1) it follows that the functor $u^{\mathbb{T}}_{(\mathcal{K}, \mathcal{E})}$ is faithful (resp. full and faithful) if and only if the composite functor $u^{\mathbb{T}'}_{(\mathcal{H}, \mathcal{E})} \circ \xi_{\mathcal{E}}$ is, when regarded as a functor with values in the subcategory $\mathrm{Im}(a_{\mathcal{E}})$ of $\mathbb{T}'$-$\mathrm{mod}(\mathcal{E})$. The interest of this reformulation lies in the alternative description of the functor

$$u^{\mathbb{T}'}_{(\mathcal{H}, \mathcal{E})} : \mathbf{Flat}(\mathcal{H}^{\mathrm{op}}, \mathcal{E}) \to \mathbb{T}'\text{-}\mathrm{mod}(\mathcal{E})$$

which is available under our hypotheses. Indeed, we have the following result:

Proposition 7.2.22 *Under the hypotheses specified above, the functor*

$$u^{\mathbb{T}'}_{(\mathcal{H}, \mathcal{E})} : \mathbf{Flat}(\mathcal{H}^{\mathrm{op}}, \mathcal{E}) \to \mathbb{T}'\text{-}\mathrm{mod}(\mathcal{E}) \simeq \mathbf{Flat}(\mathcal{P}^{\mathrm{op}}, \mathcal{E})$$

sends any flat functor $F : \mathcal{H}^{\mathrm{op}} \to \mathcal{E}$ to its restriction $F|_{\mathcal{P}^{\mathrm{op}}} : \mathcal{P}^{\mathrm{op}} \to \mathcal{E}$, and acts accordingly on the natural transformations.

Proof This follows immediately from Theorem 5.2.5 in view of the fact that $\mathcal{H}$ contains $\mathcal{P}$ and all the objects of $\mathcal{P}$ are finitely presentable $\mathbb{T}'$-models (since $\mathbb{T}'$ is by our hypotheses of presheaf type classified by the topos $[\mathcal{P}, \mathbf{Set}]$). □

As a corollary of the proposition, we immediately obtain the following result:

Theorem 7.2.23 *Under the hypotheses specified above*

(i) *$\mathbb{T}$ satisfies condition (iii)(b)-(1) of Remark 6.3.2(f) with respect to the category $\mathcal{K}$ if and only if, for any Grothendieck topos $\mathcal{E}$, every natural transformation between flat functors in the image of the functor $\xi_{\mathcal{E}}$ is determined by its restriction to the objects of $\mathcal{P}^{\mathrm{op}}$.*

(ii) *$\mathbb{T}$ satisfies condition (iii)(c) of Theorem 6.3.1 with respect to the category $\mathcal{K}$ if and only if, for any Grothendieck topos $\mathcal{E}$ and any $F, G \in \mathbf{Flat}(\mathcal{K}^{\mathrm{op}}, \mathcal{E})$, every natural transformation $\xi_{\mathcal{E}}(F)|_{\mathcal{P}^{\mathrm{op}}} \to \xi_{\mathcal{E}}(G)|_{\mathcal{P}^{\mathrm{op}}}$ in $\mathrm{Im}(a_{\mathcal{E}})$ is the image under $u^{\mathbb{T}'}_{(\mathcal{H}, \mathcal{E})} \circ \xi_{\mathcal{E}}$ of a unique natural transformation $F \to G$.*

□

Theorem 7.2.23 can be notably applied for investigating the satisfaction of condition (iii)(b) of Theorem 6.3.1 by the injectivization of a geometric theory $\mathbb{T}$ satisfying the hypotheses of Proposition 7.2.18, as shown by the following proposition. Before stating it, we need the following lemma:

Lemma 7.2.24 *Let $\mathcal{C}$ be a small category, $\mathcal{E}$ a Grothendieck topos and $F : \mathcal{C}^{op} \to \mathcal{E}$ a flat functor. Then for any epimorphism $f : a \to b$ in $\mathcal{C}$ the arrow $F(f) : F(b) \to F(a)$ is monic in $\mathcal{E}$.*

Proof The thesis follows at once from the fact that flat functors preserve all the finite limits which exist in the domain category, using the well-known characterization of monomorphisms in terms of pullbacks. □

Proposition 7.2.25 *Let $\mathbb{T}$ be the injectivization (in the sense of Definition 7.2.10) of a theory of presheaf type $\mathbb{T}'$ whose finitely presentable models are all finitely generated (cf. for instance Proposition 7.2.18) and whose finitely generated models are all quotients (in the sense of section 7.2.3) of finitely presentable $\mathbb{T}'$-models (cf. for instance Proposition 7.2.19). Then $\mathbb{T}$ satisfies condition (iii)(b)-(1) of Remark 6.3.2(f) with respect to the category of finitely generated $\mathbb{T}'$-models.*

Proof It suffices to apply Lemma 7.2.24 in conjunction with Theorem 7.2.23 by taking $\mathcal{K}$ equal to the category of finitely generated $\mathbb{T}'$-models and sortwise injective homomorphisms between them, $\mathcal{H}$ equal to the category of finitely generated $\mathbb{T}'$-models and homomorphisms between them and $\mathcal{P}$ equal to the category of finitely presentable $\mathbb{T}'$-models and sortwise injective homomorphisms between them. □

We shall now establish a general result about injectivizations of theories of presheaf type, namely that for any theory of presheaf type $\mathbb{T}$ such that all the monic arrows in the category f.p.$\mathbb{T}$-mod(**Set**) are sortwise injective, the injectivization of $\mathbb{T}$ satisfies condition (iii)(c) of Theorem 6.3.1 with respect to the category of finitely presentable $\mathbb{T}$-models and sortwise injective homomorphisms between them.

Before proving this theorem, we need a number of preliminary results.

Lemma 7.2.26 *Let A be an object of a Grothendieck topos $\mathcal{E}$, $r : R \rightarrowtail A$ a subobject of A in $\mathcal{E}$, $e : E \to A$ an arrow in $\mathcal{E}$ and $\{f_i : E_i \to E \mid i \in I\}$ an epimorphic family in $\mathcal{E}$. If for every $i \in I$ the arrow $e \circ f_i$ factors through r then e factors through r.*

Proof The arrow $\coprod_{i \in I} f_i : \coprod_{i \in I} E_i \to E$ induced by the universal property of the coproduct is an epimorphism, since by our hypothesis the family $\{f_i : E_i \to E \mid i \in I\}$ is epimorphic. The factorizations $b_i : E_i \to R$ of the arrows $e \circ f_i$ through r induce an arrow $b := \coprod_{i \in I} b_i : \coprod_{i \in I} E_i \to R$ such that $r \circ b = e \circ \coprod_{i \in I} f_i$. Consider the epi-mono factorization $b = h \circ k$ in $\mathcal{E}$ of the arrow b, where $k : \coprod_{i \in I} E_i \twoheadrightarrow U$

and $h : U \rightarrowtail R$, and the epi-mono factorization $e = m \circ n$ of the arrow e in $\mathcal{E}$, where $n : E \twoheadrightarrow T$ and $m : T \rightarrowtail A$. Clearly, the arrow $e \circ \coprod_{i \in I} f_i$ factors both as $m \circ (n \circ \coprod_{i \in I} f_i)$ and as $(r \circ h) \circ k$. Now, as m and $r \circ h$ are monic and $n \circ \coprod_{i \in I} f_i$ and k are epic, the uniqueness of the epi-mono factorization of a given arrow in a topos implies that there exists an isomorphism $i : T \cong U$ such that $r \circ h \circ i = m$ and $i \circ n \circ \coprod_{i \in I} f_i = k$. The arrow $h \circ i \circ n$ thus provides a factorization of e through r, as required. □

The notation employed in the statements and proofs of the following results is borrowed from section 5.2.2.

The proof of the following lemma is straightforward from the results of section 5.2.2.

Lemma 7.2.27 *Let $F : \mathcal{D}^{\mathrm{op}} \to \mathcal{E}$ be a flat functor from a subcategory $\mathcal{D}$ of a small category $\mathcal{C}$ to a Grothendieck topos $\mathcal{E}$ and $x : E \to \tilde{F}(d)$ a generalized element which factors through $\chi_d^F : F(d) \to \tilde{F}(d)$ as $\chi_d^F \circ x'$. Then the natural transformation $\alpha_x : \gamma^*_{\tilde{\mathcal{E}}/E} \circ y_\mathcal{C}(d) \to !^*_E \circ \tilde{F}$ corresponding to x is equal to $\widetilde{\alpha_{x'}}$, where $\alpha_{x'}$ is the natural transformation $\gamma^*_{\tilde{\mathcal{E}}/E} \circ y_\mathcal{D}(d) \to !^*_E \circ F$ corresponding to x'.* □

Given a small category $\mathcal{C}$, we shall write $\mathcal{C}_m$ for the category whose objects are the objects of $\mathcal{C}$ and whose arrows are the monic arrows in $\mathcal{C}$ between them. The following results concern extensions $\tilde{F}$ of flat functors F along the embedding $\mathcal{C}_m \hookrightarrow \mathcal{C}$.

Proposition 7.2.28 *Let $\mathcal{C}$ be a small category and $F : (\mathcal{C}_m)^{\mathrm{op}} \to \mathcal{E}$ a flat functor. Then, for any object $c \in \mathcal{C}$ and any generalized element $x : E \to \tilde{F}(c)$, x factors through $\chi_c^F : F(c) \to \tilde{F}(c)$ if and only if the corresponding natural transformation $\alpha_x : \gamma^*_{\tilde{\mathcal{E}}/E} \circ y_\mathcal{C}(c) \to !^*_E \circ \tilde{F}$ is pointwise monic.*

Proof The 'only if' direction follows at once from Lemma 7.2.27 and Theorem 7.2.8. To prove the 'if' one, thanks to the localization technique, we can suppose without loss of generality that $E = 1_\mathcal{E}$.

Suppose that the natural transformation $\alpha_x : \gamma^*_{\tilde{\mathcal{E}}} \circ y_\mathcal{C}(c) \to \tilde{F}$ corresponding to the generalized element $x : 1_\mathcal{E} \to \tilde{F}(c)$ is pointwise monic. We want to prove that x factors through $\chi_c^F : F(c) \to \tilde{F}(c)$. Recall that for any object d of $\mathcal{C}$, $\alpha_x(d) : \coprod_{f \in \mathrm{Hom}_\mathcal{C}(d,c)} 1_\mathcal{E} \to \tilde{F}(d)$ is defined as the arrow which sends the coproduct arrow indexed by f to the generalized element $\tilde{F}(f) \circ x : 1_\mathcal{E} \to \tilde{F}(d)$.

By Lemma 7.2.12, if α_x is pointwise monic then, for any object $d \in \mathcal{C}$ and arrows $f, g : d \to c$ in $\mathcal{C}$, either $f = g$ or the equalizer of $\tilde{F}(f) \circ x$ and $\tilde{F}(g) \circ x$ is zero. Consider the pullbacks of the jointly epimorphic arrows $\kappa_{(a,z)} : \tilde{F}(a) \to \tilde{F}(c)$ along the arrow $x : 1_\mathcal{E} \to \tilde{F}(c)$:

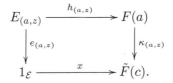

We shall prove that the epimorphic family $\{e_{(a,z)} : E_{(a,z)} \to 1_{\mathcal{E}} \mid (a,z) \in \mathcal{A}_c\}$ satisfies the condition that, for any $(a,z) \in \mathcal{A}_c$, the composite arrow $x \circ e_{(a,z)}$ factors through $\chi_c^F : F(c) \to \tilde{F}(c)$; this will imply our thesis by Lemma 7.2.26. As if $E_{(a,z)} \cong 0$ then $x \circ e_{(a,z)}$ factors through $\chi_c^F : F(c) \to \tilde{F}(c)$, it suffices to consider the case $E_{(a,z)} \not\cong 0$. Under this hypothesis, the arrow $z : c \to a$ is monic in $\mathcal{C}$. Indeed, for any two arrows $f, g : b \to c$ such that $z \circ f = z \circ g$, $\tilde{F}(f) \circ x \circ e_{(a,z)} = \tilde{F}(f) \circ \kappa_{(a,z)} \circ h_{(a,z)} = \tilde{F}(f) \circ \tilde{F}(z) \circ \chi_a^F \circ h_{(a,z)} = \tilde{F}(g) \circ \tilde{F}(z) \circ \chi_a^F \circ h_{(a,z)} = \tilde{F}(g) \circ \kappa_{(a,z)} \circ h_{(a,z)} = \tilde{F}(g) \circ x \circ e_{(a,z)}$; if the equalizer of $\tilde{F}(f) \circ x$ and $\tilde{F}(g) \circ x$ were zero then $E_{(a,z)}$ would also be isomorphic to zero (since it would admit an arrow to zero), so $f = g$. Now, z being monic, we have a factorization $\kappa_{(a,z)} = \chi_c^F \circ F(z)$ and hence $x \circ e_{(a,z)} = \kappa_{(a,z)} \circ h_{(a,z)}$ factors through χ_c^F, as required. $\square$

Remark 7.2.29 By Proposition 7.2.14, Proposition 7.2.28 can be reformulated as follows: for any object $c \in \mathcal{C}$, $F(c) \cong \text{Hom}^{\mathcal{E}}_{\mathbb{T}_{\mathcal{C}_m}\text{-mod}(\mathcal{E})}(\gamma_{\mathcal{E}}^* \circ y_{\mathcal{C}}(c), \tilde{F})$, where $\mathbb{T}_{\mathcal{C}_m}$ is the geometric theory of flat functors on $(\mathcal{C}_m)^{\text{op}}$.

For a small category $\mathcal{C}$ and a Grothendieck topos $\mathcal{E}$, let $\textbf{Flat}_m(\mathcal{C}^{\text{op}}, \mathcal{E})$ be the subcategory of $\textbf{Flat}(\mathcal{C}^{\text{op}}, \mathcal{E})$ whose objects are those of $\textbf{Flat}(\mathcal{C}^{\text{op}}, \mathcal{E})$ and whose arrows are the natural transformations between them which are pointwise monic.

Theorem 7.2.30 *Let $\mathcal{C}$ be a small category and $\mathcal{E}$ a Grothendieck topos. Then the extension functor*

$$\xi_{\mathcal{E}} : \textbf{Flat}((\mathcal{C}_m)^{\text{op}}, \mathcal{E}) \to \textbf{Flat}(\mathcal{C}^{\text{op}}, \mathcal{E})$$

along the embedding $\mathcal{C}_m \hookrightarrow \mathcal{C}$, which by Corollary 7.2.8 takes values into the subcategory $\textbf{Flat}_m(\mathcal{C}^{\text{op}}, \mathcal{E})$, is faithful and full on this latter category.

Proof We already know from Remark 5.2.2(c) that the functor $\xi_{\mathcal{E}}$ is faithful, so it remains to prove that it is full on $\textbf{Flat}_m(\mathcal{C}^{\text{op}}, \mathcal{E})$. Let F, G be flat functors $(\mathcal{C}_m)^{\text{op}} \to \mathcal{E}$, and $\beta : \tilde{F} \to \tilde{G}$ be a pointwise monic natural transformation between them. We want to prove that there exists a natural transformation $\alpha : F \to G$ such that $\beta = \tilde{\alpha}$. It suffices to show that, for any $c \in \mathcal{C}$, $\beta(c) : \tilde{F}(c) \to \tilde{G}(c)$ restricts (along the arrows $\chi_c^F : F(c) \to \tilde{F}(c)$ and $\chi_c^G : G(c) \to \tilde{G}(c)$) to an arrow $F(c) \to G(c)$. To this end, we define a function $\gamma_E : \text{Hom}_{\mathcal{E}}(E, F(c)) \to \text{Hom}_{\mathcal{E}}(E, G(c))$ natural in $E \in \mathcal{E}$. By Remark 5.2.2(c), the set $\text{Hom}_{\mathcal{E}}(E, F(c))$ (resp. the set $\text{Hom}_{\mathcal{E}}(E, G(c))$) can be identified with the set of arrows $E \to \tilde{F}(c)$ (resp. with the set of arrows $E \to \tilde{G}(c)$) which factor through χ_c^F (resp. through χ_c^G). By Proposition 7.2.28, for any flat functor $H : \mathcal{C}^{\text{op}} \to \mathcal{E}$, the arrows

$E \to \tilde{H}(c)$ which factor through χ_c^H correspond precisely to the pointwise monic natural transformations $\gamma^*_{\tilde{\mathcal{E}}/E} \circ y_\mathcal{C}(c) \to !^*_E \circ \tilde{H}$. Now, since β is pointwise monic, $!^*_E(\beta) : !^*_E \circ \tilde{F} \to !^*_E \circ \tilde{G}$ is also pointwise monic (since the functor $!^*_E$ preserves monomorphisms, it being the inverse image of a geometric morphism); hence any generalized element $E \to \tilde{F}(c)$ which factors through χ_c^F gives rise, by composition with $!^*_E(\beta)$ of the corresponding, pointwise monic, natural transformation $\gamma^*_{\tilde{\mathcal{E}}/E} \circ y_\mathcal{C}(c) \to !^*_E \circ \tilde{F}$, to a pointwise monic natural transformation $\gamma^*_{\tilde{\mathcal{E}}/E} \circ y_\mathcal{C}(c) \to !^*_E \circ \tilde{G}$, that is to a generalized element $E \to \tilde{G}(c)$ which factors through χ_c^G. But this generalized element is precisely $\beta_c \circ x$. So $\beta(c) : \tilde{F}(c) \to \tilde{G}(c)$ restricts to an arrow $F(c) \to G(c)$, as required. □

The following proposition identifies a class of theories of presheaf type $\mathbb{T}$ with the property that the monic arrows of the category f.p.$\mathbb{T}$-mod(**Set**) are sortwise injective.

Proposition 7.2.31 *Let $\mathbb{T}$ be a theory of presheaf type over a signature Σ such that, for every sort A of Σ, the formula $\{x^A . \top\}$ presents a $\mathbb{T}$-model, and let $f : M \to N$ be a homomorphism of finitely presentable $\mathbb{T}$-models M and N. Then f is monic as an arrow of* f.p.$\mathbb{T}$-mod(**Set**) *(equivalently, as an arrow of $\mathbb{T}$-mod(**Set**)) if and only if it is sortwise injective.*

Proof Let A be a sort of Σ. As $\{x^A . \top\}$ presents a $\mathbb{T}$-model F_A, for any $\mathbb{T}$-model P in **Set** we have an equivalence $\mathrm{Hom}_{\mathbb{T}\text{-mod}(\mathbf{Set})}(F_A, P) \cong PA$ natural in P. In particular, we have equivalences $\mathrm{Hom}_{\mathbb{T}\text{-mod}(\mathbf{Set})}(F_A, M) \cong MA$ and $\mathrm{Hom}_{\mathbb{T}\text{-mod}(\mathbf{Set})}(F_A, N) \cong NA$ under which the function

$$f \circ - : \mathrm{Hom}_{\mathbb{T}\text{-mod}(\mathbf{Set})}(F_A, M) \to \mathrm{Hom}_{\mathbb{T}\text{-mod}(\mathbf{Set})}(F_A, N)$$

corresponds to the function $f_A : MA \to NA$. Now, if f is monic then the function $f \circ - : \mathrm{Hom}_{\mathbb{T}\text{-mod}(\mathbf{Set})}(F_A, M) \to \mathrm{Hom}_{\mathbb{T}\text{-mod}(\mathbf{Set})}(F_A, N)$ is injective, equivalently $f_A : MA \to NA$ is injective, as required. □

Theorem 7.2.32 *Let $\mathbb{T}$ be a theory of presheaf type over a signature Σ such that every monic arrow in* f.p.$\mathbb{T}$-mod(**Set**) *is sortwise injective. Then the injectivization of $\mathbb{T}$ satisfies condition (iii)(c) of Theorem 6.3.1 with respect to the category* f.p.$\mathbb{T}$-mod(**Set**)$_m$.

Proof Since every monic arrow in f.p.$\mathbb{T}$-mod(**Set**) is sortwise injective, the category f.p.$\mathbb{T}$-mod(**Set**)$_m$ is a subcategory both of the category $\mathbb{T}_m$-mod(**Set**) and of the category f.p.$\mathbb{T}$-mod(**Set**). So, by Proposition 7.2.7, the composite functor

$$\xi_\mathcal{E} : \mathbf{Flat}(\text{f.p.}\mathbb{T}\text{-mod}(\mathbf{Set})_m^{\mathrm{op}}, \mathcal{E}) \to \mathbf{Flat}(\text{f.p.}\mathbb{T}\text{-mod}(\mathbf{Set})^{\mathrm{op}}, \mathcal{E}) \simeq \mathbb{T}\text{-mod}(\mathcal{E})$$

takes values in the subcategory $\mathbb{T}_m$-mod$(\mathcal{E})$ of $\mathbb{T}$-mod$(\mathcal{E})$. Now, $\mathbb{T}_m$ satisfies condition (iii)(c) of Theorem 6.3.1 with respect to the category f.p.$\mathbb{T}$-mod(**Set**)$_m$ if and only if the functor $\xi_{\mathcal{E}}$ is faithful and full on $\mathbb{T}_m$-mod$(\mathcal{E})$ (cf. the discussion preceding Proposition 7.2.22). But this follows from Theorem 7.2.30 in light of the fact that for any sortwise monic $\mathbb{T}$-model homomorphism $f : M \to N$ in $\mathcal{E}$, the natural transformation $\alpha_f : F_M \to F_N$ corresponding to it under the canonical Morita equivalence

$$\mathbf{Flat}(\text{f.p.}\mathbb{T}\text{-mod}(\mathbf{Set})^{\mathrm{op}}, \mathcal{E}) \simeq \mathbb{T}\text{-mod}(\mathcal{E})$$

for $\mathbb{T}$ is pointwise monic. This latter fact can be proved as follows. For any object D of the category f.p.$\mathbb{T}$-mod$(\mathbf{Set})$, the value of α_f at D can be identified with the arrow $[[\vec{x}.\phi]]_M \to [[\vec{x}.\phi]]_N$ canonically induced by f, where $\phi(\vec{x})$ is 'the' formula which presents the model D; so if f is sortwise monic then $\alpha_f(D)$ is monic in $\mathcal{E}$, it being the restriction of a monic arrow (namely $f_{A_1} \times \cdots \times f_{A_n}$, where $\vec{x} = (x^{A_1}, \ldots, x^{A_n})$) along two subobjects. □

We shall say that a geometric expansion $\mathbb{T}'$ of a geometric theory $\mathbb{T}$ is *fully faithful* if for every Grothendieck topos $\mathcal{E}$ the induced functor

$$(p_{\mathbb{T}}^{\mathbb{T}'})_{\mathcal{E}} : \mathbb{T}'\text{-mod}(\mathcal{E}) \to \mathbb{T}\text{-mod}(\mathcal{E})$$

is full and faithful.

Corollary 7.2.33 *Let $\mathbb{T}$ be a theory of presheaf type over a signature Σ. If every monic arrow in f.p.$\mathbb{T}$-mod$(\mathbf{Set})$ is sortwise injective then there is a fully faithful expansion of the injectivization of $\mathbb{T}$ which is classified by the topos $[\text{f.p.}\mathbb{T}\text{-mod}(\mathbf{Set})_m, \mathbf{Set}]$.*

Proof Consider the canonical geometric morphism

$$p_{\text{f.p.}\mathbb{T}\text{-mod}(\mathbf{Set})_m} : [\text{f.p.}\mathbb{T}\text{-mod}(\mathbf{Set})_m, \mathbf{Set}] \to \mathbf{Set}[\mathbb{T}_m]$$

(cf. section 5.2.3). By Theorem 7.1.5, this morphism is, up to equivalence, of the form $p_{\mathbb{T}_m}^{\mathbb{T}'}$ for some expansion $\mathbb{T}'$ of $\mathbb{T}_m$.

By Theorem 7.2.32, the theory $\mathbb{T}_m$ satisfies condition (iii)(c) of Theorem 6.3.1 with respect to the category f.p.$\mathbb{T}$-mod$(\mathbf{Set})_m$, whence the functor

$$(p_{\mathbb{T}_m}^{\mathbb{T}'})_{\mathcal{E}} : \mathbb{T}'\text{-mod}(\mathcal{E}) \to \mathbb{T}_m\text{-mod}(\mathcal{E})$$

is full and faithful, i.e. $\mathbb{T}'$ is a fully faithful expansion of $\mathbb{T}_m$. □

Notice that if the finitely presentable $\mathbb{T}$-models coincide with the finitely presentable $\mathbb{T}_m$-models then the expansion provided by Corollary 7.2.33 is a presheaf completion of the theory $\mathbb{T}_m$. A simple example of a theory satisfying this condition (and the hypotheses of the corollary) is the theory $\mathbb{A}$ of commutative rings with unit; indeed, the finitely presented commutative rings with unit are precisely the finitely generated ones, that is the finitely presentable models of $\mathbb{A}_m$.

7.2.5 A criterion for injectivizations

In this section we shall establish a result providing a sufficient condition for the injectivization of a theory of presheaf type of a certain form to be of presheaf type too. Before stating it, we need some preliminaries.

The following definition gives a natural topos-theoretic generalization of the standard notion of congruence on a set-based structure.

Definition 7.2.34 Let Σ be a one-sorted first-order signature and M a Σ-structure in a Grothendieck topos $\mathcal{E}$. An equivalence relation $R \rightarrowtail M \times M$ on M in $\mathcal{E}$ is said to be a *congruence* if for any function symbol f of Σ of arity n we have a commutative diagram

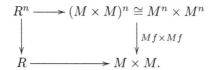

Below, for a one-sorted signature Σ, by a Σ-structure epimorphism in a topos $\mathcal{E}$ we mean a Σ-structure homomorphism whose underlying arrow is an epimorphism in $\mathcal{E}$.

Proposition 7.2.35 *Let Σ be a one-sorted first-order signature and M a Σ-structure in a Grothendieck topos $\mathcal{E}$. For any congruence R on M, there exists a Σ-structure M/R in $\mathcal{E}$ whose underlying object is the quotient in $\mathcal{E}$ of M by the relation R, and a Σ-structure epimorphism $p_R : M \to M/R$ given by the canonical projection. Conversely, for any Σ-structure epimorphism $q : M \to N$, the kernel pair R_q of q is a congruence on M such that q is isomorphic to M/R_q.*

Proof The thesis follows immediately from the exactness properties of Grothendieck toposes relating epimorphisms and equivalence relations. □

Proposition 7.2.36 *Let $\mathbb{T}$ be a geometric theory over a one-sorted signature Σ and M a $\mathbb{T}$-model in a Grothendieck topos $\mathcal{E}$. If the axioms of $\mathbb{T}$ are all of the form $(\phi \vdash_{\vec{x}} \psi)$, where ϕ is a geometric formula not containing any conjunctions, then for any congruence R on M the Σ-structure M/R is a $\mathbb{T}$-model.*

Proof The thesis can be easily proved by induction on the structure of geometric formulae over Σ, using the fact that the action of the canonical projection homomorphism $p_R : M \to M/R$ on subobjects (of powers of M) preserves the top subobject, the natural order on subobjects, image factorizations and arbitrary unions. □

The following lemma shows that one can always perform image factorizations of homomorphisms of structures in regular categories.

Lemma 7.2.37 *Let Σ be a first-order signature and $\mathcal{C}$ a regular category. Then every Σ-structure homomorphism $f : M \to N$ in $\mathcal{C}$ can be factored as $h \circ g$, where $h : N' \rightarrowtail N$ is a Σ-substructure of N and $g : M \to N'$ is sortwise a cover.*

Proof First, we notice that in any regular category finite products of covers are covers; indeed, composites of covers are covers (this can be easily proved by using the definition of a cover as an arrow orthogonal to the class of monomorphisms), and the product of two covers $f \times g$, where $f : A \to B$ and $g : C \to D$, is equal to the composite $(1_B \times g) \circ (f \times 1_C)$ of two arrows which are pullbacks of covers and hence covers.

For every sort A of Σ, we set $N'A$ equal to $\text{Im}(f_A)$, g_A equal to the canonical cover $MA \to \text{Im}(f_A)$ and h_A equal to the canonical subobject $N'A \rightarrowtail NA$. For any function symbol $\xi : A_1 \cdots A_n \to B$ of Σ, we set $N'\xi$ equal to the restriction $N'A_1 \times \cdots \times N'A_n \to N'B$ of $N\xi : NA_1 \times \cdots \times NA_n \to NB$. This restriction actually exists (and is unique) since f is a Σ-structure homomorphism and $g_{A_1} \times \cdots \times g_{A_n}$ is a cover. For any relation symbol R of Σ of type $A_1 \cdots A_n$, we set $N'R$ equal to the intersection of NR with the canonical subobject $N'A_1 \times \cdots \times N'A_n \rightarrowtail NA_1 \times \cdots \times NA_n$. Clearly, $f = h \circ g$, g is sortwise a cover and h is a Σ-substructure embedding, as required. $\square$

The following result, giving an explicit characterization of decidable objects in terms of their generalized elements, was stated in [62] as Exercise VIII.8(a).

Lemma 7.2.38 *Let $\mathcal{E}$ be a cocomplete topos and A an object of $\mathcal{E}$. Then A is decidable if and only if for any generalized elements $x, y : E \to A$, there exists an epimorphic family (which can be taken consisting of just two elements) $\{e_i : E_i \to E \mid i \in I\}$ such that for any $i \in I$, either $x \circ e_i = y \circ e_i$ or the equalizer of $x \circ e_i$ and $y \circ e_i$ is zero.*

Proof Let us suppose that A is decidable. Let $d_A : D_A \rightarrowtail A \times A$ be the complement of the diagonal subobject $\delta_A : A \rightarrowtail A \times A$. Consider the pullback of $<x, y> : E \to A \times A$ along d_A:

$$\begin{array}{ccc} E' & \xrightarrow{u} & D_A \\ {\scriptstyle s}\downarrow & & \downarrow{\scriptstyle d_A} \\ E & \xrightarrow{<x,y>} & A \times A \end{array}$$

The equalizer $i : R \rightarrowtail E'$ of $x \circ s$ and $y \circ s$ is zero; indeed, by definition of D_A, the diagram

$$\begin{array}{ccc} 0 & \longrightarrow & A \\ \downarrow & & \downarrow{\scriptstyle \delta_A} \\ D_A & \xrightarrow{d_A} & A \times A. \end{array}$$

is a pullback, and the arrows $z := x \circ s \circ i = y \circ s \circ i : R \to A$ and $u \circ i : R \to D_A$ satisfy the condition $\delta_A \circ z = d_A \circ u \circ i$.

Let us denote by $t : E'' \rightarrowtail E$ the pullback of the subobject $\delta_A : A \rightarrowtail A \times A$ along $<x, y>$. The arrows $s : E' \to E$ and $t : E'' \to E$ are jointly epimorphic, as they are pullbacks of arrows which are jointly epimorphic, namely δ_A and d_A.

Therefore, by setting $I = \{0, 1\}$, $E_0 = E'$, $E_1 = E''$, $e_0 : E_0 \to E$ equal to s and $e_1 : E_1 \to E$ equal to t, we have that the family $\{e_i : E_i \to E \mid i \in I\}$ satisfies the condition in the statement of the lemma.

Conversely, let us suppose that the condition in the statement of the lemma is satisfied. To prove that A is decidable, we have to show that the subobject $\delta_A \cup \neg \delta_A \rightarrowtail A \times A$ is an isomorphism, in other words that every subobject $<x, y> : E \rightarrowtail A \times A$ factors through it.

For any generalized elements $x', y' : E' \to A$, the arrow $<x', y'> : E' \to A \times A$ factors through $\neg \delta_A \rightarrowtail A \times A$ if and only if the equalizer of x' and y' is zero. Indeed, $<x', y'>$ factors through $\neg \delta_A$ if and only if its image does, and by definition of $\neg \delta_A$ this holds if and only if the pullback of it (or equivalently, of $<x', y'>$) along δ_A is zero, i.e. if and only if the equalizer of x' and y' is zero. Now, by our hypotheses, there exist an epimorphic family $\{e_i : E_i \to E \mid i \in I\}$ such that for every $i \in I$ either $x \circ e_i = y \circ e_i$ or the equalizer of $x \circ e_i$ and $y \circ e_i$ is zero. By Lemma 7.2.26, to prove that $<x, y>$ factors through $\delta_A \vee \neg \delta_A \rightarrowtail A \times A$ it suffices to show that for any $i \in I$, $<x \circ e_i, y \circ e_i>$ factors through $\delta_A \vee \neg \delta_A \rightarrowtail A \times A$. But by our hypothesis, for any given $i \in I$, either $x \circ e_i = y \circ e_i$ (which implies that $<x \circ e_i, y \circ e_i>$ factors through $\delta_A : A \rightarrowtail A \times A$) or the equalizer of $x \circ e_i$ and $y \circ e_i$ is zero (which implies, as we have just seen, that $<x \circ e_i, y \circ e_i>$ factors through $\neg \delta_A \rightarrowtail A \times A$); therefore, for every $i \in I$, $<x \circ e_i, y \circ e_i>$ factors through $\delta_A \cup \neg \delta_A \rightarrowtail A \times A$, as required. □

Proposition 7.2.39 *Let Σ be a one-sorted signature, c a finite Σ-structure in **Set**, M a Σ-structure in a Grothendieck topos $\mathcal{E}$ whose underlying object is decidable and E an object of $\mathcal{E}$. Then, for any Σ-structure homomorphism $f : c \to \mathrm{Hom}_{\mathcal{E}}(E, M)$, there exist an epimorphic family $\{e_i : E_i \to E \mid i \in I\}$ in $\mathcal{E}$ and for each $i \in I$ a quotient map $q_i : c \to c_i$, where c_i is a finite Σ-structure, and a disjunctive Σ-structure homomorphism (in the sense of Lemma 7.2.16) $J_i : c_i \rightarrowtail \mathrm{Hom}_{\mathcal{E}}(E_i, M)$ such that $J_i \circ q_i = \mathrm{Hom}_{\mathcal{E}}(e_i, M) \circ f$ for all $i \in I$:*

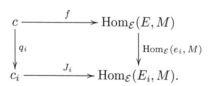

Proof Let us suppose that c has n elements $x_1, \ldots, x_n$. We know from Lemma 7.2.38 that for any pair (i, j), where $i, j \in \{1, \ldots, n\}$, there exist arrows $e_{(i,j)} : E_{(i,j)} \to E$ and $e'_{(i,j)} : E'_{(i,j)} \to E$ in $\mathcal{E}$ such that $e_{(i,j)}$ and $e'_{(i,j)}$ are jointly epimorphic, $f(x_i) \circ e_{(i,j)} = f(x_j) \circ e_{(i,j)}$ and the equalizer of $f(x_i) \circ e'_{(i,j)}$ and $f(x_j) \circ e'_{(i,j)}$ is zero. The iterated fibred product of all these epimorphic families (indexed by the pairs (i, j) such that $i, j \in \{1, \ldots, n\}$) thus yields an epimorphic family $\{u_k : U_k \to E \mid k \in K\}$ with the property that for every $k \in K$ there exists a subset $S_k \subseteq \{1, \ldots, n\} \times \{1, \ldots, n\}$ such that, for every $(i, j) \in S_k$, $f(x_i) \circ u_k = f(x_j) \circ u_k$ and, for every $(i, j) \notin S_k$, the equalizer of $f(x_i) \circ u_k$ and

$f(x_j) \circ u_k$ is zero. For each $k \in K$, consider the quotient $q_k : c \to c_q$ of c by the congruence generated by the pairs of the form (x_i, x_j) for $(i,j) \in S_k$. It is easily seen that the Σ-structure homomorphism $\mathrm{Hom}_{\mathcal{E}}(e_k, M) \circ f$ factors through q_k and that the resulting factorization is a disjunctive Σ-structure homomorphism. We have thus found a set of data satisfying the condition in the statement of the proposition, as required. □

Proposition 7.2.40 *Let $\mathbb{T}$ be a theory of presheaf type over a signature Σ. Suppose that the finitely presentable $\mathbb{T}$-models coincide with the finitely presentable $\mathbb{T}_m$-models and that the following condition is satisfied: for any Grothendieck topos $\mathcal{E}$, object E of $\mathcal{E}$ and Σ-structure homomorphism $x : c \to \mathrm{Hom}_{\mathcal{E}}(E, M)$, where c is a finitely presentable $\mathbb{T}$-model and M is a sortwise decidable $\mathbb{T}$-model, there exist an epimorphic family $\{e_i : E_i \to E \mid i \in I\}$ in $\mathcal{E}$ and for each $i \in I$ a $\mathbb{T}$-model homomorphism $f_i : c \to c_i$ in f.p.$\mathbb{T}$-mod(**Set**) and a sortwise disjunctive Σ-structure homomorphism $x_i : c_i \to \mathrm{Hom}_{\mathcal{E}}(E_i, M)$ such that $x_i \circ f_i = \mathrm{Hom}_{\mathcal{E}}(e_i, M) \circ x$. Then the injectivization $\mathbb{T}_m$ of $\mathbb{T}$ satisfies conditions (i) and (ii) of Theorem 6.3.1 with respect to its category of finitely presentable models.*

Proof The proposition represents the particular case of Theorem 7.2.3 for the faithful interpretation of $\mathbb{T}$ into its injectivization, with $\mathcal{K}$ and $\mathcal{H}$ equal to the category of finitely presentable $\mathbb{T}$-models. The hypotheses of the theorem are satisfied by Remark 7.2.4. □

Corollary 7.2.41 *Let $\mathbb{T}$ be a theory of presheaf type over a signature Σ. Suppose that the finitely presentable $\mathbb{T}$-models coincide with the finitely presentable $\mathbb{T}_m$-models, that the monic arrows of the category f.p.$\mathbb{T}$-mod(**Set**) are all sortwise injective, and that the following condition is satisfied: for any Grothendieck topos $\mathcal{E}$, object E of $\mathcal{E}$ and Σ-structure homomorphism $x : c \to \mathrm{Hom}_{\mathcal{E}}(E, M)$, where c is a finitely presentable $\mathbb{T}$-model and M is a sortwise decidable $\mathbb{T}$-model, there exist an epimorphic family $\{e_i : E_i \to E \mid i \in I\}$ in $\mathcal{E}$ and for each $i \in I$ a $\mathbb{T}$-model homomorphism $f_i : c \to c_i$ in f.p.$\mathbb{T}$-mod(**Set**) and a sortwise disjunctive Σ-structure homomorphism $x_i : c_i \to \mathrm{Hom}_{\mathcal{E}}(E_i, M)$ such that $x_i \circ f_i = \mathrm{Hom}_{\mathcal{E}}(e_i, M) \circ x$. Then the injectivization $\mathbb{T}_m$ of $\mathbb{T}$ is of presheaf type.*

Proof By Theorem 7.2.32, $\mathbb{T}_m$ satisfies condition (iii) of Theorem 6.3.1 (since all the monic arrows in the category f.p.$\mathbb{T}$-mod(**Set**) are sortwise injective), while Proposition 7.2.40 ensures that it satisfies conditions (i) and (ii) of Theorem 6.3.1. We can thus conclude that $\mathbb{T}_m$ is of presheaf type. □

Remark 7.2.42 By Propositions 7.2.18 and 7.2.20, every geometric theory $\mathbb{T}$ whose finitely presentable models are precisely the finitely generated ones and whose axioms are all of the form $(\phi \vdash_{\vec{x}} \psi)$, where ψ is a quantifier-free geometric formula, satisfies the property that the finitely presentable $\mathbb{T}$-models coincide with the finitely presentable $\mathbb{T}_m$-models.

As a consequence of Corollary 7.2.41, we obtain the following result:

Corollary 7.2.43 *Let $\mathbb{T}$ be a geometric theory over a one-sorted signature Σ such that every quotient of a finitely presentable $\mathbb{T}$-model is a $\mathbb{T}$-model (e.g. a theory whose axioms are all of the form $(\phi \vdash_{\vec{x}} \psi)$, where ϕ does not contain any conjunctions – cf. Proposition 7.2.36). Suppose that the finitely presentable $\mathbb{T}$-models are exactly the finite $\mathbb{T}$-models and that all the monic arrows in the category f.p.$\mathbb{T}$-mod(**Set**) are injective functions. Then if $\mathbb{T}$ is of presheaf type, $\mathbb{T}_m$ is of presheaf type too.*

Proof Propositions 7.2.36 and 7.2.39 ensure that all the conditions of Corollary 7.2.41 are satisfied. Therefore $\mathbb{T}_m$ is of presheaf type, as required. □

8

QUOTIENTS OF A THEORY OF PRESHEAF TYPE

In Chapter 3, we have described the classifying topos of a quotient of a geometric theory in a syntactic way. In this chapter we shall study the quotients of a given theory of presheaf type $\mathbb{T}$ by means of Grothendieck topologies that can be naturally attached to them, thereby establishing a 'semantic' representation for the classifying topos of such a quotient as a subtopos of the classifying topos of $\mathbb{T}$.

In section 8.1 we discuss such 'semantic' representations and introduce, given a theory of presheaf type $\mathbb{T}$ classified by a topos $[\mathcal{C}, \mathbf{Set}]$ and a Grothendieck topology J on the category $\mathcal{C}^{\mathrm{op}}$, the notion of J-homogeneous $\mathbb{T}$-model, showing that the models of the quotient of $\mathbb{T}$ corresponding to the subtopos $\mathbf{Sh}(\mathcal{C}^{\mathrm{op}}, J) \hookrightarrow [\mathcal{C}, \mathbf{Set}]$ via Theorem 3.2.5 are precisely the $\mathbb{T}$-models which are J-homogeneous (if the chosen Morita equivalence for $\mathbb{T}$ is canonical). We obtain explicit axiomatizations for the J-homogeneous models in terms of J and of the formulae which present the finitely presentable $\mathbb{T}$-models. We then restrict our attention to the classifying toposes of quotients with enough set-based models and establish, for any such quotient, a characterization of the Grothendieck topology corresponding to it in terms of its category of set-based models. We also discuss coherent quotients of a cartesian theory and the Grothendieck topologies which correspond to them, namely the finite-type ones; we show that the lattice operations on Grothendieck topologies naturally restrict to the collection of finite-type ones, and that one can effectively compute the lattice operations on the set of quotients of a given geometric theory by means of calculations in the associated lattices of Grothendieck topologies.

In section 8.2 we identify two main independent conditions which ensure that a quotient of a given theory of presheaf type is again of presheaf type: a finality property of a family of indexed functors naturally associated with the quotient, and a rigidity property of the Grothendieck topology corresponding to the quotient under the above-mentioned duality. We then discuss the problem of finding a geometric theory classified by a given presheaf topos $[\mathcal{K}, \mathbf{Set}]$, and prove a general theorem ensuring that if the category $\mathcal{K}$ can be realized as a full subcategory of the category of finitely presentable models of a theory of presheaf type $\mathbb{T}$, there exists a quotient of $\mathbb{T}$ classified by the topos $[\mathcal{K}, \mathbf{Set}]$, which can be explicitly described in terms of $\mathbb{T}$ and $\mathcal{K}$ provided that some natural conditions are satisfied.

Theories, Sites, Toposes. Olivia Caramello.
© Olivia Caramello 2018. Published 2018 by Oxford University Press.

8.1 Studying quotients through the associated Grothendieck topologies

Given a geometric theory $\mathbb{T}$ classified by a presheaf topos via an equivalence

$$[\mathcal{C}, \mathbf{Set}] \simeq \mathbf{Sh}(\mathcal{C}_{\mathbb{T}}, J_{\mathbb{T}}),$$

we have an induced bijective correspondence between the subtoposes of $[\mathcal{C}, \mathbf{Set}]$ and the subtoposes of $\mathbf{Sh}(\mathcal{C}_{\mathbb{T}}, J_{\mathbb{T}})$, and hence between the Grothendieck topologies on the category $\mathcal{C}^{\mathrm{op}}$ and the quotients of the theory $\mathbb{T}$ (see Theorem 3.2.5):

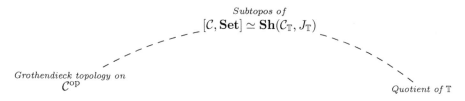

Under this correspondence, each quotient $\mathbb{T}'$ of $\mathbb{T}$ corresponds to a Grothendieck topology J on $\mathcal{C}^{\mathrm{op}}$ such that the topos $\mathbf{Sh}(\mathcal{C}^{\mathrm{op}}, J)$ classifies $\mathbb{T}'$.

It is natural to wonder if we can characterize the models of $\mathbb{T}'$ among those of $\mathbb{T}$ directly in terms of $\mathbb{T}$ and of J, without any reference to flat functors. As we shall see below, the technique of 'Yoneda representation of flat functors' described in section 5.3 provides us with a means for solving this problem.

8.1.1 The notion of J-homogeneous model

This section is based on [17].

By the results of section 6.1.1, we can suppose the Cauchy-completion $\hat{\mathcal{C}}$ of $\mathcal{C}$ to be equal to f.p.$\mathbb{T}$-mod($\mathbf{Set}$) without loss of generality, and that we have a canonical Morita equivalence

$$\xi^{\mathcal{E}} : \mathbf{Flat}(\hat{\mathcal{C}}^{\mathrm{op}}, \mathcal{E}) \to \mathbb{T}\text{-mod}(\mathcal{E}).$$

The restriction $\tau_{\xi} : \hat{\mathcal{C}} \to$ f.p.$\mathbb{T}$-mod($\mathbf{Set}$) of the equivalence $\xi^{\mathbf{Set}} : \mathbf{Flat}(\hat{\mathcal{C}}^{\mathrm{op}}, \mathbf{Set}) \simeq \mathbb{T}$-mod($\mathbf{Set}$) is thus isomorphic to the identity, and we have a commutative diagram

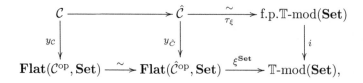

where $y_{\mathcal{C}} : \mathcal{C} \to \mathbf{Flat}(\mathcal{C}^{\mathrm{op}}, \mathbf{Set})$ and $y_{\hat{\mathcal{C}}} : \hat{\mathcal{C}} \to \mathbf{Flat}(\hat{\mathcal{C}}^{\mathrm{op}}, \mathbf{Set})$ are the Yoneda embeddings and the horizontal arrows in the left-hand square are the canonical ones.

Recall from Corollary 5.3.2 that every flat functor $F \in \mathbf{Flat}(\mathcal{C}^{\mathrm{op}}, \mathcal{E})$ can be represented as
$$F \cong \mathrm{Hom}^{\mathcal{E}}_{\mathbf{Flat}(\mathcal{C}^{\mathrm{op}},\mathcal{E})}(\overline{y_{\mathcal{C}}}(-), F),$$
where $\overline{y_{\mathcal{C}}} : \mathcal{C} \to \mathbf{Flat}(\mathcal{C}^{\mathrm{op}}, \mathcal{E})$ is the functor given by the composite
$$\mathcal{C} \xrightarrow{y_{\mathcal{C}}} \mathbf{Flat}(\mathcal{C}^{\mathrm{op}}, \mathbf{Set}) \xrightarrow{\gamma_{\mathcal{E}}^{*} \circ -} \mathbf{Flat}(\mathcal{C}^{\mathrm{op}}, \mathcal{E})$$
and $\mathrm{Hom}^{\mathcal{E}}_{\mathbf{Flat}(\mathcal{C},\mathcal{E})}(\overline{y_{\mathcal{C}}}(-), F)$ is the functor given by the composite
$$\mathcal{C}^{\mathrm{op}} \xrightarrow{\overline{y_{\mathcal{C}}}^{\mathrm{op}} \times \Delta F} \mathbf{Flat}(\mathcal{C}^{\mathrm{op}}, \mathcal{E})^{\mathrm{op}} \times \mathbf{Flat}(\mathcal{C}^{\mathrm{op}}, \mathcal{E}) \xrightarrow{\mathrm{Hom}^{\mathcal{E}}_{\mathbf{Flat}(\mathcal{C}^{\mathrm{op}},\mathcal{E})}} \mathcal{E}.$$

We want to rewrite this Yoneda representation in terms of the theory $\mathbb{T}$.

Recall from section 6.2.1 that the $\mathcal{E}$-indexed category $\mathbb{T}\text{-mod}(\mathcal{E})$ is locally small, with object of morphisms
$$\mathrm{Hom}^{\mathcal{E}}_{\mathbb{T}\text{-mod}(\mathcal{E})}(M, N)$$
from M to N in $\mathbb{T}\text{-mod}(\mathcal{E})$. The naturality in $\mathcal{E}$ of the Morita equivalence between $\mathbb{T}$ and the theory of flat functors on $\mathcal{C}^{\mathrm{op}}$ implies the commutativity of the following diagram:

$$\begin{array}{ccc} \mathbf{Flat}(\mathcal{C}^{\mathrm{op}}, \mathbf{Set}) & \xrightarrow{\sim} & \mathbb{T}\text{-mod}(\mathbf{Set}) \\ {\scriptstyle \gamma_{\mathcal{E}}^{*} \circ -} \downarrow & & \downarrow {\scriptstyle \gamma_{\mathcal{E}}^{*}(-)} \\ \mathbf{Flat}(\mathcal{C}^{\mathrm{op}}, \mathcal{E}) & \xrightarrow{\sim} & \mathbb{T}\text{-mod}(\mathcal{E}). \end{array}$$

From the commutativity of the two diagrams above we deduce the following representation for F:
$$F \cong \mathrm{Hom}^{\mathcal{E}}_{\mathbb{T}\text{-mod}(\mathcal{E})}(\gamma_{\mathcal{E}}^{*}(i_{\mathcal{C}}(-)), M_F),$$
where M_F is the $\mathbb{T}$-model in $\mathcal{E}$ corresponding to $F \in \mathbf{Flat}(\mathcal{C}^{\mathrm{op}}, \mathcal{E})$ via the above Morita equivalence and $i_{\mathcal{C}} : \mathcal{C} \to \mathbb{T}\text{-mod}(\mathbf{Set})$ is the canonical induced functor. This motivates the following definition:

Definition 8.1.1 (cf. Definition 2.17 [17]) Let $\mathcal{C}$ be a small category, $\mathcal{E}$ a Grothendieck topos and $\mathbb{T}$ a theory of presheaf type classified by the topos $[\mathcal{C}, \mathbf{Set}]$.

(a) Given a cosieve S in $\mathcal{C}$ on an object c, a model $M \in \mathbb{T}\text{-mod}(\mathcal{E})$ is said to be S-homogeneous if the family of arrows

$$\mathrm{Hom}^{\mathcal{E}}_{\mathbb{T}\text{-mod}(\mathcal{E})}(\gamma_{\mathcal{E}}^{*}(i_{\mathcal{C}}(d)), M) \xrightarrow{\mathrm{Hom}^{\mathcal{E}}_{\mathbb{T}\text{-mod}(\mathcal{E})}(\gamma_{\mathcal{E}}^{*}(i_{\mathcal{C}}(f)), M)} \mathrm{Hom}^{\mathcal{E}}_{\mathbb{T}\text{-mod}(\mathcal{E})}(\gamma_{\mathcal{E}}^{*}(i_{\mathcal{C}}(c)), M)$$

for $f \in S$, is epimorphic in $\mathcal{E}$.

(b) Given a Grothendieck topology J on $\mathcal{C}^{\mathrm{op}}$, a model $M \in \mathbb{T}\text{-mod}(\mathcal{E})$ is said to be J-homogeneous if M is S-homogeneous for every $S \in J(c)$.

Remark 8.1.2 It is immediate to see that if J is the atomic topology on $\mathcal{C}^{\mathrm{op}}$ then a model $M \in \mathbb{T}\text{-mod}(\mathcal{E})$ is J-homogeneous if and only if, for every arrow $f : c \to d$ in $\mathcal{C}$, the arrow

$$\mathrm{Hom}^{\mathcal{E}}_{\mathbb{T}\text{-mod}(\mathcal{E})}(\gamma^*_{\mathcal{E}}(i_{\mathcal{C}}(d)), M) \xrightarrow{\mathrm{Hom}^{\mathcal{E}}_{\mathbb{T}\text{-mod}(\mathcal{E})}(\gamma^*_{\mathcal{E}}(i_{\mathcal{C}}(f)), M)} \mathrm{Hom}^{\mathcal{E}}_{\mathbb{T}\text{-mod}(\mathcal{E})}(\gamma^*_{\mathcal{E}}(i_{\mathcal{C}}(c)), M)$$

is an epimorphism in $\mathcal{E}$. In this case we shall simply say 'homogeneous' instead of 'J-homogeneous'.

Since the model M_F is J-homogeneous if and only if the functor F is J-continuous, we have the following result:

Theorem 8.1.3 (cf. Theorem 2.19 [17]). *Let $\mathbb{T}$ be a theory of presheaf type classified by a topos $[\mathcal{C}, \mathbf{Set}]$ and J a Grothendieck topology on $\mathcal{C}^{\mathrm{op}}$. Then the subtopos $\mathbf{Sh}(\mathcal{C}^{\mathrm{op}}, J) \hookrightarrow [\mathcal{C}, \mathbf{Set}]$ classifies the J-homogeneous $\mathbb{T}$-models; that is, for any quotient $\mathbb{T}'$ of $\mathbb{T}$, the $\mathbb{T}'$-models are exactly the J-homogeneous $\mathbb{T}$-models in any Grothendieck topos if and only if $\mathbb{T}'$ is classified by the topos $\mathbf{Sh}(\mathcal{C}^{\mathrm{op}}, J)$ and the diagram in $\mathfrak{CAT}$*

$$\begin{array}{ccc}
\mathbb{T}'\text{-mod}(\mathcal{E}) & \xrightarrow{\simeq} & \mathbf{Geom}(\mathcal{E}, \mathbf{Sh}(\mathcal{C}^{\mathrm{op}}, J)) \\
\downarrow i^{\mathcal{E}}_{\mathbb{T}'} & & \downarrow j \circ - \\
\mathbb{T}\text{-mod}(\mathcal{E}) & \xrightarrow[\chi^{\mathcal{E}}]{\simeq} & \mathbf{Geom}(\mathcal{E}, [\mathcal{C}, \mathbf{Set}])
\end{array}$$

commutes (up to isomorphism) naturally in $\mathcal{E} \in \mathfrak{BTop}$, where χ is a canonical Morita equivalence for the theory $\mathbb{T}$,

$$j : \mathbf{Sh}(\mathcal{C}^{\mathrm{op}}, J) \hookrightarrow [\mathcal{C}, \mathbf{Set}]$$

is the canonical geometric inclusion and

$$i^{\mathcal{E}}_{\mathbb{T}'} : \mathbb{T}'\text{-mod}(\mathcal{E}) \hookrightarrow \mathbb{T}\text{-mod}(\mathcal{E})$$

is the inclusion functor. □

Specializing the theorem to the case of the atomic topology gives the following result:

Corollary 8.1.4 (cf. Corollary 2.20 [17]). *Let $(\mathcal{C}^{\mathrm{op}}, J)$ be an atomic site and $\mathbb{T}$ a theory classified by a topos $[\mathcal{C}, \mathbf{Set}]$. Then the subtopos $\mathbf{Sh}(\mathcal{C}^{\mathrm{op}}, J) \hookrightarrow [\mathcal{C}, \mathbf{Set}]$ classifies the homogeneous $\mathbb{T}$-models.* □

Let us now rephrase in more explicit terms what it means for a model to be J-homogeneous. For this, we first express the condition that a given family of arrows as in Definition 8.1.1 be epimorphic as the validity of a sentence written in the internal language of the topos, and then use Kripke-Joyal semantics to spell out what it means for that sentence to be valid in terms of generalized elements.

Recall that if $\mathcal{E}$ is a Grothendieck topos and $\{f_i : C_i \to C \mid i \in I\}$ is a family of arrows in it indexed by a set I, then this family is epimorphic if and only if the logical sentence $(\forall y^{\ulcorner C \urcorner})(\bigvee_{i \in I}(\exists x^{\ulcorner C_i \urcorner}(\ulcorner f_i \urcorner(x) = y)))$ written in the internal language $\Sigma_\mathcal{E}$ of the topos $\mathcal{E}$ (cf. section 1.3.1.1) holds in the tautological $\Sigma_\mathcal{E}$-structure in $\mathcal{E}$. Given a separating family of objects $\mathcal{S}$ for $\mathcal{E}$, the validity in $\mathcal{E}$ of this sentence is in turn equivalent, by Kripke-Joyal semantics, to the following condition: for any object $E \in \mathcal{S}$ and arrow $y : E \to C$, there exist an epimorphic family $\{r_i : E_i \to E \mid i \in I, E_i \in \mathcal{S}\}$ in $\mathcal{E}$ and generalized elements $\{x_i : E_i \to C_i \mid i \in I\}$ such that $y \circ r_i = f_i \circ x_i$ for each $i \in I$. Applying this to the families of arrows in Definition 8.1.1 and recalling that the objects $\mathrm{Hom}^\mathcal{E}_{\mathbb{T}\text{-mod}(\mathcal{E})}(\gamma_\mathcal{E}^*(i_\mathcal{C}(d)), M)$ are the objects of morphisms from $\gamma_\mathcal{E}^*(i_\mathcal{C}(d))$ to M in $\mathbb{T}\text{-mod}(\mathcal{E})$, we obtain the characterization expressed by the following result:

Theorem 8.1.5 (cf. Theorem 2.21 and Corollary 2.22 [17]). *Let $\mathcal{E}$ be a Grothendieck topos with a separating family of objects $\mathcal{S}$ and $\mathbb{T}$ a theory of presheaf type classified by a topos $[\mathcal{C}, \mathbf{Set}]$. Given a Grothendieck topology J on $\mathcal{C}^{\mathrm{op}}$, a model $M \in \mathbb{T}\text{-mod}(\mathcal{E})$ is J-homogeneous if and only if for every cosieve $S \in J(c)$ in $\mathcal{C}$, object $E \in \mathcal{S}$ and arrow $y :!_E^*(\gamma_\mathcal{E}^*(i_\mathcal{C}(c))) \to !_E^*(M)$ in $\mathbb{T}\text{-mod}(\mathcal{E}/E)$, there exist an epimorphic family $\{p_f : E_f \to E, f \in S, E_f \in \mathcal{S}\}$ in $\mathcal{E}$ and for each arrow $f : c \to d$ in S an arrow $u_f :!_{E_f}^*(\gamma_\mathcal{E}^*(i_\mathcal{C}(d))) \to !_{E_f}^*(M)$ in $\mathbb{T}\text{-mod}(\mathcal{E}/E_f)$ such that $p_f^*(y) = u_f \circ !_{E_f}^*(\gamma_\mathcal{E}^*(i_\mathcal{C}(f)))$.*

In particular, if $\mathcal{E} = \mathbf{Set}$ then a model $M \in \mathbb{T}\text{-mod}(\mathbf{Set})$ is J-homogeneous if and only if for every cosieve $S \in J(c)$ in $\mathcal{C}$ and arrow $y : i_\mathcal{C}(c) \to M$ in $\mathbb{T}\text{-mod}(\mathbf{Set})$, there exist an arrow $f : c \to d$ in $\mathcal{C}$ belonging to S and an arrow $u_f : i_\mathcal{C}(d) \to M$ in $\mathbb{T}\text{-mod}(\mathbf{Set})$ such that $y = u_f \circ i_\mathcal{C}(f)$. □

By specializing the theorem to the case of the atomic topology, we immediately obtain the following result:

Corollary 8.1.6 (cf. Corollaries 2.23 and 2.24 [17]). *Let $\mathcal{E}$ be a Grothendieck topos with a separating family of objects $\mathcal{S}$ and $\mathbb{T}$ a theory of presheaf type classified by a topos $[\mathcal{C}, \mathbf{Set}]$. If the category $\mathcal{C}^{\mathrm{op}}$ satisfies the right Ore condition then a model $M \in \mathbb{T}\text{-mod}(\mathcal{E})$ is homogeneous if and only if for every arrow $f : c \to d$ in $\mathcal{C}$, object $E \in \mathcal{S}$ and arrow $y :!_E^*(\gamma_\mathcal{E}^*(i_\mathcal{C}(c))) \to !_E^*(M)$ in $\mathbb{T}\text{-mod}(\mathcal{E}/E)$, there exist an object $E_f \in \mathcal{S}$, an epimorphism $p_f : E_f \twoheadrightarrow E$ in $\mathcal{E}$ and an arrow $u_f :!_{E_f}^*(\gamma_\mathcal{E}^*(i_\mathcal{C}(d))) \to !_{E_f}^*(M)$ in $\mathbb{T}\text{-mod}(\mathcal{E}/E_f)$ such that $p_f^*(y) = u_f \circ !_{E_f}^*(\gamma_\mathcal{E}^*(i_\mathcal{C}(f)))$.*

In particular, if $\mathcal{E} = \mathbf{Set}$ then a model $M \in \mathbb{T}\text{-mod}(\mathbf{Set})$ is homogeneous if and only if for any arrows $f : c \to d$ in $\mathcal{C}$ and $y : i_{\mathcal{C}}(c) \to M$ in $\mathbb{T}\text{-mod}(\mathbf{Set})$, there exists an arrow $u_f : i_{\mathcal{C}}(d) \to M$ in $\mathbb{T}\text{-mod}(\mathbf{Set})$ such that $y = u_f \circ i_{\mathcal{C}}(f)$:

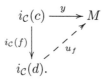

□

Remarks 8.1.7 (a) Our notion of homogeneous model essentially specializes to that of (weakly) homogeneous model in classical model theory (cf. Chapter 7 of [49]); in fact, this link has been exploited in [23] to obtain a topos-theoretic interpretation and generalization of Fraïssé's theorem on countably categorical theories (cf. section 10.4 below).

(b) Under the hypotheses of Definition 8.1.1, for any Grothendieck topos $\mathcal{E}$ and any object $E \in \mathcal{E}$, there is an isomorphism

$$!_E^*(\mathrm{Hom}_{\mathbb{T}\text{-mod}(\mathcal{E})}^{\mathcal{E}}(\gamma_{\mathcal{E}}^*(i_{\mathcal{C}}(c)), M)) \cong \mathrm{Hom}_{\mathbb{T}\text{-mod}(\mathcal{E}/E)}^{\mathcal{E}/E}(\gamma_{\mathcal{E}/E}^*(i_{\mathcal{C}}(c)), !_E^*(M))$$

natural in $c \in \mathcal{C}$. Hence, if $M \in \mathbb{T}\text{-mod}(\mathcal{E})$ is J-homogeneous then $!_E^*(M) \in \mathbb{T}\text{-mod}(\mathcal{E}/E)$ is also J-homogeneous. This implies that, when dealing with theories $\mathbb{T}'$ that one wants to prove to satisfy the conditions of Theorem 8.1.5, it suffices to deal with generalized elements defined on $1_{\mathcal{E}}$ by the localization principle. This is illustrated by the example in section 8.1.1.1.

8.1.1.1 Dense linear orders without endpoints as homogeneous models

In this section, we recover, as an application of Corollaries 8.1.4 and 8.1.6, the well-known representation of the classifying topos of the theory of dense linearly ordered objects without endpoints as the category $\mathbf{Sh}(\mathbf{Ord}_{\mathrm{fm}}^{\mathrm{op}}, J_{\mathrm{at}})$ of atomic sheaves on the opposite of the category $\mathbf{Ord}_{\mathrm{fm}}$ of finite ordinals and order-preserving injections between them (cf. Example D3.4.11 [55]).

The theory $\mathbb{L}'$ of dense linearly ordered objects without endpoints is defined over a one-sorted signature consisting of one relation symbol $<$ apart from equality, and has the following axioms:

$$((x < y) \wedge (y < x) \vdash_{x,y} \bot);$$
$$(\top \vdash_{x,y} (x = y) \vee (x < y) \vee (y < x));$$
$$((x < y) \wedge (y < z) \vdash_{x,y,z} x < z);$$
$$(\top \vdash_{[]} (\exists x)\top);$$
$$((x < y) \vdash_{x,y} (\exists z)((x < z) \wedge (z < y)));$$

$$(x \vdash_x (\exists y, z)((y < x) \wedge (x < z))).$$

The first three axioms give the theory $\mathbb{L}$ of (decidably) linearly ordered objects. It is well known that this theory is of presheaf type (cf. for instance section 9.1 below); the category of finitely presentable $\mathbb{L}$-models can be identified with $\mathbf{Ord}_{\mathrm{fm}}$, so the classifying topos of $\mathbb{L}$ is equivalent to the presheaf topos $[\mathbf{Ord}_{\mathrm{fm}}, \mathbf{Set}]$. Clearly, the category $\mathbf{Ord}_{\mathrm{fm}}^{\mathrm{op}}$ satisfies the right Ore condition, so we can equip it with the atomic topology J_{at}.

We shall now proceed to verify that the models of $\mathbb{L}'$ are precisely the homogeneous $\mathbb{L}$-models in every Grothendieck topos $\mathcal{E}$.

A model $M \in \mathbb{L}\text{-mod}(\mathcal{E})$ consists of a pair (I, R) where I is an object of $\mathcal{E}$ and R is a relation on I satisfying the diagrammatic forms of the three axioms of $\mathbb{L}$. We will prove that for each Grothendieck topos $\mathcal{E}$, a model $M = (I, R) \in \mathbb{L}\text{-mod}(\mathcal{E})$ is homogeneous if and only if it is a model of $\mathbb{L}'$, that is if and only if (I, R) is non-empty, dense and without endpoints.

In one direction, let us suppose that M is homogeneous and prove that (I, R) is dense. For each object $E \in \mathcal{E}$, we denote by $<_E$ the order induced by R on $\mathrm{Hom}_{\mathcal{E}}(E, I)$. By the localizing principle (cf. Remark 8.1.7(b)), it is enough to prove that if $x, y : 1 \to I$ are two generalized elements of I with $x <_1 y$ then there exist an object $E \in \mathcal{E}$, an epimorphism $p : E \twoheadrightarrow 1$ and an arrow $z : E \to I$ such that $x \circ p <_E z <_E y \circ p$. Consider the arrow $f : 2 = \{0, 1\} \to 3 = \{0, 1, 2\}$ in $\mathbf{Ord}_{\mathrm{fm}}$ defined by $f(0) = 0$ and $f(1) = 2$; the arrows x and y induce, via the assignment $(0 \to x, 1 \to y)$ and the universal property of the coproduct $\gamma_{\mathcal{E}}^*(2)$, an arrow $s : \gamma_{\mathcal{E}}^*(2) \to I$ in $\mathbb{L}\text{-mod}(\mathcal{E})$. The homogeneity of M thus ensures the existence of an object $E \in \mathcal{E}$, an epimorphism $p : E \twoheadrightarrow 1$ and an arrow $t : !_E^*(\gamma_{\mathcal{E}}^*(3)) \to !_E^*(I)$ in $\mathbb{L}\text{-mod}(\mathcal{E}/E)$ such that $t \circ !_E^*(\gamma_{\mathcal{E}}^*(f)) = !_E^*(s)$. Then the composite arrow

$$E \cong !_E^*(\gamma_{\mathcal{E}}^*(1)) \xrightarrow{!_E^*(\gamma_{\mathcal{E}}^*(u))} !_E^*(\gamma_{\mathcal{E}}^*(3)) \xrightarrow{t} !_E^*(I) \xrightarrow{\pi_I} I,$$

where $u : 1 \to 3$ is the arrow in $\mathbf{Ord}_{\mathrm{fm}}$ which picks out the element $1 \in 3 = \{0, 1, 2\}$, gives an arrow $z : E \to I$ with the required properties. The proofs that (I, R) is non-empty and without endpoints are similar and left to the reader.

Conversely, let us prove that if $M \in \mathbb{L}'\text{-mod}(\mathcal{E})$ then M is homogeneous. Again, by the localization principle, this amounts to showing that, for any arrows $f : n \to m$ in $\mathbf{Ord}_{\mathrm{fm}}$ and $s : \gamma_{\mathcal{E}}^*(n) \to I$ in $\mathbb{L}\text{-mod}(\mathcal{E})$, there exist an object $E \in \mathcal{E}$, an epimorphism $p : E \twoheadrightarrow 1$ in $\mathcal{E}$ and an arrow $t : !_E^*(\gamma_{\mathcal{E}}^*(m)) \to !_E^*(I)$ in $\mathbb{L}\text{-mod}(\mathcal{E}/E)$ such that $t \circ !_E^*(\gamma_{\mathcal{E}}^*(f)) = !_E^*(s)$. The arrow s can be identified, via the universal property of the coproduct $\gamma_{\mathcal{E}}^*(n)$, with a family $\{h_i : 1 \to I \mid i \in n\}$ of generalized elements of I. To find an arrow t satisfying the required property, we inductively use the fourth or fifth axioms of the theory $\mathbb{L}'$ to obtain, starting from the arrows h_i, an object $E \in \mathcal{E}$, an epimorphism $p : E \twoheadrightarrow 1$ in $\mathcal{E}$ and m generalized elements $z_j : E \to I$ (for $j \in m$) such that, for any $i \in n$, $z_{f(i)} = h_i \circ p$ and, for any $j, j' \in m$, $(j < j')$ implies $(z_j <_E z_{j'})$. The family $\{z_j : E \to I \mid j \in m\}$ thus yields an arrow $t : !_E^*(\gamma_{\mathcal{E}}^*(m)) \to !_E^*(I)$ in $\mathbb{L}\text{-mod}(\mathcal{E}/E)$ with the required property.

8.1.2 Axiomatizations for the J-homogeneous models

Given a quotient $\mathbb{T}'$ of a theory of presheaf type $\mathbb{T}$, the Grothendieck topology J on f.p.$\mathbb{T}$-mod$(\mathbf{Set})^{\mathrm{op}}$ such that the subtopos

$$\mathbf{Sh}(\text{f.p.}\mathbb{T}\text{-mod}(\mathbf{Set})^{\mathrm{op}}, J) \hookrightarrow [\text{f.p.}\mathbb{T}\text{-mod}(\mathbf{Set}), \mathbf{Set}]$$

corresponds via the canonical equivalence

$$[\text{f.p.}\mathbb{T}\text{-mod}(\mathbf{Set}), \mathbf{Set}] \simeq \mathbf{Sh}(\mathcal{C}_{\mathbb{T}}, J_{\mathbb{T}})$$

to the subtopos of $\mathbf{Sh}(\mathcal{C}_{\mathbb{T}}, J_{\mathbb{T}})$ associated with $\mathbb{T}'$ via Theorem 3.2.5, will be called the *induced $\mathbb{T}$-topology of $\mathbb{T}'$*; note that, by Remark 3.2.6, the canonical Morita equivalence

$$\mathbb{T}\text{-mod}(\mathcal{E}) \simeq \mathbf{Flat}(\text{f.p.}\mathbb{T}\text{-mod}(\mathbf{Set})^{\mathrm{op}}, \mathcal{E})$$

restricts to a Morita equivalence

$$\mathbb{T}'\text{-mod}(\mathcal{E}) \simeq \mathbf{Flat}_J(\text{f.p.}\mathbb{T}\text{-mod}(\mathbf{Set})^{\mathrm{op}}, \mathcal{E}).$$

The following theorem shows that the J-homogeneous models can be axiomatized by geometric sequents over the signature of $\mathbb{T}$.

Theorem 8.1.8 *Let $\mathbb{T}$ be a theory of presheaf type and J a Grothendieck topology on f.p.$\mathbb{T}$-mod$(\mathbf{Set})^{\mathrm{op}}$. Then there exists a (unique up to syntactic equivalence) quotient $\mathbb{T}'$ of $\mathbb{T}$ such that the $\mathbb{T}'$-models are exactly the J-homogeneous $\mathbb{T}$-models in every Grothendieck topos.*

Proof Via the canonical equivalence $[\text{f.p.}\mathbb{T}\text{-mod}(\mathbf{Set}), \mathbf{Set}] \simeq \mathbf{Sh}(\mathcal{C}_{\mathbb{T}}, J_{\mathbb{T}})$ of classifying toposes for $\mathbb{T}$, the geometric inclusion

$$\mathbf{Sh}(\text{f.p.}\mathbb{T}\text{-mod}(\mathbf{Set})^{\mathrm{op}}, J) \hookrightarrow [\text{f.p.}\mathbb{T}\text{-mod}(\mathbf{Set}), \mathbf{Set}]$$

corresponds to a subtopos of $\mathbf{Sh}(\mathcal{C}_{\mathbb{T}}, J_{\mathbb{T}})$; the quotient of $\mathbb{T}$ corresponding to this inclusion via the duality of Theorem 3.2.5 thus axiomatizes the J-homogeneous $\mathbb{T}$-models by Remark 3.2.6 and the discussion preceding Theorem 8.1.8. $\square$

Remark 8.1.9 If the category f.p.$\mathbb{T}$-mod$(\mathbf{Set})^{\mathrm{op}}$ satisfies the right Ore condition and J_{at} is the atomic topology on it, then the geometric inclusion

$$\mathbf{Sh}(\text{f.p.}\mathbb{T}\text{-mod}(\mathbf{Set})^{\mathrm{op}}, J_{\mathrm{at}}) \hookrightarrow [\text{f.p.}\mathbb{T}\text{-mod}(\mathbf{Set}), \mathbf{Set}]$$

corresponds to the subtopos $\mathbf{sh}_{\neg\neg}(\mathbf{Sh}(\mathcal{C}_{\mathbb{T}}, J_{\mathbb{T}}))$ of $\mathbf{Sh}(\mathcal{C}_{\mathbb{T}}, J_{\mathbb{T}})$, and hence the J_{at}-homogeneous models are axiomatized by the Booleanization of $\mathbb{T}$ (cf. section 4.2.3). Analogously, the quotient of $\mathbb{T}$ corresponding to the De Morgan topology on the category f.p.$\mathbb{T}$-mod$(\mathbf{Set})^{\mathrm{op}}$ (cf. section 2 of [13] for the definition of this topology) is precisely the DeMorganization of $\mathbb{T}$ (as defined in section 4.2.3).

The following results provide, for any theory of presheaf type $\mathbb{T}$, explicit descriptions of the induced $\mathbb{T}$-topology of a given quotient of $\mathbb{T}$ and, reciprocally, of the quotient of $\mathbb{T}$ corresponding to a given Grothendieck topology on the category f.p.$\mathbb{T}$-mod$(\mathbf{Set})^{\mathrm{op}}$.

Given a sequent σ of the form $(\phi \vdash_{\vec{x}} \bigvee_{i\in I}(\exists \vec{y_i})\theta_i)$, where, for any $i \in I$, $[\theta_i]:\{\vec{y_i}.\psi\} \to \{\vec{x}.\phi\}$ is an arrow in $\mathcal{C}_{\mathbb{T}}$ and $\phi(\vec{x})$, $\psi(\vec{y_i})$ are formulae presenting respectively $\mathbb{T}$-models $M_{\{\vec{x}.\phi\}}$ and $M_{\{\vec{y_i}.\psi_i\}}$, we shall denote by S_σ the sieve on the category f.p.$\mathbb{T}$-mod$(\mathbf{Set})^{\mathrm{op}}$ defined as follows. For each $i \in I$, $[[\vec{y_i},\vec{x}.\theta_i]]_{M_{\{\vec{y_i}.\psi_i\}}}$ is the graph of a map $[[\vec{y_i}.\psi_i]]_{M_{\{\vec{y_i}.\psi_i\}}} \to [[\vec{x}.\phi]]_{M_{\{\vec{y_i}.\psi_i\}}}$; the image under it of the generators of $M_{\{\vec{y_i}.\psi_i\}}$ is an element of $[[\vec{x}.\phi]]_{M_{\{\vec{y_i}.\psi_i\}}}$ and this in turn determines, by definition of $M_{\{\vec{x}.\phi\}}$, a unique arrow $s_i : M_{\{\vec{x}.\phi\}} \to M_{\{\vec{y_i}.\psi_i\}}$ in $\mathbb{T}$-mod$(\mathbf{Set})$. We define S_σ to be the sieve in f.p.$\mathbb{T}$-mod$(\mathbf{Set})^{\mathrm{op}}$ on $M_{\{\vec{x}.\phi\}}$ generated by the arrows s_i as i varies in I. Notice that S_σ is the sieve in f.p.$\mathbb{T}$-mod$(\mathbf{Set})^{\mathrm{op}}$ corresponding under the equivalence f.p.$\mathbb{T}$-mod$(\mathbf{Set})^{\mathrm{op}} \simeq \mathcal{C}_{\mathbb{T}}^{\mathrm{irr}}$ of Theorem 6.1.13 to the sieve in $\mathcal{C}_{\mathbb{T}}^{\mathrm{irr}}$ generated by the arrows $[\theta_i] : \{\vec{y_i}.\psi\} \to \{\vec{x}.\phi\}$ (for $i \in I$); indeed, for each $i \in I$, the arrow s_i defined above coincides with the arrow $M_{[\theta_i]}$ defined in the proof of Theorem 6.1.13.

Theorem 8.1.10 *Let $\mathbb{T}$ be a theory of presheaf type over a signature Σ and $\mathbb{T}'$ a quotient of $\mathbb{T}$ obtained from $\mathbb{T}$ by adding axioms σ of the form $(\phi \vdash_{\vec{x}} \bigvee_{i\in I}(\exists \vec{y_i})\theta_i)$ where, for each $i \in I$, $[\theta_i] : \{\vec{y_i}.\psi\} \to \{\vec{x}.\phi\}$ is an arrow in $\mathcal{C}_{\mathbb{T}}$ and $\phi(\vec{x})$, $\psi(\vec{y_i})$ are geometric formulae over Σ presenting respectively $\mathbb{T}$-models $M_{\{\vec{x}.\phi\}}$ and $M_{\{\vec{y_i}.\psi_i\}}$ (note that, by Theorem 3.2.5, every quotient of $\mathbb{T}$ has an axiomatization of this form, cf. also Theorem 8.1.12 below). With the above notation, the induced $\mathbb{T}$-topology of $\mathbb{T}'$ is generated by the sieves S_σ.*

Proof Let $F :$ f.p.$\mathbb{T}$-mod$(\mathbf{Set})^{\mathrm{op}} \to \mathcal{E}$ be a flat functor; if $M_{\{\vec{x}.\phi\}}$ is a finitely presented $\mathbb{T}$-model then $F(M_{\{\vec{x}.\phi\}}) = [[\vec{x}.\phi]]_{M_F}$ where M_F is the $\mathbb{T}$-model in $\mathcal{E}$ corresponding to F via the canonical Morita equivalence for $\mathbb{T}$. Indeed, denoting by $g : \mathcal{E} \to \mathbf{Set}[\mathbb{T}] \simeq [$f.p.$\mathbb{T}$-mod$(\mathbf{Set}),\mathbf{Set}]$ the geometric morphism corresponding to F via the universal property of the classifying topos of $\mathbb{T}$, we have that $F = g^* \circ y$ where $y :$ f.p.$\mathbb{T}$-mod$(\mathbf{Set})^{\mathrm{op}} \to [$f.p.$\mathbb{T}$-mod$(\mathbf{Set}),\mathbf{Set}]$ is the Yoneda embedding. But $M_F = g^*(N_{\mathbb{T}})$ where $N_{\mathbb{T}}$ is the universal model of $\mathbb{T}$ lying in the classifying topos $\mathbf{Set}[\mathbb{T}] \simeq [$f.p.$\mathbb{T}$-mod$(\mathbf{Set}),\mathbf{Set}]$ (as in Theorem 6.1.1), and the representable $\mathrm{Hom}_{\text{f.p.}\mathbb{T}\text{-mod}(\mathbf{Set})}(M_{\{\vec{x}.\phi\}},-) \in [$f.p.$\mathbb{T}$-mod$(\mathbf{Set}),\mathbf{Set}]$ is clearly isomorphic to $[[\vec{x}.\phi]]_{N_{\mathbb{T}}}$. Since inverse image functors of geometric morphisms preserve the interpretations of geometric formulae, we may conclude that $F(M_{\{\vec{x}.\phi\}}) = [[\vec{x}.\phi]]_{M_F}$, as required. It is also immediate to see that, for any arrow $[\theta_i] : \{\vec{y_i}.\psi\} \to \{\vec{x}.\phi\}$ in $\mathcal{C}_{\mathbb{T}}$, the graph of the arrow $F(M_{[\theta_i]})$ is precisely $[[\vec{y_i},\vec{x}.\theta_i]]_{M_F}$.

Now, for any sequent σ as in the statement of the theorem, a model $N \in \mathbb{T}$-mod$(\mathcal{E})$ is S_σ-homogeneous if and only if σ holds in N; indeed, N is S_σ-homogeneous if and only if F_N sends S_σ to an epimorphic family, that is, by the above discussion, if and only if σ holds in N. So, if J is the induced $\mathbb{T}$-topology of $\mathbb{T}'$, from the equivalence

$$\mathbb{T}'\text{-mod}(\mathcal{E}) \simeq \mathbf{Flat}_J(\text{f.p.}\mathbb{T}\text{-mod}(\mathbf{Set})^{\text{op}}, \mathcal{E})$$

it follows that for any $\mathcal{E} \in \mathfrak{BTop}$ and any $F \in \mathbf{Flat}_J(\text{f.p.}\mathbb{T}\text{-mod}(\mathbf{Set})^{\text{op}}, \mathcal{E})$, F sends S_σ to an epimorphic family for any axiom σ of $\mathbb{T}'$ as in the statement of the theorem. This implies, by Lemma 3 at p. 393 of [63] applied to the canonical geometric inclusion

$$\mathbf{Sh}(\text{f.p.}\mathbb{T}\text{-mod}(\mathbf{Set})^{\text{op}}, J) \hookrightarrow [\text{f.p.}\mathbb{T}\text{-mod}(\mathbf{Set}), \mathbf{Set}],$$

that for any such σ the associated sheaf functor

$$a_J : [\text{f.p.}\mathbb{T}\text{-mod}(\mathbf{Set}), \mathbf{Set}] \to \mathbf{Sh}(\text{f.p.}\mathbb{T}\text{-mod}(\mathbf{Set})^{\text{op}}, J)$$

sends the monomorphism $S_\sigma \hookrightarrow \text{Hom}_{\text{f.p.}\mathbb{T}\text{-mod}(\mathbf{Set})}(M_{\{\vec{x}.\phi\}}, -)$ to an isomorphism i.e. that S_σ is J-covering. Thus the Grothendieck topology J' generated by the sieves S_σ is contained in J. To prove that $J' = J$, it is equivalent to verify, by Diaconescu's equivalence and the 2-dimensional Yoneda lemma (cf. section 3.1.1), that for any Grothendieck topos $\mathcal{E}$, $\mathbf{Flat}_{J'}(\text{f.p.}\mathbb{T}\text{-mod}(\mathbf{Set})^{\text{op}}, \mathcal{E}) = \mathbf{Flat}_J(\text{f.p.}\mathbb{T}\text{-mod}(\mathbf{Set})^{\text{op}}, \mathcal{E})$. Now, since $J' \subseteq J$, $\mathbf{Flat}_J(\text{f.p.}\mathbb{T}\text{-mod}(\mathbf{Set})^{\text{op}}, \mathcal{E}) \subseteq \mathbf{Flat}_{J'}(\text{f.p.}\mathbb{T}\text{-mod}(\mathbf{Set})^{\text{op}}, \mathcal{E})$. To prove the other inclusion, observe that, since J' contains all the sieves S_σ, the $\mathbb{T}$-model corresponding to a functor $F \in \mathbf{Flat}_{J'}(\text{f.p.}\mathbb{T}\text{-mod}(\mathbf{Set})^{\text{op}}, \mathcal{E})$ belongs to $\mathbb{T}'\text{-mod}(\mathcal{E})$ and hence, by the equivalence $\mathbb{T}'\text{-mod}(\mathcal{E}) \simeq \mathbf{Flat}_J(\text{f.p.}\mathbb{T}\text{-mod}(\mathbf{Set})^{\text{op}}, \mathcal{E})$, F is J-continuous, as required. □

Remarks 8.1.11 (a) Theorem 8.1.10 generalizes the method of construction of the classifying topos of a quotient of a cartesian theory given by Propositions D3.1.7 and D3.1.10 [55] (cf. also the theorem at p. 25 of [39] for the particular case of a Horn theory); indeed, it is well known that the opposite of the category of finitely presentable models of a cartesian theory is equivalent to its cartesian syntactic category (cf. Theorem 2.1.21).
(b) By Theorem 6.1.13, every sieve on f.p.$\mathbb{T}$-mod$(\mathbf{Set})^{\text{op}}$ is of the form S_σ for some geometric sequent σ as in the statement of Theorem 8.1.10.
(c) Given a quotient $\mathbb{T}'$ of a theory of presheaf type $\mathbb{T}$, one can obtain an axiomatization of $\mathbb{T}'$ of the form required in Theorem 8.1.10 by replacing any axiom $(\phi \vdash_{\vec{x}} \psi)$ of $\mathbb{T}'$, where $\phi(\vec{x})$ and $\psi(\vec{x})$ are geometric formulae over the signature of $\mathbb{T}$, with the collection of sequents corresponding to the sieves S on the category $\mathcal{C}_\mathbb{T}^{\text{irr}}$ obtained as follows. By Corollary 6.1.10, there is a $J_\mathbb{T}$-covering sieve T on $\{\vec{x}.\phi\}$ in $\mathcal{C}_\mathbb{T}$ generated by arrows f whose domain is a $\mathbb{T}$-irreducible formula. For each such arrow f, consider the object A_f of $\mathcal{C}_\mathbb{T}$ given by the pullback in $\mathcal{C}_\mathbb{T}$ of f along the canonical monomorphism $\{\vec{x}.\phi \wedge \psi\} \hookrightarrow \{\vec{x}.\phi\}$; then, again by Corollary 6.1.10, there exists a $J_\mathbb{T}$-covering sieve on A_f generated by arrows k^f whose domains are $\mathbb{T}$-irreducible geometric formulae.

The sieves S in $\mathcal{C}_{\mathbb{T}}^{\text{irr}}$ on the objects $\text{dom}(f)$ generated by the composites of the arrows k^f with the canonical arrow $A_f \to \text{dom}(f)$ satisfy the required condition since they are all covering for a Grothendieck topology on $\mathcal{C}_{\mathbb{T}}$ containing $J_{\mathbb{T}}$ if and only if the sieve generated by the canonical monomorphism $\{\vec{x} . \phi \wedge \psi\} \rightarrowtail \{\vec{x} . \phi\}$ is.

Theorem 8.1.12 *Let $\mathbb{T}$ be a theory of presheaf type over a signature Σ and J a Grothendieck topology on the category* f.p.$\mathbb{T}$-mod(**Set**)^{op}. *Then the unique (up to syntactic equivalence) quotient $\mathbb{T}'$ of $\mathbb{T}$ whose induced $\mathbb{T}$-topology is J (and hence whose models in every Grothendieck topos are exactly the J-homogeneous $\mathbb{T}$-models) is axiomatized by the sequents σ over Σ of the form $(\phi \vdash_{\vec{x}} \bigvee_{i \in I} (\exists \vec{y_i}) \theta_i)$ (where, for each $i \in I$, $[\theta_i] : \{\vec{y_i} . \psi\} \to \{\vec{x} . \phi\}$ is an arrow in $\mathcal{C}_{\mathbb{T}}$ and $\phi(\vec{x})$, $\psi(\vec{y_i})$ are geometric formulae over Σ which present $\mathbb{T}$-models) such that the sieve S_{σ} is J-covering. In fact, for any sequent σ of the above form, σ is provable in $\mathbb{T}'$ if and only if S_{σ} is J-covering.*

Proof By Theorem 8.1.3, the quotient $\mathbb{T}'$ of $\mathbb{T}$ whose induced $\mathbb{T}$-topology is J is characterized by the property that its models in every Grothendieck topos are the J-homogeneous $\mathbb{T}$-models. From the proof of Theorem 8.1.10 and Remark 8.1.11(b) it thus follows that $\mathbb{T}'$ is axiomatized by the sequents σ whose associated sieves S_{σ} are J-covering.

It thus remains to prove that, for any sequent σ of the form specified above, σ is provable in $\mathbb{T}'$ if and only if S_{σ} is J-covering. Since universal models are preserved by inverse image functors of geometric inclusions (cf. Lemma 2.1 [19]), this follows from the fact that the canonical monomorphism $S_{\sigma} \rightarrowtail \text{Hom}_{\text{f.p.}\mathbb{T}\text{-mod}(\mathbf{Set})}(M_{\{\vec{x} . \phi\}}, -)$ is isomorphic to $[[\vec{x} . \bigvee_{i \in I} (\exists \vec{y_i}) \theta_i)]]_{N_{\mathbb{T}}} \rightarrowtail [[\vec{x} . \phi]]_{N_{\mathbb{T}}}$, where $N_{\mathbb{T}}$ is the universal model of $\mathbb{T}$ defined in Theorem 6.1.1 (cf. Theorem 3.3 [19]), in light of the conservativity of universal models. □

Remark 8.1.13 Theorem 8.1.12, combined with the duality of Theorem 3.2.5, yields a proof-theoretic equivalence between the classical deduction system for geometric logic over $\mathbb{T}$ and the proof system $\mathcal{T}_{\text{f.p.}\mathbb{T}\text{-mod}(\mathbf{Set})^{\text{op}}}^{T}$ (where T is the trivial Grothendieck topology) introduced in section 3.2.2 (cf. also Remarks 8.1.11(b)-(c)). This generalizes Coste-Lombardi-Roy's correspondence [40] between dynamical theories (viewed as coherent quotients of a Horn theory) and the finite-type Grothendieck topologies associated with them.

Corollary 8.1.14 *Let $\mathbb{T}$ be a theory of presheaf type such that* f.p.$\mathbb{T}$-mod(**Set**)^{op} *satisfies the right Ore condition. Then the quotient $\mathbb{T}'$ of $\mathbb{T}$ whose induced $\mathbb{T}$-topology is the atomic topology on* f.p.$\mathbb{T}$-mod(**Set**)^{op} *is the theory obtained from $\mathbb{T}$ by adding all the axioms of the form $(\phi \vdash_{\vec{x}} (\exists \vec{y}) \theta)$, where $[\theta] : \{\vec{y} . \psi\} \to \{\vec{x} . \phi\}$ is any arrow in $\mathcal{C}_{\mathbb{T}}$ and $\phi(\vec{x})$, $\psi(\vec{y})$ are geometric formulae over the signature of $\mathbb{T}$ which present $\mathbb{T}$-models.*

Proof The thesis follows from Theorem 8.1.12 in light of Remark 8.1.11(b), by observing that the collection of principal sieves on f.p.$\mathbb{T}$-mod(**Set**)^{op} generates the atomic topology on f.p.$\mathbb{T}$-mod(**Set**)^{op}. □

Remark 8.1.15 If $\mathbb{T}$ is cartesian then the hypotheses of Corollary 8.1.14 are satisfied; indeed, the equivalence f.p.$\mathbb{T}$-mod(**Set**)$^{\mathrm{op}} \simeq \mathcal{C}_{\mathbb{T}}^{\mathrm{cart}}$ of Theorem 2.1.21 shows that the category f.p.$\mathbb{T}$-mod(**Set**)^{op} is cartesian and hence *a fortiori* satisfies the right Ore condition.

8.1.3 Quotients with enough set-based models

In this section we focus on quotients of theories of presheaf type having enough set-based models. We will extend some ideas and results from sections 3.2 and 3.4 of [68], rewriting them in a general topos-theoretic context. This will lead in particular to a representation for the classifying topos of such a quotient in terms of its category of set-based models, holding under the hypothesis that it has enough points.

Let us first review some basic facts about ind-completions which will be useful for our analysis; we refer the reader to section C4.2 of [55] for more details.

Recall that a finitely accessible category $\mathcal{L}$ is a category which is equivalent to the ind-completion Ind-$\mathcal{C}$ of a small category $\mathcal{C}$; Ind-$\mathcal{C}$ is defined to be the full subcategory of $[\mathcal{C}^{\mathrm{op}}, \mathbf{Set}]$ on the flat functors $F : \mathcal{C}^{\mathrm{op}} \to \mathbf{Set}$. Every representable functor is flat, so the Yoneda embedding $y : \mathcal{C} \to [\mathcal{C}^{\mathrm{op}}, \mathbf{Set}]$ factors through the embedding Ind-$\mathcal{C} \hookrightarrow [\mathcal{C}^{\mathrm{op}}, \mathbf{Set}]$; we will denote this factorization by $y_{\mathcal{C}} : \mathcal{C} \to$ Ind-$\mathcal{C}$. Moreover, the inclusion Ind-$\mathcal{C} \hookrightarrow [\mathcal{C}^{\mathrm{op}}, \mathbf{Set}]$ creates filtered colimits.

Given a finitely accessible category $\mathcal{L}$, we write f.p.$\mathcal{L}$ for the full subcategory of $\mathcal{L}$ on the finitely presentable objects; then the embedding f.p.$\mathcal{L} \hookrightarrow \mathcal{L}$ corresponds to the embedding $y_{\mathrm{f.p.}\mathcal{L}} : \mathrm{f.p.}\mathcal{L} \to$ Ind-$\mathcal{L}$ under the canonical equivalence $\mathcal{L} \simeq$ Ind-$\mathcal{L}$ (cf. Proposition C4.2.2 and Corollary A1.1.9 [55]).

The ind-completion Ind-$\mathcal{C}$ of a small category $\mathcal{C}$ is the free filtered-colimit completion of $\mathcal{C}$ (cf. Corollary C4.2.6 [55]); that is, for any category $\mathcal{D}$ with filtered colimits, any functor $H : \mathcal{C} \to \mathcal{D}$ extends, via $y_{\mathcal{C}} : \mathcal{C} \to$ Ind-$\mathcal{C}$, uniquely up to canonical isomorphism, to a filtered-colimit-preserving functor $\overline{H} :$ Ind-$\mathcal{C} \to \mathcal{D}$. Notice that if $H = \mathrm{Hom}_{\mathcal{C}}(c, -)$ then $\overline{H}$ is naturally isomorphic to $\mathrm{Hom}_{\mathrm{Ind}\text{-}\mathcal{C}}(y_{\mathcal{C}}(c), -)$.

Generalizing from section 3.2 of [67], given a small category $\mathcal{C}$, we shall construct correspondences between the collection $\mathcal{S}_{\mathrm{Ind}\text{-}\mathcal{C}}$ of full subcategories of Ind-$\mathcal{C}$ and the collection $\mathcal{G}(\mathcal{C})$ of Grothendieck topologies on $\mathcal{C}^{\mathrm{op}}$. Given a sieve S in $\mathcal{C}^{\mathrm{op}}$ on an object $c \in \mathcal{C}$, we denote by $\overline{S}$ the extension Ind-$\mathcal{C} \to \mathbf{Set}$ of $S : \mathcal{C} \to \mathbf{Set}$ (regarded as a subfunctor of the representable $\mathrm{Hom}_{\mathcal{C}}(c, -)$) along $y_{\mathcal{C}}$ as above.

Let us define two maps $\mathcal{K} : \mathcal{G}(\mathcal{C}) \to \mathcal{S}_{\mathrm{Ind}\text{-}\mathcal{C}}$ and $\mathcal{H} : \mathcal{S}_{\mathrm{Ind}\text{-}\mathcal{C}} \to \mathcal{G}(\mathcal{C})$ as follows.

Given a Grothendieck topology J on $\mathcal{C}^{\mathrm{op}}$, $\mathcal{K}(J)$ is the full subcategory of Ind-$\mathcal{C}$ defined by

$$d \in \mathcal{K}(J) \text{ iff } \overline{S}(d) = \mathrm{Hom}_{\mathrm{Ind}\text{-}\mathcal{C}}(y_{\mathcal{C}}(c), d) \text{ for all } S \in J(c),$$

for any $d \in$ Ind-$\mathcal{C}$. Here by the equality

$$\overline{S}(d) = \mathrm{Hom}_{\mathrm{Ind}\text{-}\mathcal{C}}(y_{\mathcal{C}}(c), d)$$

we mean that the canonical map $\overline{S}(d) \to \mathrm{Hom}_{\mathrm{Ind}\text{-}\mathcal{C}}(y_{\mathcal{C}}(c), d)$ is an isomorphism, equivalently that whenever $d = \mathrm{colim}(y_{\mathcal{C}} \circ G)$ in Ind-$\mathcal{C}$ for a functor $G : \mathcal{I} \to \mathcal{C}$ defined on a small filtered category $\mathcal{I}$, for any arrow $r : c \to G(i)$ there exist objects $j, k \in \mathcal{I}$ and arrows $s : c \to G(j)$ in S and $\chi : i \to k$, $\xi : j \to k$ in $\mathcal{I}$ such that $G(\chi) \circ r = G(\xi) \circ s$.

In the converse direction, given a full subcategory $\mathcal{D}$ of Ind-$\mathcal{C}$, we define $\mathcal{H}(\mathcal{D})$ by

$$S \in \mathcal{H}(\mathcal{D})(c) \text{ iff } \overline{S}(d) = \mathrm{Hom}_{\mathrm{Ind}\text{-}\mathcal{C}}(y_{\mathcal{C}}(c), d) \text{ for all } d \in \mathcal{D},$$

for any sieve S in $\mathcal{C}^{\mathrm{op}}$ on an object c.

It is easy to verify that for any full subcategory $\mathcal{D}$ of Ind-$\mathcal{C}$, $\mathcal{H}(\mathcal{D})$ is indeed a Grothendieck topology on $\mathcal{C}^{\mathrm{op}}$; we provide the details for the reader's convenience. It is clear that the maximality axiom holds. Let us verify the stability axiom. Given an arrow $f : c \to c'$ in $\mathcal{C}$ and a sieve $S \in \mathcal{H}(\mathcal{D})(c)$, we want to prove that $f^*(S) \in \mathcal{H}(\mathcal{D})(c')$, i.e. that for any functor $G : \mathcal{I} \to \mathcal{C}$ defined on a small filtered category $\mathcal{I}$ and any arrow $r : c' \to G(i)$, there exist $j, k \in \mathcal{I}$ and arrows $s : c' \to G(j)$ in $f^*(S)$ and $\chi : i \to k$, $\xi : j \to k$ in $\mathcal{I}$ such that $G(\chi) \circ r = G(\xi) \circ s$. Consider the arrow $r \circ f$; since $S \in \mathcal{H}(\mathcal{D})(c)$, there exist $j', k' \in \mathcal{I}$ and arrows $s' : c \to G(j')$ in S and $\chi' : i \to k'$, $\xi' : j' \to k'$ in $\mathcal{I}$ such that $G(\chi') \circ r \circ f = G(\xi') \circ s'$. Then if we take $i = i'$, $j = k = k'$, $\chi = \chi'$, $\xi = 1_k$ and $s = G(\chi) \circ r$, we have that $s \in f^*(S)$ and hence our thesis is satisfied. It remains to verify the transitivity axiom. Given a sieve R in $\mathcal{C}^{\mathrm{op}}$ on an object c and a sieve $S \in \mathcal{H}(\mathcal{D})(c)$ such that $f^*(R) \in \mathcal{H}(\mathcal{D})(c')$ for any arrow $f : c \to c'$ in $\mathcal{C}$ belonging to S, we want to prove that $R \in \mathcal{H}(\mathcal{D})(c)$. Since $S \in \mathcal{H}(\mathcal{D})(c)$, given an arrow $r : c \to G(i)$ in $\mathcal{C}$ there exist $j, k \in \mathcal{I}$ and arrows $f' : c \to G(j)$ in S and $\chi : i \to k$, $\xi : j \to k$ in $\mathcal{I}$ such that $G(\chi) \circ r = G(\xi) \circ f'$; now, since $f'^*(R) \in \mathcal{H}(\mathcal{D})(c')$, there exist $j', k' \in \mathcal{I}$ and arrows $g : G(j) \to G(j')$ in $f'^*(R)$ and $\chi' : k \to k'$, $\xi' : j' \to k'$ in $\mathcal{I}$ such that $G(\chi') \circ G(\xi) = G(\xi') \circ g$; so $G(\chi' \circ \chi) \circ r = G(\xi') \circ g \circ f'$ and our thesis is satisfied.

Next, we note that $\mathcal{S}_{\mathrm{Ind}\text{-}\mathcal{C}}$ and $\mathcal{G}(\mathcal{C})$ are naturally equipped with partial orders (respectively the inclusion ordering between full subcategories of $\mathcal{S}_{\mathrm{Ind}\text{-}\mathcal{C}}$ and the inclusion ordering between Grothendieck topologies on $\mathcal{C}^{\mathrm{op}}$) and if we regard them as poset categories then the correspondences $\mathcal{H}$ and $\mathcal{K}$ become contravariant functors; moreover, it is immediate to see that they form a Galois connection between $\mathcal{S}_{\mathrm{Ind}\text{-}\mathcal{C}}$ and $\mathcal{G}(\mathcal{C})$. From the formal theory of Galois connections, it thus follows that $\mathcal{H}(\mathcal{K}(\mathcal{H}(\mathcal{D}))) = \mathcal{H}(\mathcal{D})$ for any full subcategory $\mathcal{D}$ of Ind-$\mathcal{C}$ and $\mathcal{K}(\mathcal{H}(\mathcal{K}(J))) = \mathcal{K}(J)$ for any Grothendieck topology J on $\mathcal{C}^{\mathrm{op}}$.

The following lemma represents the extension of Lemma 3.11 [68] to the context of finitely accessible categories.

Lemma 8.1.16 *Let J be a Grothendieck topology on a small category $\mathcal{C}^{\mathrm{op}}$. Then, with the above notation, the functor $\mathrm{Hom}_{\mathrm{Ind}\text{-}\mathcal{C}}(y_{\mathcal{C}}(-), d) : \mathcal{C}^{\mathrm{op}} \to \mathbf{Set}$ is J-continuous if and only if $d \in \mathcal{K}(J)$, for any $d \in \mathrm{Ind}\text{-}\mathcal{C}$.*

Proof Recall from section 1.1.5 that via the equivalence

$$\mathbf{Geom}(\mathbf{Set}, [\mathcal{C}, \mathbf{Set}]) \simeq \mathbf{Flat}(\mathcal{C}^{\mathrm{op}}, \mathbf{Set})$$

a flat functor $F : \mathcal{C}^{\mathrm{op}} \to \mathbf{Set}$ corresponds to the geometric morphism having as inverse image $F \otimes_{\mathcal{C}} - \cong - \otimes_{\mathcal{C}^{\mathrm{op}}} F : [\mathcal{C}, \mathbf{Set}] \to \mathbf{Set}$ (cf. section 1.1.5 for the notation); so F is J-continuous (for a Grothendieck topology J on $\mathcal{C}^{\mathrm{op}}$) if and only if for every $S \in J(c)$, $F \otimes_{\mathcal{C}} -$ sends the monomorphism $S \rightarrowtail \mathrm{Hom}_{\mathcal{C}}(c, -)$ to an isomorphism. The functor $\mathrm{Hom}_{\mathrm{Ind}\text{-}\mathcal{C}}(y_{\mathcal{C}}(-), d) : \mathcal{C}^{\mathrm{op}} \to \mathbf{Set}$ is flat for any $d \in \mathrm{Ind}\text{-}\mathcal{C}$ (by the Yoneda representation of flat functors, cf. section 5.3), so it is J-continuous if and only if for every $S \in J(c)$, $\mathrm{Hom}_{\mathrm{Ind}\text{-}\mathcal{C}}(y_{\mathcal{C}}(-), d) \otimes_{\mathcal{C}} -$ sends $S \rightarrowtail \mathrm{Hom}_{\mathcal{C}}(c, -)$ to an isomorphism.

Now, if $d = \mathrm{colim}(y_{\mathcal{C}} \circ G)$ in $\mathrm{Ind}\text{-}\mathcal{C}$ for a functor $G : \mathcal{I} \to \mathcal{C}$ defined on a small filtered category $\mathcal{I}$ then

$$\mathrm{Hom}_{\mathrm{Ind}\text{-}\mathcal{C}}(y_{\mathcal{C}}(-), d) \cong \mathrm{colim}_{[\mathcal{C}^{\mathrm{op}}, \mathbf{Set}]} \mathrm{Hom}_{\mathrm{Ind}\text{-}\mathcal{C}}(y_{\mathcal{C}}(-), G(-)),$$

since all the objects of $\mathcal{C}$ are finitely presentable in $\mathrm{Ind}\text{-}\mathcal{C}$ and colimits in functor categories are computed pointwise. So, for any functor $P : \mathcal{C} \to \mathbf{Set}$, since $(- \otimes_{\mathcal{C}} P)$ preserves filtered colimits (having a right adjoint) and $\mathrm{Hom}_{\mathcal{C}}(-, c) \otimes_{\mathcal{C}} P \cong P(c)$ for any $c \in \mathcal{C}$ (cf. the proof of Theorem 1.1.31), we have that $\mathrm{Hom}_{\mathrm{Ind}\text{-}\mathcal{C}}(y_{\mathcal{C}}(-), d) \otimes_{\mathcal{C}} P \cong \mathrm{colim}(P \circ G) = \overline{P}(d)$. Hence $\mathrm{Hom}_{\mathrm{Ind}\text{-}\mathcal{C}}(y_{\mathcal{C}}(-), d) \otimes_{\mathcal{C}} -$ sends $S \rightarrowtail \mathrm{Hom}_{\mathcal{C}}(c, -)$ to the monomorphism $\overline{S}(d) \rightarrowtail \mathrm{Hom}_{\mathrm{Ind}\text{-}\mathcal{C}}(y_{\mathcal{C}}(c), d)$, from which our thesis follows. □

Remark 8.1.17 The proof of Lemma 8.1.16 actually shows that for any small category $\mathcal{C}$, sieve S in $\mathcal{C}^{\mathrm{op}}$ and object d of $\mathrm{Ind}\text{-}\mathcal{C}$, the equality

$$\overline{S}(d) = \mathrm{Hom}_{\mathrm{Ind}\text{-}\mathcal{C}}(y_{\mathcal{C}}(c), d)$$

holds if and only if d (regarded as a set-based model of the theory of flat functors on $\mathcal{C}^{\mathrm{op}}$) is S-homogeneous in the sense of Definition 8.1.1(a). It follows that for any Grothendieck topology J on $\mathcal{C}^{\mathrm{op}}$, $\mathcal{K}(J)$ is the full subcategory of $\mathrm{Ind}\text{-}\mathcal{C}$ on the J-homogeneous objects, and that for any full subcategory $\mathcal{D}$ of $\mathrm{Ind}\text{-}\mathcal{C}$, the $\mathcal{H}(\mathcal{D})$-covering sieves are precisely the sieves S in $\mathcal{C}^{\mathrm{op}}$ such that every object of $\mathcal{D}$ is S-homogeneous.

Proposition 8.1.18 *Let $\mathbb{T}$ be a theory of presheaf type and $\mathbb{T}'$ a quotient of $\mathbb{T}$. Then, denoting by $\mathbb{T}'$-$\mathrm{mod}(\mathbf{Set})$ the full subcategory of $\mathbb{T}$-$\mathrm{mod}(\mathbf{Set})$ on the $\mathbb{T}'$-models, we have $\mathcal{K}(\mathcal{H}(\mathbb{T}'\text{-}\mathrm{mod}(\mathbf{Set}))) = \mathbb{T}'\text{-}\mathrm{mod}(\mathbf{Set})$.*

Proof By Theorems 3.2.5 and 8.1.3, there exists a (unique) Grothendieck topology J on f.p.$\mathbb{T}$-$\mathrm{mod}(\mathbf{Set})^{\mathrm{op}}$ such that the $\mathbb{T}'$-models are exactly the J-homogeneous ones in every Grothendieck topos (cf. also the proof of Theorem 8.1.8); but by Lemma 8.1.16 a $\mathbb{T}$-model M in $\mathbf{Set}$ is J-homogeneous if and only if $M \in \mathcal{K}(J)$, whence $\mathbb{T}'$-$\mathrm{mod}(\mathbf{Set}) = \mathcal{K}(J)$. Our thesis thus follows from the discussion preceding Lemma 8.1.16. □

Remark 8.1.19 By Theorem 8.1.8 and Remark 8.1.17, every full subcategory of $\mathbb{T}$-mod(**Set**) of the form $\mathcal{K}(J)$ (for a Grothendieck topology J on the category f.p.$\mathbb{T}$-mod(**Set**)^{op}) is of the form $\mathbb{T}'$-mod(**Set**) for a quotient $\mathbb{T}'$ of $\mathbb{T}$. It thus follows from Proposition 8.1.18 and the discussion preceding Lemma 8.1.16 that, in the case of categories $\mathcal{C}$ of the form f.p.$\mathbb{T}$-mod(**Set**) for a theory of presheaf type $\mathbb{T}$, the 'closed' full subcategories of our Galois correspondence are precisely the categories of set-based models of quotients of $\mathbb{T}$.

We are now ready to prove the main result of this section.

Theorem 8.1.20 *Let $\mathbb{T}$ be a theory of presheaf type and $\mathbb{T}'$ be a quotient of $\mathbb{T}$ having enough set-based models. Then the topos*

$$\mathbf{Sh}(\text{f.p.}\mathbb{T}\text{-mod}(\mathbf{Set})^{\mathrm{op}}, \mathcal{H}(\mathbb{T}'\text{-mod}(\mathbf{Set})))$$

classifies $\mathbb{T}'$ (via Theorem 3.2.5), provided that it has enough points.

Proof From Theorem 8.1.8 we know that there exists a quotient $\mathbb{T}''$ of $\mathbb{T}$ classified by the topos $\mathbf{Sh}(\text{f.p.}\mathbb{T}\text{-mod}(\mathbf{Set})^{\mathrm{op}}, \mathcal{H}(\mathbb{T}'\text{-mod}(\mathbf{Set})))$ whose models in any Grothendieck topos are exactly the $\mathcal{H}(\mathbb{T}'\text{-mod}(\mathbf{Set}))$-homogeneous $\mathbb{T}$-models. Now, if the topos $\mathbf{Sh}(\text{f.p.}\mathbb{T}\text{-mod}(\mathbf{Set})^{\mathrm{op}}, \mathcal{H}(\mathbb{T}'\text{-mod}(\mathbf{Set})))$ has enough points then $\mathbb{T}''$ has enough set-based models (cf. Proposition 4.2.20). But $\mathbb{T}''$ has the same set-based models as the theory $\mathbb{T}'$ by Lemma 8.1.16 and Proposition 8.1.18 and $\mathbb{T}'$ has enough set-based models by our hypotheses, so we may conclude that $\mathbb{T}'$ and $\mathbb{T}''$ are syntactically equivalent and hence that they have equivalent classifying toposes; in particular, the topos

$$\mathbf{Sh}(\text{f.p.}\mathbb{T}\text{-mod}(\mathbf{Set})^{\mathrm{op}}, \mathcal{H}(\mathbb{T}'\text{-mod}(\mathbf{Set})))$$

classifies $\mathbb{T}'$ via Theorem 3.2.5, as required. □

From the proof of Theorem 8.1.20, we can extract the following result:

Proposition 8.1.21 *Let $\mathbb{T}$ be a theory of presheaf type and J a Grothendieck topology on the category f.p.$\mathbb{T}$-mod(**Set**)^{op} such that both the toposes*

$$\mathbf{Sh}(\text{f.p.}\mathbb{T}\text{-mod}(\mathbf{Set})^{\mathrm{op}}, J)$$

and

$$\mathbf{Sh}(\text{f.p.}\mathbb{T}\text{-mod}(\mathbf{Set})^{\mathrm{op}}, \mathcal{H}(\mathcal{K}(J)))$$

have enough points. Then $J = \mathcal{H}(\mathcal{K}(J))$.

Proof By the results of section 1.3.4, $J = \mathcal{H}(\mathcal{K}(J))$ if and only if there exists an equivalence

$$\mathbf{Sh}(\text{f.p.}\mathbb{T}\text{-mod}(\mathbf{Set})^{\mathrm{op}}, J) \simeq \mathbf{Sh}(\text{f.p.}\mathbb{T}\text{-mod}(\mathbf{Set})^{\mathrm{op}}, \mathcal{H}(\mathcal{K}(J)))$$

which commutes (in the obvious sense) with the canonical geometric inclusions

$$\mathbf{Sh}(\text{f.p.}\mathbb{T}\text{-mod}(\mathbf{Set})^{\mathrm{op}}, J) \hookrightarrow [\text{f.p.}\mathbb{T}\text{-mod}(\mathbf{Set}), \mathbf{Set}]$$

and

$$\mathbf{Sh}(\text{f.p.}\mathbb{T}\text{-mod}(\mathbf{Set})^{\mathrm{op}}, \mathcal{H}(\mathcal{K}(J))) \hookrightarrow [\text{f.p.}\mathbb{T}\text{-mod}(\mathbf{Set}), \mathbf{Set}];$$

but this is equivalent, by the 2-dimensional Yoneda lemma (cf. section 3.1.1) and the universal property of classifying toposes, to saying that the quotients $\mathbb{T}'$ and $\mathbb{T}''$ of $\mathbb{T}$ classified respectively by

$$\mathbf{Sh}(\text{f.p.}\mathbb{T}\text{-mod}(\mathbf{Set})^{\mathrm{op}}, J)$$

and

$$\mathbf{Sh}(\text{f.p.}\mathbb{T}\text{-mod}(\mathbf{Set})^{\mathrm{op}}, \mathcal{H}(\mathcal{K}(J)))$$

via Theorem 8.1.8 have exactly the same models in every Grothendieck topos. Now, if both $\mathbb{T}'$ and $\mathbb{T}''$ have enough set-based models then this happens precisely when they have exactly the same set-based models, equivalently (by Lemma 8.1.16) when $\mathcal{K}(J) = \mathcal{K}(\mathcal{H}(\mathcal{K}(J)))$; but this holds by the formal properties of Galois connections (cf. the discussion preceding Lemma 8.1.16). □

Remark 8.1.22 Theorem 8.1.20 and Proposition 8.1.21 can be notably applied in the case of a coherent quotient of a cartesian theory $\mathbb{T}$. Indeed, assuming the axiom of choice, if $\mathbb{T}'$ is a coherent quotient of $\mathbb{T}$ then it readily follows that $\mathcal{H}(\mathbb{T}'\text{-mod}(\mathbf{Set}))$ is a topology of finite type in the sense of Definition 8.1.23 (cf. the proof of Theorem 8.1.10 and Theorem 10.8.6(iii)). Conversely, if J is a topology of finite type on f.p.$\mathbb{T}$-mod$(\mathbf{Set})^{\mathrm{op}}$ then there exists a (unique up to syntactic equivalence, assuming the axiom of choice) coherent quotient $\mathbb{T}'$ of $\mathbb{T}$ such that $\mathbb{T}'\text{-mod}(\mathbf{Set}) = \mathcal{K}(J)$. See also Proposition 8.1.27 below.

Finally, let us discuss how our results relate to those in sections 3.2 and 3.4 of [68]. There the authors dealt with the case of embeddings $\mathcal{C} \hookrightarrow \text{Ind-}\mathcal{C}$ of the form f.p.$\mathcal{L} \hookrightarrow \mathcal{L}$ for a locally finitely presentable category $\mathcal{L}$. It is well known that these categories $\mathcal{L}$ are precisely the categories of set-based models of cartesian theories and that for any such category $\mathcal{L}$ the category f.p.$\mathcal{L}$ admits a syntactic description as the opposite of the syntactic category of the relevant cartesian theory (cf. Theorem 2.1.21). This fact is exploited in an essential way to derive some results in [68], for example Proposition 3.5. Instead, we have arrived at Proposition 8.1.18, which generalizes Proposition 3.5, by using the theory of classifying toposes and Theorem 3.2.5. Also, our Lemma 8.1.16 generalizes Lemma 3.11 [68], whose proof relied on the locally finite presentability of the category $\mathcal{L}$, and our Proposition 8.1.21 implies Proposition 3.12 [68] (by Theorem 2.1.12, Proposition 3.4(a) [68] and Remark 8.1.22).

8.1.4 Coherent quotients and topologies of finite type

In this section we shall study the lattice structure on the collection of coherent quotients of a given cartesian theory. Recall from section 2.2.4 that the notion of provability in a cartesian (resp. regular, coherent) theory $\mathbb{T}$ with respect to cartesian (resp. regular, coherent) logic coincides with the notion of provability in $\mathbb{T}$ with respect to geometric logic.

Recall that a Grothendieck topology on $\mathcal{C}$ is said to be generated by a given collection $\mathcal{F}$ of presieves on $\mathcal{C}$ if it is the smallest Grothendieck topology J on $\mathcal{C}$ such that all the sieves generated by presieves in $\mathcal{F}$ are J-covering (cf. section 3.1.2).

Definition 8.1.23 A Grothendieck topology J on a category $\mathcal{C}$ is said to be *of finite type* if it is generated by a collection of finite presieves on $\mathcal{C}$.

Proposition 8.1.24 *Let $\mathcal{C}$ be a category and J a Grothendieck topology on $\mathcal{C}$. Then J is of finite type if and only if there exists an assignment K sending to each object $c \in \mathcal{C}$ a collection $K(c)$ of finite presieves in $\mathcal{C}$ on c which satisfies the following properties:*

(i) *If $R \in K(c)$ then for any arrow $g : d \to c$ in $\mathcal{C}$ there exists a presieve $S \in K(d)$ such that, for each arrow f in S, $g \circ f \in R$.*

(ii) *If $\{f_i : c_i \to c \mid i \in I\} \in K(c)$ and for each $i \in I$ we have a presieve $\{g_{ij} : d_{ij} \to c_i \mid j \in J_i\} \in K(c_i)$ then there exists a presieve $P \in K(c)$ such that $P \subseteq \{f_i \circ g_{ij} : d_{ij} \to c \mid i \in I, j \in J_i\}$.*

(iii) *For any sieve S on $c \in \mathcal{C}$, $S \in J(c)$ if and only if $S \supseteq T$ for some $T \in K(c)$.*

Proof The 'if' part of the proposition follows immediately from Definition 3.1.3. Let us prove the 'only if' part. We define K by stipulating that, for any presieve V on an object $c \in \mathcal{C}$, $V \in K(c)$ if and only if V is finite and the sieve generated by it is J-covering. By Definition 3.1.3, K satisfies properties (i) and (ii) of the proposition. Let us now define K' as follows: for any sieve R on an object $c \in \mathcal{C}$, $R \in K'(c)$ if and only if $R \supseteq T$ for some $T \in K(c)$. We want to prove that $J = K'$. By Definition 3.1.3, K' is a Grothendieck topology and, clearly, K' is contained in J. But the fact that J is of finite type implies that $J \subseteq K'$, whence $J = K'$, as required. $\square$

Remark 8.1.25 The proof of Proposition 8.1.24 shows that a Grothendieck topology J on a small category $\mathcal{C}$ is of finite type if and only if every J-covering sieve contains a J-covering sieve generated by a finite presieve.

Proposition 8.1.26 *Let $\mathcal{C}$ be a small category and J_1, J_2 Grothendieck topologies on $\mathcal{C}$.*

(i) *If J_1, J_2 are of finite type then $J_1 \wedge J_2$ is of finite type.*

(ii) *If J_1, J_2 are of finite type then $J_1 \vee J_2$ is of finite type.*

Proof

(i) For any sieve S on an object $c \in \mathcal{C}$, $S \in (J_1 \wedge J_2)(c)$ if and only if $S \in J_1(c)$ and $S \in J_2(c)$. Let us denote by K_1 (resp. K_2) the collection of finite presieves which generate a J_1-covering (resp. a J_2-covering) sieve, and let K be defined as follows: for any presieve V on $c \in \mathcal{C}$, $V \in K(c)$ if and only if there exist $V_1 \in K_1(c)$ and $V_2 \in K_2(c)$ such that $V = V_1 \cup V_2$. It is immediate to see that K satisfies conditions (i) and (ii) of Proposition 8.1.24. Further, for any sieve S on $c \in \mathcal{C}$, $S \in (J_1 \wedge J_2)(c)$ if and only if $S \supseteq T$ for some $T \in K(c)$. So $J_1 \wedge J_2$ is of finite type by Proposition 8.1.24.

(ii) Since $J_1 \vee J_2$ is the smallest Grothendieck topology on $\mathcal{C}$ which contains both J_1 and J_2, our thesis follows at once by observing that we can get a collection of finite presieves generating $J_1 \vee J_2$ by taking the union of any two collections of finite presieves generating J_1 and J_2.

□

Notice that, given two representations $\mathbf{Sh}(\mathcal{C}, J) \simeq \mathbf{Sh}(\mathcal{C}', J')$ of the same Grothendieck topos in terms of small sites, we may construct a bijection between the class $\mathfrak{Groth}_J^{\mathcal{C}}$ of Grothendieck topologies on $\mathcal{C}$ which contain J and the class $\mathfrak{Groth}_{J'}^{\mathcal{C}'}$ of Grothendieck topologies on $\mathcal{C}'$ which contain J'. Indeed, it is well known (cf. section 1.3.4) that the Grothendieck topologies on $\mathcal{C}$ (resp. $\mathcal{C}'$) which contain J (resp. J') are in bijection with the (isomorphism classes of) geometric inclusions into the topos $\mathbf{Sh}(\mathcal{C}, J)$ (resp. $\mathbf{Sh}(\mathcal{C}', J')$), whence composing geometric inclusions with the equivalence $\mathbf{Sh}(\mathcal{C}, J) \simeq \mathbf{Sh}(\mathcal{C}', J')$ yields a bijection between $\mathfrak{Groth}_J^{\mathcal{C}}$ and $\mathfrak{Groth}_{J'}^{\mathcal{C}'}$. Since the natural order between geometric inclusions corresponds to the canonical order between the associated Grothendieck topologies, our bijection between $\mathfrak{Groth}_J^{\mathcal{C}}$ and $\mathfrak{Groth}_{J'}^{\mathcal{C}'}$ is order-preserving and hence a Heyting algebra isomorphism. This fact can be notably exploited for obtaining explicit descriptions of the lattice operations between theories, as will be illustrated by the example in section 8.1.5. For instance, if $\mathbb{T}$ is a theory of presheaf type then its classifying topos can be represented both as $\mathbf{Sh}(\mathcal{C}_\mathbb{T}, J_\mathbb{T})$ and as the presheaf topos $[\text{f.p.}\mathbb{T}\text{-mod}(\mathbf{Set}), \mathbf{Set}]$; thus, by Theorem 3.2.5, there is a Heyting algebra isomorphism

$$\mathfrak{Th}_\Sigma^\mathbb{T} \cong \mathfrak{Groth}_T^{\text{f.p.}\mathbb{T}\text{-mod}(\mathbf{Set})^{\text{op}}}$$

(where T is the trivial Grothendieck topology on f.p.$\mathbb{T}$-mod$(\mathbf{Set})^{\text{op}}$) between the (syntactic-equivalence classes of) quotients $\mathbb{T}'$ of $\mathbb{T}$ and the Grothendieck topologies J on the category f.p.$\mathbb{T}$-mod$(\mathbf{Set})^{\text{op}}$, with the property that for any such $\mathbb{T}'$ the topos $\mathbf{Sh}(\text{f.p.}\mathbb{T}\text{-mod}(\mathbf{Set})^{\text{op}}, J)$ of sheaves with respect to the corresponding Grothendieck topology J classifies $\mathbb{T}'$ (cf. section 8.1.2):

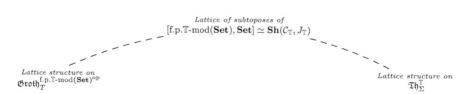

Proposition 8.1.27 *Let $\mathbb{T}$ be a cartesian theory over a signature Σ and $\mathcal{C}_\mathbb{T}^{\text{cart}}$ its cartesian syntactic category. Then the bijection between the (syntactic-equivalence*

classes of) quotients of $\mathbb{T}$ and the Grothendieck topologies on $\mathcal{C}_{\mathbb{T}}^{\text{cart}}$ induced by Theorem 3.2.5 restricts to a bijection between the (syntactic-equivalence classes of) coherent quotients of $\mathbb{T}$ and the finite-type Grothendieck topologies on $\mathcal{C}_{\mathbb{T}}^{\text{cart}}$.

Proof The equivalence of classifying toposes $\mathbf{Sh}(\mathcal{C}_{\mathbb{T}}, J_{\mathbb{T}}) \simeq [\mathcal{C}_{\mathbb{T}}^{\text{cart}}, \mathbf{Set}]$ induces by Theorem 3.2.5 a bijection between the (syntactic-equivalence classes of) geometric quotients of $\mathbb{T}$ and the Grothendieck topologies on $\mathcal{C}_{\mathbb{T}}^{\text{cart}}$, which we may describe as follows. Given a Grothendieck topology J on $\mathcal{C}_{\mathbb{T}}^{\text{cart}}$, the corresponding theory is axiomatized by the geometric sequents over Σ of the form $(\psi \vdash_{\vec{y}} \bigvee_{i \in I} (\exists \vec{x_i}) \theta_i)$, where $\{[\theta_i] \mid i \in I\}$ is any family of morphisms

$$\{\vec{x_i} . \phi_i\} \xrightarrow{[\theta_i]} \{\vec{y} . \psi\}$$

in $\mathcal{C}_{\mathbb{T}}^{\text{cart}}$ forming a J-covering sieve. Conversely, by Proposition D1.3.10 [55], any geometric (resp. coherent) theory over Σ can be axiomatized by sequents of the form $(\psi \vdash_{\vec{y}} \bigvee_{i \in I} (\exists \vec{x_i}) \theta_i)$, where ψ and the θ_i are cartesian formulae over Σ such that for any $i \in I$ the sequent $(\theta_i \vdash_{\vec{x_i}, \vec{y}} \psi)$ is provable in geometric logic (where I may be taken finite if $\mathbb{T}$ is coherent); the corresponding Grothendieck topology on $\mathcal{C}_{\mathbb{T}}^{\text{cart}}$ is thus generated by the sieves

$$\{\vec{x_i}, \vec{y'} . \theta_i\} \xrightarrow{[\theta_i \wedge \vec{y} = \vec{y'}]} \{\vec{y} . \psi\}$$

as i varies in I.

Our thesis then follows from Remark 3.2.9 or from Theorem 8.1.10 and Remark 8.1.11(b). $\square$

Remark 8.1.28 The analogue of Proposition 8.1.27 for regular theories also holds; specifically, the regular quotients of a cartesian theory $\mathbb{T}$ correspond to the Grothendieck topologies J on $\mathcal{C}_{\mathbb{T}}^{\text{cart}}$ such that every J-covering sieve contains a J-covering sieve generated by a single arrow.

Corollary 8.1.29 Let $\mathbb{T}$ be a cartesian theory over a signature Σ. Then the collection of (syntactic-equivalence classes of) coherent quotients of $\mathbb{T}$ form a sublattice of the collection $\mathfrak{Th}_{\Sigma}^{\mathbb{T}}$ of (syntactic-equivalence classes of) quotients of $\mathbb{T}$.

Proof This follows immediately from Proposition 8.1.27, Theorem 3.2.5 and Proposition 8.1.26. $\square$

Corollary 8.1.29 implies that, more generally, for any geometric theory $\mathbb{T}$ over a signature Σ, the class of coherent theories in $\mathfrak{Th}_{\Sigma}^{\mathbb{T}}$ is closed under finite meets and joins in $\mathfrak{Th}_{\Sigma}^{\mathbb{T}}$; indeed, by the discussion preceding Proposition 8.1.27, the meets and joins of subtoposes of $\mathbf{Sh}(\mathcal{C}_{\mathbb{T}}, J_{\mathbb{T}}) \simeq \mathbf{Sh}(\mathcal{C}_{\mathbb{O}_\Sigma}, J_{\mathbb{T}}^{\mathbb{O}_\Sigma})$ (where $\mathbb{O}_\Sigma$ is the empty theory over Σ) are the same as those calculated in the lattice of subtoposes of $\mathbf{Sh}(\mathcal{C}_{\mathbb{O}_\Sigma}, J_{\mathbb{O}_\Sigma})$.

Remark 8.1.30 By Theorem 2.2.2, the order relation between coherent theories in $\mathfrak{Th}_\Sigma^{\mathbb{T}}$ is equivalent to the natural ordering on them defined as follows: $\mathbb{T}_1 \leq \mathbb{T}_2$ if and only if every (coherent) axiom of $\mathbb{T}_1$ is provable in $\mathbb{T}_2$ using coherent logic. Moreover, assuming the axiom of choice, the classical completeness theorem for coherent logic (Theorem 1.4.18) implies that this order relation also coincides with the well-known ordering on first-order theories, the condition $\mathbb{T}_1 \leq \mathbb{T}_2$ being equivalent to the requirement 'for any Σ-structure M in **Set**, if M is a $\mathbb{T}_2$-model then M is a $\mathbb{T}_1$-model'.

8.1.5 An example

In this section, as an application of the results obtained in section 8.1.4, we calculate the meet of the theory of local rings and that of integral domains in the lattice of coherent (equivalently, of geometric) theories over the signature of the theory of commutative rings with unit.

Let Σ be the one-sorted signature consisting of two binary function symbols $+$ and $\cdot$, one unary function symbol $-$ and two constants 0 and 1, and let $\mathbb{T}$ be the algebraic theory over Σ axiomatizing the notion of commutative ring with unit. The category f.p.$\mathbb{T}$-mod(**Set**) coincides with the category **Rng**$_{\text{f.g.}}$ of finitely generated commutative rings with unit. The theory $\mathbb{T}_1$ of local rings is obtained from $\mathbb{T}$ by adding the sequents

$$(0 = 1 \vdash_{[]} \bot)$$

and

$$((\exists z)((x + y) . z = 1) \vdash_{x,y} (\exists z)(x . z = 1) \vee (\exists z)(y . z = 1)),$$

while the theory $\mathbb{T}_2$ of integral domains is obtained from $\mathbb{T}$ by adding the sequents

$$(0 = 1 \vdash_{[]} \bot)$$

and

$$(x . y = 0 \vdash_{x,y} (x = 0) \vee (y = 0)).$$

Consider the Grothendieck topologies J_1 and J_2 on f.p.$\mathbb{T}$-mod(**Set**)^{op} corresponding respectively to $\mathbb{T}_1$ and $\mathbb{T}_2$. By Example D3.1.11(a) [55] and the calculation preceding Lemma 6.6 [13], we have the following descriptions of J_1 and J_2: for any $A \in$ f.p.$\mathbb{T}$-mod(**Set**) and any sieve S on A in f.p.$\mathbb{T}$-mod(**Set**)^{op},

- $S \in J_1(A)$ if and only if S contains a finite family $\{\xi_i : A \to A[s_i^{-1}] \mid 1 \leq i \leq n\}$ of canonical inclusions $\xi_i : A \to A[s_i^{-1}]$ in **Rng**$_{\text{f.g.}}$ where $\{s_1, \ldots, s_n\}$ is any set of elements of A which is not contained in any proper ideal of A;
- $S \in J_2(A)$ if and only if either A is the zero ring and S is the empty sieve on it or S contains a non-empty finite family $\{\pi_{a_i} : A \to A/(a_i) \mid 1 \leq i \leq n\}$ of canonical projections $\pi_{a_i} : A \to A/(a_i)$ in **Rng**$_{\text{f.g.}}$ where $\{a_1, \ldots, a_n\}$ is any set of elements of A such that $a_1 \cdot \ldots \cdot a_n = 0$.

We may identify the polynomials with integer coefficients in a finite number of variables with the R-equivalence classes of terms over Σ, where R is the equivalence relation given by '$t_1 \mathrel{R} t_2$ if and only if $(\top \vdash t_1 = t_2)$ is provable in $\mathbb{T}$'.

The meet $\mathbb{T}_1 \wedge \mathbb{T}_2$ in $\mathfrak{Th}_\Sigma^\mathbb{T}$ (or, equivalently, in $\mathfrak{Th}_\Sigma^{\mathbb{O}_\Sigma}$) corresponds to the intersection $J_1 \wedge J_2$ in $\mathfrak{Groth}_T^{\mathbf{Rng}_{\mathrm{f.g.}}^{\mathrm{op}}}$, where T is the trivial Grothendieck topology on $\mathbf{Rng}_{\mathrm{f.g.}}^{\mathrm{op}}$ (cf. the discussion preceding Proposition 8.1.27). Now, the topology $J_1 \wedge J_2$ is generated by the collection of presieves of the form $V_1 \cup V_2$ where V_1 and V_2 are finite presieves on the same object of $\mathbf{Rng}_{\mathrm{f.g.}}$ which respectively generate a J_1-covering sieve and a J_2-covering sieve (cf. the proof of Proposition 8.1.26(i)), and any object of $\mathbf{Rng}_{\mathrm{f.g.}}$ has the form $\mathbb{Z}[x_1,\ldots,x_n]/(P_1,\ldots,P_s)$ where $P_1,\ldots,P_s$ are polynomials in $\mathbb{Z}[x_1,\ldots,x_n]$. It thus follows from Theorem 8.1.12 and Remark 8.1.11(b) that $\mathbb{T}_1 \wedge \mathbb{T}_2$ is the theory obtained from $\mathbb{T}$ by adding the sequents

$$(0 = 1 \vdash_{[]} \bot)$$

and

$$\left(\bigwedge_{1 \leq s \leq m} P_s(\vec{x}) = 0 \vdash_{\vec{x}} \bigvee_{1 \leq i \leq k} (\exists y)(G_i(\vec{x}) \cdot y = 1) \vee \bigvee_{1 \leq j \leq l} H_j(\vec{x}) = 0 \right),$$

for any polynomials P_s, G_i and H_j in a finite string $\vec{x}$ of variables with the property that if $\vec{x} = (x_1,\ldots,x_n)$ then $\{P_1,\ldots,P_s,G_1,\ldots,G_k\}$ is any set of elements of $\mathbb{Z}[x_1,\ldots,x_n]$ which is not contained in any proper ideal of $\mathbb{Z}[x_1,\ldots,x_n]$ and $\prod_{1 \leq j \leq l} H_j \in (P_1,\ldots,P_s)$ in $\mathbb{Z}[x_1,\ldots,x_n]$.

8.2 Presheaf-type quotients

In this section we shall prove some results which show that it is often useful, in investigating whether a given geometric theory $\mathbb{S}$ is of presheaf type, to consider it in relation to a theory $\mathbb{T}$ of presheaf type of which $\mathbb{S}$ is a quotient.

8.2.1 Finality conditions

The following result is a corollary of Theorem 7.2.3 and Proposition 7.2.5.

Corollary 8.2.1 *Let $\mathbb{T}$ be a theory of presheaf type over a signature Σ and $\mathbb{S}$ a quotient of $\mathbb{T}$ such that all the finitely presentable $\mathbb{S}$-models are finitely presentable as $\mathbb{T}$-models (e.g. $\mathbb{T}$ can be the empty theory over a finite signature and $\mathbb{S}$ any geometric theory over Σ whose finitely presentable models are all finite, cf. Proposition 9.1.1 below). Suppose, moreover, that for any object E of a Grothendieck topos $\mathcal{E}$ and any Σ-structure homomorphism $x : c \to \mathrm{Hom}_\mathcal{E}(E,M)$, where c is a finitely presentable $\mathbb{T}$-model and M is a $\mathbb{S}$-model in $\mathcal{E}$, there exist an epimorphic family $\{e_i : E_i \to E \mid i \in I\}$ in $\mathcal{E}$ and for each $i \in I$ a $\mathbb{T}$-model homomorphism $f_i : c \to c_i$, where c_i is a finitely presentable $\mathbb{S}$-model, and a Σ-structure homomorphism $x_i : c_i \to \mathrm{Hom}_\mathcal{E}(E_i,M)$ such that $x_i \circ f_i = \mathrm{Hom}_\mathcal{E}(e_i,M) \circ x$ for all $i \in I$. Then $\mathbb{S}$ is of presheaf type.*

Proof Condition (iii) of Theorem 6.3.1 is satisfied by Proposition 6.3.12, while the fact that conditions (i) and (ii) of Theorem 6.3.1 hold follows from Theorem 7.2.3 in view of Proposition 7.2.5 (in its statement take $\mathbb{T}$ equal to $\mathbb{S}$, $\mathbb{T}'$ equal to $\mathbb{T}$, $\mathcal{K}$ equal to the category of finitely presentable $\mathbb{S}$-models and $\mathcal{H}$ equal to the category of finitely presentable $\mathbb{T}$-models – the fact that $\int a$ is $\mathcal{E}$-full readily follows from the fact that $\mathbb{S}$ is a quotient of $\mathbb{T}$). □

In the sequel we shall refer to a family of homomorphisms x_i as in the statement of Corollary 8.2.1 as a 'localization' of the homomorphism x.

Remarks 8.2.2 (a) Corollary 8.2.1, when applied to the empty theory $\mathbb{T}$ over a finite signature and to a geometric theory $\mathbb{S}$ over Σ whose finitely presentable models are all finite, generalizes Theorem 3.6 [73], whose hypotheses expressed syntactically in terms of geometric logic, once interpreted in topos-theoretic semantics, are stronger than those of the corollary.

(b) Assuming the first of the hypotheses of Corollary 8.2.1, the second hypothesis is not only sufficient but also necessary for $\mathbb{S}$ to be of presheaf type. Indeed, if $\mathbb{S}$ is of presheaf type then every model M of $\mathbb{S}$ in a Grothendieck topos $\mathcal{E}$ is canonically an $\mathcal{E}$-filtered colimit of a diagram of 'constant' models associated with the finitely presentable $\mathbb{S}$-models (cf. sections 6.2.3 and 6.3.3), which by the first hypothesis are also finitely presentable as $\mathbb{T}$-models, whence the fact that the second hypothesis of the corollary is satisfied follows from Propositions 6.2.13 and 6.2.15.

(c) Corollary 8.2.1 can often be applied to pairs of the form $(\mathbb{S}, \mathbb{T} = \mathbb{S}_c)$, where $\mathbb{S}$ is a geometric theory over a signature Σ and $\mathbb{T}$ is equal to its cartesianization $\mathbb{S}_c$. It is not known if it is true in general that every finitely presentable model of a theory of presheaf type $\mathbb{T}$ is finitely presentable also as a $\mathbb{T}_c$-model. Nonetheless, this property is satisfied by all the examples of theories of presheaf type considered in this book (see Chapter 9 below).

Corollary 8.2.1 can also be applied to pairs consisting of a geometric theory over a signature Σ and the empty theory over the same signature, provided that the former satisfies appropriate hypotheses. In order to apply the corollary in this context, we need the following lemma. Below in this section, given a finite signature Σ, when we say that a Σ-structure c is *finite* we mean that cA is finite for every sort A of Σ.

Lemma 8.2.3 *Let Σ be a finite signature and $\mathbb{T}$ a geometric theory over Σ with a finite number of axioms each of which is of the form $(\top \vdash_{\vec{x}} \bigvee\limits_{i \in I} \phi_i)$, where the ϕ_i are finite conjunctions of atomic formulae. Then, for any $\mathbb{T}$-model M in a Grothendieck topos $\mathcal{E}$, any object E of $\mathcal{E}$ and any Σ-structure homomorphism $f : c \to \mathrm{Hom}_{\mathcal{E}}(E, M)$, where c is a finite Σ-structure, there exist an epimorphic family $\{e_i : E_i \to E \mid i \in I\}$ in $\mathcal{E}$ and for each $i \in I$ a finite Σ-substructure c_i of $\mathrm{Hom}_{\mathcal{E}}(E_i, M)$ which is a model of $\mathbb{T}$ and a Σ-structure homomorphism $f_i : c \to c_i$ such that $\mathrm{Hom}_{\mathcal{E}}(e_i, M) \circ f = j_i \circ f_i$ (where j_i is the canonical inclusion of c_i into $\mathrm{Hom}_{\mathcal{E}}(E_i, M)$).*

Proof For each axiom σ of the form $(\top \vdash_{\vec{x}_\sigma} \bigvee_{i \in I_\sigma} \phi_i^\sigma)$ and any finite tuple $\vec{\xi}$ of elements of c of the same type as $\vec{x}_\sigma$, since M is by our hypotheses a $\mathbb{T}$-model, there exists an epimorphic family $\{e_i^{\vec{\xi}} : E_i^{\vec{\xi}} \to E \mid i \in I_\sigma\}$ in $\mathcal{E}$ such that for each $i \in I_\sigma$, $f(\vec{\xi}) \circ e_i^{\vec{\xi}}$ factors through $[[\vec{x}_\sigma . \phi_i]]_M$. Since there is only a finite number of axioms of $\mathbb{T}$ and of elements of c, and the fibred product of a finite number of epimorphic families is again an epimorphic family, there exists an epimorphic family $\{e_k : E_k \to E \mid k \in K\}$ in $\mathcal{E}$ such that, for any $k \in K$, any axiom σ of $\mathbb{T}$ and any tuple $\vec{\xi}$ of elements of c of the appropriate sorts, the element $f(\vec{\xi}) \circ e_k \in \mathrm{Hom}_\mathcal{E}(E_k, M)$ satisfies the formula on the right-hand side of the sequent σ. Hence if we consider, for each $k \in K$, the surjection-inclusion factorization $j_k \circ f_k$ of the homomorphism $\mathrm{Hom}_\mathcal{E}(e_k, M) \circ f : c \to \mathrm{Hom}_\mathcal{E}(E_k, M)$ (in the sense of Lemma 7.2.37), we obtain that $c_k = \mathrm{Im}(f_k)$ is finite, since f_k is surjective, and that all the axioms of $\mathbb{T}$ are satisfied in c_k (since the formulae ϕ_i are finite conjunctions of atomic formulae and c_k is a Σ-substructure of $\mathrm{Hom}_\mathcal{E}(E_k, M)$). This proves our thesis. $\square$

Corollary 8.2.4 *Let Σ be a finite signature and $\mathbb{T}$ a geometric theory over Σ with a finite number of axioms each of which is of the form $(\top \vdash_{\vec{x}} \bigvee_{i \in I} \phi_i)$, where the ϕ_i are finite conjunctions of atomic formulae. Suppose that for any $\mathbb{T}$-model M in a Grothendieck topos $\mathcal{E}$ and any object E of $\mathcal{E}$, every finitely generated Σ-substructure of $\mathrm{Hom}_\mathcal{E}(E, M)$ has sortwise only a finite number of elements besides the constants (e.g. when the signature Σ does not contain function symbols except for a finite number of constants). Then $\mathbb{T}$ is of presheaf type, classified by the topos of covariant set-valued functors on the category of finite models of $\mathbb{T}$.*

Proof As Σ is finite, every $\mathbb{T}$-model that sortwise contains only finitely many elements besides the constants is finitely presentable as a model of the empty theory $\mathbb{O}_\Sigma$ over Σ (cf. Proposition 9.1.1 below). This remark ensures, by Lemma 8.2.3, that the theory $\mathbb{T}$ satisfies the hypotheses of Corollary 8.2.1 with respect to $\mathbb{O}_\Sigma$. Since the latter theory is of presheaf type (as it is cartesian), it follows that $\mathbb{T}$ is of presheaf type as well, as required. $\square$

8.2.2 Rigid topologies

In this section we shall analyse the presheaf-type quotients of a theory of presheaf type $\mathbb{T}$ in terms of the associated subtoposes of the classifying topos of $\mathbb{T}$.

Recall from section 6.1.3 that a Grothendieck topology J on a category $\mathcal{C}$ is said to be rigid if for every object c of $\mathcal{C}$, the family of all the arrows to c in $\mathcal{C}$ from J-irreducible objects of $\mathcal{C}$ (i.e. objects d with the property that the only J-covering sieve on d is the maximal one), generates a J-covering sieve.

Let us consider a theory of presheaf type $\mathbb{T}$ and a quotient $\mathbb{T}'$ of $\mathbb{T}$ corresponding to a Grothendieck topology J on f.p.$\mathbb{T}$-mod$(\mathbf{Set})^{\mathrm{op}}$ as in section 8.1.2. The following result gives a characterization of the models of $\mathbb{T}'$ which are finitely presentable as models of $\mathbb{T}$ in terms of the topology J; it is useful for recognizing the quotients of a theory of presheaf type which are again of presheaf type.

Theorem 8.2.5 *Let $\mathbb{T}'$ be a quotient of a theory of presheaf type $\mathbb{T}$ corresponding via the duality of Theorem 3.2.5 to a Grothendieck topology J on f.p.$\mathbb{T}$-mod($\mathbf{Set}$)^{op} (as in section 8.1.2), and let M be a finitely presentable $\mathbb{T}$-model. Then M is a model of $\mathbb{T}'$ if and only if it is J-irreducible as an object of f.p.$\mathbb{T}$-mod($\mathbf{Set}$)^{op}.*

Proof As shown in section 8.1.2, the theory $\mathbb{T}'$ can be identified with the quotient of $\mathbb{T}$ obtained by adding all the sequents σ of the form $(\phi \vdash_{\vec{x}} \bigvee_{i \in I} (\exists \vec{y_i}) \theta_i)$ where, for any $i \in I$, $[\theta_i] : \{\vec{y_i}.\psi_i\} \to \{\vec{x}.\phi\}$ is an arrow in $\mathcal{C}_{\mathbb{T}}$, $\phi(\vec{x})$ and $\psi_i(\vec{y_i})$ are formulae presenting respectively $\mathbb{T}$-models $M_{\{\vec{x}.\phi\}}$ and $M_{\{\vec{y_i}.\psi_i\}}$ and the sieve S_σ on $M_{\{\vec{x}.\phi\}}$ in the category f.p.$\mathbb{T}$-mod($\mathbf{Set}$)^{op} generated by the arrows $M_{[\theta_i]}$ (using the notation in the proof of Theorem 6.1.13) is J-covering. On the other hand, the Grothendieck topology J associated with the quotient $\mathbb{T}'$ consists of all the sieves of the form S_σ for a sequent σ provable in $\mathbb{T}'$ of the form $(\phi \vdash_{\vec{x}} \bigvee_{i \in I} (\exists \vec{y_i}) \theta_i)$, where, for any $i \in I$, $[\theta_i] : \{\vec{y_i}.\psi_i\} \to \{\vec{x}.\phi\}$ is an arrow in $\mathcal{C}_{\mathbb{T}}$ and $\phi(\vec{x})$, $\psi(\vec{y_i})$ are formulae presenting respectively $\mathbb{T}$-models $M_{\{\vec{x}.\phi\}}$ and $M_{\{\vec{y_i}.\psi_i\}}$.

Let us establish the following claim, from which our thesis will follow easily: for any sequent $\sigma := (\phi \vdash_{\vec{x}} \bigvee_{i \in I} (\exists \vec{y_i}) \theta_i)$ of the above form and any set-based $\mathbb{T}$-model N, σ holds in N if and only if every arrow $f : M_{\{\vec{x}.\phi\}} \to N$ in $\mathbb{T}$-mod($\mathbf{Set}$) factors through an arrow in S_σ.

Clearly, σ holds in N if and only if for any $\vec{a} \in [[\vec{x}.\phi]]_N$ there exist $i \in I$ and $\vec{b} \in [[\vec{y_i}.\psi_i]]_N$ such that $(\vec{b}, \vec{a}) \in [[\vec{y_i}, \vec{x}.\theta_i]]_N$. By the universal property of the model $M_{\{\vec{x}.\phi\}}$, the tuples $\vec{a} \in [[\vec{x}.\phi]]_N$ can be identified with the arrows $M_{\{\vec{x}.\phi\}} \to N$ in $\mathbb{T}$-mod($\mathbf{Set}$) via the correspondence which assigns to $\vec{a} \in [[\vec{x}.\phi]]_N$ the unique $\mathbb{T}$-model homomorphism $f_{\vec{a}} : M_{\{\vec{x}.\phi\}} \to N$ such that $f_{\vec{a}}(\vec{\xi_\phi}) = \vec{a}$, and similarly for the tuples $\vec{b} \in [[\vec{y_i}.\psi_i]]_N$. We can easily see that the property that $(\vec{b}, \vec{a}) \in [[\vec{y_i}, \vec{x}.\theta_i]]_N$ is exactly equivalent to the requirement that $f_{\vec{b}} \circ M_{[\theta_i]} = f_{\vec{a}}$; indeed, the equality $f_{\vec{b}} \circ M_{[\theta_i]} = f_{\vec{a}}$ holds if and only if $(f_{\vec{b}} \circ M_{[\theta_i]})(\vec{\xi_\phi}) = f_{\vec{a}}(\vec{\xi_\phi})$ and, since $M_{[\theta_i]}(\vec{\xi_\phi}) = [[\vec{y_i}, \vec{x}.\theta_i]]_{M_{\{\vec{y_i}.\psi_i\}}}(\vec{\xi_{\psi_i}})$, we have that $(f_{\vec{b}} \circ M_{[\theta_i]})(\vec{\xi_\phi}) = f_{\vec{b}}([[\vec{y_i}, \vec{x}.\theta_i]]_{M_{\{\vec{y_i}.\psi_i\}}}(\vec{\xi_{\psi_i}})) = [[\vec{y_i}, \vec{x}.\theta_i]]_N(f_{\vec{b}}(\vec{\xi_\phi})) = [[\vec{y_i}, \vec{x}.\theta_i]]_N(\vec{b})$ (where we have identified $[[\vec{y_i}, \vec{x}.\theta_i]]_{M_{\{\vec{y_i}.\psi_i\}}}$ and $[[\vec{y_i}, \vec{x}.\theta_i]]_N$ with the maps of which they are the graphs). This proves our claim.

Now, for any sequent σ of the above form which is provable in $\mathbb{T}'$, the associated sieve S_σ is J-covering whence for any arrow $f : M_{\{\vec{x}.\phi\}} \to M$ in f.p.$\mathbb{T}$-mod($\mathbf{Set}$) the pullback sieve $f^*(S_\sigma)$ is J-covering on M and hence maximal (i.e. f belongs to S_σ) if M is J-irreducible. The above claim thus shows that if M is J-irreducible then M is a model of $\mathbb{T}'$. In the converse direction, if M is a model of $\mathbb{T}'$, which we can suppose of the form $M_{\{\vec{x}.\phi\}}$ for some geometric formula $\phi(\vec{x})$ over the signature of $\mathbb{T}$ by Theorem 6.1.13, then for any J-covering sieve $S := S_\sigma$ on $M = M_{\{\vec{x}.\phi\}}$ in the category f.p.$\mathbb{T}$-mod($\mathbf{Set}$)^{op} the sequent $\sigma := (\phi \vdash_{\vec{x}} \bigvee_{i \in I} (\exists \vec{y_i})\theta_i)$ is provable in $\mathbb{T}'$ and hence valid in M, so by the above claim the identity on M belongs to S_σ, i.e. S_σ is maximal. □

We can characterize the Grothendieck topologies J such that the corresponding subtopos $\mathbf{Sh}(\text{f.p.}\mathbb{T}\text{-mod}(\mathbf{Set})^{\mathrm{op}}, J) \hookrightarrow [\text{f.p.}\mathbb{T}\text{-mod}(\mathbf{Set}), \mathbf{Set}]$ is an essential geometric inclusion (that is, a geometric inclusion whose inverse image admits a left adjoint) by using the following site characterizations:

(1) A canonical geometric inclusion $\mathbf{Sh}(\mathcal{C}, J) \hookrightarrow [\mathcal{C}^{\mathrm{op}}, \mathbf{Set}]$, where $\mathcal{C}$ is a small Cauchy-complete category and the topos $\mathbf{Sh}(\mathcal{C}, J)$ is equivalent to a presheaf topos, is essential if and only if the Grothendieck topology J is rigid.

(2) A geometric morphism (resp. a geometric inclusion) $[\mathcal{C}, \mathbf{Set}] \to [\mathcal{D}, \mathbf{Set}]$, where $\mathcal{D}$ is a Cauchy-complete category, is essential if and only if it is induced by a functor (resp. by a full and faithful functor) $\mathcal{C} \to \mathcal{D}$.

The latter characterization is well-known (cf. Lemma A4.1.5 and Example A4.2.12(b) [55]), while the former can be proved as follows. If J is rigid then, denoting by $\mathcal{D}$ the full subcategory of $\mathcal{C}$ on the J-irreducible objects, the Comparison Lemma yields an equivalence $\mathbf{Sh}(\mathcal{C}, J) \simeq [\mathcal{D}^{\mathrm{op}}, \mathbf{Set}]$ which makes the canonical geometric inclusion $\mathbf{Sh}(\mathcal{C}, J) \hookrightarrow [\mathcal{C}^{\mathrm{op}}, \mathbf{Set}]$ isomorphic to the canonical geometric inclusion $[\mathcal{D}^{\mathrm{op}}, \mathbf{Set}] \hookrightarrow [\mathcal{C}^{\mathrm{op}}, \mathbf{Set}]$ induced by the full embedding of categories $\mathcal{D}^{\mathrm{op}} \hookrightarrow \mathcal{C}^{\mathrm{op}}$; in particular, the geometric morphism $\mathbf{Sh}(\mathcal{C}, J) \hookrightarrow [\mathcal{C}^{\mathrm{op}}, \mathbf{Set}]$ is essential. Conversely, if the canonical geometric inclusion $i_J : \mathbf{Sh}(\mathcal{C}, J) \hookrightarrow [\mathcal{C}^{\mathrm{op}}, \mathbf{Set}]$ is essential and $\mathbf{Sh}(\mathcal{C}, J)$ is equivalent to a presheaf topos $[\mathcal{D}^{\mathrm{op}}, \mathbf{Set}]$, by (2) we have that i_J is isomorphic to a geometric inclusion $[\mathcal{D}^{\mathrm{op}}, \mathbf{Set}] \to [\mathcal{C}^{\mathrm{op}}, \mathbf{Set}]$ induced by a full embedding $\mathcal{D}^{\mathrm{op}} \hookrightarrow \mathcal{C}^{\mathrm{op}}$. Now, the topology $J_\mathcal{D}$ on $\mathcal{C}$ defined by saying that a sieve R on an object c of $\mathcal{C}$ is J-covering if and only if it contains all the morphisms from objects of $\mathcal{D}$ to c is clearly rigid, and the Comparison Lemma yields an equivalence $\mathbf{Sh}(\mathcal{C}, J_\mathcal{D}) \simeq [\mathcal{D}^{\mathrm{op}}, \mathbf{Set}]$ which makes the geometric morphism $[\mathcal{D}^{\mathrm{op}}, \mathbf{Set}] \to [\mathcal{C}^{\mathrm{op}}, \mathbf{Set}]$ isomorphic to the canonical geometric inclusion $\mathbf{Sh}(\mathcal{C}, J_\mathcal{D}) \hookrightarrow [\mathcal{C}^{\mathrm{op}}, \mathbf{Set}]$ (cf. Example A2.2.4(d) [55]). It then follows that $J = J_\mathcal{D}$; in particular, J is rigid.

These site characterizations define the arches of a 'bridge' (in the sense of section 2.2) leading to the following result:

Theorem 8.2.6 *Let $\mathbb{T}'$ be a quotient of a theory of presheaf type $\mathbb{T}$, corresponding to a Grothendieck topology J on the category* f.p.$\mathbb{T}$-mod($\mathbf{Set}$)^{op} *under the duality of Theorem 3.2.5 (as in section 8.1.2). Then the topology J is rigid if and only if $\mathbb{T}'$ is of presheaf type and every finitely presentable $\mathbb{T}'$-model is finitely presentable also as a $\mathbb{T}$-model.*

Proof If $\mathbb{T}'$ is of presheaf type and every finitely presentable $\mathbb{T}'$-model is finitely presentable also as a $\mathbb{T}$-model then the geometric inclusion corresponding via the duality of Theorem 3.2.5 to the quotient $\mathbb{T}'$ of $\mathbb{T}$ is induced by the embedding

$$\text{f.p.}\mathbb{T}'\text{-mod}(\mathbf{Set}) \hookrightarrow \text{f.p.}\mathbb{T}\text{-mod}(\mathbf{Set})$$

(cf. Theorem 8.2.5). From the above characterizations, it thus follows that the Grothendieck topology J is rigid. Conversely, if J is rigid then $\mathbf{Sh}(\text{f.p.}\mathbb{T}\text{-mod}(\mathbf{Set})^{\text{op}}, J)$ is equivalent to the presheaf topos $[\mathcal{D}, \mathbf{Set}]$, where $\mathcal{D}$ is the opposite of the full subcategory of f.p.$\mathbb{T}$-mod$(\mathbf{Set})^{\text{op}}$ on the J-irreducible objects. So $\mathbb{T}'$ is of presheaf type. Clearly, $\mathcal{D}$ is Cauchy-complete (as f.p.$\mathbb{T}$-mod$(\mathbf{Set})$ is Cauchy-complete and the property of J-irreducibility is stable under retracts) and hence, since $\mathbb{T}'$ is of presheaf type, it is equal to the category of finitely presentable $\mathbb{T}'$-models; in particular, every finitely presentable $\mathbb{T}'$-model is finitely presentable as a $\mathbb{T}$-model. □

Remarks 8.2.7 (a) The proof of Theorem 8.2.6 shows that if $\mathbb{T}'$ is of presheaf type and every finitely presentable $\mathbb{T}'$-model is finitely presentable as a $\mathbb{T}$-model, the full subcategory f.p.$\mathbb{T}'$-mod$(\mathbf{Set})$ of f.p.$\mathbb{T}$-mod$(\mathbf{Set})$ and the topology J can be defined in terms of each other as follows:
- the objects of f.p.$\mathbb{T}'$-mod$(\mathbf{Set})$ are precisely the J-irreducible objects of f.p.$\mathbb{T}$-mod$(\mathbf{Set})$;
- a sieve S in f.p.$\mathbb{T}$-mod$(\mathbf{Set})^{\text{op}}$ on an object c is J-covering if and only if it contains all the arrows in f.p.$\mathbb{T}$-mod$(\mathbf{Set})$ from finitely presentable $\mathbb{T}'$-models to c.

(b) If $\mathbb{T}'$ is a presheaf-type quotient of a theory of presheaf type $\mathbb{T}$ then, for any finitely presentable $\mathbb{T}$-model c, any $\mathbb{T}'$-model M and any $\mathbb{T}$-model homomorphism $f : c \to M$, there exist a $\mathbb{T}$-model homomorphism $g : c \to c'$ to a finitely presentable $\mathbb{T}'$-model c' and a $\mathbb{T}$-model homomorphism $h : c' \to M$ such that $h \circ g = f$. Indeed, M can be expressed as a filtered colimit of finitely presentable $\mathbb{T}'$-models c'; therefore, as c is finitely presentable as a $\mathbb{T}$-model, f necessarily factors through a colimit arrow $c' \to M$ (cf. Remark 8.2.2(b)).

(c) Under the hypothesis that every finitely presentable $\mathbb{T}'$-model is finitely presentable also as a $\mathbb{T}$-model, the theory $\mathbb{T}'$ can be characterized as the quotient of $\mathbb{T}$ obtained by adding all the sequents of the form $(\phi(\vec{x}) \vdash_{\vec{x}} \bigvee_{i \in I} \exists \vec{x}_i \theta_i(\vec{x}_i, \vec{x}))$ where $\phi(\vec{x})$ is a $\mathbb{T}$-irreducible formula (equivalently, a formula presenting a $\mathbb{T}$-model) and $\{\theta_i(\vec{x}_i, \vec{x}) : \{\vec{x}_i . \phi_i\} \to \{\vec{x} . \phi\} \mid i \in I\}$ is the (set of representatives of formulae in the) family of $\mathbb{T}$-provably functional formulae from a formula $\phi_i(\vec{x}_i)$ presenting a $\mathbb{T}'$-model to $\phi(\vec{x})$ (cf. the proof of Theorem 8.2.5 and Remark 8.2.7(a)).

(d) Theorem 8.2.6 has been applied in [32] (cf. Theorem 6.20 therein) to prove that the theory of local MV-algebras in a proper variety of MV-algebras is of presheaf type.

The following result identifies a natural class of presheaf-type quotients of presheaf-type theories satisfying the condition in Theorem 8.2.6.

Theorem 8.2.8 *Let $\mathbb{T}$ be a theory of presheaf type over a signature Σ. Then any quotient $\mathbb{T}'$ of $\mathbb{T}$ obtained from $\mathbb{T}$ by adding sequents of the form $(\phi \vdash_{\vec{x}} \bot)$, where $\phi(\vec{x})$ is a geometric formula over Σ, is classified by the topos $[\mathcal{T}, \mathbf{Set}]$, where $\mathcal{T}$ is the full subcategory of* f.p.$\mathbb{T}$-mod$(\mathbf{Set})$ *on the $\mathbb{T}'$-models.*

Proof Since by Corollary 6.1.10 we can cover each $\phi(\vec{x})$ with $\mathbb{T}$-irreducible formulae in the syntactic category $\mathcal{C}_{\mathbb{T}}$ of $\mathbb{T}$, we can suppose without loss of generality all the $\phi(\vec{x})$ to be $\mathbb{T}$-irreducible, that is to present a $\mathbb{T}$-model $M_{\{\vec{x}.\phi\}}$. Then, by Theorem 8.1.10, $\mathbb{T}'$ is classified by the topos $\mathbf{Sh}(\text{f.p.}\mathbb{T}\text{-mod}(\mathbf{Set})^{\text{op}}, J)$, where J is the smallest Grothendieck topology on f.p.$\mathbb{T}$-mod$(\mathbf{Set})^{\text{op}}$ which contains all the empty sieves on the models $M_{\{\vec{x}.\phi\}}$ presented by the formulae $\phi(\vec{x})$ occurring in the axioms $(\phi \vdash_{\vec{x}} \bot)$ added to $\mathbb{T}$ to form $\mathbb{T}'$. Let $\mathcal{T}$ be the full subcategory of f.p.$\mathbb{T}$-mod$(\mathbf{Set})$ on the $\mathbb{T}'$-models. Then $\mathcal{T}$ is J-dense; indeed, any object not in $\mathcal{T}$ admits an arrow in f.p.$\mathbb{T}$-mod$(\mathbf{Set})$ from a model of the form $M_{\{\vec{x}.\phi\}}$ and hence is J-covered by the empty sieve in f.p.$\mathbb{T}$-mod$(\mathbf{Set})^{\text{op}}$. Further, for any object of $\mathcal{T}$, the J-covering sieves on it are exactly the maximal ones; therefore, the Comparison Lemma yields an equivalence $\mathbf{Sh}(\text{f.p.}\mathbb{T}\text{-mod}(\mathbf{Set})^{\text{op}}, J) \simeq [\mathcal{T}, \mathbf{Set}]$, as required. $\square$

Remark 8.2.9 Theorem 8.2.8 could alternatively be deduced from Corollary 8.2.1 by observing that if $\mathbb{T}$ and $\mathbb{T}'$ are theories as in the hypotheses of the theorem then the domain of any $\mathbb{T}$-model homomorphism having as codomain a $\mathbb{T}'$-model is also a $\mathbb{T}'$-model whence the hypotheses of the corollary are trivially satisfied.

8.2.3 Finding theories classified by a given presheaf topos

The following theorem provides a method for constructing theories of presheaf type whose category of finitely presentable models is equivalent, up to Cauchy completion, to a given small category of structures.

Theorem 8.2.10 *Let $\mathbb{T}$ be a theory of presheaf type and $\mathcal{A}$ a full subcategory of* f.p.$\mathbb{T}$-mod$(\mathbf{Set})$. *Then the $\mathcal{A}$-completion $\mathbb{T}_{\mathcal{A}}$ of $\mathbb{T}$ (i.e. the theory consisting of all the geometric sequents over the signature of $\mathbb{T}$ which are valid in all the models in $\mathcal{A}$) is of presheaf type classified by the topos $[\mathcal{A}, \mathbf{Set}]$; in particular, the presentable $\mathbb{T}_{\mathcal{A}}$-models are precisely the retracts of models in $\mathcal{A}$.*

Proof Since every model in $\mathcal{A}$ is finitely presentable as a $\mathbb{T}$-model, we have a geometric inclusion $i : [\mathcal{A}, \mathbf{Set}] \hookrightarrow \mathbf{Set}[\mathbb{T}] \simeq [\text{f.p.}\mathbb{T}\text{-mod}(\mathbf{Set}), \mathbf{Set}]$ induced by the canonical inclusion $\mathcal{A} \hookrightarrow \text{f.p.}\mathbb{T}\text{-mod}(\mathbf{Set})$. This subtopos corresponds, by the duality of Theorem 3.2.5, to a quotient of $\mathbb{T}$ classified by the topos $[\mathcal{A}, \mathbf{Set}]$, which can be characterized as the collection of geometric sequents which hold in every model in $\mathcal{A}$, that is as the $\mathcal{A}$-completion $\mathbb{T}_{\mathcal{A}}$ of $\mathbb{T}$ (cf. Remark 6.1.2(c)).

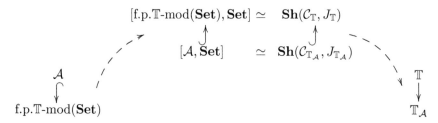

Therefore $\mathbb{T}_{\mathcal{A}}$ is of presheaf type classified by the topos $[\mathcal{A}, \mathbf{Set}]$. In particular, the finitely presentable $\mathbb{T}_{\mathcal{A}}$-models are the finitely presentable objects of Ind-$\mathcal{A}$, that is the retracts of objects of $\mathcal{A}$ in Ind-$\mathcal{A} \simeq \mathbb{T}_{\mathcal{A}}\text{-mod}(\mathbf{Set})$. □

Remarks 8.2.11 (a) Theorem 8.2.10 is a generalization of Joyal-Wraith's recognition theorem (Theorem 1.1 [5]), as well as of Proposition 5.3 [51]. Indeed, Theorem 1.1 [5] can be obtained as the particular case when $\mathbb{T}_{\mathcal{A}}$ has enough set-based models and every set-based model of $\mathbb{T}_{\mathcal{A}}$ is a filtered colimit of finitely presentable models of $\mathbb{T}_{\mathcal{A}}$ which are also finitely presentable as $\mathbb{T}$-models, while Proposition 5.3 [51] can be obtained as the specialization of this result to the case when $\mathbb{T}_{\mathcal{A}}$ is a disjunctive theory and $\mathbb{T}$ is a cartesian theory (in view of the fact that, by Theorem 4.1 [51], the category of set-based models of a disjunctive theory $\mathbb{S}$ is multiply finitely presentable and hence every set-based model of $\mathbb{S}$ can be expressed as a filtered colimit of finitely presentable $\mathbb{S}$-models).

(b) It is worth comparing Theorem 8.2.10 with Theorem 6.3.1: if $\mathbb{T}'$ is a quotient of a theory of presheaf type $\mathbb{T}$ and $\mathcal{K}$ is a small category of set-based $\mathbb{T}'$-models such that every model in $\mathcal{K}$ is finitely presentable as a $\mathbb{T}$-model then, in order to conclude that $\mathbb{T}'$ is of presheaf type classified by the topos $[\mathcal{K}, \mathbf{Set}]$, we can either verify conditions (i) and (ii) of Theorem 6.3.1 or prove that the models in $\mathcal{K}$ are jointly conservative for $\mathbb{T}'$. Indeed, as observed in Remark 6.3.2(e), conditions (i) and (ii) of Theorem 6.3.1 together imply that the models in $\mathcal{K}$ are jointly conservative for $\mathbb{T}'$.

Corollary 8.2.12 *Assuming the axiom of choice, every coherent theory (or, more generally, any theory in a countable fragment of geometric logic as in the hypotheses of Theorem 5.1.7 [64]) over a finite relational signature whose axioms do not involve existential quantifications is of presheaf type.*

Proof Let $\mathbb{T}$ be a theory satisfying the hypotheses of the Corollary. Since the axioms of $\mathbb{T}$ do not involve existential quantifications, every substructure of a model of $\mathbb{T}$ is a model of $\mathbb{T}$. Moreover, since the signature of $\mathbb{T}$ is relational, every finitely generated structure over the signature of $\mathbb{T}$ contains only a finite number of elements besides the constants. So every finitely presentable model of $\mathbb{T}$ contains only a finite number of elements besides the constants (as any model of $\mathbb{T}$ is a filtered union of its finite submodels). On the other hand, since the signature of $\mathbb{T}$ is finite, every such model is finitely presentable as a model of the empty theory over the signature of $\mathbb{T}$ (cf. Proposition 9.1.1 below). Therefore condition

(iii) of Theorem 6.3.1 is satisfied by the theory $\mathbb{T}$ with respect to the category of its finite models and homomorphisms between them (cf. Proposition 6.3.12(i)). Now, assuming the axiom of choice, every theory satisfying the hypotheses of Theorem 5.1.7 [64], in particular any coherent theory, has enough set-based models, whence our thesis follows from Theorem 8.2.10. □

In light of Theorem 8.2.10, a natural first step in constructing a geometric theory classified by a presheaf topos $[\mathcal{K}, \mathbf{Set}]$ consists in finding a theory of presheaf type $\mathbb{T}$ such that the category $\mathcal{K}$ can be identified with a full subcategory of the category of finitely presentable $\mathbb{T}$-models. Indeed, under these hypotheses, Theorem 8.2.10 ensures the existence of a quotient $\mathbb{T}_\mathcal{K}$ of $\mathbb{T}$ classified by the topos $[\mathcal{K}, \mathbf{Set}]$, which can be characterized as the theory consisting of all the geometric sequents over the signature of $\mathbb{T}$ which are valid in every model in $\mathcal{K}$. In most cases, if one has a natural candidate $\mathbb{S}$ for a theory classified by the topos $[\mathcal{K}, \mathbf{Set}]$, the theory $\mathbb{T}$ can be chosen to be the Horn part of $\mathbb{S}$ (i.e. the theory consisting of all the Horn sequents which are provable in $\mathbb{S}$) or the cartesianization $\mathbb{S}_c$ of $\mathbb{S}$.

Of course, the abstract characterization of $\mathbb{T}_\mathcal{K}$ as the theory consisting of all the geometric sequents over the signature of $\mathbb{T}$ which are valid in every model in $\mathcal{K}$ is not very useful in specific contexts where one looks for an axiomatization of $\mathbb{T}_\mathcal{K}$ that is as simple and 'economical' as possible. To address this issue, we observe that if every $\mathbb{T}$-model M in $\mathcal{K}$ is strongly finitely presented as a model of $\mathbb{T}$ as well as finitely generated (in the sense that for any sort A of the signature of $\mathbb{T}$ the elements of the set MA are precisely given by the interpretations in M of terms $t(\vec{x})$ whose output sort is A, where $\vec{x}$ are the generators of M as a strongly finitely presented model of $\mathbb{T}$) then we dispose of an explicit axiomatization of the theory $\mathbb{T}_\mathcal{K}$, as shown by the following result:

Theorem 8.2.13 *Let $\mathbb{T}$ be a geometric theory over a signature Σ and $\mathcal{K}$ a full subcategory of the category of set-based $\mathbb{T}$-models such that every $\mathbb{T}$-model in $\mathcal{K}$ is both strongly finitely presentable and finitely generated (with respect to the same generators). Then the following sequents, where we denote by $\mathcal{P}$ the set of (representatives of) geometric formulae over Σ which strongly present a $\mathbb{T}$-model in $\mathcal{K}$, added to the axioms of $\mathbb{T}$, yield an axiomatization of a quotient of $\mathbb{T}$ classified by the topos $[\mathcal{K}, \mathbf{Set}]$ via the Morita equivalence induced by the canonical geometric morphism $p_\mathcal{K} : [\mathcal{K}, \mathbf{Set}] \to \mathbf{Sh}(\mathcal{C}_\mathbb{T}, J_\mathbb{T})$:*

(i) *the sequent*

$$\left(\top \vdash_{[]} \bigvee_{\phi(\vec{x}) \in \mathcal{P}} (\exists \vec{x})\phi(\vec{x})\right);$$

(ii) *for any formulae $\phi(\vec{x})$ and $\psi(\vec{y})$ in $\mathcal{P}$, where $\vec{x} = (x_1^{A_1}, \ldots, x_n^{A_n})$ and $\vec{y} = (y_1^{B_1}, \ldots, y_m^{B_m})$, the sequent*

$$\left(\phi(\vec{x}) \wedge \psi(\vec{y}) \vdash_{\vec{x},\vec{y}}\right.$$

$$\bigvee_{\substack{\chi(\vec{z})\in\mathcal{P},\, t_1^{A_1}(\vec{z}),\ldots,t_n^{A_n}(\vec{z}) \\ s_1^{B_1}(\vec{z}),\ldots,s_m^{B_m}(\vec{z})}} (\exists \vec{z})\left(\chi(\vec{z}) \wedge \bigwedge_{\substack{i\in\{1,\ldots,n\}, \\ j\in\{1,\ldots,m\}}} (x_i = t_i(\vec{z}) \wedge y_j = s_j(\vec{z}))\right)\bigg),$$

where the disjunction is taken over all the formulae $\chi(\vec{z})$ in $\mathcal{P}$ and all the tuples of terms $t_1^{A_1}(\vec{z}),\ldots,t_n^{A_n}(\vec{z})$ and $s_1^{B_1}(\vec{z}),\ldots,s_m^{B_m}(\vec{z})$ such that

$$(t_1(\vec{\xi_\chi}),\ldots,t_n(\vec{\xi_\chi})) \in [[\vec{x}.\phi]]_{M_{\{\vec{z}.\chi\}}}$$

and

$$(s_1(\vec{\xi_\chi}),\ldots,s_m(\vec{\xi_\chi})) \in [[\vec{y}.\psi]]_{M_{\{\vec{z}.\chi\}}};$$

(iii) for any formulae $\phi(\vec{x})$ and $\psi(\vec{y})$ in $\mathcal{P}$, where $\vec{x} = (x_1^{A_1},\ldots,x_n^{A_n})$ and $\vec{y} = (y_1^{B_1},\ldots,y_m^{B_m})$, and any terms $t_1^{A_1}(\vec{y}), s_1^{A_1}(\vec{y}),\ldots,t_n^{A_n}(\vec{y}), s_n^{A_n}(\vec{y})$ such that $(t_1(\vec{\xi_\phi}),\ldots,t_n(\vec{\xi_\phi})) \in [[\vec{y}.\phi]]_{M_{\{\vec{y}.\psi\}}}$ and $(s_1(\vec{\xi_\phi}),\ldots,s_n(\vec{\xi_\phi})) \in [[\vec{y}.\phi]]_{M_{\{\vec{y}.\psi\}}}$, the sequent

$$\left(\bigwedge_{i\in\{1,\ldots,n\}} (t_i(\vec{y}) = s_i(\vec{y})) \wedge \psi(\vec{y}) \vdash_{\vec{y}}\right.$$

$$\bigvee_{\substack{\chi(\vec{z})\in\mathcal{P}, \\ u_1^{B_1}(\vec{z}),\ldots,u_m^{B_m}(\vec{z})}} (\exists \vec{z})\left(\chi(\vec{z}) \wedge \bigwedge_{j\in\{1,\ldots,m\}} (y_j = u_j(\vec{z}))\right)\bigg),$$

where the disjunction is taken over all the formulae $\chi(\vec{z})$ in $\mathcal{P}$ and all the tuples of terms $u_1^{B_1}(\vec{z}),\ldots,u_m^{B_m}(\vec{z})$ such that

$$(u_1(\vec{\xi_\chi}),\ldots,u_m(\vec{\xi_\chi})) \in [[\vec{y}.\psi]]_{M_{\{\vec{z}.\chi\}}}$$

and

$$t_i(u_1(\vec{\xi_\chi}),\ldots,u_m(\vec{\xi_\chi})) = s_i(u_1(\vec{\xi_\chi}),\ldots,u_m(\vec{\xi_\chi})) \text{ in } M_{\{\vec{z}.\chi\}}$$

for all $i \in \{1,\ldots,n\}$;

(iv) for any sort A of Σ, the sequent

$$\left(\top \vdash_{x^A} \bigvee_{\chi(\vec{z})\in\mathcal{P},\, t^A(\vec{z})} (\exists \vec{z})(\chi(\vec{z}) \wedge x = t(\vec{z}))\right),$$

where the disjunction is taken over all the formulae $\chi(\vec{z})$ in $\mathcal{P}$ and all the terms $t^A(\vec{z})$;

(v) *for any sort A of Σ, any formulae $\phi(\vec{x})$ and $\psi(\vec{y})$ in $\mathcal{P}$, where $\vec{x} = (x_1^{A_1}, \ldots, x_n^{A_n})$ and $\vec{y} = (y_1^{B_1}, \ldots, y_m^{B_m})$, and any terms $t^A(\vec{x})$ and $s^A(\vec{y})$, the sequent*

$$\left(\phi(\vec{x}) \wedge \psi(\vec{y}) \wedge t(\vec{x}) = s(\vec{y}) \vdash_{\vec{x},\vec{y}} \right.$$

$$\left. \bigvee_{\substack{\chi(\vec{z}) \in \mathcal{P},\, p_1^{A_1}(\vec{z}),\ldots,p_n^{A_n}(\vec{z}) \\ q_1^{B_1}(\vec{z}),\ldots,q_m^{B_m}(\vec{z})}} (\exists \vec{z}) \left(\chi(\vec{z}) \wedge \bigwedge_{\substack{i \in \{1,\ldots,n\}, \\ j \in \{1,\ldots,m\}}} (x_i = p_i(\vec{z}) \wedge y_j = q_j(\vec{z})) \right) \right),$$

where the disjunction is taken over all the formulae $\chi(\vec{z})$ in $\mathcal{P}$ and all the tuples of terms $p_1^{A_1}(\vec{z}), \ldots, p_n^{A_n}(\vec{z})$ and $q_1^{B_1}(\vec{z}), \ldots, q_m^{B_m}(\vec{z})$ such that

$$(p_1(\vec{\xi_\chi}), \ldots, p_n(\vec{\xi_\chi})) \in [[\vec{x}.\phi]]_{M_{\{\vec{z}.\chi\}}},$$

$$(q_1(\vec{\xi_\chi}), \ldots, q_m(\vec{\xi_\chi})) \in [[\vec{y}.\psi]]_{M_{\{\vec{z}.\chi\}}}$$

and

$$t(p_1(\vec{\xi_\chi}), \ldots, p_n(\vec{\xi_\chi})) = s(q_1(\vec{\xi_\chi}), \ldots, q_m(\vec{\xi_\chi})) \text{ in } M_{\{\vec{z}.\chi\}}.$$

If moreover $\mathbb{T}$ is of presheaf type (in which case every finitely presentable $\mathbb{T}$-model is strongly finitely presentable – cf. Corollary 6.2.8) then the quotient axiomatized by these sequents coincides with the theory $\mathbb{T}_\mathcal{K}$ defined in Theorem 8.2.10.

Proof Let $\mathbb{R}$ be the geometric theory obtained from $\mathbb{T}$ by adding all the sequents specified above. The objects of $\mathcal{K}$ are clearly models of $\mathbb{R}$ (cf. Remarks 6.3.7(a) and 6.3.11); since they are strongly finitely presentable as $\mathbb{T}$-models, it follows from Proposition 6.3.12(ii) that condition (iii) of Theorem 6.3.1 is satisfied by the theory $\mathbb{R}$ with respect to the category $\mathcal{K}$. By Remarks 6.3.7(a) and 6.3.11, $\mathbb{R}$ also satisfies conditions (i) and (ii) of Theorem 6.3.1 with respect to $\mathcal{K}$. Theorem 6.3.1 thus implies that $\mathbb{R}$ is of presheaf type classified by the topos $[\mathcal{K}, \mathbf{Set}]$ via a Morita equivalence induced by the canonical geometric morphism $p_\mathcal{K} : [\mathcal{K}, \mathbf{Set}] \to \mathbf{Sh}(\mathcal{C}_\mathbb{T}, J_\mathbb{T})$; this ensures in particular that if $\mathbb{T}$ is of presheaf type then $\mathbb{R}$ is equal to the theory $\mathbb{T}_\mathcal{K}$. $\square$

Remark 8.2.14 Theorem 8.2.13 has been applied in [33] to describe geometric theories classified by Connes' cyclic topos ([35]), Connes-Consani's epicyclic topos ([37]) and Connes-Consani's arithmetic topos ([36]). It has also been used in [32] to obtain a simple axiomatization for the geometric theory of finite MV-chains (cf. Theorem 9.7 therein).

9

EXAMPLES OF THEORIES OF PRESHEAF TYPE

In this chapter we discuss some classical as well as new examples of theories of presheaf type from the perspective of the theory developed in the previous chapters. We revisit in particular well-known examples of theories of presheaf type whose finitely presentable models are all finite, notably including the geometric theory of finite sets, and give fully constructive proofs of the fact that Moerdijk's theory of abstract circles and Johnstone's theory of Diers fields are of presheaf type. Next, we introduce new examples of theories of presheaf type, including the theory of algebraic extensions of a given field, the theory of locally finite groups, the theory of vector spaces with linear independence predicates and the theory of abelian ℓ-groups with strong unit. We also show that the injectivization of the algebraic theory of groups is not of presheaf type and explicitly describe a presheaf completion for it (in the sense of section 7.1.3).

9.1 Theories whose finitely presentable models are finite

Geometric theories whose finitely presentable models are all finite often happen to be of presheaf type. Corollary 8.2.4 provides a theoretical justification for this fact, while Corollary 9.1.3 below brings alternative evidence in this respect.

The following proposition shows that set-theoretic finiteness implies finite presentability in categories of set-based models of geometric theories over finite signatures. Recall that a model of a geometric theory over a finite signature Σ is said to be finite if the disjoint union of the MA for A a sort of Σ is a finite set.

Proposition 9.1.1 *Let $\mathbb{T}$ be a geometric theory over a finite signature. Then every finite model M of $\mathbb{T}$ is finitely presented (and hence finitely presentable).*

Proof For any $\mathbb{T}$-models M and N in **Set**, a bunch of functions $\xi_A : MA \to NA$ indexed by the sorts A of Σ is the set of components of a $\mathbb{T}$-model homomorphism $M \to N$ if and only if the following conditions are satisfied: (1) for any function symbol $f : A_1 \cdots A_n \to B$ of Σ and any tuple $(x_1, \ldots, x_n) \in MA_1 \times \cdots \times MA_n$, if $y = Mf(x_1, \ldots, x_n)$ then $\xi_B(y) = Nf(\xi_{A_1}(x_1), \ldots, \xi_{A_n}(x_n))$ and (2) for any relation symbol $R \rightarrowtail A_1 \cdots A_n$ of Σ and any $(x_1, \ldots, x_n) \in MR$, $(\xi_{A_1}(x_1), \ldots, \xi_{A_n}(x_n)) \in NR$.

Given a finite $\mathbb{T}$-model M, let us enumerate its elements (i.e. the elements of MA for some A) $a_1, \ldots, a_n$ in such a way that $a_i \neq a_j$ for $i \neq j$. Let us define a presentation of M as a model of $\mathbb{T}$ as follows: the generators $(x_1, \ldots, x_n)$ are the elements $(a_1, \ldots, a_n)$ of M, while the relations are given by the formulae which are either of the form $x_i = f(x_j, \ldots, x_k)$ (for any tuple $(a_i, a_j, \ldots, a_k)$ of elements of M and function symbol f of the signature of $\mathbb{T}$ such that $a_i = Mf(a_j, \ldots, a_k)$) or of the form $R(x_j, \ldots, x_k)$ (for any tuple $(a_j, \ldots, a_k)$ of elements of M such that $(a_j, \ldots, a_k) \in MR$). By the above remarks, this is indeed a presentation for M, and it is finite since the signature of $\mathbb{T}$ is finite. □

Remark 9.1.2 Proposition 9.1.1 specializes to Example 1.2 (4) [1] in the case where the signature of $\mathbb{T}$ contains no function symbols.

Proposition 9.1.1, combined with Theorem 8.2.10, yields the following result:

Corollary 9.1.3 *Let Σ be a finite signature and $\mathcal{A}$ a category of finite Σ-structures and Σ-structure homomorphisms between them. Then the geometric theory consisting of all the geometric sequents over Σ which are valid in all the structures in $\mathcal{A}$ (that is, the $\mathcal{A}$-completion of the empty theory $\mathbb{O}_\Sigma$ over Σ in the sense of Theorem 8.2.10) is of presheaf type, classified by the topos $[\mathcal{A}, \mathbf{Set}]$, and its finitely presentable models are precisely the finite ones.*

Remark 9.1.4 Corollary 9.1.3 can be notably applied to theories $\mathbb{T}$ with enough set-based models such that every model of $\mathbb{T}$ can be expressed as a filtered colimit of finite models of $\mathbb{T}$. Recall that, assuming the axiom of choice, every coherent theory has enough set-based models; it thus follows from the corollary that, assuming the axiom of choice, the injectivization of any finitary theory of presheaf type whose finitely presentable models are all finite is again of presheaf type.

Let us now discuss a few concrete examples of theories of presheaf type whose finitely presentable models are finite.

As an application of Corollary 8.2.4, we can recover at once the well-known results that the following theories are of presheaf type:

- The theory of decidable objects (cf. [56] and p. 907 of [55]);
- The theory of decidable Boolean algebras (cf. Example D3.4.12 [55]);
- The theory of linear orders;
- The theory of intervals (cf. Definition 2.1.27 and Proposition 2.1.28).

The fact that these theories are of presheaf type also follows from Corollary 9.1.3 by assuming the axiom of choice.

Notice that the former two theories are injectivizations of two cartesian theories, namely the empty theory over a signature consisting of just one sort and the algebraic theory of Boolean algebras. The latter two theories satisfy the hypotheses of Corollary 7.2.43, whence their injectivizations are of presheaf type too.

9.2 The theory of abstract circles

The *theory $\mathbb{C}$ of abstract circles* has been introduced by I. Moerdijk and shown in [66] to have the property that the points of *Connes' topos* of cyclic sets (introduced in [35]) can be identified with the set-based models of $\mathbb{C}$; in [66] it is also stated that Connes' topos actually classifies $\mathbb{C}$, but the argument given therein appears incomplete.

We shall prove in this section, by using Corollary 8.2.1, that $\mathbb{C}$ is of presheaf type classified by Connes' topos. Specifically, we shall show that $\mathbb{C}$ satisfies the hypotheses of the corollary with respect to its Horn part (i.e. the Horn theory consisting of the collection of all Horn sequents which are provable in $\mathbb{C}$). We will also prove that the injectivization of $\mathbb{C}$ is of presheaf type as well.

The signature Σ of the theory $\mathbb{C}$ consists of two sorts P and S (variables of type P will be denoted by letters $x, y, \ldots$, while variables of type S will be denoted by letters $a, b, \ldots$), two unary function symbols $0 : P \to S$ and $1 : P \to S$, two unary function symbols $\delta_0 : S \to P$ and $\delta_1 : S \to P$, one unary function symbol $* : S \to S$ and a ternary predicate R of type S. The individuals of sort P should be thought of as *points* and those of sort S as *segments*; the predicate R formalizes the partial operation of *concatenation* of segments.

The axioms of $\mathbb{C}$ can be formulated as follows (using the terminology of [66]):

1. *Non-triviality axioms*:
$$(\top \vdash_{[]} (\exists x)(x = x));$$
$$(\top \vdash_{x,y} (\exists a)(\delta_0(a) = x \wedge \delta_1(a) = y));$$
$$(0(x) = 1(x) \vdash_x \bot);$$

2. *'Equational' axioms*:
$$(\top \vdash_a a^{**} = a);$$
$$(\top \vdash_a \delta_0(a^*) = \delta_1(a));$$
$$(\top \vdash_x \delta_0(0(x)) = x \wedge \delta_1(0(x)) = x);$$
$$(\top \vdash_x 0(x)^* = 1(x));$$
$$(\delta_0(a) = x \wedge \delta_1(a) = x \vdash_{x,a} a = 0(x) \vee a = 1(x));$$

3. *Axioms for concatenation*:
$$(R(a,b,c) \wedge R(a,b,c') \vdash_{a,b,c,c'} c = c');$$
$$(R(a,b,c) \vdash_{a,b,c} \delta_0(c) = \delta_0(a) \wedge \delta_1(c) = \delta_1(b));$$
$$(R(a,b,c) \vdash_{a,b,c} R(c^*, a, b^*));$$
$$(R(a,b,d) \wedge R(d,c,e) \vdash_{a,b,c,d,e} (\exists e')(R(b,c,e') \wedge R(a,e',e)));$$
$$(R(a,b,0(x)) \vdash_{a,b,x} a = 0(x));$$
$$(\delta_0(a) = x \vdash_{a,x} R(0(x), a, a));$$
$$(\delta_1(a) = \delta_0(b) \vdash_{a,b} (\exists c) R(a,b,c) \vee (\exists d) R(b^*, a^*, d)).$$

A model of $\mathbb{C}$ is said to be an *abstract circle*. Any set of points P of the circle S^1 defines an abstract circle S_P whose segments $a \in S$ such that $\delta_0(a) = x$ and $\delta_1(a) = y$ are the oriented arcs on S^1 from the point x to the point y. For any natural number $n > 0$ there is exactly one abstract circle, up to isomorphism, whose set of points has n elements; we shall denote it by the symbol C_n.

To prove that $\mathbb{C}$ is of presheaf type, we first notice that the sequent

$$(\delta_0(a) = \delta_0(b) \wedge \delta_1(a) = \delta_1(b) \vdash_{a,b} a = b)$$

is provable in $\mathbb{C}$. This sequent can be easily deduced as a consequence of the seventh axiom of group 3 and the fifth axiom of group 2. We shall refer to it as the 'uniqueness axiom'.

It follows that, modulo the non-triviality axioms and the uniqueness axiom, we can replace any expression (equivalent to one) of the form $(\exists c)(\phi(c) \wedge \delta_0(c) = x \wedge \delta_1(c) = y)$ arising in an axiom of $\mathbb{C}$ with the requirement that the unique segment d such that $\delta_0(d) = x$ and $\delta_1(d) = y$ satisfies ϕ. In particular, the fourth axiom of group 3 is provably equivalent, modulo the non-triviality axioms and the uniqueness axiom, to the following sequent:

$$(\delta_0(u) = \delta_0(b) \wedge \delta_1(u) = \delta_1(c) \wedge R(a,b,d) \wedge$$
$$\wedge R(d,c,e) \vdash_{a,b,c,d,e,u} R(b,c,u) \wedge R(a,u,e)).$$

Similarly, the seventh axiom of group 3 is provably equivalent, modulo the non-triviality axioms and the uniqueness axiom, to the following sequent:

$$(\delta_0(c) = \delta_0(a) \wedge \delta_1(c) = \delta_1(b) \wedge \delta_1(a) = \delta_0(b) \vdash_{a,b,c} R(a,b,c) \vee R(b^*, a^*, c^*)).$$

From these remarks it follows that $\mathbb{C}$ admits a presentation in which none of the axioms involve existential quantifications except for the first and second of group 1.

Let us show that $\mathbb{C}$ satisfies the hypotheses of Corollary 8.2.1 with respect to its Horn part and the category of finite $\mathbb{C}$-models.

Given a homomorphism $f : c \to \text{Hom}_{\mathcal{E}}(E, M)$, where c is a finite model of the Horn part of $\mathbb{C}$, M is a model of $\mathbb{C}$ in a Grothendieck topos $\mathcal{E}$ and E is an object of $\mathcal{E}$, we can 'localize' f (in the sense of section 8.2.1) a finite number of times (once for the first axiom of group 1 and once for each pair of points of c for the second axiom of group 1) to obtain Σ-substructure homomorphisms $f_i : c_i \hookrightarrow \text{Hom}_{\mathcal{E}}(E_i, M)$ such that the structures c_i are finite models of the Horn part of $\mathbb{C}$ satisfying the non-triviality axioms (notice that any substructure of a structure of the form $\text{Hom}_{\mathcal{E}}(E_i, M)$, where M is a model of $\mathbb{C}$ in $\mathcal{E}$, satisfies all the Horn sequents provable in $\mathbb{C}$). We can clearly further 'localize' each of these homomorphisms so as to obtain the satisfaction of all the other axioms of $\mathbb{C}$; since this can be done without modifying their domains, the final result will be a family of homomorphisms whose domains are structures satisfying all the axioms of $\mathbb{C}$.

Next, we notice that for any set-based model M of the Horn part of $\mathbb{C}$, any point $x \in MP$ and any segment $a \in MS$, there exists a finite substructure N of M such that NP contains x and NS contains a (take N equal to the substructure of M given by the sets
$$NP = \{x, \delta_0(a), \delta_1(a)\}$$
and
$$NS = \{a, 0(x), 1(x), 0(\delta_0(a)), 0(\delta_1(a)), 1(\delta_0(a)), 1(\delta_1(a)))\}).$$
Moreover, any two finite substructures N_1 and N_2 of M are contained in a common substructure N of M (take N equal to the substructure of M given by $NP = N_1P \cup N_2P$ and $NS = N_1S \cup N_2S$). This, combined with the above argument (specialized to the case $\mathcal{E} = \mathbf{Set}$), shows that every set-based model of $\mathbb{C}$ is a directed union of finite models of $\mathbb{C}$. It follows in particular that every finitely presentable $\mathbb{C}$-model is finite (it being a retract of a finite model). On the other hand, every finite model of $\mathbb{C}$ is finitely presentable as a model of the Horn part of $\mathbb{C}$ (cf. Proposition 9.1.1). We can thus conclude that the hypotheses of Corollary 8.2.1 are satisfied, whence $\mathbb{C}$ is of presheaf type classified by the topos of covariant set-valued functors on the category of finite models of $\mathbb{C}$.

Let us now show that the injectivization $\mathbb{C}_m$ of $\mathbb{C}$ is of presheaf type as well.

Due to the presence of conjunctions in the premises of some axioms of $\mathbb{C}$, we cannot directly apply Corollary 7.2.43 to conclude that $\mathbb{C}_m$ is of presheaf type. We shall instead apply Corollary 7.2.41. Since every $\mathbb{C}$-model in $\mathbf{Set}$ is a directed union of finite $\mathbb{C}$-models, the finitely presentable $\mathbb{C}_m$-models are exactly the finite $\mathbb{C}$-models. Moreover, by Proposition 7.2.31, the monic $\mathbb{C}$-model homomorphisms in $\mathbf{Set}$ are precisely the homomorphisms which are sortwise injective; indeed, the formulae $\{x^P \, . \, \top\}$ and $\{x^S \, . \, \top\}$ (strongly) present respectively the $\mathbb{C}$-models C_1 and C_2. To show that the hypotheses of Corollary 7.2.41 are satisfied, it remains to verify that for every Grothendieck topos $\mathcal{E}$, object E of $\mathcal{E}$ and Σ-structure homomorphism $x : c \to \mathrm{Hom}_\mathcal{E}(E, M)$, where c is a finite $\mathbb{C}$-model and M is a sortwise decidable $\mathbb{C}$-model, there exist an epimorphic family $\{e_i : E_i \to E \mid i \in I\}$ in $\mathcal{E}$ and for each $i \in I$ a $\mathbb{C}$-model homomorphism $f_i : c \to c_i$ of finite $\mathbb{C}$-models and a sortwise disjunctive Σ-structure homomorphism $x_i : c_i \to \mathrm{Hom}_\mathcal{E}(E_i, M)$ (in the sense of Lemma 7.2.16) such that $x_i \circ f_i = \mathrm{Hom}_\mathcal{E}(e_i, M) \circ x$. To this end, we make $\mathbb{C}$ into a one-sorted theory by identifying points x with the segments 0_x and rewriting its axioms appropriately, so that we can apply Proposition 7.2.39. The proposition yields an epimorphic family $\{e_i : E_i \to E \mid i \in I\}$ in $\mathcal{E}$ and for each $i \in I$ a surjective homomorphism $q_i : c \to c_i$, where c_i is a finite Σ-structure, and a disjunctive Σ-structure homomorphism $J_i : c_i \rightarrowtail \mathrm{Hom}_\mathcal{E}(E_i, M)$ such that $J_i \circ q_i = \mathrm{Hom}_\mathcal{E}(e_i, M) \circ f$. Now, since c is a $\mathbb{C}$-model and q_i is surjective, the structure c_i satisfies the non-triviality axioms. We can clearly suppose that $E_i \not\cong 0$ without loss of generality, and hence that the arrows J_i are injective. If we consider the image factorizations of the Σ-structure homomorphisms J_i (in the sense of Lemma 7.2.37), we thus obtain substructures $c'_i \rightarrowtail \mathrm{Hom}_\mathcal{E}(E_i, M)$ whose underlying sets are isomorphic to those of the c_i (since the J_i are injective) and Σ-structure homomorphisms $q'_i : c \to c'_i$. Since the c'_i have underlying sets isomorphic to those of the c_i, they all satisfy the non-triviality axiom, and, at the cost of refining the epimorphic family $\{e_i : E_i \to E \mid i \in I\}$, we can suppose that they satisfy all the other axioms of $\mathbb{C}$ (cf. the argument given above for showing that $\mathbb{C}$ satisfies the hypotheses of Corollary 8.2.1 with respect to its Horn part and the category of finite $\mathbb{C}$-models). So the hypotheses of Corollary 7.2.41 are satisfied, and we can conclude that $\mathbb{C}_m$ is of presheaf type classified by the topos of covariant set-valued functors on the category of finite models of $\mathbb{C}$ and injective homomorphisms between them.

Remark 9.2.1 In [33] we have described another theory, written over the signature of oriented groupoids (i.e. groupoids with a positivity predicate on arrows which is preserved by composition), which is classified by the cyclic topos, as well as a related theory classified by *Connes-Consani's epicyclic topos* (cf. section 4 of [37]). The advantage of working with oriented groupoids instead of abstract circles is that the composition of two segments (i.e. arrows of the groupoid) is always defined provided that the codomain of the first coincides with the domain of the second.

9.3 The geometric theory of finite sets

In this section we shall revisit, from the point of view of the theory developed in section 8.2.1, a well-known example of a theory of presheaf type, namely the geometric theory $\mathbb{T}$ of finite sets. Recall from Example D1.1.7(k) [55] that the signature Σ of $\mathbb{T}$ consists of one sort A and an n-ary relation symbol R_n for each $n > 0$. The axioms of $\mathbb{T}$ are as follows. For each n, we have the axiom

$$\sigma_n \equiv \left(R_n(x_1, \ldots, x_n) \vdash_{x_1,\ldots,x_n,y} \bigvee_{1 \leq i \leq n} y = x_i \right),$$

expressing the requirement that if an n-tuple of individuals satisfies the relation R_n then it exhausts the members of (the set interpreting) the sort A. (The case $n = 0$ of this axiom is $(R_0 \vdash_{[]} \bot)$, which says that if R_0 holds then the interpretation of the sort A must be empty.) We also have the axiom

$$\left(\top \vdash_{[]} \bigvee_{n \in \mathbb{N}} (\exists x_1) \cdots (\exists x_n) R_n(x_1, \ldots, x_n) \right).$$

Finally, to ensure that the interpretations of the R_n are uniquely determined by that of the sort A (i.e. that R_n holds for all the n-tuples which exhaust the elements of the interpretation of the sort A, and not just for some of them), one adds the axiom

$$(R_n(x_1, \ldots, x_n) \vdash_{x_1 \ldots, x_n} R_m(x_{f(1)}, \ldots, x_{f(m)}))$$

for any surjection $f : \{1, 2, \ldots, m\} \to \{1, 2, \ldots, n\}$, and

$$(R_n(x_1,\ldots,x_n) \wedge x_i = x_j \vdash_{x_1\ldots,x_n} R_{n-1}(x_1,\ldots,x_{i-1},x_{i+1},\ldots,x_n))$$

whenever $1 \leq i < j \leq n$.

We can deduce that $\mathbb{T}$ is of presheaf type as an application of Corollary 8.2.1.

Notice that the set-based models of $\mathbb{T}$ can be identified with the finite sets, while the $\mathbb{T}$-model homomorphisms are precisely the surjective functions between them.

Let us regard $\mathbb{T}$ as a quotient of its Horn part (i.e. the theory $\mathbb{T}_H$ consisting of all the Horn sequents over the signature of $\mathbb{T}$ which are provable in $\mathbb{T}$). The criterion for finite presentability given by Lemma 6.1.12 ensures that every finite set, regarded as a model of $\mathbb{T}$, is finitely presentable as a model of $\mathbb{T}_H$; indeed, the finiteness of the structure immediately implies that the second condition of the lemma is satisfied, while the satisfaction of the first condition follows from the fact that every function from a finite $\mathbb{T}_H$-model of cardinality n to a set-based $\mathbb{T}_H$-model M which preserves the predicate R_n preserves the predicate R_m for any $m > n$ (by the 'introduction' and 'elimination' rules expressed by the last two groups of axioms of $\mathbb{T}$).

To apply Corollary 8.2.1, it thus remains to prove that the second condition in the statement of the corollary is satisfied. Given a homomorphism $f : a \to \mathrm{Hom}_\mathcal{E}(E,M)$, where a is a finite model $\mathbb{T}_H$ and M is a $\mathbb{T}$-model in a Grothendieck topos $\mathcal{E}$, if the cardinality of aA is n then for every $m > n$ the sequent σ_m is satisfied in a provided that σ_n is. Indeed, if $m \geq n$ then for any m-tuple $(y_1,\ldots,y_m)$ of elements of aA there exists a sub-n-tuple of elements of aA obtained by removing $m - n$ repetitions, which satisfies the relation R_n if the m-tuple $(y_1,\ldots,y_m)$ satisfies the relation R_m, since a, as a model of $\mathbb{T}_H$, satisfies the Horn sequent expressing the 'elimination' rule (i.e. the last group of axioms of $\mathbb{T}$). Similarly, by invoking the 'introduction' rules (i.e. the last but one group of axioms of $\mathbb{T}$), one can prove that if the cardinality of aA is n then for every $k < n$ the sequent σ_k is satisfied in a if σ_n is.

Let us show that we can inductively 'localize' (in the sense of section 8.2.1) the homomorphism f to eventually arrive at Σ-structure homomorphisms $f_i : a_i \to \mathrm{Hom}_\mathcal{E}(E_i, M)$ defined on Σ-structures a_i which satisfy all the axioms of $\mathbb{T}$.

Starting from a Σ-structure homomorphism $f : a \to \mathrm{Hom}_\mathcal{E}(E,M)$, where a is a finite $\mathbb{T}_H$-model, from the fact that M is a model of $\mathbb{T}$ it follows that there exist an epimorphic family $\{e_i : E_i \to E \mid i \in I\}$ in $\mathcal{E}$ and for each $i \in I$ a natural number n_i and an n_i-tuple $(\xi_1,\ldots,\xi_{n_i})$ of generalized elements $E_i \to MA$ such that $< \xi_1,\ldots,\xi_{n_i} >$ factors through the interpretation of R_{n_i} in M. By taking a_i to be the Σ-substructure of $\mathrm{Hom}_\mathcal{E}(E_i, M)$ on the finite subset consisting of the elements $\xi_1,\ldots,\xi_{n_i}$ and all the elements in the image of the homomorphism $\mathrm{Hom}_\mathcal{E}(e_i, M) \circ f$, we clearly obtain a structure satisfying the second axiom of $\mathbb{T}$ and also the last two groups of axioms of $\mathbb{T}$ (the validity of Horn sequents being inherited by substructures). Now, in order to obtain from this family of Σ-structure homomorphisms a 'localization' such that the domains of its homomorphisms satisfy all the axioms of $\mathbb{T}$, it suffices to 'localize' each f_i

a finite number of times (one for each (n_i+1)-tuple $\vec{u}$ of elements of a_i, where n_i is the cardinality of a_i), endowing the sets d_i arising in the surjection-inclusion factorizations of the homomorphisms with the quotient structure induced by the domains a_i via the relevant quotient map q_i (in the sense that the interpretation of each relation symbol R of Σ in such a set d_i is defined to be equal to the image of the interpretation of R in a_i under the quotient map q_i).

Since the Σ-structure homomorphisms $q_i : a_i \to d_i$ in such a 'localization' are quotient maps, the fact that the d_i satisfy the last two groups of axioms of $\mathbb{T}$ follows automatically from the fact that the a_i do. By the above considerations, it will thus be enough to show that each d_i satisfies the sequent σ_{n_i}, where n_i is the cardinality of d_iA, and the second axiom of $\mathbb{T}$. But the fact that each d_i satisfies the second axiom of $\mathbb{T}$ follows from the fact that a_i does. So the structures d_i satisfy all the axioms of $\mathbb{T}$.

This completes the proof that the hypotheses of Corollary 8.2.1 are satisfied by the theory $\mathbb{T}$ with respect to its Horn part and the category of (finite) $\mathbb{T}$-models; therefore $\mathbb{T}$ is of presheaf type classified by the topos $[\mathbf{S}, \mathbf{Set}]$ where $\mathbf{S}$ is the category of finite sets and surjective functions between them.

9.4 The theory of Diers fields

As observed in [51] (cf. section 5 therein), the coherent theory $\mathbb{T}$ of fields is not of presheaf type. In fact, one can easily identify two properties which are preserved by homomorphisms of fields but which are not definable in finitely generated fields by geometric formulae over the signature of $\mathbb{T}$ (cf. Theorem 6.1.4): the property of a field to have characteristic 0, and the property of a tuple $(x_1, \ldots, x_n, x_{n+1})$ of elements of a field that the element x_{n+1} is transcendental over the subfield generated by the elements $x_1, \ldots, x_n$. This can be easily seen by arguing as follows. Assuming the axiom of choice, the theory $\mathbb{T}$ has enough set-based models (it being coherent). Hence if the property of having characteristic 0 were definable by a geometric sentence ϕ over the signature of fields then the sequent $(\top \vdash_{[]} \phi \vee \bigvee_{p \in \mathbb{P}} \phi_p)$, where $\mathbb{P}$ is the set of prime numbers and ϕ_p (for each $p \in \mathbb{P}$) is the sentence $p.1 = 0$ expressing the property of having characteristic p, would be provable in $\mathbb{T}$. But, $\mathbb{T}$ being coherent, the sequent $(\top \vdash_{[]} \phi \vee \bigvee_{p \in \mathbb{P}'} \phi_p)$ would be provable in $\mathbb{T}$ for some finite subset $\mathbb{P}' \subseteq \mathbb{P}$ (cf. Theorem 10.8.6(iii) below), which is absurd as it would imply that the set of all possible characteristics of a field is finite. A similar argument works for the 'transcendence' property which, like the former, is the complement of a property definable by a strictly infinitary geometric formula.

In order to make such properties definable and obtain a presheaf completion of the theory $\mathbb{T}$, it is thus necessary to enlarge the signature of $\mathbb{T}$ with new relation symbols. In fact, Johnstone introduces in [51] a 0-ary predicate R_0, expressing the property of a field to have characteristic 0, and for each natural number $n \geq 0$ an $(n+1)$-predicate $R_{n+1}(x_1, \ldots, x_n, x_{n+1})$ expressing the property of x_{n+1} to be transcendental over the subfield generated by the elements $x_1, \ldots, x_n$. Let Σ'

be the resulting signature. Formally, one has to impose the following axioms over Σ' to ensure that these predicates have indeed the required meaning (below we use the abbreviation $\mathrm{Inv}(z)$ for the formula $(\exists x)(x \cdot z = 1)$):

$$\left(\top \vdash R_0 \vee \bigvee_{p \in \mathbb{P}} p.1 = 0\right);$$

$$(R_0 \wedge p.1 = 0 \vdash \bot)$$

for each $p \in \mathbb{P}$;

$$\left(\top \vdash_{x_1,\ldots,x_{n+1}} R_n(x_1,\ldots,x_n,x_{n+1}) \vee \right.$$

$$\left. \left(\bigvee_{m \in \mathbb{N},\, \vec{c}_0^{\vec{n}},\ldots,\vec{c}_m^{\vec{n}}} \left(\sum_{j=0}^{m} P_{\vec{c}_j^{\vec{n}}} x_{n+1}^j = 0\right) \wedge \bigvee_{i \in \{0,1,\ldots,m\}} \mathrm{Inv}(P_{\vec{c}_i^{\vec{n}}})\right)\right)$$

for each natural number $n \geq 0$, where the former disjunction is taken over all the natural numbers $m \geq 0$ and all the tuples $\vec{c}_i^{\vec{n}}$ (for $i \in \{0,\ldots,m\}$) of integer coefficients (i.e. coefficients of the form $1 + \cdots + 1$ an integer number of times) of polynomials in n variables $x_1,\ldots,x_n$ of degree $\leq m$ and the expression $P_{\vec{c}_i^{\vec{n}}}$ (for each $i \in \{0,\ldots,m\}$) denotes the polynomial term in the variables $x_1,\ldots,x_n$ corresponding to the tuple $\vec{c}_i^{\vec{n}}$, and

$$\left(R_n(x_1,\ldots,x_n,x_{n+1}) \wedge \left(\bigvee_{m \in \mathbb{N},\, \vec{c}_0^{\vec{n}},\ldots,\vec{c}_m^{\vec{n}}} \left(\left(\sum_{j=0}^{m} P_{\vec{c}_j^{\vec{n}}} x_{n+1}^j = 0\right) \wedge \right.\right.\right.$$

$$\left.\left.\left. \bigvee_{i \in \{0,1,\ldots,m\}} \mathrm{Inv}(P_{\vec{c}_i^{\vec{n}}})\right)\right) \vdash_{x_1,\ldots,x_{n+1}} \bot\right).$$

Let $\mathbb{D}$ be the theory, called in [51] the *theory of Diers fields*, obtained from $\mathbb{T}$ by adding these new predicates and the above-mentioned axioms.

Note that the geometric formula

$$\bigvee_{m \in \mathbb{N},\, \vec{c}_0^{\vec{n}},\ldots,\vec{c}_m^{\vec{n}}} \left(\sum_{j=0}^{m} P_{\vec{c}_j^{\vec{n}}} x_{n+1}^j = 0\right) \wedge \bigvee_{i \in \{0,1,\ldots,m\}} \mathrm{Inv}(P_{\vec{c}_i^{\vec{n}}})$$

is $\mathbb{D}$-provably equivalent to a disjunction of geometric formulae, namely, the formulae $\left(\sum_{j=0}^{m} P_{\vec{c}_j^{\vec{n}}} x_{n+1}^j = 0\right) \wedge \mathrm{Inv}(P_{\vec{c}_i^{\vec{n}}})$ (for each m, $\vec{c}_0^{\vec{n}},\ldots,\vec{c}_m^{\vec{n}}$ and $i \in \{0,\ldots,m\}$), each of which has a $\mathbb{D}$-provable complement, namely the formula

$$\mathrm{Inv}\left(\sum_{j=0}^{m} P_{\vec{c}_j^{\vec{n}}} x_{n+1}^j\right) \vee P_{\vec{c}_i^{\vec{n}}} = 0.$$

Notice also that, for any tuples $\vec{c_i^n}$ (for $i \in \{0, \ldots, m\}$) of integer coefficients of a polynomial expression $P_{\vec{c_i^n}}$ in the variables $x_1, \ldots, x_n$, the cartesian sequent

$$(*) \qquad \left(R_n(x_1, \ldots, x_n, x_{n+1}) \wedge \operatorname{Inv}(P_{\vec{c_i^n}}) \vdash_{x_1,\ldots,x_{n+1}} \operatorname{Inv}\left(\sum_{j=0}^{m} P_{\vec{c_j^n}} x_{n+1}^j \right) \right)$$

is provable in $\mathbb{D}$.

Following [51], we observe that the theory $\mathbb{D}$ satisfies the property that every finitely presentable $\mathbb{D}$-model, i.e. every finitely generated field, is finitely presented as a model of its cartesianization. This will follow from Lemma 7.1.7 once we have proved that every finitely generated field F is presented by a finite set of generators, in the weak sense of the lemma, by a formula over the signature of $\mathbb{D}$. To this end, we regard $\mathbb{T}$ as axiomatized over the signature of von Neumann regular rings, which contains a unary function for the operation of pseudoinverse; over this signature, every finitely generated field (in the sense of field theory) becomes finitely generated (in the sense of model theory, cf. section 7.2.3). Notice that $\mathbb{D}$ satisfies the first set of hypotheses of Lemma 7.1.7 with respect to the theory $\mathbb{T}$.

We shall prove that every field F with a finite set of generators is presented as a model of the cartesianization of $\mathbb{D}$, in the weak sense of Lemma 7.1.7, by a geometric formula over the signature of $\mathbb{D}$ by induction on the number n of generators of F. If $n = 0$ then F is equal to its prime field, so either F has characteristic p, in which case it is equal to $\mathbb{F}_p$, or F has characteristic 0, in which case it is equal to $\mathbb{Q}$. Now, the field $\mathbb{F}_p$ is clearly (weakly) presented by the formula $p.1 = 0$, while the field $\mathbb{Q}$ is (weakly) presented by the formula R_0 since the sequent $(R_0 \vdash \operatorname{Inv}(n.1))$ is provable in the cartesianization of $\mathbb{D}$ for each non-zero natural number n. Now, consider a field F generated by $n+1$ elements $x_1, \ldots, x_n, x_{n+1}$. If we denote by F_0 its prime field, we have that $F = F_0(x_1, \ldots, x_n)(x_{n+1})$. Suppose that $F_0(x_1, \ldots, x_n)$ is (weakly) presented by a formula in n variables $\phi(x_1, \ldots, x_n)$ with the elements $x_1, \ldots, x_n$ as generators. There are two cases: either the element x_{n+1} is transcendental over the field $F_0(x_1, \ldots, x_n)$ or not.

In the first case, F is isomorphic to the field of rational functions in one variable with coefficients in $F_0(x_1, \ldots, x_n)$, and it is (weakly) presented by the formula $\{x_1, \ldots, x_n, x_{n+1} . \phi \wedge R_{n+1}\}$; in other words, for any model $(A, \{AR_n \mid n \in \mathbb{N}\})$ of the cartesianization of $\mathbb{D}$, the function which assigns a ring homomorphism $f : F \to A$ with the property that $f(x_1, \ldots, x_n, x_{n+1}) \in AR_{n+1} \cap [[\vec{x} . \phi]]_A$ to the element $f(x_1, \ldots, x_n, x_{n+1})$ is injective and surjective on $AR_{n+1} \cap [[\vec{x} . \phi]]_A$. This can be proved as follows. Since F is generated by the elements $x_1, \ldots, x_{n+1}$, the injectivity is clear, so it remains to prove the surjectivity, i.e. that for any $(n+1)$-tuple $(a_1, \ldots, a_n, a_{n+1}) \in AR_{n+1} \cap [[\vec{x} . \phi]]_A$ there exists a ring homomorphism $F \to A$ which sends x_i to a_i for each $i \in \{1, \ldots, n+1\}$. By the induction hypothesis, since $(a_1, \ldots, a_n) \in [[\vec{x} . \phi]]_A$, there exists a unique ring homomorphism

$g : F_0(x_1, \ldots, x_n) \to A$ such that $g(x_i) = a_i$ for each $i \in \{1, \ldots, n\}$. By definition of F, there exists a ring homomorphism $f : F \to A$ which extends g and sends x_{n+1} to a_{n+1} if and only if for every polynomial P with a non-zero coefficient $P_{c_i^{\vec{n}}}(x_1, \ldots, x_n)$ in $F_0(x_1, \ldots, x_n)$, $\sum_{j=0}^{m} g(P_{c_j^{\vec{n}}})a^j$ is invertible in A. But since $P_{c_i^{\vec{n}}}(x_1, \ldots, x_n)$ is non-zero (equivalently, invertible) in the field $F_0(x_1, \ldots, x_n)$, its image $g(P_{c_i^{\vec{n}}}(x_1, \ldots, x_n)) = P_{c_i^{\vec{n}}}(g(x_1), \ldots, g(x_n))$ under the homomorphism g is invertible in A; therefore, since the sequent $(*)$ holds in A, the condition $(f(x_1), \ldots, f(x_n), a) \in AR_{n+1}$ entails the fact that $\sum_{j=0}^{m} g(P_{c_j^{\vec{n}}})a^j$ is invertible in A, as required.

In the second case, consider the minimal polynomial P for x_{n+1} over the field $F_0(x_1, \ldots, x_n)$; then F is isomorphic to the quotient of $F_0(x_1, \ldots, x_n)[X]$ by the ideal generated by the polynomial P. It is immediate to see that F is (weakly) presented by the formula in $n+1$ variables $\phi \wedge P(x_{n+1}) = 0$ with generators $x_1, \ldots, x_n, x_{n+1}$.

These arguments show that the theories $\mathbb{T}$ and $\mathbb{D}$ satisfy the hypotheses of Lemma 7.1.7. It follows that all the finitely generated fields are finitely presented models of the cartesianization of $\mathbb{D}$, as required. Condition (iii) of Theorem 6.3.1 is therefore satisfied by $\mathbb{D}$ with respect to the category of finitely generated fields (cf. Proposition 6.3.12(i)). We could have alternatively deduced this from Corollary 6.3.17 and Theorem 6.3.22.

In [51], it is shown that $\mathbb{D}$ is a theory of presheaf type by assuming a form of the axiom of choice in order to ensure that $\mathbb{D}$ has enough set-based models. Theorem 6.3.1 allows us to prove that $\mathbb{D}$ is of presheaf type directly, without assuming any non-constructive principles. Having already proved that condition (iii) of Theorem 6.3.1 is satisfied, it remains to show that conditions (i) and (ii) hold as well. To prove this, we shall verify that conditions (ii)(a), (ii)(b) and (ii)(c) of Theorem 6.3.4 and conditions (ii)(a) and (ii)(b) of Theorem 6.3.8 are satisfied.

The fact that condition (ii)(a) of Theorem 6.3.8 holds follows immediately from the above discussion in view of Remarks 6.3.9(b) and 6.2.6(c). Condition (ii)(c) of Theorem 6.3.4 is trivially satisfied while condition (ii)(a) of Theorem 6.3.4 follows from condition (ii)(a) of Theorem 6.3.8 (cf. Remarks 6.3.5(b) and 6.3.5(c)). By Remark 6.3.9(a), condition (ii)(b) of Theorem 6.3.4 implies condition (ii)(b) of Theorem 6.3.8. So, to complete the proof that $\mathbb{D}$ is of presheaf type, it remains to show that condition (ii)(b) of Theorem 6.3.4 holds, i.e. that for any finitely generated fields c and d, any $\mathbb{D}$-model M in a Grothendieck topos $\mathcal{E}$ and any Σ'-structure homomorphisms $f : c \to \mathrm{Hom}_{\mathcal{E}}(E, M)$ and $g : d \to \mathrm{Hom}_{\mathcal{E}}(E, M)$, there exist an epimorphic family $\{e_i : E_i \to E \mid i \in I\}$ in $\mathcal{E}$ and for each $i \in I$ a finitely generated field b_i, field homomorphisms $u_i : c \to b_i$, $v_i : d \to b_i$ and a Σ'-structure homomorphism $z_i : b_i \to \mathrm{Hom}_{\mathcal{E}}(E_i, M)$ such that $\mathrm{Hom}_{\mathcal{E}}(e_i, M) \circ f = z_i \circ u_i$ and $\mathrm{Hom}_{\mathcal{E}}(e_i, M) \circ g = z_i \circ v_i$.

We can prove this by induction on the sum n of the minimal numbers of generators of c and d.

Before proceeding further, it is convenient to remark the following fact: for any model $(A, \{AR_n \mid n \in \mathbb{N}\})$ of the cartesianization of $\mathbb{D}$ where A is not the trivial ring (for instance, a Σ'-structure of the form $\mathrm{Hom}_{\mathcal{E}}(E, M)$, where M is a model of $\mathbb{D}$ in a topos $\mathcal{E}$ and E is a non-zero object of $\mathcal{E}$) and any field e considered as a model of $\mathbb{D}$, all the Σ'-structure homomorphisms $e \to A$ reflect the satisfaction of the relations R_n, i.e. $f(x_1, \ldots, x_n) \in AR_n$ implies $(x_1, \ldots, x_n) \in eR_n$. This easily follows from the disjunctive axioms of $\mathbb{D}$ defining R_n and the cartesian axiom (*). Note also that such homomorphisms are always injective (since their domain is a field and their codomain is a non-zero ring).

For proving that our condition is satisfied, we can suppose without loss of generality all the objects E arising in Σ'-structure homomorphisms to structures of the form $\mathrm{Hom}_{\mathcal{E}}(E, M)$ to be non-zero (since removing zero arrows from an epimorphic family leaves the family epimorphic).

If $n = 0$ (that is, if both c and d are equal to their prime fields) then c and d have the same characteristic; indeed, equalities of the form $p.1 = 0$ are preserved and reflected by the homomorphisms f and g. So c and d are isomorphic, whence the condition is trivially satisfied. Let us now assume that the condition is true for all $k \leq n$ and prove it for $n + 1$. We can represent c as $c'(x)$ in such a way that a set of generators for c with minimal cardinality can be obtained by adding an element x to a set of generators for c'; then by the induction hypothesis there exist an epimorphic family $\{e_i : E_i \to E \mid i \in I\}$ in $\mathcal{E}$ and for each $i \in I$ a finitely generated field u_i, field homomorphisms $f_i : c' \to u_i$ and $g_i : d \to u_i$ and a Σ'-structure homomorphism $r_i : u_i \to \mathrm{Hom}_{\mathcal{E}}(E_i, M)$ such that $r_i \circ f_i = \mathrm{Hom}_{\mathcal{E}}(e_i, M) \circ f|_{c'}$ and $r_i \circ g_i = \mathrm{Hom}_{\mathcal{E}}(e_i, M) \circ g$. Now, consider for each $i \in I$ the element $f(x) \circ e_i \in \mathrm{Hom}_{\mathcal{E}}(E_i, M)$. Suppose that c' has m generators $\chi_1, \ldots, \chi_m$; then for each $i \in I$, by the disjunctive axiom of $\mathbb{D}$ involving R_m, there exists an epimorphic family $\{f_{i,j} : F_{i,j} \to E_i \mid j \in J_i\}$ in $\mathcal{E}$ such that for any $j \in J_i$, either $R_{m+1}((r_i \circ f_i)(\chi_1) \circ e_i \circ f_{i,j}, \ldots, (r_i \circ f_i)(\chi_m) \circ e_i \circ f_{i,j}, f(x) \circ e_i \circ f_{i,j})$ or $f(x) \circ e_i \circ f_{i,j}$ is the root of a non-zero polynomial P with coefficients belonging to the von Neumann regular subring of $\mathrm{Hom}_{\mathcal{E}}(F_{i,j}, M)$ generated by the elements $(r_i \circ f_i)(\chi_1) \circ e_i \circ f_{i,j}, \ldots, (r_i \circ f_i)(\chi_m) \circ e_i \circ f_{i,j}$. In the first case, $f(x) \circ e_i \circ f_{i,j}$ is transcendental over c' via the embedding $\mathrm{Hom}_{\mathcal{E}}(f_{i,j}, M) \circ r_i \circ f_i$ and over u_i via the embedding $\mathrm{Hom}_{\mathcal{E}}(f_{i,j}, M) \circ r_i$ (apply the above remarks to these two embeddings); it follows that the homomorphism f_i extends to a homomorphism from $c = c'(x)$ to u_i. In the second case, the homomorphism $\mathrm{Hom}_{\mathcal{E}}(e_i \circ f_{i,j}, M) \circ f$ being injective, there exists a non-zero polynomial P with coefficients in c' such that $f(x) \circ e_i \circ f_{i,j}$ is a root of the image of P under $\mathrm{Hom}_{\mathcal{E}}(f_{i,j}, M) \circ r_i \circ f_i = \mathrm{Hom}_{\mathcal{E}}(e_i \circ f_{i,j}, M) \circ f|_{c'}$. It follows that x is a root of P in c. We can thus clearly suppose P to be irreducible without loss of generality and represent $c = c'(x)$ as $c = c'[z]/P(z)$, via an isomorphism sending z to x; denoting by P' the image of P under the homomorphism f_i, the arrow f_i thus yields an arrow $c = c'[z]/P(z) \to u_i[w]/P'(w)$ and the homomorphism $\mathrm{Hom}_{\mathcal{E}}(f_{i,j}, M) \circ r_i$ factors through the quotient map $u_i \to u_i[w]/P'(w)$ yielding a ring homomorphism which is in fact a Σ'-structure homomorphism (by Lemma 7.1.7, cf. the above argument for proving that every finitely generated field is finitely presented as a model of the cartesianization of $\mathbb{D}$).

This completes our proof that $\mathbb{D}$ is of presheaf type.

Notice that we could have alternatively shown that $\mathbb{D}$ satisfies condition (ii)(b) of Theorem 6.3.4 either by using Remark 6.3.5(d) or by applying Theorem 6.3.6.

9.5 The theory of algebraic extensions of a given field

Let F be a field. We define the theory $\mathbb{T}_F$ of *algebraic extensions* of F as the expansion of the coherent theory of fields obtained by adding one constant symbol $\bar{a}$ for each element $a \in F$ and the following axioms:

$$(\top \vdash \overline{1_F} = 1);$$

$$(\top \vdash \overline{0_F} = 0);$$

$$(\top \vdash \overline{a} + \overline{b} = \overline{a +_F b})$$

for any elements $a, b \in F$ (where the symbol $+_F$ denotes the addition operation in F);

$$(\top \vdash \overline{a} \cdot \overline{b} = \overline{a \cdot_F b})$$

for any elements $a, b \in F$ (where the symbol $\cdot_F$ denotes the multiplication operation in F), plus the algebraicity axiom

$$\left(\top \vdash_x \bigvee_{n \in \mathbb{N},\, a_0,\ldots,a_{n-1},a_n \in F} \overline{a_n} \cdot x^n + \overline{a_{n-1}} \cdot x^{n-1} + \cdots + \overline{a_0} = 0 \right).$$

Let us prove that $\mathbb{T}_F$ is of presheaf type. Clearly, the finitely presentable models of $\mathbb{T}_F$ are exactly the finitely generated algebraic extensions of F, that is the finite extensions of F. One can prove, by adapting the argument used in the proof that every finitely generated field is finitely presented as a model of the cartesianization of the theory of Diers fields established in section 9.4, that every finite extension of F is finitely presented as a model of the cartesianization of $\mathbb{T}_F$. Specifically, every finite extension $F(x_1,\ldots,x_n)$ of F is presented by the conjunction of the formulae of the form $P_i(x_1,\ldots,x_i)(x_{i+1}) = 0$ (for $i = 0,\ldots,n-1$), where P_i is the minimal polynomial of the element x_{i+1} over the field $F(x_1,\ldots,x_i)$.

Condition (iii) of Theorem 6.3.1 is thus satisfied by the theory $\mathbb{T}_F$ with respect to the category of finite extensions of F (cf. Proposition 6.3.12(i)). In verifying that conditions (i) and (ii) of Theorem 6.3.1 are satisfied, one is reduced as in section 9.4 to check that condition (ii)(b) of Theorem 6.3.4 holds; again, this can be done by an argument similar to the one given in section 9.4. We can thus conclude that $\mathbb{T}_F$ is of presheaf type classified by the topos of covariant set-valued functors on the category of finite extensions of F.

Next, let us consider the theory $\mathbb{S}_F$ of $\mathbb{T}_F$ of *separable extensions* of F, that is the quotient of $\mathbb{T}_F$ obtained by adding the following sequent:

$$\left(\top \vdash_x \bigvee_{n \in \mathbb{N},\, (a_0,\ldots,a_{n-1},a_n) \in \mathcal{S}_F^n} \overline{a_n} \cdot x^n + \overline{a_{n-1}} \cdot x^{n-1} + \cdots + \overline{a_0} = 0 \right),$$

where $\mathcal{S}_F^n$ is the set of $(n+1)$-tuples of elements $a_0, \ldots, a_n$ of F such that the polynomial $a_n Z^n + a_{n-1} Z^{n-1} + \cdots + a_0 \in F[Z]$ is irreducible and separable.

Clearly, the finitely presentable $\mathbb{S}_F$-models are precisely the finite separable extensions of F. In particular, every finitely presentable $\mathbb{S}_F$-model is finitely presentable also as a $\mathbb{T}_F$-model. In fact, by Artin's primitive element theorem, every finite separable extension of F is presented by a formula of the form $\{x \,.\, \overline{a_n} \cdot x^n + \overline{a_{n-1}} \cdot x^{n-1} + \cdots + \overline{a_0} = 0\}$.

The hypotheses of Corollary 8.2.1 are trivially satisfied by the quotient $\mathbb{S}_F$ of $\mathbb{T}_F$; indeed, for any $\mathbb{T}_F$-model homomorphism $M \to N$ (in an arbitrary Grothendieck topos), if N is separable (i.e. a model of $\mathbb{S}_F$) then M is *a fortiori* separable as well. We can thus conclude that the theory $\mathbb{S}_F$ is also of presheaf type, classified by the category of covariant set-valued functors on the category of finite separable extensions of F.

Remark 9.5.1 The theory of fields of fixed finite characteristic p which are algebraic over their prime field, which is shown in [29] to be of presheaf type classified by the topos of covariant set-valued functors on the category of finite fields of characteristic p, is (trivially) Morita-equivalent to the theory $\mathbb{T}_{\mathbb{F}_p}$ introduced above (where $\mathbb{F}_p$ is the field with p elements).

9.6 Groups with decidable equality

In this section we shall study the injectivization $\mathbb{G}_m$ of the (algebraic) theory $\mathbb{G}$ of groups and describe a presheaf completion for it (in the sense of section 7.1.3).

Clearly, the finitely presentable $\mathbb{G}_m$-models are precisely the finitely generated groups. Even if $\mathbb{G}_m$ satisfies condition (iii) of Theorem 6.3.1 with respect to the category of finitely generated groups and injective homomorphisms between them (by Theorem 6.3.22 and Corollary 6.3.17), $\mathbb{G}_m$ is not of presheaf type. To see this, consider the property of an element x of a group G to be non-nilpotent. This property is clearly preserved by injective homomorphisms of groups, so if $\mathbb{G}_m$ were of presheaf type it would be definable in finitely generated groups by a geometric formula $\phi(x)$ over the signature of $\mathbb{G}_m$ (by Theorem 6.1.4) and hence the sequent

$$\left(\top \vdash_x \phi(x) \vee \bigvee_{n \in \mathbb{N}} (x^n = 1) \right)$$

would be provable in $\mathbb{G}_m$ (since every theory of presheaf type has enough set-based models and this sequent is valid in every set-based $\mathbb{G}_m$-model by definition of ϕ). As $\mathbb{G}_m$ is coherent, the disjunction on the right-hand side could be replaced by a finite subdisjunction (cf. Theorem 10.8.6(iii)); but this is absurd since it would imply that there exists a natural number n such that every nilpotent element x of a group satisfies $x^n = 1$.

To obtain a presheaf completion of the theory $\mathbb{G}_m$, we add to the signature of $\mathbb{G}_m$ a relation symbol R_N^n for each natural number n and any normal subgroup N of the free group F_n on n generators, and the following axioms (where the symbol $\neq$ denotes the predicate of $\mathbb{G}_m$ which is $\mathbb{G}_m$-provably complemented to the equality relation):

$$\left(\top \vdash_{\vec{x}} R_N^n(\vec{x}) \vee \left(\bigvee_{w,w' \in F_n \mid ww'^{-1} \in N} w(\vec{x}) \neq w'(\vec{x})\right) \vee \left(\bigvee_{w,w' \in F_n \mid ww'^{-1} \notin N} w(\vec{x}) = w'(\vec{x})\right)\right)$$

and

$$\left(R_N^n(\vec{x}) \wedge \left(\bigvee_{w,w' \in F_n \mid ww'^{-1} \in N} w(\vec{x}) \neq w'(\vec{x}) \vee \bigvee_{w,w' \in F_n \mid ww'^{-1} \notin N} w(\vec{x}) = w'(\vec{x})\right) \vdash_{\vec{x}} \bot\right)$$

for any $n \in \mathbb{N}$ and any normal subgroup N of F_n, and

$$\left(\top \vdash_{\vec{x}} \bigvee_{N \in \mathcal{N}_n} R_N^n(\vec{x})\right),$$

where $\mathcal{N}_n$ is the set of normal subgroups of F_n (for any $n \in \mathbb{N}$).

Let $\mathbb{G}_p$ be the resulting theory; we shall prove that it is of presheaf type.

The theories $\mathbb{G}_m$ and $\mathbb{G}_p$ clearly satisfy the first set of hypotheses of Lemma 7.1.7. Let us show that every finitely generated group is finitely presented as a model of the cartesianization of $\mathbb{G}_p$. By Lemma 7.1.7, to prove that F_n/N is presented by the formula $R_N^n(\vec{x})$ as a model of the cartesianization of $\mathbb{G}_p$, it suffices to verify that for any set-based model $(G', \{G'R_N^n \mid n \in \mathbb{N}, N \in \mathcal{N}_n\})$ of the cartesianization of $\mathbb{G}_p$, denoting by $\vec{\xi} = (\xi_1, \ldots, \xi_n)$ the generators of the group F_n, the group homomorphisms $f : F_n/N \to G'$ such that $f(\vec{\xi}) \in [[\vec{x} \cdot R_N^n]]_{G'}$ correspond exactly, via the assignment $f \to f(\vec{\xi})$, to the n-tuples $\vec{y}$ of elements of G' which belong to the interpretation in G' of the formula R_N^n. Given an n-tuple $\vec{y}$ of elements of G' which belong to the interpretation of R_N^n in G', we can define a function $f : F_n/N \to G'$ by setting $f([w]) = w_{G'}(\vec{y})$ (where $w_{G'}$ is the interpretation in G' of the term $w \in F_n$). This is a well-defined injective group homomorphism since the following sequents are provable in the cartesianization of $\mathbb{G}_p$ and hence are valid in G':

$$(R_N^n \vdash_{\vec{x}} w = w')$$

for any $w, w' \in F_n$ such that $ww'^{-1} \in N$, and

$$(R_N^n \vdash_{\vec{x}} w \neq w')$$

for any $w, w' \in F_n$ such that $ww'^{-1} \notin N$. Since every finitely generated group is, up to isomorphism, of the form F_n/N for some natural number n and some normal subgroup N of F_n, we can conclude that the theory $\mathbb{G}_p$ satisfies condition (iii) of Theorem 6.3.1 with respect to the category $\mathbf{Grp}_{\text{f.g.}}$ of finitely generated groups and injective homomorphisms between them (cf. Proposition 6.3.12(i)). We could have alternatively proved this by invoking Corollary 6.3.17 and Theorem 6.3.22.

As in the case of the theory of Diers fields treated in section 9.4, in order to verify that the theory $\mathbb{G}_p$ satisfies conditions (i) and (ii) of Theorem 6.3.1 one is reduced to show that it satisfies condition (ii)(b) of Theorem 6.3.4; but this follows from Remark 6.3.5(d). The theory $\mathbb{G}_p$ is thus classified by the topos $[\mathbf{Grp}_{\text{f.g.}}, \mathbf{Set}]$.

Notice that the category $\mathbf{Grp}_{\text{f.g.}}$ is cocartesian; indeed, it has an initial object (namely, the trivial group) and pushouts (given by the free product with amalgamation construction, cf. [69]). It follows that the topos $[\mathbf{Grp}_{\text{f.g.}}, \mathbf{Set}]$ is coherent. The theory $\mathbb{G}_p$, in spite of being infinitary, is thus classified by a coherent topos. This fact has various implications for $\mathbb{G}_p$. For instance, the coherence of the classifying topos $\mathbf{Set}[\mathbb{G}_p]$ of $\mathbb{G}_p$ implies that every formula over the signature of $\mathbb{G}_p$ presenting a finitely generated group (regarded as a $\mathbb{G}_p$-model) is not only $\mathbb{G}_p$-irreducible, but also a coherent object of $\mathbf{Set}[\mathbb{G}_p]$; in particular, for any $\mathbb{G}_p$-compact formula $\{\vec{x}.\phi\}$ over the signature of $\mathbb{G}_p$, the formula $\{\vec{x}, \vec{y}.\phi(\vec{x}) \wedge \phi(\vec{y})\}$, where $\vec{x}$ and $\vec{y}$ are two disjoint contexts of the same type, is also $\mathbb{G}_p$-compact (cf. Definition 10.8.4(b) below for the notion of $\mathbb{T}$-compact formula for a geometric theory $\mathbb{T}$). Semantically speaking, a formula $\{\vec{x}.\phi\}$ is $\mathbb{G}_p$-compact if whenever $\{S_i \mid i \in I\}$ is a family of assignments $G \to S_i^G \subseteq G^n$ sending each finitely generated group G to a subset $S_i^G \subseteq [[\vec{x}.\phi]]_G$ in such a way that every injective homomorphism $f : G \to G'$ of finitely generated groups sends tuples in S_i^G to tuples in $S_i^{G'}$, if $[[\vec{x}.\phi]]_G = \bigcup_{i \in I} S_i^G$ then there exists a finite subset $J \subseteq I$ such that $[[\vec{x}.\phi]]_G = \bigcup_{i \in J} S_i^G$ for all G. Assuming the axiom of choice, by Deligne's theorem (Theorem 2.1.12) the fact that the category $\mathbf{Grp}_{\text{f.g.}}$ is cocartesian also implies the existence of enough set-based models for any quotient of $\mathbb{G}_p$ whose associated Grothendieck topology on $(\mathbf{Grp}_{\text{f.g.}})^{\text{op}}$ is of finite type.

9.7 Locally finite groups

Let $\mathbb{G}$ be the algebraic theory of groups. Since every finite group is finitely presented as a $\mathbb{G}$-model (cf. Proposition 9.1.1) and $\mathbb{G}$ is of presheaf type, Theorem

8.2.10 ensures that the quotient $\mathbb{U}$ of $\mathbb{G}$ consisting of all the geometric sequents over the signature of $\mathbb{G}$ that are valid in every finite group is classified by the topos $[\mathbf{Grp}_{\text{fin}}, \mathbf{Set}]$, where $\mathbf{Grp}_{\text{fin}}$ is the category of finite groups and homomorphisms between them. Since $\mathbb{U}$ is classified by the topos $[\mathbf{Grp}_{\text{fin}}, \mathbf{Set}]$, the set-based models of $\mathbb{U}$ are exactly the groups which can be expressed as filtered colimits of finite groups. This yields the following characterization of locally finite groups (recall that a group is said to be *locally finite* if all its finitely generated subgroups are finite).

Proposition 9.7.1 *The locally finite groups are exactly the groups which satisfy all the geometric sequents over the signature of the (algebraic) theory of groups which hold in all finite groups.*

Proof In view of the above discussion, it remains to verify that a group is locally finite if and only if it is a filtered colimit of finite groups. This can be proved as follows. If a group is locally finite then it is the filtered union of all its finitely generated (whence finite) subgroups. Conversely, suppose that G is a filtered colimit of finite groups. Then G is the directed union of the images in G of these finite groups, which are again finite. It follows that every finitely generated subgroup H of G is contained in one of the groups in this directed union (since the colimit is filtered, there exists one of them which contains all the generators of H) and hence it is *a fortiori* finite, as required. □

The injectivization of $\mathbb{U}$ is also of presheaf type (by Corollary 7.2.43) and can be characterized as the quotient of the injectivization $\mathbb{G}_m$ of $\mathbb{G}$ consisting of all the geometric sequents over its signature which hold in all finite groups.

9.8 Vector spaces

Let $\mathbb{V}_K$ be the expansion of the algebraic theory of vector spaces over a field K obtained by adding, for each natural number n, an n-ary predicate R_n expressing the property of an n-tuple of elements to be linearly independent, that is the following sequents:

$$\left(\top \vdash_{\vec{x}} R_n(\vec{x}) \vee \bigvee_{\substack{(k_1,\ldots,k_n)\in K^n \\ k_i \neq 0 \text{ for some } i}} k_1 x_1 + \cdots + k_n x_n = 0 \right)$$

and

$$\left(R_n(\vec{x}) \wedge \left(\bigvee_{\substack{(k_1,\ldots,k_n)\in K^n \\ k_i \neq 0 \text{ for some } i}} k_1 x_1 + \cdots + k_n x_n = 0 \right) \vdash_{\vec{x}} \bot \right).$$

The category of set-based models of the theory $\mathbb{V}_K$ has as objects the vector spaces over K and as arrows the injective homomorphisms between them. The finitely presentable $\mathbb{V}_K$-models are precisely the finite-dimensional vector spaces over K.

By using techniques analogous to those employed in section 9.4, one can prove that $\mathbb{V}_K$ is of presheaf type. Also, by using Corollary 8.2.1, one can easily show that, for any fixed natural number n, the expansion of the theory $\mathbb{V}_K$ obtained by adding the sequent

$$\left(\top \vdash_{x_1,\ldots,x_{n+1}} \bigvee_{\substack{(k_1,\ldots,k_{n+1}) \in K^{n+1} \\ k_i \neq 0 \text{ for some } i}} k_1 x_1 + \cdots + k_{n+1} x_{n+1} = 0\right)$$

is of presheaf type; note that the set-based models of this theory are precisely the vector spaces over K of dimension $\leq n$.

9.9 The theory of abelian ℓ-groups with strong unit

Recall that an abelian ℓ-group with strong unit is a lattice-ordered group $(G, 0, \leq)$ with a distinguished element u, called the unit of the group, such that for any element $x \in G$ such that $x \geq 0$ there exists a natural number n such that $x \leq nu$, where $nu = u + \cdots + u$ n times. The reader may consult Chapter 2 of [70] for an introduction to the theory of lattice-ordered groups (also called ℓ-groups).

We can axiomatize the theory $\mathbb{L}_u$ of abelian ℓ-groups with strong unit over a signature Σ consisting of four binary function symbols $+$, $-$, inf, sup, two constants 0 and u and a binary relation symbol $\leq$, by using Horn sequents to formalize the notion of abelian ℓ-group and the following geometric sequent to express the property of being a strong unit:

$$\left(x \geq 0 \vdash_x \bigvee_{n \in \mathbb{N}} x \leq nu\right).$$

The following lemma will be useful for proving that the theory $\mathbb{L}_u$ is of presheaf type.

Lemma 9.9.1 *Let G be an abelian group with a distinguished element u and generators $x_1, \ldots, x_n$. If for every $i \in \{1, \ldots, n\}$ there exists a natural number k_i such that $|x_i| \leq k_i u$ then u is a strong unit for G.*

Proof Recall that the absolute value $|x|$ of an element x of an abelian ℓ-group $(G, 0, +, -, \leq, \inf, \sup)$ is the element $\sup(x, -x)$. For any $x \in G$, $|x| \geq 0$, $|x| = |-x|$, and for any $x, y \in G$ the triangular inequality $|x + y| \leq |x| + |y|$ holds.

Since G is generated by elements $x_1, \ldots, x_n$, every element x of G can be expressed as the interpretation $t(x_1, \ldots, x_n)$ of a term t over the signature Σ. Let us prove that there exists a natural number n such that $|x| \leq nu$ by induction on the structure of t; this will clearly imply our thesis, since if $x \geq 0$ then $|x| = x$. If t is a variable then the claim is clearly true by our hypotheses. If $x = x' + x''$ with $|x'| \leq n'u$ and $|x''| \leq n''u$ then by the triangular inequality we have $|x| \leq n' + n''$, and similarly for the subtraction, inf and sup cases. □

Let us verify that the theory $\mathbb{L}_u$ satisfies the hypotheses of Corollary 8.2.1 with respect to its Horn part (i.e. the theory $\mathbb{H}$ consisting of all the Horn sequents which are provable in $\mathbb{L}_u$).

We have to prove that for any finitely presentable $\mathbb{H}$-model c, any model G of $\mathbb{L}_u$ in a Grothendieck topos $\mathcal{E}$, any object E of $\mathcal{E}$ and any Σ-structure homomorphism $f : c \to \mathrm{Hom}_{\mathcal{E}}(E, G)$, there exist an epimorphic family $\{e_i : E_i \to E \mid i \in I\}$ in $\mathcal{E}$ and for each $i \in I$ a finitely presentable model c_i of $\mathbb{L}_u$ and Σ-structure homomorphisms $f_i : c \to c_i$ and $u_i : c_i \to \mathrm{Hom}_{\mathcal{E}}(E_i, G)$ such that $\mathrm{Hom}_{\mathcal{E}}(e_i, G) \circ f = u_i \circ f_i$. Let us suppose that c is presented as a $\mathbb{H}$-model by a cartesian formula $\phi(x_1, \ldots, x_n)$ with generators $\xi_1, \ldots, \xi_n$. Since G is an ℓ-group with strong unit, there exist an epimorphic family $\{e_i : E_i \to E \mid i \in I\}$ in $\mathcal{E}$ and for each $k \in \{1, \ldots, n\}$ and $i \in I$ a natural number $m_{k,i}$ such that $f(|\xi_k|) \circ e_i \leq m_{k,i} u_{\mathrm{Hom}_{\mathcal{E}}(E_i, G)}$ (where $u_{\mathrm{Hom}_{\mathcal{E}}(E_i, G)}$ denotes the unit of the ℓ-group $\mathrm{Hom}_{\mathcal{E}}(E_i, G)$). For each $i \in I$, let c_i be the $\mathbb{H}$-model presented by the cartesian formula $\phi(x_1, \ldots, x_n) \wedge |x_1| \leq m_{1,i} \wedge \cdots \wedge |x_n| \leq m_{n,i}$. For each $i \in I$, we have a natural quotient homomorphism $f_i : c \to c_i$ through which $\mathrm{Hom}_{\mathcal{E}}(e_i, G) \circ f$ factors, and the resulting factorization u_i satisfies the required property that $\mathrm{Hom}_{\mathcal{E}}(e_i, G) \circ f = u_i \circ f_i$. Since $\mathbb{H}$ is a Horn theory, each c_i is generated by the n-tuple $(f_i(\xi_1), \ldots, f_i(\xi_n))$ which presents it as a $\mathbb{H}$-model (cf. Remark 6.1.18(a)). Therefore the c_i are ℓ-groups with strong unit by Lemma 9.9.1. This argument also shows that the finitely presentable $\mathbb{L}_u$-models are exactly the finitely presented $\mathbb{H}$-models whose unit is strong (cf. Theorem 8.2.6).

Therefore all the hypotheses of Corollary 8.2.1 are satisfied and we can conclude that the theory $\mathbb{L}_u$ is of presheaf type. In fact, $\mathbb{L}_u$ is shown in [30] to be Morita-equivalent to the (algebraic) theory of MV-algebras.

10

SOME APPLICATIONS

In this section we describe some applications of the theory developed in the book in a variety of different mathematical contexts. These applications are meant to give the reader a flavour of the kind of insight that classifying toposes can offer on 'concrete' questions arising in classical mathematics, and by no means exhaust the scope of applicability of the general theory. In fact, the main methodology that we shall use to generate such applications, namely the 'bridge technique' described in section 2.2, can, by virtue of its generality, be applied in a great number of different situations.

We shall in particular discuss restrictions of Morita equivalences to quotients of the two theories, give a solution to a problem of Lawvere concerning the boundary operator on subtoposes, establish syntax-semantics 'bridges' for quotients of theories of presheaf type, present topos-theoretic interpretations and generalizations of Fraïssé's theorem in model theory on countably categorical theories and of topological Galois theory, develop a notion of maximal spectrum of a commutative ring with unit and investigate generalized compactness conditions for geometric theories allowing us to identify theories lying in smaller fragments of geometric logic.

10.1 Restrictions of Morita equivalences

It follows as an immediate consequence of the duality of Theorem 3.2.5 that every Morita equivalence between two geometric theories yields, by restriction, Morita equivalences between their quotients. More specifically, we have the following result:

Theorem 10.1.1 *Let $\mathbb{T}_1$ and $\mathbb{T}_2$ be two Morita-equivalent theories. Then for any quotient $\mathbb{S}_1$ of $\mathbb{T}_1$ there exists a unique quotient $\mathbb{S}_2$ of $\mathbb{T}_2$ (up to syntactic equivalence) such that the Morita equivalence between $\mathbb{T}_1$ and $\mathbb{T}_2$ restricts to a Morita equivalence between $\mathbb{S}_1$ and $\mathbb{S}_2$:*

$$\begin{array}{ccc} \mathbb{T}_1\text{-mod}(\mathcal{E}) & \simeq & \mathbb{T}_2\text{-mod}(\mathcal{E}) \\ \uparrow & & \uparrow \\ \mathbb{S}_1\text{-mod}(\mathcal{E}) & \simeq & \mathbb{S}_2\text{-mod}(\mathcal{E}). \end{array}$$

Theories, Sites, Toposes. Olivia Caramello.
© Olivia Caramello 2018. Published 2018 by Oxford University Press.

Proof The theorem arises from the following 'bridge', whose arches are provided by Theorem 3.2.5:

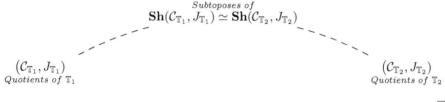

Remark 10.1.2 This theorem would be trivial if the notion of Morita equivalence 'collapsed' to bi-interpretability (as defined in section 2.1.3) but, as we have remarked in section 2.2.2, this is not the case. In fact, there are many pairs of theories which are Morita-equivalent without being bi-interpretable; in such situations, the theories cannot be related to each other directly, that is by using a syntactic 'dictionary' allowing us to translate formulae written in the language of one theory into formulae written in the language of the other, but Theorem 10.1.1, as well as other results obtained by applying the 'bridge' technique in connection with other invariants of classifying toposes, apply, allowing a non-trivial transfer of properties and results across them. For concrete illustrations of this remark, we refer the reader to [30] and [31], which provide natural examples of theories which are Morita-equivalent without being bi-interpretable.

10.2 A solution to the boundary problem for subtoposes

The third of the open problems in topos theory proposed by W. Lawvere in [61] concerns an operation on the lattice of subtoposes of a given topos, introduced in [60] under the name of *boundary* of a subtopos. The problem asks to describe the effect of the boundary operator on the classes of models classified by these subtoposes, both syntactically and semantically.

Recall from [60] that the boundary of an element a in a co-Heyting algebra C is given by the meet between a and the smallest element c of C such that $a \vee c = 1$. The canonical example of a co-Heyting algebra is the lattice of closed subsets of a topological space; in this case, the notion of boundary specializes to the ordinary topological one (that is, the boundary of a closed subset C is the set of elements of C which do not belong to its interior).

Thanks to the explicit description of the Heyting algebra operations on quotients of a given geometric theory obtained in sections 4.1.2 and 4.1.3, and to the description of the Heyting operations on Grothendieck topologies established in section 4.1.1, we can give an answer to both of the above-mentioned questions.

Indeed, the operation of boundary on subtoposes of the classifying topos of a geometric theory $\mathbb{T}$ over a signature Σ corresponds, via the duality of Theorem 3.2.5, to the operation $\mathbb{T}' \to \mathbb{T}' \vee \neg \mathbb{T}'$ in the lattice $\mathfrak{Th}_\Sigma^\mathbb{T}$. The explicit description of the lattice and Heyting operations on $\mathfrak{Th}_\Sigma^\mathbb{T}$ obtained in sections 4.1.2 and 4.1.3

thus provides an explicit axiomatization of the quotient $\partial \mathbb{T}'$ of $\mathbb{T}$ corresponding to the boundary of the subtopos $\mathbf{Set}[\mathbb{T}']$ of $\mathbf{Set}[\mathbb{T}]$ associated with a quotient $\mathbb{T}'$ of $\mathbb{T}$ via Theorem 3.2.5, as follows: $\partial \mathbb{T}'$ is obtained from $\mathbb{T}'$ by adding all the geometric sequents $(\psi \vdash_{\vec{y}} \psi')$ over Σ with the property that $(\psi' \vdash_{\vec{y}} \psi)$ is provable in $\mathbb{T}$ and for any $\mathbb{T}$-provably functional geometric formula $\theta(\vec{x}, \vec{y})$ from a geometric formula-in-context $\{\vec{x} \,.\, \phi\}$ to $\{\vec{y} \,.\, \psi\}$ and any geometric formula χ in the context $\vec{x}$ such that $(\chi \vdash_{\vec{x}} \phi)$ is provable in $\mathbb{T}$, the conjunction of the facts

(i) $(\phi \vdash_{\vec{x}} \chi)$ is provable in $\mathbb{T}'$
(ii) $((\exists \vec{y})(\theta(\vec{x}, \vec{y}) \wedge \psi'(\vec{y})) \vdash_{\vec{x}} \chi)$ is provable in $\mathbb{T}$

implies that $(\phi \vdash_{\vec{x}} \chi)$ is provable in $\mathbb{T}$.

Concerning the problem of getting a semantic description of the models of the quotient $\mathbb{T}'$ of a theory of presheaf type $\mathbb{T}$ corresponding to the boundary of a given subtopos $\mathcal{F}$ of the classifying topos $\mathbf{Set}[\mathbb{T}] \simeq [\text{f.p.}\mathbb{T}\text{-mod}(\mathbf{Set}), \mathbf{Set}]$ of $\mathbb{T}$, we observe that if $\mathcal{F}$ corresponds to a Grothendieck topology J on the category f.p.$\mathbb{T}$-mod$(\mathbf{Set})^{\mathrm{op}}$ as in section 8.1.2 then the models of $\mathbb{T}'$ are precisely the $(J \vee \neg J)$-homogeneous $\mathbb{T}$-models (in the sense of Definition 8.1.1).

Of course, these are abstract solutions which work uniformly for any geometric theory (resp. for any theory of presheaf type) $\mathbb{T}$. In practice one is often interested in finding more 'economical' axiomatizations which exploit the specific features of the theories under consideration; in this respect, the duality between quotients and Grothendieck topologies established in Chapter 8 and the explicit formulae for the lattice operations on Grothendieck topologies obtained in section 4.1.1 are clearly relevant (cf. for instance section 8.1.5).

10.3 Syntax-semantics 'bridges'

In this section we establish several results on theories of presheaf type and their quotients by transferring certain invariant properties of (sub)toposes across two different representations of a given classifying topos, according to the technique 'toposes as bridges' described in section 2.2.

We have seen in section 6.1.2 that if $\mathbb{T}$ is a theory of presheaf type then, by choosing a canonical Morita equivalence for $\mathbb{T}$ (in the sense of section 6.1.1), we have an equivalence of classifying toposes

$$[\text{f.p.}\mathbb{T}\text{-mod}(\mathbf{Set}), \mathbf{Set}] \simeq \mathbf{Sh}(\mathcal{C}_{\mathbb{T}}, J_{\mathbb{T}})$$

sending the universal model $N_{\mathbb{T}}$ of $\mathbb{T}$ in the topos $[\text{f.p.}\mathbb{T}\text{-mod}(\mathbf{Set}), \mathbf{Set}]$ (as in Theorem 6.1.1) into the universal model $U_{\mathbb{T}}$ of $\mathbb{T}$ in $\mathbf{Sh}(\mathcal{C}_{\mathbb{T}}, J_{\mathbb{T}})$ (as in Remark 2.1.9). In particular, if $M \in \text{f.p.}\mathbb{T}\text{-mod}(\mathbf{Set})$ is a $\mathbb{T}$-model presented by a formula $\phi(\vec{x})$ over Σ then, denoting by

$$y : \text{f.p.}\mathbb{T}\text{-mod}(\mathbf{Set})^{\mathrm{op}} \to [\text{f.p.}\mathbb{T}\text{-mod}(\mathbf{Set}), \mathbf{Set}]$$

the Yoneda embedding, we have that $y(M)$ is isomorphic to $[[\vec{x} \,.\, \phi]]_{N_{\mathbb{T}}}$ and hence corresponds, under the equivalence

$$[\text{f.p.}\mathbb{T}\text{-mod}(\mathbf{Set}), \mathbf{Set}] \simeq \mathbf{Sh}(\mathcal{C}_\mathbb{T}, J_\mathbb{T}),$$

to the sheaf $y_\mathbb{T}(\{\vec{x}.\phi\}) = [[\vec{x}.\phi]]_{U_\mathbb{T}}$, where $y_\mathbb{T} : \mathcal{C}_\mathbb{T} \to \mathbf{Sh}(\mathcal{C}_\mathbb{T}, J_\mathbb{T})$ is the Yoneda embedding.

Now, if $\mathbb{T}'$ is a quotient of $\mathbb{T}$ then the subtopos of $\mathbf{Sh}(\mathcal{C}_\mathbb{T}, J_\mathbb{T})$ corresponding to it via Theorem 3.2.5 transfers via the equivalence

$$[\text{f.p.}\mathbb{T}\text{-mod}(\mathbf{Set}), \mathbf{Set}] \simeq \mathbf{Sh}(\mathcal{C}_\mathbb{T}, J_\mathbb{T})$$

to a subtopos

$$\mathbf{Sh}(\text{f.p.}\mathbb{T}\text{-mod}(\mathbf{Set})^{\text{op}}, J) \hookrightarrow [\text{f.p.}\mathbb{T}\text{-mod}(\mathbf{Set}), \mathbf{Set}]$$

of $[\text{f.p.}\mathbb{T}\text{-mod}(\mathbf{Set}), \mathbf{Set}]$, where J is the induced $\mathbb{T}$-topology of $\mathbb{T}'$ (as defined in section 8.1.2). This gives rise to an equivalence

$$\mathbf{Sh}(\text{f.p.}\mathbb{T}\text{-mod}(\mathbf{Set})^{\text{op}}, J) \simeq \mathbf{Sh}(\mathcal{C}_{\mathbb{T}'}, J_{\mathbb{T}'})$$

of classifying toposes of $\mathbb{T}'$ which sends $l_J^{\text{f.p.}\mathbb{T}\text{-mod}(\mathbf{Set})^{\text{op}}}(M)$ to $y_{\mathbb{T}'}(\{\vec{x}.\phi\}) = [[\vec{x}.\phi]]_{U_{\mathbb{T}'}}$, where $l_J^{\text{f.p.}\mathbb{T}\text{-mod}(\mathbf{Set})^{\text{op}}}$ is the composite of y with the associated sheaf functor

$$[\text{f.p.}\mathbb{T}\text{-mod}(\mathbf{Set}), \mathbf{Set}] \to \mathbf{Sh}(\text{f.p.}\mathbb{T}\text{-mod}(\mathbf{Set})^{\text{op}}, J)$$

and $y_{\mathbb{T}'} : \mathcal{C}_{\mathbb{T}'} \to \mathbf{Sh}(\mathcal{C}_{\mathbb{T}'}, J_{\mathbb{T}'})$ is the Yoneda embedding.

We shall establish links between 'geometrical' properties of J and syntactic properties of $\mathbb{T}'$ by transferring suitable invariant properties of toposes or of objects of toposes across the equivalence

$$\mathbf{Sh}(\text{f.p.}\mathbb{T}\text{-mod}(\mathbf{Set})^{\text{op}}, J) \simeq \mathbf{Sh}(\mathcal{C}_{\mathbb{T}'}, J_{\mathbb{T}'}).$$

In order to perform these 'translations' of properties of one site into properties of the other across this equivalence, we shall need characterizations relating properties of (objects of) sites and properties of (objects of) the corresponding toposes, which will play the role of the 'arches' of our bridges (in the sense of section 2.2.3).

As 'global' invariant properties of toposes we shall use the notions of atomic topos, locally connected topos and the invariant property of a topos to be equivalent to a presheaf topos, while as 'local' invariant properties of objects of toposes we shall use the notion of atom, of indecomposable object and of irreducible object.

Before proceeding further, let us recall the precise meaning of the above-mentioned terms. An *atom* of a topos is a non-zero object which does not have any proper subobjects. An *indecomposable object* of a Grothendieck topos is an object which does not admit any non-trivial coproduct decomposition. An *irreducible object* of a Grothendieck topos is an object such that every epimorphic family to it contains a split epimorphism. A Grothendieck topos is said to be *atomic* if all its subobject lattices are complete atomic Boolean algebras, equivalently if it has a separating set of atoms. A Grothendieck topos is *locally connected* (resp. equivalent to a presheaf topos) if and only if it has a separating set of indecomposable (resp. irreducible) objects. For a proof of these characterizations we refer the reader to [20].

These invariants admit natural well-known site characterizations of the kind 'if a site $(\mathcal{C}, J)$ satisfies a certain property then the topos $\mathbf{Sh}(\mathcal{C}, J)$ satisfies the given invariant' (cf. section C3 of [55]), which can be combined with criteria going in the other direction of the form 'if the classifying topos of a geometric theory $\mathbb{T}$ satisfies the given invariant then the theory $\mathbb{T}$ satisfies a certain property' to form 'bridges' which allow us to deduce properties of a quotient $\mathbb{T}'$ from properties of the associated Grothendieck topology J:

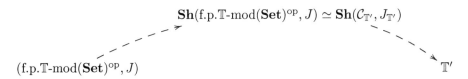

$$\mathbf{Sh}(\text{f.p.}\mathbb{T}\text{-mod}(\mathbf{Set})^{\mathrm{op}}, J) \simeq \mathbf{Sh}(\mathcal{C}_{\mathbb{T}'}, J_{\mathbb{T}'})$$

$(\text{f.p.}\mathbb{T}\text{-mod}(\mathbf{Set})^{\mathrm{op}}, J) \qquad\qquad\qquad\qquad\qquad \mathbb{T}'$

In order to state the following result, obtained by applying this methodology, it is convenient to introduce some terminology.

Definition 10.3.1 Let $\mathbb{T}$ be a geometric theory over a signature Σ and $\phi(\vec{x})$ a geometric formula-in-context over Σ.

(a) We say that $\phi(\vec{x})$ is $\mathbb{T}$-*complete* if the sequent $(\phi \vdash_{\vec{x}} \bot)$ is not provable in $\mathbb{T}$, and for every geometric formula χ in the same context either $(\chi \wedge \phi \vdash_{\vec{x}} \bot)$ or $(\phi \vdash_{\vec{x}} \chi)$ is provable in $\mathbb{T}$ (cf. p. 928 of [55]).

(b) We say that $\phi(\vec{x})$ is $\mathbb{T}$-*indecomposable* if for any family $\{\psi_i(\vec{x}) \mid i \in I\}$ of geometric formulae in the same context such that, for each i, ψ_i $\mathbb{T}$-provably implies ϕ and, for any distinct $i, j \in I$, $(\psi_i \wedge \psi_j \vdash_{\vec{x}} \bot)$ is provable in $\mathbb{T}$, we have that if $(\phi \vdash_{\vec{x}} \bigvee_{i \in I} \psi_i)$ is provable in $\mathbb{T}$ then $(\phi \vdash_{\vec{x}} \psi_i)$ is provable in $\mathbb{T}$ for some $i \in I$ (cf. Definition 3.1(b) [20]).

(c) We say that $\phi(\vec{x})$ is $\mathbb{T}$-*irreducible* if for any family $\{\theta_i \mid i \in I\}$ of $\mathbb{T}$-provably functional geometric formulae $\{\vec{x}_i, \vec{x} . \theta_i\}$ from $\{\vec{x}_i . \phi_i\}$ to $\{\vec{x} . \phi\}$ such that $(\phi \vdash_{\vec{x}} \bigvee_{i \in I} (\exists \vec{x}_i)\theta_i)$ is provable in $\mathbb{T}$, there exist $i \in I$ and a $\mathbb{T}$-provably functional geometric formula $\{\vec{x}, \vec{x}_i . \theta'\}$ from $\{\vec{x} . \phi\}$ to $\{\vec{x}_i . \phi_i\}$ such that the bisequent $(\phi(\vec{x}) \wedge \vec{x} = \vec{x}' \dashv\vdash_{\vec{x}, \vec{x}'} (\exists \vec{x}_i)(\theta'(\vec{x}, \vec{x}_i) \wedge \theta_i(\vec{x}_i, \vec{x}')))$ is provable in $\mathbb{T}$ (cf. Definition 6.1.9).

As shown by the following lemma, the above-mentioned notions arise precisely by expressing in terms of the syntactic site $(\mathcal{C}_\mathbb{T}, J_\mathbb{T})$ the property of the image of a geometric formula $\{\vec{x} . \phi\}$ under the Yoneda embedding $y_\mathbb{T} : \mathcal{C}_\mathbb{T} \hookrightarrow \mathbf{Sh}(\mathcal{C}_\mathbb{T}, J_\mathbb{T})$ to be an atom (resp. an indecomposable object, an irreducible object) of the topos $\mathbf{Sh}(\mathcal{C}_\mathbb{T}, J_\mathbb{T})$.

Lemma 10.3.2 (cf. Lemma 3.2 [20]). *Let $\mathbb{T}$ be a geometric theory over a signature Σ and $\phi(\vec{x})$ a geometric formula-in-context over Σ. Then*
  (i) *$\phi(\vec{x})$ is $\mathbb{T}$-complete if and only if $y_{\mathbb{T}}(\{\vec{x}.\phi\})$ is an atom of $\mathbf{Sh}(\mathcal{C}_{\mathbb{T}}, J_{\mathbb{T}})$.*
  (ii) *$\phi(\vec{x})$ is $\mathbb{T}$-indecomposable if and only if $y_{\mathbb{T}}(\{\vec{x}.\phi\})$ is an indecomposable object of $\mathbf{Sh}(\mathcal{C}_{\mathbb{T}}, J_{\mathbb{T}})$.*
  (iii) *$\phi(\vec{x})$ is $\mathbb{T}$-irreducible if and only if $y_{\mathbb{T}}(\{\vec{x}.\phi\})$ is an irreducible object of $\mathbf{Sh}(\mathcal{C}_{\mathbb{T}}, J_{\mathbb{T}})$.*

The following site characterizations will provide the other 'arches' of our 'bridges':

Proposition 10.3.3 (cf. Proposition 2.5 [20]). *Let $(\mathcal{C}, J)$ be a small site and $l_J^{\mathcal{C}} : \mathcal{C} \to \mathbf{Sh}(\mathcal{C}, J)$ the composite of the Yoneda embedding $\mathcal{C} \hookrightarrow [\mathcal{C}^{\mathrm{op}}, \mathbf{Set}]$ with the associated sheaf functor.*
  (i) *If J is trivial then, for any $c \in \mathcal{C}$, $l_J^{\mathcal{C}}(c)$ is an irreducible object of $\mathbf{Sh}(\mathcal{C}, J)$.*
  (ii) *If $(\mathcal{C}, J)$ is locally connected (in the sense that every J-covering sieve S on an object c of $\mathcal{C}$ is connected as a full subcategory of $\mathcal{C}/c$, cf. p. 657 of [55]) then, for each $c \in \mathcal{C}$, $l_J^{\mathcal{C}}(c)$ is an indecomposable object of $\mathbf{Sh}(\mathcal{C}, J)$; in particular, the topos $\mathbf{Sh}(\mathcal{C}, J)$ is locally connected.*
  (iii) *If $(\mathcal{C}, J)$ is atomic (in the sense that $\mathcal{C}$ satisfies the right Ore condition and J is the atomic topology on it) then, for each $c \in \mathcal{C}$, $l_J^{\mathcal{C}}(c)$ is an atom of $\mathbf{Sh}(\mathcal{C}, J)$; in particular, the topos $\mathbf{Sh}(\mathcal{C}, J)$ is atomic.*

We are now ready to state the main theorem of this section.

Theorem 10.3.4 (cf. [20]). *Let $\mathbb{T}$ be a theory of presheaf type over a signature Σ and $\mathbb{T}'$ a quotient of $\mathbb{T}$ with induced $\mathbb{T}$-topology J on f.p.$\mathbb{T}$-mod$(\mathbf{Set})^{\mathrm{op}}$ (as in section 8.1.2).*
  (i) *If J is trivial then $\mathbb{T}'(=\mathbb{T})$ satisfies the property that for any geometric formula $\{\vec{y}.\psi\}$ over Σ there exist $\mathbb{T}$-irreducible formulae-in-context $\{\vec{x}_i.\phi_i\}$ (for $i \in I$) and $\mathbb{T}$-provably functional geometric formulae $\{\vec{x}_i, \vec{y}.\theta_i\}$ from $\{\vec{x}_i.\phi_i\}$ to $\{\vec{y}.\psi\}$ such that $(\psi \vdash_{\vec{y}} \bigvee_{i \in I}(\exists \vec{x}_i)\theta_i)$ is provable in $\mathbb{T}'$ (cf. Corollary 6.1.10).*
  Moreover, any formula which presents a $\mathbb{T}$-model is $\mathbb{T}$-irreducible (cf. Theorem 6.1.13).
  (ii) *If the site (f.p.$\mathbb{T}$-mod$(\mathbf{Set})^{\mathrm{op}}, J)$ is locally connected then $\mathbb{T}'$ satisfies the property that every geometric formula-in-context over Σ is $\mathbb{T}'$-provably equivalent to a disjunction of $\mathbb{T}'$-indecomposable geometric formulae in the same context which are $\mathbb{T}'$-provably disjoint from each other.*
  Moreover, every formula which presents a $\mathbb{T}$-model is $\mathbb{T}'$-indecomposable.
  (iii) *If the site (f.p.$\mathbb{T}$-mod$(\mathbf{Set})^{\mathrm{op}}, J)$ is atomic then $\mathbb{T}'$ satisfies the property that every geometric formula-in-context over Σ is $\mathbb{T}'$-provably equivalent to a disjunction of $\mathbb{T}'$-complete geometric formulae in the same context.*
  Moreover, every formula which presents a $\mathbb{T}$-model is $\mathbb{T}'$-complete.

Proof

(i) The first assertion about $\mathbb{T}$ follows from Corollary 6.1.10. Given a geometric formula $\phi(\vec{x})$ over Σ presenting a $\mathbb{T}$-model M, by Lemma 10.3.2(iii) $\phi(\vec{x})$ is $\mathbb{T}$-irreducible if and only if $y_{\mathbb{T}}(\{\vec{x}\,.\,\phi\})$ is an irreducible object of $\mathbf{Sh}(\mathcal{C}_{\mathbb{T}}, J_{\mathbb{T}})$. But this is equivalent to saying that $y(M)$ is irreducible in $[\text{f.p.}\mathbb{T}\text{-mod}(\mathbf{Set}), \mathbf{Set}]$ (cf. the discussion preceding Definition 10.3.1), and this is true since all representables in a presheaf topos are irreducible objects of it (cf. Proposition 10.3.3).

(ii) and (iii) Given a geometric formula $\phi(\vec{x})$ over Σ presenting a $\mathbb{T}$-model M, by Lemma 10.3.2(ii) (resp. Lemma 10.3.2(i)) $\phi(\vec{x})$ is $\mathbb{T}'$-indecomposable (resp. $\mathbb{T}'$-complete) if and only if $y_{\mathbb{T}'}(\{\vec{x}\,.\,\phi\})$ is an indecomposable object (resp. an atom) of $\mathbf{Sh}(\mathcal{C}_{\mathbb{T}'}, J_{\mathbb{T}'})$; but this means that $l_J^{\text{f.p.}\mathbb{T}\text{-mod}(\mathbf{Set})^{\text{op}}}(M)$ is an indecomposable object (resp. an atom) of $\mathbf{Sh}(\text{f.p.}\mathbb{T}\text{-mod}(\mathbf{Set})^{\text{op}}, J)$, and this is true by Proposition 10.3.3. The above-mentioned characterizations of the invariant property of a Grothendieck topos to be locally connected (resp. atomic) thus yield our thesis.

□

Remark 10.3.5 As shown in [21], one can easily introduce new 'geometric' invariant properties of toposes other than the ones considered above, defined for instance by requiring the existence of separating sets of objects satisfying particular properties, and establish site characterizations for them yielding 'bridges' such as the ones in the proof of Theorem 10.3.4. In other words, Theorem 10.3.4 provides just a few examples of insights relating the syntactic properties of a quotient $\mathbb{T}'$ of a theory of presheaf type $\mathbb{T}$ and the 'geometric' properties of the associated Grothendieck topology J which are obtainable by applying the 'bridge' technique.

10.4 Topos-theoretic Fraïssé theorem

In this section, which is based on [23], we present a topos-theoretic interpretation and wide generalization of the well-known result (Theorem 7.4.1(a) [49]) providing the link between Fraïssé's construction and countably categorical theories. The key concepts involved in Fraïssé's construction (i.e. amalgamation and joint embedding properties, homogeneous structures and atomicity and completeness of the theory axiomatizing them) will be seen to naturally arise from the expression of topos-theoretic invariants on a classifying topos in terms of different sites of definition for it, and the connections between them to result from 'bridges' (in the sense of section 2.2) involving these different representations.

The context in which we shall formulate our topos-theoretic interpretation of Fraïssé's theorem is that of theories of presheaf type, which we comprehensively studied in Chapter 6. Recall that this class of theories is very extensive and includes all the cartesian theories as well as many other interesting theories arising in different fields of mathematics (cf. also Chapter 9).

The following notions, which will play a central role in our analysis, are natural categorical generalizations of the concepts involved in the classical Fraïssé construction.

Definition 10.4.1 A category $\mathcal{C}$ is said to satisfy the *amalgamation property* (AP) if for any objects $a, b, c \in \mathcal{C}$ and arrows $f : a \to b$, $g : a \to c$ in $\mathcal{C}$ there exist an object $d \in \mathcal{C}$ and arrows $f' : b \to d$, $g' : c \to d$ in $\mathcal{C}$ such that $f' \circ f = g' \circ g$:

$$\begin{array}{ccc} a & \xrightarrow{f} & b \\ g \downarrow & & \downarrow f' \\ c & \xrightarrow{g'} & d. \end{array}$$

The amalgamation property on a category is also called the left Ore condition, and its dual the right Ore condition. Notice that if $\mathcal{C}$ satisfies AP then we can equip $\mathcal{C}^{\mathrm{op}}$ with the atomic topology (that is, the Grothendieck topology whose covering sieves are exactly the non-empty ones). This point will be a fundamental ingredient of our topos-theoretic interpretation of Fraïssé's theorem.

Definition 10.4.2 A category $\mathcal{C}$ is said to satisfy the *joint embedding property* (JEP) if for every pair of objects $a, b \in \mathcal{C}$ there exist an object $c \in \mathcal{C}$ and arrows $f : a \to c$, $g : b \to c$ in $\mathcal{C}$:

$$\begin{array}{ccc} & & a \\ & & \downarrow f \\ b & \xrightarrow{g} & c. \end{array}$$

Notice that if $\mathcal{C}$ has a weakly initial object then AP on $\mathcal{C}$ implies JEP on $\mathcal{C}$; however, in general the two notions are distinct from each other.

Definition 10.4.3 Let $\mathcal{C} \hookrightarrow \mathcal{D}$ be an embedding of categories.

(a) An object $u \in \mathcal{D}$ is said to be $\mathcal{C}$-*homogeneous* if for any objects $a, b \in \mathcal{C}$ and arrows $j : a \to b$ in $\mathcal{C}$ and $\chi : a \to u$ in $\mathcal{D}$ there exists an arrow $\tilde{\chi} : b \to u$ in $\mathcal{D}$ such that $\tilde{\chi} \circ j = \chi$:

$$\begin{array}{ccc} a & \xrightarrow{\chi} & u \\ j \downarrow & \nearrow \tilde{\chi} & \\ b. & & \end{array}$$

(b) An object $u \in \mathcal{D}$ is said to be $\mathcal{C}$-*ultrahomogeneous* if for any objects $a, b \in \mathcal{C}$ and arrows $j : a \to b$ in $\mathcal{C}$ and $\chi_1 : a \to u$, $\chi_2 : b \to u$ in $\mathcal{D}$ there exists an isomorphism $\check{j} : u \to u$ in $\mathcal{D}$ such that $\check{j} \circ \chi_1 = \chi_2 \circ j$:

$$\begin{array}{ccc} a & \xrightarrow{\chi_1} & u \\ j \downarrow & & \downarrow \check{j} \\ b & \xrightarrow{\chi_2} & u. \end{array}$$

(c) An object $u \in \mathcal{D}$ is said to be $\mathcal{C}$-*universal* if for every $a \in \mathcal{C}$ there exists an arrow $\chi : a \to u$ in $\mathcal{D}$:
$$a \xrightarrow{\chi} u.$$

Remarks 10.4.4 (a) Any $\mathcal{C}$-ultrahomogeneous and $\mathcal{C}$-universal object is $\mathcal{C}$-homogeneous.

(b) In verifying that an object u in $\mathcal{D}$ is $\mathcal{C}$-ultrahomogeneous one can clearly suppose, without loss of generality, that the arrow j in Definition 10.4.3(b) is an identity.

Let $\mathbb{T}$ be a theory of presheaf type over a signature Σ. If the category f.p.$\mathbb{T}$-mod(**Set**) of its finitely presentable models satisfies the amalgamation property then we can put on f.p.$\mathbb{T}$-mod(**Set**)^{op} the atomic topology J_{at}, obtaining a subtopos

$$\mathbf{Sh}(\text{f.p.}\mathbb{T}\text{-mod}(\mathbf{Set})^{\mathrm{op}}, J) \hookrightarrow [\text{f.p.}\mathbb{T}\text{-mod}(\mathbf{Set}), \mathbf{Set}]$$

of the classifying topos of $\mathbb{T}$ which, once a canonical Morita equivalence for $\mathbb{T}$ (in the sense of section 6.1.1) is chosen, corresponds by the duality of Theorem 3.2.5 to a unique quotient $\mathbb{T}'$ of $\mathbb{T}$ classified by it. This quotient can be characterized, by Corollary 8.1.14, as the theory over Σ obtained from $\mathbb{T}$ by adding all the sequents of the form $(\psi \vdash_{\vec{y}} (\exists \vec{x}) \theta(\vec{x}, \vec{y}))$, where $\phi(\vec{x})$ and $\psi(\vec{y})$ are formulae which present a $\mathbb{T}$-model and $\theta(\vec{x}, \vec{y})$ is a $\mathbb{T}$-provably functional formula over Σ from $\{\vec{x}.\phi\}$ to $\{\vec{y}.\psi\}$.

Since $\mathbf{Sh}(\text{f.p.}\mathbb{T}\text{-mod}(\mathbf{Set})^{\mathrm{op}}, J_{\mathrm{at}})$ classifies the theory $\mathbb{T}'$, by the syntactic method for constructing classifying toposes (cf. Theorem 2.1.8), our canonical Morita equivalence for $\mathbb{T}$ induces an equivalence

$$\mathbf{Sh}(\text{f.p.}\mathbb{T}\text{-mod}(\mathbf{Set})^{\mathrm{op}}, J_{\mathrm{at}}) \simeq \mathbf{Sh}(\mathcal{C}_{\mathbb{T}'}, J_{\mathbb{T}'}),$$

which can be used for building 'bridges' between the two sites by considering appropriate topos-theoretic invariants on the classifying topos of $\mathbb{T}'$:

$$\mathbf{Sh}(\text{f.p.}\mathbb{T}\text{-mod}(\mathbf{Set})^{\mathrm{op}}, J_{\mathrm{at}}) \simeq \mathbf{Sh}(\mathcal{C}_{\mathbb{T}'}, J_{\mathbb{T}'})$$

(f.p.$\mathbb{T}$-mod(**Set**)^{op}, J_{at}) $\qquad\qquad\qquad\qquad\qquad$ $(\mathcal{C}_{\mathbb{T}'}, J_{\mathbb{T}'})$

First, we observe that, by Theorem 8.1.3 and Remark 8.1.2, the theory $\mathbb{T}'$ axiomatizes the homogeneous $\mathbb{T}$-models (in the sense of Remark 8.1.2) in every Grothendieck topos $\mathcal{E}$:

Geometric morphism from $\mathcal{E}$ to
$$\mathbf{Sh}(\text{f.p.}\mathbb{T}\text{-mod}(\mathbf{Set})^{\mathrm{op}}, J_{\mathrm{at}}) \simeq \mathbf{Sh}(\mathcal{C}_{\mathbb{T}'}, J_{\mathbb{T}'})$$

(f.p.$\mathbb{T}$-mod(**Set**)^{op}, J_{at}) $\qquad\qquad\qquad\qquad\qquad$ $(\mathcal{C}_{\mathbb{T}'}, J_{\mathbb{T}'})$
homogeneous $\mathbb{T}$-model in $\mathcal{E}$ $\qquad\qquad\qquad\qquad\qquad$ $\mathbb{T}'$-*model in $\mathcal{E}$*

In particular, taking $\mathcal{E} = \mathbf{Set}$, we obtain that the set-based $\mathbb{T}'$-models are the set-based models M of $\mathbb{T}$ such that for any arrow $y : c \to M$ in $\mathbb{T}$-mod(**Set**) and any arrow $f : c \to d$ in f.p.$\mathbb{T}$-mod(**Set**) there exists an arrow u in $\mathbb{T}$-mod(**Set**) such that $u \circ f = y$:

For this reason, we shall call $\mathbb{T}'$ the 'theory of homogeneous $\mathbb{T}$-models'. Notice that a set-based $\mathbb{T}$-model M is homogeneous if and only if it is f.p.$\mathbb{T}$-mod(**Set**)-homogeneous as an object of the category $\mathbb{T}$-mod(**Set**) in the sense of Definition 10.4.3.

Next, let us consider the invariant property of the classifying topos of $\mathbb{T}'$ to be atomic. The fact that (f.p.$\mathbb{T}$-mod(**Set**)^{op}, J_{at}) is an atomic site implies, by a well-known site characterization (cf. also Proposition 10.3.3), that the topos $\mathbf{Sh}$(f.p.$\mathbb{T}$-mod(**Set**)^{op}, J_{at}) is atomic. By looking at this invariant from the point of view of the other site $(\mathcal{C}_{\mathbb{T}'}, J_{\mathbb{T}'})$ we obtain, as an application of the classical characterization of the geometric theories whose classifying topos is atomic (cf. Proposition D3.4.13 [50]), that the theory $\mathbb{T}'$ is atomic in a logical sense (cf. Theorem 10.3.4(iii)), i.e. that every geometric formula-in-context over its signature is $\mathbb{T}'$-provably equivalent to a disjunction of $\mathbb{T}'$-complete formulae in the same context (cf. Definition 10.3.1(a)):

Moreover, recall from Theorem 10.3.4(iii) that every geometric formula over Σ which presents a $\mathbb{T}$-model is $\mathbb{T}'$-complete. Indeed, denoting by

$$y_{\mathbb{T}'} : \mathcal{C}_{\mathbb{T}'} \to \mathbf{Sh}(\mathcal{C}_{\mathbb{T}'}, J_{\mathbb{T}'})$$

the Yoneda embedding and by

$$l_{J_{\mathrm{at}}} : \text{f.p.}\mathbb{T}\text{-mod}(\mathbf{Set})^{\mathrm{op}} \to \mathbf{Sh}((\text{f.p.}\mathbb{T}\text{-mod}(\mathbf{Set}))^{\mathrm{op}}, J_{\mathrm{at}})$$

the composite of the Yoneda embedding

$$\text{f.p.}\mathbb{T}\text{-mod}(\mathbf{Set})^{\mathrm{op}} \to [\text{f.p.}\mathbb{T}\text{-mod}(\mathbf{Set}), \mathbf{Set}]$$

with the associated sheaf functor

$$[\text{f.p.}\mathbb{T}\text{-mod}(\mathbf{Set}), \mathbf{Set}] \to \mathbf{Sh}(\text{f.p.}\mathbb{T}\text{-mod}(\mathbf{Set})^{\mathrm{op}}, J_{\mathrm{at}}),$$

we have that for any finitely presentable $\mathbb{T}$-model M the objects $l_{J_{\mathrm{at}}}(M)$ and $y_{\mathbb{T}'}(\{\vec{x}.\phi\})$, where $\phi(\vec{x})$ is 'the' formula presenting M, correspond to each other under the Morita equivalence for $\mathbb{T}'$ considered above; now, the fact that J_{at} is the atomic topology implies that $l_{J_{at}}(M)$ is an atom (cf. Proposition 10.3.3), while the property of $y_{\mathbb{T}'}(\{\vec{x}.\phi\})$ to be an atom is precisely equivalent to the condition of $\mathbb{T}'$-completeness of the formula $\phi(\vec{x})$ (cf. Lemma 10.3.2(i)). In general, it is not true that all the atoms of the classifying topos of $\mathbb{T}'$ have this form (for example, for any discrete group G, regarded as a one-object category, the topos $[G^{op}, \mathbf{Set}]$ is atomic and its atoms can be identified with the transitive G-sets, while there is only one representable, namely the functor sending the unique object of G to the underlying set of G), but the fact that in an atomic topos every atom can be covered by a representable ensures that if an object of the form $y_{\mathbb{T}'}(\{\vec{y}.\psi\})$ is an atom, then there exists a $\mathbb{T}'$-provably functional formula $\theta(\vec{x}, \vec{y})$ from a formula $\{\vec{x}.\phi\}$ presenting a $\mathbb{T}$-model to $\{\vec{y}.\psi\}$ with the property that the sequent $(\psi \vdash_{\vec{y}} (\exists \vec{x})\theta(\vec{x}, \vec{y}))$ is provable in $\mathbb{T}'$. We shall characterize the subcanonical atomic sites such that every atom of the corresponding topos comes from the site (up to isomorphism) in section 10.5 (cf. Theorem 10.5.6).

Let us now consider the invariant property of a topos to be *two-valued* (in the sense that its only two subterminal objects are the zero and the identity one, and they are distinct from each other) in connection with the above Morita equivalence for $\mathbb{T}'$. By using the characterization of the subterminal objects of a topos $\mathbf{Sh}(\mathcal{C}, J)$ in terms of J-ideals on $\mathcal{C}$ provided by Remark 1.1.25, it is easy to see that for any atomic site $(\mathcal{C}, J_{at})$ such that $\mathcal{C}$ is non-empty, the topos $(\mathcal{C}, J_{at})$ is two-valued if and only if $\mathcal{C}^{op}$ satisfies the joint embedding property (cf. Theorem 3.6 and Lemma 3.7 [23]). On the other hand, applying this characterization in the case of the classifying topos of a geometric theory $\mathbb{S}$, represented as the topos of sheaves on its geometric syntactic site, yields the following criterion: the topos $\mathbf{Sh}(\mathcal{C}_{\mathbb{S}}, J_{\mathbb{S}})$ is two valued if and only if the theory $\mathbb{S}$ is *complete*, in the sense that every geometric sentence over the signature of $\mathbb{S}$ is either provably false or provably true, but not both – cf. Definition 4.2.21 (notice that the subterminal objects of a topos $\mathbf{Sh}(\mathcal{C}_{\mathbb{S}}, J_{\mathbb{S}})$ correspond precisely to the $\mathbb{S}$-provable equivalence classes of geometric sentences over the signature of $\mathbb{S}$). Notice that if a theory is complete then all its set-based models satisfy the same geometric sentences; the converse holds if the theory has enough set-based models. This notion of completeness for geometric theories notably agrees with the classical model-theoretic one for (finitary) first-order theories when applied to atomic theories (cf. Remark 10.4.7 below).

By applying these criteria in the context of the Morita equivalence

$$\mathbf{Sh}(\text{f.p.}\mathbb{T}\text{-mod}(\mathbf{Set})^{op}, J_{at}) \simeq \mathbf{Sh}(\mathcal{C}_{\mathbb{T}'}, J_{\mathbb{T}'}),$$

we obtain the following logical equivalence: the category f.p.$\mathbb{T}$-mod($\mathbf{Set}$), if it is non-empty, satisfies JEP if and only if the theory $\mathbb{T}'$ is complete:

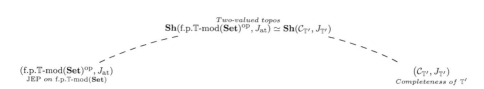

Finally, recall from [18] (cf. Theorem 3.15 therein) that, assuming the axiom of countable choice, every geometric theory which is atomic and complete is countably categorical, that is any two of its set-based countable models are isomorphic.

Summarizing, we have the following

Theorem 10.4.5 (Theorem 3.8 [23]). *Let $\mathbb{T}$ be a theory of presheaf type such that the category f.p.$\mathbb{T}$-mod(**Set**) is non-empty and satisfies the amalgamation and joint embedding properties. Then (any theory Morita-equivalent to) the theory $\mathbb{T}'$ of homogeneous $\mathbb{T}$-models is complete and atomic; in particular, assuming the axiom of countable choice, any two countable homogeneous $\mathbb{T}$-models in **Set** are isomorphic.*

Moreover, every geometric formula which presents a $\mathbb{T}$-model is $\mathbb{T}'$-complete.
□

Remark 10.4.6 If the category f.p.$\mathbb{T}$-mod(**Set**) satisfies AP then each of its connected components satisfies AP as well as JEP. The toposes of sheaves on these subcategories with respect to the atomic topology are precisely the classifying toposes of the *completions* of the theory $\mathbb{T}'$, i.e. of the (complete) quotients of $\mathbb{T}'$ obtained by adding an axiom of the form $(\top \vdash_{[]} \phi)$ for a $\mathbb{T}'$-complete geometric sentence ϕ.

Theorem 10.4.5 is a vast generalization of the well-known result (Theorem 7.4.1(a) [49]) allowing one to build countably categorical theories through Fraïssé's method (the classical result can be recovered as the particular case of the theorem when $\mathbb{T}$ is the quotient of the empty theory over a finite signature corresponding as in Corollary 9.1.3 to a uniformly finite collection of finitely presented models of it satisfying the hereditary property). In fact, as argued in Chapter 6, the context of theories of presheaf type is a very general and natural one both from a mathematical and a model-theoretic perspective; moreover, as we have seen in the previous chapters, there are many effective means for generating theories of presheaf type or testing whether a certain theory is of presheaf type.

Note that Theorem 10.4.5 represents a faithful application of the 'bridge' technique (cf. section 2.2) in the context of the canonical Morita equivalence induced by a theory of presheaf type, the invariants used in the proof being the notion of subtopos, the notion of geometric morphism to a classifying topos, the properties of a topos to be atomic and two-valued and the notion of atom of a topos:

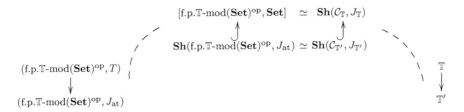

The notion of homogeneous $\mathbb{T}$-model is most relevant when all the $\mathbb{T}$-model homomorphisms in **Set** are injective; the notion of injectivization of a geometric theory, which we introduced in section 7.2.2, thus plays an important role in this context.

Examples of theories to which Theorem 10.4.5 can be applied include the following:

(a) The theory $\mathbb{I}$ of decidable objects (that is, the injectivization of the empty theory over a one-sorted signature, in the sense of Definition 7.2.10) is of presheaf type, and its category of finitely presentable models is the category $\mathbf{I}$ of finite sets and injections between them. The theory of homogeneous $\mathbb{I}$-models is classified by the topos $\mathbf{Sh}(\mathbf{I}^{\mathrm{op}}, J_{\mathrm{at}})$, also known as the *Schanuel topos*. Notice that the homogeneous $\mathbb{I}$-models in **Set** are precisely the infinite sets.

(b) The theory of decidable Boolean algebras (that is, the injectivization of the algebraic theory of Boolean algebras) is of presheaf type (cf. section 9.1). Its finitely presentable models are the finite Boolean algebras, while its homogeneous set-based models are the atomless Boolean algebras. The well-known result that any two countable atomless Boolean algebras are isomorphic thus follows from Theorem 10.4.5 assuming the axiom of countable choice.

(c) The theory of (decidably) linearly ordered objects (that is, the injectivization of the coherent theory of linear orders) is of presheaf type (cf. section 9.1), and its category of finitely presentable models coincides with the category of finite linear orders and order-preserving injections between them. Its homogeneous models are precisely the dense linearly ordered objects without endpoints (cf. section 8.1.1.1). Theorem 10.4.5 thus ensures that the theory of dense linearly ordered objects without endpoints is atomic and complete.

(d) As shown in [29] (cf. Propositions 2.3 and 2.4 therein), the (infinitary) geometric theory of fields of finite characteristic which are algebraic over their prime field is of presheaf type and the theory of its homogeneous models can be identified with the theory of fields of finite characteristic which are algebraic over their prime field and algebraically closed; the completions of this theory are obtained precisely by adding, in each case, the axiom fixing the characteristic of the field (i.e. the sequent $(\top \vdash_{[]} p.1 = 0)$ for a prime number p). Theorem 10.4.5 thus implies, assuming the countable axiom of choice, the (well-known) fact that any two algebraic closures of a given finite field are isomorphic and, without assuming any form of the axiom of choice, the fact that any two algebraic closures of a given finite field satisfy the same first-order sentences (cf. Remark 10.4.7) written in the language of fields; in fact, this property is true more generally for any base field (the theory of algebraic extensions of a base field is of presheaf type – cf. section 9.5 – and satisfies the hypotheses of Theorem 10.4.5).

Given a theory of presheaf type $\mathbb{T}$ whose category f.p.$\mathbb{T}$-mod(**Set**) of finitely presentable models satisfies AP then, by Remark 8.1.9, the theory $\mathbb{T}'$ of homogeneous $\mathbb{T}$-models can be identified with the *Booleanization* of the theory $\mathbb{T}$ defined in section 4.2.3, that is as the quotient of $\mathbb{T}$ obtained by adding the sequent

$$(\top \vdash_{\vec{y}} \psi)$$

for any $\mathbb{T}$-stably consistent formula-in-context $\psi(\vec{y})$.

Remark 10.4.7 As every atomic topos is Boolean (since every atomic frame is isomorphic to a powerset), every atomic theory is Boolean. Every Boolean geometric theory satisfies the following key property: every first-order formula-in-context over its signature is provably equivalent in the theory, using classical logic, to a geometric formula in the same context (cf. Remark 4.2.19). This property has two immediate but remarkable consequences:

- Every Boolean geometric theory is complete as a geometric theory (in the sense of Definition 4.2.21) if and only if it is complete in the sense of classical model theory (that is, every first-order sentence over its signature is provably false or provably true using classical logic, but not both). In general, such an equivalence between the two notions of completeness only holds at the cost of replacing the given first-order theory with a geometric theory over a larger signature, e.g. its Morleyization (cf. Remark 2.1.3).
- Every Boolean geometric theory (in particular, every theory of homogeneous models as in Theorem 10.4.5) is model-complete (in the sense of classical model theory), that is every embedding of set-based models of it is an elementary embedding.

There is a kind of converse to Theorem 10.4.5, relating the amalgamation property on the category f.p.$\mathbb{T}$-mod(**Set**) with a semantic property of the theory $\mathbb{T}'$. In fact, the theory $\mathbb{T}'$ of homogeneous $\mathbb{T}$-models can be defined even if f.p.$\mathbb{T}$-mod(**Set**)^{op} does not satisfy the right Ore condition, by replacing the atomic topology J_{at} with the Grothendieck topology on it generated by the family of all non-empty sieves.

Theorem 10.4.8 (Theorem 3.12 [23]). *Let $\mathbb{T}$ be a theory of presheaf type and $\mathbb{T}'$ the theory of homogeneous $\mathbb{T}$-models. Suppose that $\mathbb{T}'$ has enough set-based models and that all the arrows in $\mathbb{T}$-mod(**Set**) are monic. Then the following conditions are equivalent:*

(i) *The category f.p.$\mathbb{T}$-mod(**Set**) satisfies the amalgamation property.*

(ii) *For every finitely presentable $\mathbb{T}$-model c there exist a set-based $\mathbb{T}'$-model (that is, a homogeneous $\mathbb{T}$-model in* **Set***) M and a $\mathbb{T}$-model homomorphism $c \to M$.*

We can naturally expect the consideration of invariants other than the ones used in the proof of Theorem 10.4.5 in connection with the same Morita equivalence for $\mathbb{T}'$ to lead to other relationships between the geometry of the category f.p.$\mathbb{T}$-mod(**Set**) and the syntactic properties of the theory $\mathbb{T}'$. For example, the consideration of the invariant notion of arrow between two atoms leads to the following result:

Proposition 10.4.9 *Let $\mathbb{T}$ be a theory of presheaf type such that the category f.p.$\mathbb{T}$-mod(**Set**) satisfies the amalgamation property, and let $\mathbb{T}'$ be the theory of homogeneous $\mathbb{T}$-models. Then, if all the arrows in f.p.$\mathbb{T}$-mod(**Set**) are strict monomorphisms (in the sense of condition (i) of Theorem 10.5.5 below), every $\mathbb{T}'$-provably functional geometric formula from a formula $\{\vec{x}.\phi\}$ which presents a $\mathbb{T}$-model to another formula $\{\vec{y}.\psi\}$ which presents a $\mathbb{T}$-model is $\mathbb{T}$-provably functional from $\{\vec{x}.\phi\}$ to $\{\vec{y}.\psi\}$ (equivalently, corresponds to a $\mathbb{T}$-model homomorphisms $M_{\{\vec{y}.\psi\}} \to M_{\{\vec{x}.\phi\}}$).*

Proof Using the above notation, given two formulae $\phi(\vec{x})$ and $\psi(\vec{y})$ presenting respectively $\mathbb{T}$-models $M_{\{\vec{x}.\phi\}}$ and $M_{\{\vec{y}.\psi\}}$, the objects $l_{J_{\mathrm{at}}}(M_{\{\vec{x}.\phi\}})$ and $y_{\mathbb{T}'}(\{\vec{x}.\phi\})$ (resp. $l_{J_{\mathrm{at}}}(M_{\{\vec{y}.\psi\}})$ and $y_{\mathbb{T}'}(\{\vec{y}.\psi\}))$ correspond to each other under the Morita equivalence

$$\mathbf{Sh}(\text{f.p.}\mathbb{T}\text{-mod}(\mathbf{Set})^{\mathrm{op}}, J_{\mathrm{at}}) \simeq \mathbf{Sh}(\mathcal{C}_{\mathbb{T}'}, J_{\mathbb{T}'}).$$

Therefore the arrows $l_{J_{\mathrm{at}}}(M_{\{\vec{x}.\phi\}}) \to l_{J_{\mathrm{at}}}(M_{\{\vec{y}.\psi\}})$ in $\mathbf{Sh}(\text{f.p.}\mathbb{T}\text{-mod}(\mathbf{Set})^{\mathrm{op}}, J_{\mathrm{at}})$ correspond bijectively to the arrows $y_{\mathbb{T}'}(\{\vec{x}.\phi\}) \to y_{\mathbb{T}'}(\{\vec{y}.\psi\})$ in $\mathbf{Sh}(\mathcal{C}_{\mathbb{T}'}, J_{\mathbb{T}'})$. But these arrows in turn correspond, $y'_{\mathbb{T}}$ being full and faithful, to the arrows $\{\vec{x}.\phi\} \to \{\vec{y}.\psi\}$ in $\mathcal{C}_{\mathbb{T}'}$, that is to the $\mathbb{T}'$-provably functional formulae $\{\vec{x}.\phi\} \to \{\vec{y}.\psi\}$.

Now, one can easily prove that the topology J_{at} is subcanonical if and only if every arrow in f.p.$\mathbb{T}$-mod(**Set**) is a strict monomorphism (in the sense of satisfying condition (i) of Theorem 10.5.5). So, if this condition holds then the arrows $l_{J_{\mathrm{at}}}(M_{\{\vec{x}.\phi\}}) \to l_{J_{\mathrm{at}}}(M_{\{\vec{y}.\psi\}})$ in $\mathbf{Sh}(\text{f.p.}\mathbb{T}\text{-mod}(\mathbf{Set})^{\mathrm{op}}, J_{\mathrm{at}})$ correspond precisely to the $\mathbb{T}$-model homomorphism $M_{\{\vec{y}.\psi\}} \to M_{\{\vec{x}.\phi\}}$ (equivalently, by Theorem 6.1.13, to the $\mathbb{T}$-provably functional formulae $\{\vec{x}.\phi\} \to \{\vec{y}.\psi\}$). In other words, we have the following 'bridge':

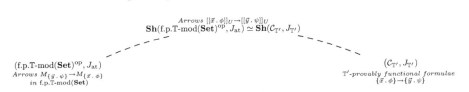

(where U denotes 'the' universal model of $\mathbb{T}'$ in its classifying topos). □

We shall reconsider the same invariant in section 10.5 in the context of Galois-type representations for the classifying topos of $\mathbb{T}'$ (cf. Theorem 10.5.3 and Remark 10.5.4(b)).

Another natural application of the 'bridge technique' in the context of the above Morita equivalence for $\mathbb{T}'$ is obtained by considering the invariant property of a topos to have enough points. It is interesting to consider this invariant in combination with the (invariant) property of a topos to be non-trivial (recall that a topos is said to be non-trivial if its initial and terminal objects are non-isomorphic) in the context of the investigation of the existence of points of a given topos; indeed, if a non-trivial topos has enough points then it has at least a point. Now, one can profitably exploit the geometry of sites of definition of a topos to investigate these kinds of issues. As an example, by Theorem 3.3 [6] the theory $\mathbb{T}^{\mathcal{C}}$ of flat functors on a small category $\mathcal{C}$ (cf. the proof of Theorem 2.1.11) can be axiomatized coherently over its signature if and only if $\mathcal{C}$ has all *finite fc-limits* (that is, for any finite diagram with values in $\mathcal{C}$ there is a finite family of cones over it such that every cone over the diagram factors through one in the family). Since $\mathbb{T}^{\mathcal{C}}$ is classified by the presheaf topos $[\mathcal{C}^{\mathrm{op}}, \mathbf{Set}]$, if the category $\mathcal{C}$ satisfies the right Ore condition then the theory obtained from $\mathbb{T}^{\mathcal{C}}$ by adding the axiom

$$(\top \vdash_{y^{\ulcorner d \urcorner}} (\exists x^{\ulcorner c \urcorner})(\ulcorner f \urcorner(x) = y))$$

for any arrow $f : c \to d$ in $\mathcal{C}$, is classified by the topos $\mathbf{Sh}(\mathcal{C}, J_{\mathrm{at}})$ and hence it is coherent if the theory $\mathbb{T}^{\mathcal{C}}$ is. So, if the category $\mathcal{C}$ has all finite fc-limits and satisfies the right Ore condition, the topos $\mathbf{Sh}(\mathcal{C}, J_{\mathrm{at}})$ is coherent and hence, assuming the axiom of choice, it has enough points. Now, it is easy to see that the topos $\mathbf{Sh}(\mathcal{C}, J_{\mathrm{at}})$ is non-trivial if and only if the category $\mathcal{C}$ is non-empty. Applying these remarks in connection with the Morita equivalence

$$\mathbf{Sh}(\mathrm{f.p.}\mathbb{T}\text{-mod}(\mathbf{Set})^{\mathrm{op}}, J_{\mathrm{at}}) \simeq \mathbf{Sh}(\mathcal{C}_{\mathbb{T}'}, J_{\mathbb{T}'})$$

then yields the following existence theorem for homogeneous models:

Theorem 10.4.10 (Theorem 3.4 [23]). *Let $\mathbb{T}$ be a theory of presheaf type whose category* f.p.$\mathbb{T}$-mod($\mathbf{Set}$) *of finitely presentable models satisfies the amalgamation property. If the topos* $\mathbf{Sh}(\mathrm{f.p.}\mathbb{T}\text{-mod}(\mathbf{Set})^{\mathrm{op}}, J_{\mathrm{at}})$ *has enough points (e.g. if the category* f.p.$\mathbb{T}$-mod($\mathbf{Set}$) *has all finite fc-colimits) and there exists at least one $\mathbb{T}$-model in $\mathbf{Set}$, then, assuming the axiom of choice, there exists at least one homogeneous $\mathbb{T}$-model in $\mathbf{Set}$.* □

Notice that this theorem represents a consistency result for logical theories arising from purely combinatorial/geometrical considerations.

10.5 Maximal theories and Galois representations

In this section, which is based on [27], we establish, under some natural hypotheses, Galois-type representations for the atomic two-valued toposes considered in section 10.4 as toposes of continuous actions of a topological group of automorphisms of a suitable structure. This leads to a framework which generalizes Grothendieck's theory of Galois categories (cf. [47]) and allows one to build Galois-type theories in a great variety of new mathematical contexts.

Theorem 10.5.1 (cf. Corollary 3.7 [27]). *Let $\mathbb{T}$ be a theory of presheaf type whose category f.p.$\mathbb{T}$-mod(**Set**) of finitely presentable models satisfies* AP *and* JEP, *and let M be a f.p.$\mathbb{T}$-mod(**Set**)-universal and f.p.$\mathbb{T}$-mod(**Set**)-ultrahomogeneous model of $\mathbb{T}$ (in the sense of Definition 10.4.3). Then we have an equivalence of toposes*

$$\mathbf{Sh}(\text{f.p.}\mathbb{T}\text{-mod}(\mathbf{Set})^{\text{op}}, J_{\text{at}}) \simeq \mathbf{Cont}(\text{Aut}(M)),$$

where $\text{Aut}(M)$ *is endowed with the topology of pointwise convergence (in which a basis of open neighbourhoods of the identity is given by the sets of the form $\{f : M \cong M \mid f(\vec{a}) = \vec{a}\}$ for a tuple $\vec{a}$ of elements of M), which is induced by the functor*

$$F : \text{f.p.}\mathbb{T}\text{-mod}(\mathbf{Set})^{\text{op}} \to \mathbf{Cont}(\text{Aut}(M))$$

*sending any model c of f.p.$\mathbb{T}$-mod(**Set**) to the set $\text{Hom}_{\mathbb{T}\text{-mod}(\mathbf{Set})}(c, M)$ (endowed with the obvious action by $\text{Aut}(M)$) and any arrow $f : c \to d$ in f.p.$\mathbb{T}$-mod(**Set**) to the $\text{Aut}(M)$-equivariant map*

$$- \circ f : \text{Hom}_{\mathbb{T}\text{-mod}(\mathbf{Set})}(d, M) \to \text{Hom}_{\mathbb{T}\text{-mod}(\mathbf{Set})}(c, M).$$

Remark 10.5.2 Theorem 10.5.1 also admits a purely categorical formulation (not involving the concept of theory of presheaf type), obtained by replacing the category f.p.$\mathbb{T}$-mod(**Set**) with an arbitrary small category $\mathcal{C}$, the category $\mathbb{T}$-mod(**Set**) with the ind-completion Ind-$\mathcal{C}$ of $\mathcal{C}$, the model M with an object u of Ind-$\mathcal{C}$ and the group $\text{Aut}(M)$ with the group of automorphisms of u in the category Ind-$\mathcal{C}$, topologized in such a way that a basis of open neighbourhoods of the identity is given by the sets of the form $\{f : u \cong u \mid f \circ \chi = \chi\}$ for an arrow $\chi : c \to u$ in Ind-$\mathcal{C}$ from an object c of $\mathcal{C}$ to u (see Theorem 3.5 [27]).

Under the hypotheses of Theorem 10.5.1, let $\mathbb{T}'$ be the theory of homogeneous $\mathbb{T}$-models (as defined in section 10.4). Then the model M, endowed with the (continuous) canonical action of $\text{Aut}(M)$, is a universal model of $\mathbb{T}'$ in the topos $\mathbf{Cont}(\text{Aut}(M))$ (cf. the proof of Theorem 3.1 [27]). So for any tuple $A_1, \ldots, A_n$ of sorts of the signature of $\mathbb{T}$, we have a 'bridge'

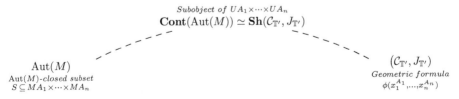

(where U is 'the' universal model of $\mathbb{T}'$ in its classifying topos), which yields the following result:

Theorem 10.5.3 *Let $\mathbb{T}'$ be the theory of homogeneous $\mathbb{T}$-models for a theory $\mathbb{T}$ satisfying the hypotheses of Theorem 10.5.1.*

(i) *For any subset $S \subseteq MA_1 \times \cdots \times MA_n$ which is closed under the action of $\text{Aut}(M)$, there exists a (unique up to $\mathbb{T}'$-provable equivalence) geometric formula $\phi(\vec{x})$ over the signature of $\mathbb{T}$ (where $\vec{x} = (x_1^{A_1}, \ldots, x_n^{A_n})$) such that $S = [[\vec{x}.\phi]]_M$.*

(ii) *For any $\text{Aut}(M)$-equivariant map $f : S \to T$ between $\text{Aut}(M)$-closed subsets S and T as in (i) there exists a (unique up to $\mathbb{T}'$-provable equivalence) $\mathbb{T}'$-provably functional geometric formula $\theta(\vec{x}, \vec{y})$ from $\phi(\vec{x})$ to $\psi(\vec{y})$, where $S = [[\vec{x}.\phi]]_M$ and $T = [[\vec{y}.\psi]]_M$, whose interpretation $[[\vec{x}, \vec{y}.\theta]]_M$ coincides with the graph of f.*

Remarks 10.5.4 (a) It easily follows from Theorem 10.5.3 that for any finite tuple $A_1, \ldots, A_n$ of sorts of the signature of $\mathbb{T}$, the orbits of the action of $\text{Aut}(M)$ on $MA_1 \times \cdots \times MA_n$ coincide with the interpretations $[[\vec{x}.\phi]]_M$ in M of $\mathbb{T}$-complete formulae $\phi(\vec{x})$, where $\vec{x} = (x_1^{A_1}, \ldots, x_n^{A_n})$, and hence correspond precisely to the $\mathbb{T}$-provable equivalence classes of $\mathbb{T}$-complete formulae in the context $\vec{x}$ (in the sense of Definition 10.3.1(a)).

(b) If in part (ii) of Theorem 10.5.3 the formulae $\phi(\vec{x})$ and $\psi(\vec{y})$ present respectively $\mathbb{T}$-models $M_{\{\vec{x}.\phi\}}$ and $M_{\{\vec{y}.\psi\}}$ and all the arrows in f.p.$\mathbb{T}$-mod(**Set**) are strict monomorphisms (cf. Theorem 10.5.5 below), then the formula θ can be taken to be $\mathbb{T}$-provably functional and hence to induce (by Theorem 6.1.13) a $\mathbb{T}$-model homomorphism $z : M_{\{\vec{y}.\psi\}} \to M_{\{\vec{x}.\phi\}}$ such that the map $[[\vec{x}.\phi]]_M \to [[\vec{y}.\psi]]_M$ whose graph is $[[\vec{x}, \vec{y}.\theta]]_M$ corresponds, under the identifications

$$\text{Hom}_{\mathbb{T}\text{-mod}(\mathbf{Set})}(M_{\{\vec{x}.\phi\}}, M) \cong [[\vec{x}.\phi]]_M$$

and

$$\text{Hom}_{\mathbb{T}\text{-mod}(\mathbf{Set})}(M_{\{\vec{y}.\psi\}}, M) \cong [[\vec{y}.\psi]]_M,$$

to the map $\text{Hom}_{\mathbb{T}\text{-mod}(\mathbf{Set})}(z, M)$.

Concerning the existence of models M satisfying the hypotheses of Theorem 10.5.1, we should remark that ultrahomogeneous structures naturally arise in many different mathematical contexts. Their existence can be proved either directly by means of an explicit construction or through abstract logical arguments. A general method for building countable ultrahomogeneous structures is provided by Fraïssé's construction in model theory (cf. Chapter 7 of [48]), while the categorical generalization established in section 2 of [23] allows one to construct ultrahomogeneous structures of arbitrary cardinality. As examples of models M satisfying the hypotheses of Theorem 10.5.1 in relation to the theories of presheaf type considered in section 10.4, we mention:

- The set $\mathbb{N}$ of natural numbers, with respect to the theory of decidable objects;

- The ordered set $(\mathbb{Q}, <)$ of rational numbers, with respect to the theory $\mathbb{L}$ of decidably linearly ordered objects considered in section 8.1.1.1;
- The unique countable atomless Boolean algebra, with respect to the theory of decidable Boolean algebras.
- Any Galois extension F' of a given field F, with respect to the $\mathcal{L}_F^{F'}$-completion (in the sense of Theorem 8.2.10) of the theory $\mathbb{T}_F$ of algebraic extensions of F (defined in section 9.5), where $\mathcal{L}_F^{F'}$ is the category of finite extensions of F which are embeddable in F' and field homomorphisms between them.

There is a natural link between ultrahomogeneity and the property of a model of an atomic complete theory to be *special*, which can be exploited to construct models satisfying the hypotheses of Theorem 10.5.1. Recall that a model M of an atomic complete theory $\mathbb{T}$ is said to be *special* if every $\mathbb{T}$-complete formula $\phi(\vec{x})$ is realized in M and, for any tuples $\vec{a}$ and $\vec{b}$ of elements of M which satisfy the same $\mathbb{T}$-complete formula, there is an automorphism of M which sends $\vec{a}$ to $\vec{b}$. Any model M satisfying the hypotheses of Theorem 10.5.1 is clearly special as a model of the theory of homogeneous $\mathbb{T}$-models. Conversely, given a special model M of an atomic complete theory $\mathbb{S}$, we have, by the Comparison Lemma (cf. Theorem 1.1.8), an equivalence $\mathbf{Sh}(\mathcal{C}_\mathbb{S}, J_\mathbb{S}) \simeq \mathbf{Sh}(\mathcal{C}_\mathbb{S}^{\mathrm{at}}, J_{\mathrm{at}})$, where $\mathcal{C}_\mathbb{S}^{\mathrm{at}}$ is the full subcategory of $\mathcal{C}_\mathbb{S}$ on the $\mathbb{S}$-complete formulae, which shows that $\mathbb{S}$ is Morita-equivalent to the theory of homogeneous $\mathbb{T}$-models for a theory of presheaf type $\mathbb{T}$ with respect to which the model M satisfies the hypotheses of Theorem 10.5.1.

Let us now discuss under which conditions the Morita equivalence of Theorem 10.5.1 restricts to an equivalence of sites.

Given an arrow $\chi : c \to M$ in $\mathbb{T}\text{-mod}(\mathbf{Set})$, where c is in f.p.$\mathbb{T}$-mod($\mathbf{Set}$), we denote by
$$\mathcal{I}_\chi := \{f : M \cong M \mid f \circ \chi = \chi\}$$
the open subgroup of $\mathrm{Aut}(M)$ consisting of the automorphisms which fix χ.

The following theorem gives necessary and sufficient conditions for the functor F of Theorem 10.5.1 to be full and faithful.

Theorem 10.5.5 (cf. Proposition 4.1 [27]). *Under the hypotheses of Theorem 10.5.1, the following conditions are equivalent:*

(i) *Every arrow $f : d \to c$ in f.p.$\mathbb{T}$-mod($\mathbf{Set}$) is a strict monomorphism, i.e. for any arrow $g : e \to c$ in f.p.$\mathbb{T}$-mod($\mathbf{Set}$) such that $h \circ g = k \circ g$ whenever $h \circ f = k \circ f$, g factors uniquely through f:*

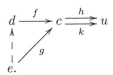

(ii) *The functor*

$$F: \text{f.p.}\mathbb{T}\text{-mod}(\mathbf{Set})^{\mathrm{op}} \to \mathbf{Cont}(\mathrm{Aut}(M))$$

of Theorem 10.5.1 is full and faithful.

(iii) For any models $c, d \in \text{f.p.}\mathbb{T}\text{-mod}(\mathbf{Set})$ and any arrows $\chi : c \to M$ and $\xi : d \to M$ in $\mathbb{T}\text{-mod}(\mathbf{Set})$, $\mathcal{I}_\xi \subseteq \mathcal{I}_\chi$ (that is, for any automorphism f of M, $f \circ \xi = \xi$ implies $f \circ \chi = \chi$) if and only if there exists exactly one arrow $f : c \to d$ in $\text{f.p.}\mathbb{T}\text{-mod}(\mathbf{Set})$ such that $\chi = \xi \circ f$:

(iv) The Grothendieck topology J_{at} is subcanonical.

Let $\mathrm{Sgr}(\mathrm{Aut}(M))$ be the preorder category consisting of the open subgroups of $\mathrm{Aut}(M)$ and the inclusion ordering between them, and let $\text{f.p.}\mathbb{T}\text{-mod}(\mathbf{Set})^{\mathrm{op}}/M$ be the category whose objects are the arrows $c \to M$ in $\mathbb{T}\text{-mod}(\mathbf{Set})$, where c is an object of $\text{f.p.}\mathbb{T}\text{-mod}(\mathbf{Set})$, and whose arrows $(\chi : c \to M) \to (\xi : d \to M)$ are the arrows $f : d \to c$ in $\text{f.p.}\mathbb{T}\text{-mod}(\mathbf{Set})$ such that $\chi \circ f = \xi$. Then the functor F of Theorem 10.5.1 yields a functor

$$\tilde{F}: \text{f.p.}\mathbb{T}\text{-mod}(\mathbf{Set})^{\mathrm{op}}/M \to \mathrm{Sgr}(\mathrm{Aut}(M))$$

sending any object $\chi : c \to M$ of $\text{f.p.}\mathbb{T}\text{-mod}(\mathbf{Set})^{\mathrm{op}}/M$ to the subgroup $\mathcal{I}_\chi$ of $\mathrm{Aut}(M)$.

Note that if the conditions of Theorem 10.5.5 are satisfied then the slice category $\text{f.p.}\mathbb{T}\text{-mod}(\mathbf{Set})^{\mathrm{op}}/M$ is a preorder. Indeed, if all the arrows of the category $\text{f.p.}\mathbb{T}\text{-mod}(\mathbf{Set})$ are monic then all the arrows of the category $\mathbb{T}\text{-mod}(\mathbf{Set})$ are monic too (cf. the proof of Theorem 7.2.32).

It is natural to wonder under which conditions the functor $\tilde{F}$ yields a bijection between the objects of the category $\text{f.p.}\mathbb{T}\text{-mod}(\mathbf{Set})^{\mathrm{op}}/M$ and the open subgroups of the topological group $\mathrm{Aut}(M)$, as in the case of classical Galois theory. The following theorem provides an answer to this question.

Theorem 10.5.6 (cf. Theorem 4.17 [27]). *Assuming that the equivalent conditions of Theorem 10.5.5 are satisfied, the following conditions are equivalent:*

(i) *The functor*

$$F: \text{f.p.}\mathbb{T}\text{-mod}(\mathbf{Set})^{\mathrm{op}} \to \mathbf{Cont}(\mathrm{Aut}(M))$$

is a categorical equivalence onto the full subcategory of $\mathbf{Cont}(\mathrm{Aut}(M))$ on the non-empty transitive actions.

(ii) *The map*

$$\tilde{F}: \text{f.p.}\mathbb{T}\text{-mod}(\mathbf{Set})^{\mathrm{op}}/M \to \mathrm{Sgr}(\mathrm{Aut}(M))$$

is a bijection onto the set of open subgroups of $\mathrm{Aut}(M)$.

(iii) *The category* f.p.$\mathbb{T}$-mod(**Set**) *has equalizers and arbitrary intersections of subobjects of a given object and for any pair of arrows* $h, k : c \to e$ *in* f.p.$\mathbb{T}$-mod(**Set**) *with equalizer* $m : d \rightarrowtail c$ *we have that for any pair of arrows* $l, n : c \to e'$, $l \circ m = n \circ m$ *if and only if there exists an arrow* $s : e' \to e''$ *such that* $(s \circ l, s \circ n)$ *belongs to the equivalence relation on* $\mathrm{Hom}_{\mathrm{f.p.}\mathbb{T}\text{-mod}(\mathbf{Set})}(c, e'')$ *generated by the relation consisting of the pairs of the form* $(t \circ h, t \circ k)$ *for an arrow* $t : e \to e''$.

(iv) *Every atom of the topos* $\mathbf{Sh}(\mathrm{f.p.}\mathbb{T}\text{-mod}(\mathbf{Set})^{\mathrm{op}}, J_{\mathrm{at}})$ *is, up to isomorphism, of the form* $l_{J_{\mathrm{at}}}(c)$ *for some object* c *of* f.p.$\mathbb{T}$-mod(**Set**), *where* $l_{J_{\mathrm{at}}}$ *is the functor*

$$\mathrm{f.p.}\mathbb{T}\text{-mod}(\mathbf{Set}) \to \mathbf{Sh}(\mathrm{f.p.}\mathbb{T}\text{-mod}(\mathbf{Set})^{\mathrm{op}}, J_{\mathrm{at}})$$

given by the composite of the Yoneda embedding with the associated sheaf functor.

Remark 10.5.7 If the category f.p.$\mathbb{T}$-mod(**Set**) satisfies the hypotheses of Theorem 10.5.6, it is always possible to 'complete' it to a category satisfying the equivalent conditions of the theorem and having an equivalent associated topos, by means of an elementary process which is a sort of completion by the addition of imaginaries (in the sense of model theory). This process is described in detail in [27] (cf. the proof of Theorem 4.15 therein) and applies to a number of classical categories such as the category of Boolean algebras and embeddings, the category of finite groups and embeddings, the category of finite graphs and embeddings, etc.

Notice that both Theorem 10.5.5 and Theorem 10.5.6 naturally arise from topos-theoretic 'bridges', the former by considering the invariant notion of arrow between two given objects, and the latter by considering the notion of atom of a topos.

10.6 A characterization theorem for geometric logic

As an application of the duality theorem of Chapter 3, we can derive a criterion for deciding whether a class of structures is the class of models of a geometric theory inside Grothendieck toposes; this criterion has led in particular to a characterization of the infinitary first-order theories which are geometric in terms of their models in Grothendieck toposes, solving a problem posed by I. Moerdijk in 1989 (cf. [15]).

The proof of our criterion relies on the following technical lemma:

Lemma 10.6.1 (Lemma 1 [15]). *Let* $\{f_i : \mathcal{E}_i \to \mathcal{E} \mid i \in I\}$ *be a family of geometric morphisms of (elementary) toposes with common codomain* $\mathcal{E}$. *Then* $\{f_i : \mathcal{E}_i \to \mathcal{E} \mid i \in I\}$ *is jointly surjective if and only if* $\mathcal{E} = \bigcup_{i \in I} \mathcal{E}'_i$, *where for each* $i \in I$, $\mathcal{E}_i \twoheadrightarrow \mathcal{E}'_i \hookrightarrow \mathcal{E}$ *is the surjection-inclusion factorization of* f_i.

Theorem 10.6.2 (cf. Theorem 2 [15]). *Let* Σ *be a first-order signature and* $\mathcal{S}$ *a collection of* Σ*-structures in Grothendieck toposes closed under isomorphisms* of structures. Then $\mathcal{S}$ is the collection of all models in Grothendieck toposes of a geometric theory over Σ if and only if it satisfies the following two conditions:

(i) For any geometric morphism $f : \mathcal{F} \to \mathcal{E}$, if $M \in \Sigma\text{-str}(\mathcal{E})$ is in $\mathcal{S}$ then $f^*(M)$ is in $\mathcal{S}$.

(ii) For any (set-indexed) jointly surjective family $\{f_i : \mathcal{E}_i \to \mathcal{E} \mid i \in I\}$ of geometric morphisms and any Σ-structure M in $\mathcal{E}$, if $f_i^*(M)$ is in $\mathcal{S}$ for every $i \in I$ then M is in $\mathcal{S}$.

Proof The 'only if' part of the theorem is trivial and well-known. Let us prove the 'if' part. Let $\mathbb{O}_\Sigma$ be the empty (geometric) theory over Σ. The $\mathbb{O}_\Sigma$-models in any Grothendieck topos $\mathcal{E}$ are precisely the Σ-structures in $\mathcal{E}$ and hence, for any Grothendieck topos $\mathcal{E}$, the geometric morphisms $\mathcal{E} \to \mathbf{Set}[\mathbb{O}_\Sigma]$ correspond to the Σ-structures in $\mathcal{E}$. The geometric morphism corresponding to a Σ-structure M will be denoted by $f_M : \mathcal{E}_M \to \mathbf{Set}[\mathbb{O}_\Sigma]$; recall that if U is a universal model of $\mathbb{O}_\Sigma$ in $\mathbf{Set}[\mathbb{O}_\Sigma]$ then $M \cong f_M^*(U)$.

Let $\mathcal{E} \hookrightarrow \mathbf{Set}[\mathbb{O}_\Sigma]$ be the subtopos of $\mathbf{Set}[\mathbb{O}_\Sigma]$ given by the union of all the subtoposes of $\mathbf{Set}[\mathbb{O}_\Sigma]$ arising as the inclusion parts of the surjection-inclusion factorizations of the geometric morphisms $f_M : \mathcal{E}_M \to \mathbf{Set}[\mathbb{O}_\Sigma]$ for M in $\mathcal{S}$, and let $a : \mathbf{Set}[\mathbb{O}_\Sigma] \to \mathcal{E}$ be the corresponding associated sheaf functor. By Theorem 3.2.5 the subtopos $\mathcal{E} \hookrightarrow \mathbf{Set}[\mathbb{O}_\Sigma]$ of $\mathbf{Set}[\mathbb{O}_\Sigma]$ corresponds to a (unique up to syntactic equivalence) geometric quotient $\mathbb{T}$ of $\mathbb{O}_\Sigma$ such that if $U_{\mathbb{O}_\Sigma}$ is a universal model of $\mathbb{O}_\Sigma$ in $\mathbf{Set}[\mathbb{O}_\Sigma]$ then $U_\mathbb{T} := a(U_{\mathbb{O}_\Sigma})$ is a universal model of $\mathbb{T}$ in $\mathcal{E}$. Let us show that $\mathbb{T}$ axiomatizes our class of structures $\mathcal{S}$.

For any structure $M \in \Sigma\text{-str}(\mathcal{E}_M)$ in $\mathcal{S}$, the subtopos $\mathcal{E}'_M \hookrightarrow \mathbf{Set}[\mathbb{O}_\Sigma]$ arising in the surjection-inclusion factorization of $f_M : \mathcal{E}_M \to \mathbf{Set}[\mathbb{O}_\Sigma]$ factors as the canonical inclusion $\mathcal{E}'_M \hookrightarrow \mathcal{E}$ followed by the inclusion $\mathcal{E} \hookrightarrow \mathbf{Set}[\mathbb{O}_\Sigma]$. Now, composing the former inclusion with the surjection part of the surjection-inclusion factorization of f_M yields a geometric morphism $h_M : \mathcal{E}_M \to \mathcal{E}$ which, when composed with $\mathcal{E} \hookrightarrow \mathbf{Set}[\mathbb{O}_\Sigma]$, is isomorphic to f_M. But $M \cong f_M^*(U_{\mathbb{O}_\Sigma})$, from which it follows that $h_M^*(U_\mathbb{T}) \cong M$ and hence that M is a model of $\mathbb{T}$. This shows that every structure in $\mathcal{S}$ is a model of $\mathbb{T}$. Let us prove that, conversely, any $\mathbb{T}$-model in a Grothendieck topos lies in $\mathcal{S}$. By Lemma 10.6.1 the family of geometric morphisms h_M for M in $\mathcal{S}$ is jointly surjective; so condition (ii) implies that $U_\mathbb{T}$ lies in $\mathcal{S}$. Now, by the universal property of the classifying topos $\mathcal{E}$ of $\mathbb{T}$, every model N of $\mathbb{T}$ in a Grothendieck topos $\mathcal{F}$ is of the form $g^*(U_\mathbb{T})$ for some geometric morphism $g : \mathcal{F} \to \mathcal{E}$, so condition (i) implies our thesis.

Note in passing that $\mathbb{T}$ can be described as the collection of geometric sequents over Σ which are valid in every structure M in $\mathcal{S}$ (cf. Theorem 4.2.13). □

10.7 The maximal spectrum of a commutative ring

The purpose of this section is to show that, by using the 'bridge' technique (cf. section 2.2), we can design natural analogues of the Zariski spectrum of a commutative ring with unit which may be useful in different contexts, for instance in connection with the problem of building an associated theory for the maximal spectrum of a ring.

As shown in [16], the different constructions of the Zariski spectrum of a commutative ring with unit A can be interpreted topos-theoretically as different representations of the topos of sheaves on it. For instance, we have a Morita equivalence

$$\mathbf{Sh}(\mathrm{Spec}(A)) \simeq \mathbf{Sh}(\mathrm{Rad}(A), J_A),$$

where $\mathrm{Rad}(A)$ is the frame of radical ideals of A (with the subset-inclusion ordering) and J_A is the canonical topology on it, and a Morita equivalence

$$\mathbf{Sh}(\mathrm{Spec}(A)) \simeq \mathbf{Sh}(\mathcal{C}_{\mathbb{P}_A}, J_{\mathbb{P}_A}),$$

where $(\mathcal{C}_{\mathbb{P}_A}, J_{\mathbb{P}_A})$ is the geometric syntactic site of the theory $\mathbb{P}_A$ of prime filters on A defined in section 2.1.6.1. In fact, the Zariski spectrum of A is the space of points of the classifying topos of $\mathbb{P}_A$ (in the sense of Definition 1.1.23).

The first Morita equivalence can be established by using the well-known isomorphism between the frame $\mathcal{O}(\mathrm{Spec}(A))$ of open sets of $\mathrm{Spec}(A)$ and the frame $\mathrm{Rad}(A)$, which sends every radical ideal I of A to the complement in $\mathrm{Spec}(A)$ of the set $\{P \in \mathrm{Spec}(A) \mid P \supseteq I\}$.

The second Morita equivalence follows from the prime ideal theorem for distributive lattices (a theorem which is equivalent to a weak form of the axiom of choice), which ensures that, the theory $\mathbb{P}_A$ being coherent, its classifying topos has enough points and hence can be represented topologically as the topos of sheaves on its space of points.

The syntactic method for constructing classifying toposes of coherent theories reviewed in section 2.1.2 yields a further representation

$$\mathbf{Sh}(\mathrm{Spec}(A)) \simeq \mathbf{Sh}(L(A), J_A^{\mathrm{coh}})$$

of the topos $\mathbf{Sh}(\mathrm{Spec}(A))$ as the topos of coherent sheaves on a distributive lattice, namely the coherent syntactic category $L(A)$ of the propositional theory $\mathbb{P}_A$.

By the Comparison Lemma (Theorem 1.1.8), the site $(L(A), J_A^{\mathrm{coh}})$ can be cut down to a smaller site $(S(A), C_A)$ generating the same topos (up to equivalence), given by the full subcategory $S(A)$ of $L(A)$ on the elements of the form $[a]$ for $a \in A$, with the induced topology C_A. This site can be intrinsically characterized as follows: $S(A)$ is the meet-semilattice given by the quotient of the underlying monoid of A by the smallest congruence which identifies a and a^2 for all elements a of A, with the order given by $[a] \leq [b]$ if and only if $a \cdot b = a$ (where $[a]$ denotes the equivalence class of a in $S(A)$), while C_A is the Grothendieck topology on $S(A)$ generated by the covering families of the form $\emptyset \in C_A([0_A])$ and of the form $[a_i] \to [a]$ (for $i = 1, \ldots, n$), where $[a_i] \leq [a]$ and $[a_1 + \cdots + a_n] = [a] = x$ (cf. sections 8.7 of [16] and V3.1 of [53]).

Each of these different representations of the small Zariski topos of A yields a different point of view on it; depending on the aspects of the spectrum that one wishes to investigate, one point of view can be more natural or convenient than another.

Now that we have reviewed the topos-theoretic interpretation of the different constructions of the Zariski spectrum of A, let us proceed to discuss how to design a suitable analogue of it for the set of maximal ideals of A.

Since every maximal ideal is prime, it is natural to put a topology on the set $\mathrm{Max}(A)$ of maximal ideals of A such that the inclusion $\mathrm{Max}(A) \hookrightarrow \mathrm{Spec}(A)$ induces a geometric inclusion $\mathbf{Sh}(\mathrm{Max}(A)) \to \mathbf{Sh}(\mathrm{Spec}(A))$. Now, for the inclusion map $\mathrm{Max}(A) \hookrightarrow \mathrm{Spec}(A)$ to be continuous it is necessary for the topology on $\mathrm{Max}(A)$ to contain the subspace topology. On the other hand, any subspace inclusion $i : Y \hookrightarrow X$ induces a geometric inclusion $\mathbf{Sh}(i) : \mathbf{Sh}(Y) \hookrightarrow \mathbf{Sh}(X)$ (cf. Proposition C1.5.1(i) [55]). So, if we endow $\mathrm{Max}(A)$ with the topology induced by that of $\mathrm{Spec}(A)$, the topos $\mathbf{Sh}(\mathrm{Max}(A))$ gets naturally identified with a subtopos of $\mathbf{Sh}(\mathrm{Spec}(A))$, which allows us to 'localize' all the representations of $\mathbf{Sh}(\mathrm{Spec}(A))$ to representations of $\mathbf{Sh}(\mathrm{Max}(A))$. We shall now proceed to describe and analyse in detail such localizations, starting from the logical representation

$$\mathbf{Sh}(\mathrm{Spec}(A)) \simeq \mathbf{Sh}(\mathcal{C}_{\mathbb{P}_A}, J_{\mathbb{P}_A}).$$

By Theorem 3.2.5, there exists exactly one quotient $\mathbb{P}_A^{\mathrm{max}}$ of $\mathbb{P}_A$ classified by the subtopos $\mathbf{Sh}(\mathrm{Max}(A)) \hookrightarrow \mathbf{Sh}(\mathrm{Spec}(A))$; that is, we have a 'bridge'

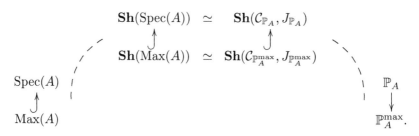

By considering the topos-theoretic invariant notion of point of a topos in connection with the Morita equivalence $\mathbf{Sh}(\mathrm{Max}(A)) \simeq \mathbf{Sh}(\mathcal{C}_{\mathbb{P}_A^{\mathrm{max}}}, J_{\mathbb{P}_A^{\mathrm{max}}})$ and recalling the well-known site characterization 'the points of the topos $\mathbf{Sh}(X)$ of sheaves on a topological space X correspond precisely to the points of the sobrification of X' (holding for any X), we obtain a 'bridge' yielding a bijective correspondence between the set-based models of the theory $\mathbb{P}_A^{\mathrm{max}}$ and the elements of the sobrification of the space $\mathrm{Max}(A)$:

It follows in particular that the sobrification of the space $\mathrm{Max}(A)$ can be realized as a space of prime ideals of A; we shall characterize these ideals more explicitly in Theorem 10.7.1. Another consequence of the above Morita equivalence is the fact that every open set of $\mathrm{Max}(A)$ is of the form $\{M \in \mathrm{Max}(A) \mid M \vDash \phi\}$ for a geometric sentence ϕ over the signature of the theory $\mathbb{P}_A$.

By considering the topos-theoretic invariant property 'to be a dense subobject with respect to a subtopos' in connection with the same Morita equivalence, we obtain a syntactic description of the theory $\mathbb{P}_A^{\max}$, as follows. Clearly, a geometric sequent $(\phi \vdash_{\vec{x}} \psi)$ over the signature of $\mathbb{P}_A$ is provable in $\mathbb{P}_A^{\max}$ if and only if the subobject $[[\vec{x}.\phi \wedge \psi]]_U \rightarrowtail [[\vec{x}.\phi]]_U$, where U is 'the' universal model of $\mathbb{P}_A$ in $\mathbf{Sh}(\mathcal{C}_{\mathbb{P}_A}, J_{\mathbb{P}_A})$, is dense with respect to the subtopos $\mathbf{Sh}(\mathcal{C}_{\mathbb{P}_A^{\max}}, J_{\mathbb{P}_A^{\max}})$. Now, since the points of the topos $\mathbf{Sh}(\mathrm{Max}(A))$ corresponding to the maximal ideals of A are jointly surjective (that is, their inverse image functors jointly reflect isomorphisms), this subobject is dense if and only if the sequent $(\phi \vdash_{\vec{x}} \psi)$ is satisfied in every prime filter of A whose complement is a maximal ideal of A. We can thus conclude that the quotient $\mathbb{P}_A^{\max}$ is obtained from the theory $\mathbb{P}_A$ by adding all the sequents over the signature of $\mathbb{P}_A$ which are satisfied by all the complements of the maximal ideals of A.

Let us now consider the problem of getting an algebraic presentation of the maximal spectrum of A, as defined above; this will lead in particular to an alternative axiomatization of the theory $\mathbb{P}_A^{\max}$. By considering the invariant notion of subtopos in connection with the Morita equivalence

$$\mathbf{Sh}(\mathrm{Spec}(A)) \simeq \mathbf{Sh}(L(A), J_A^{\mathrm{coh}}),$$

we obtain that the subtopos $\mathbf{Sh}(\mathrm{Max}(A)) \hookrightarrow \mathbf{Sh}(\mathrm{Spec}(A))$ corresponds to a unique Grothendieck topology $J_A^{\max}$ on $L(A)$ such that the canonical geometric inclusion

$$\mathbf{Sh}(L(A), J_A^{\max}) \hookrightarrow \mathbf{Sh}(L(A), J_A^{\mathrm{coh}})$$

induces an equivalence of subtoposes

$$\begin{array}{ccc} \mathbf{Sh}(\mathrm{Spec}(A)) & \simeq & \mathbf{Sh}(L(A), J_A^{\mathrm{coh}}) \\ \uparrow & & \uparrow \\ \mathbf{Sh}(\mathrm{Max}(A)) & \simeq & \mathbf{Sh}(L(A), J_A^{\max}). \end{array}$$

Theorem 3.2.5 then implies, in view of the Morita equivalence

$$\mathbf{Sh}(\mathcal{C}_{\mathbb{P}_A}, J_{\mathbb{P}_A}) \simeq \mathbf{Sh}(S(A), C_A),$$

that the quotient $\mathbb{P}_A^{\max}$ is obtained by adding all the sequents over the signature of $\mathbb{P}_A$ of the form

$$\left(P_c \vdash \bigvee_{i \in I} P_{c_i} \right)$$

for $J_A^{\max}$-covering sieves $\{[P_{c_i}] \to [P_c] \mid i \in I\}$ (notice that for each $a \in A$, $[a] \in S(A)$ and $[P_a] \in L(A)$ correspond to each other under the above equivalence $\mathbf{Sh}(L(A), J_A^{\mathrm{coh}}) \simeq \mathbf{Sh}(S(A), C_A)$).

An explicit description of the topology $J_A^{\max}$ will thus provide us with an alternative axiomatization of the theory $\mathbb{P}_A^{\max}$. In order to obtain such a description, we shall use the characterization of the J-covering sieves S on a small category $\mathcal{C}$ (for a Grothendieck topology J on $\mathcal{C}$) as the subobjects $S \rightarrowtail \mathcal{C}(-,c)$ which are dense with respect to the closure operation on $[\mathcal{C}^{\mathrm{op}}, \mathbf{Set}]$ associated with J. To calculate the closure operation on $\mathbf{Sh}(\mathrm{Spec}(A))$ corresponding to the geometric inclusion $\mathbf{Sh}(\mathrm{Max}(A)) \hookrightarrow \mathbf{Sh}(\mathrm{Spec}(A))$, we observe that for any subspace Y of a topological space X, the action on subterminals of the inverse image of the subtopos inclusion $\mathbf{Sh}(Y) \hookrightarrow \mathbf{Sh}(X)$ can be identified with the map sending any open set U of X to the largest open set of X whose intersection with Y is contained in U. Next, we observe that, since the maximal ideals of A can be characterized in an invariant way as the points of the Zariski topos $\mathbf{Sh}(\mathrm{Spec}(A))$ which are minimal with respect to the specialization ordering, and the points of the topos $\mathbf{Sh}(L(A), J_A^{\mathrm{coh}})$ correspond to the prime ideals of the lattice $L(A)$, the Morita equivalence

$$\mathbf{Sh}(\mathrm{Spec}(A)) \simeq \mathbf{Sh}(L(A), J_A^{\mathrm{coh}})$$

restricts to an equivalence

$$\mathbf{Sh}(\mathrm{Max}(A)) \simeq \mathbf{Sh}(\mathrm{Max}(L(A))),$$

where $\mathrm{Max}(L(A))$ is the subspace of the prime spectrum $\mathrm{Spec}(L(A))$ of $L(A)$ on the maximal ideals of $L(A)$.

Now, for any distributive lattice D, the subspace $\mathrm{Max}(D)$ of the prime spectrum $\mathrm{Spec}(D)$ of D consisting of the maximal ideals of D corresponds to a unique Grothendieck topology $J_D^{\max}$ on D yielding an equivalence of subtoposes

$$\begin{array}{ccc} \mathbf{Sh}(\mathrm{Spec}(D)) & \simeq & \mathbf{Sh}(D, J_D^{\mathrm{coh}}) \\ \uparrow & & \uparrow \\ \mathbf{Sh}(\mathrm{Max}(D)) & \simeq & \mathbf{Sh}(D, J_D^{\max}). \end{array}$$

For any ideal I of D, the open subset of $\mathrm{Spec}(D)$ corresponding to it is given by $\{P \in \mathrm{Spec}(D) \mid P \not\supseteq I\}$. The closure operation on $\mathbf{Sh}(D, J_D^{\mathrm{coh}})$ corresponding to the geometric inclusion $\mathbf{Sh}(D, J_D^{\max}) \hookrightarrow \mathbf{Sh}(D, J_D^{\mathrm{coh}})$ can thus be described, in terms of ideals of D, as the assignment sending any ideal I of D to the intersection of all the maximal ideals of D containing I. Under the hypothesis of the maximal ideal theorem for D (i.e. the assertion that every non-trivial ideal of D is contained in a maximal ideal), this can in turn be identified with the map sending any ideal I of D to the ideal

$$\{d \in D \mid (\forall b \in D)(b \vee d = 1) \Rightarrow (\exists c \in I)(b \vee c = 1)\}$$

(cf. section 1 of [54]). Indeed, for any element d of D, d does not belong to the intersection of all the maximal ideals of D containing I if and only if there exist a maximal ideal M of D containing I and an element $c \in M$ such that

$c \vee d = 1$; in other words, d belongs to the intersection of all the maximal ideals of D containing I if and only if for every element $b \in D$, $b \vee d = 1$ implies that for every maximal ideal M of D containing I, $b \notin M$. But, by the maximal ideal theorem for D, this latter condition means precisely that the ideal generated by b and I is trivial i.e. that there exists an element $c \in I$ such that $b \vee c = 1$.

This characterization yields the following explicit description of the topology $J_D^{\max}$: a sieve $\{c_i \to c \mid i \in I\}$ on $c \in D$ is $J_D^{\max}$-covering if and only if for every $d \in D$, $d \vee c = 1$ implies that there exists a finite subset $J \subseteq I$ such that $\bigvee_{i \in J} c_i \vee d = 1$. Notice in particular that every $J_D^{\max}$-covering sieve on the top element 1_D of D is generated by a finite family of arrows (take $d = 0$). This implies, in view of Proposition 10.8.1(iv), that (assuming the maximal ideal theorem for D) the space $\mathrm{Max}(D)$ is compact:

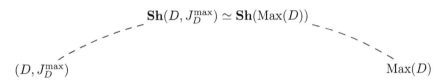

$(D, J_D^{\max})$ $\qquad\qquad\qquad\qquad\qquad\qquad$ $\mathrm{Max}(D)$

By applying this description of the topology $J_D^{\max}$ in the particular case $D = L(A)$, in conjunction with the fact that for any elements $a, b \in A$, $[P_a] \leq [P_b]$ in $L(A)$ if and only if a power of a is a multiple of b in A (cf. the technical lemma V3.2 [53]), we easily arrive at the following axiomatization of the theory $\mathbb{P}_A^{\max}$ (below we indicate by a list of elements in parentheses the ideal generated by them): the theory $\mathbb{P}_A^{\max}$ is obtained from $\mathbb{P}_A$ by adding all the sequents of the form

$$\left(P_c \vdash \bigvee_{i \in I} P_{c_i}\right)$$

for any elements $\{c_i \mid i \in I\}$ such that for any $i \in I$ a power of c_i is a multiple of c in A and for any finite set of elements $d_1, \ldots, d_n \in A$ such that $(c, d_1, \ldots, d_n) = A$ there exists a finite subset $J \subseteq I$ such that the ideal generated by the d_i (for $i = 1, \ldots, n$) and the c_i (for $i \in J$) is the whole of A.

Notice that the compactness of the space $\mathrm{Max}(A)$ can be expressed in logical terms as the assertion that, for any set $\{\phi_i \mid i \in I\}$ of geometric sentences over the signature of $\mathbb{P}_A$, if for every maximal ideal M of A there is an element $i \in I$ such that $M \vDash \phi_i$ then there exists a finite subset $J \subseteq I$ such that for every maximal ideal M of A, $M \vDash \phi_i$ for some $i \in J$.

The following result, which is a corollary of the above arguments, provides an explicit description of the sobrification of $\mathrm{Max}(A)$ as a subspace of $\mathrm{Spec}(A)$.

Theorem 10.7.1 *Let A be a commutative ring with unit. Then a prime ideal P of A belongs to the sobrification of $\mathrm{Max}(A)$ (resp. is maximal, if $\mathrm{Max}(A)$ is sober) if and only if for any elements c and $\{c_i \mid i \in I\}$ of A such that for every $i \in I$ a power of c_i is a multiple of c and for any finite set of elements $d_1, \ldots, d_n \in A$ such that $(c, d_1, \ldots, d_n) = A$ there exists a finite subset $J \subseteq I$*

such that the ideal generated by the d_j (for $j = 1, \ldots, n$) and the c_i (for $i \in J$) is the whole of A, if $c_i \in P$ for all $i \in I$ then $c \in P$. □

Remark 10.7.2 Theorem 10.7.1 notably applies to any commutative C^*-algebra (since its Gelfand spectrum $\mathrm{Max}(A)$ is sober being Hausdorff), giving a characterization of its maximal ideals.

We can 'functorialize' Theorem 10.7.1 in a natural way, obtaining a characterization of the ring homomorphisms $f : A \to B$ such that the induced continuous map $\mathrm{Spec}(B) \to \mathrm{Spec}(A)$ restricts to a (continuous) map $\mathrm{Max}(B) \to \mathrm{Max}(A)$.

Let us work once again in the more general context of morphisms of distributive lattices.

Theorem 10.7.3 Let $f : D \to D'$ be a morphism of distributive lattices. If

$$\mathrm{Spec}(f) : \mathrm{Spec}(D') \to \mathrm{Spec}(D)$$

restricts to a (necessarily continuous map)

$$\mathrm{Max}(D') \to \mathrm{Max}(D)$$

then f is a morphism of sites

$$(D, J_D^{\mathrm{max}}) \to (D', J_{D'}^{\mathrm{max}}).$$

The converse holds if $\mathrm{Max}(D)$ is sober.

Proof The thesis arises from the consideration of the following 'bridge' in light of the general theory of morphisms of sites:

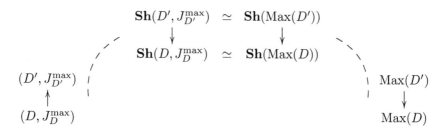

If $\mathrm{Spec}(f) : \mathrm{Spec}(D') \to \mathrm{Spec}(D)$ restricts to a (necessarily continuous map) $\mathrm{Spec}(f)| : \mathrm{Max}(D') \to \mathrm{Max}(D)$ then the following diagram (where the vertical functors are the canonical ones) commutes:

$$\begin{array}{ccc} D & \xrightarrow{f} & D' \\ \downarrow & & \downarrow \\ \mathbf{Sh}(\mathrm{Max}(D)) & \xrightarrow{\mathbf{Sh}(\mathrm{Spec}(f)|)^*} & \mathbf{Sh}(\mathrm{Max}(D')). \end{array}$$

Composing with the equivalences

$$\mathbf{Sh}(D, J_D^{\max}) \simeq \mathbf{Sh}(\mathrm{Max}(D))$$

and

$$\mathbf{Sh}(D', J_{D'}^{\max}) \simeq \mathbf{Sh}(\mathrm{Max}(D')),$$

we obtain the commutativity of the diagram

$$\begin{array}{ccc} D & \xrightarrow{f} & D' \\ \downarrow & & \downarrow \\ \mathbf{Sh}(D, J_D^{\max}) & \longrightarrow & \mathbf{Sh}(D', J_{D'}^{\max}), \end{array}$$

where the vertical functors are the canonical ones. This in turn implies that f is cover-preserving, whence a morphism of sites $(D, J_D^{\max}) \to (D', J_{D'}^{\max})$.

Conversely, if f is a morphism of sites $(D, J_D^{\max}) \to (D', J_{D'}^{\max})$ then it induces a geometric morphism

$$\mathbf{Sh}(f) : \mathbf{Sh}(D', J_{D'}^{\max}) \to \mathbf{Sh}(D, J_D^{\max}),$$

which corresponds under the Morita equivalences

$$\mathbf{Sh}(D, J_D^{\max}) \simeq \mathbf{Sh}(\mathrm{Max}(D))$$

and

$$\mathbf{Sh}(D', J_{D'}^{\max}) \simeq \mathbf{Sh}(\mathrm{Max}(D'))$$

to a geometric morphism

$$\mathbf{Sh}(\mathrm{Max}(D')) \to \mathbf{Sh}(\mathrm{Max}(D)).$$

This morphism in turn corresponds, if $\mathrm{Max}(D)$ is sober, to a unique continuous map $\mathrm{Max}(D') \to \mathrm{Max}(D)$ given by the restriction of $\mathrm{Spec}(f) : \mathrm{Spec}(D') \to \mathrm{Spec}(D)$ to $\mathrm{Max}(D')$ and $\mathrm{Max}(D)$. □

Applying this result in the context of distributive lattices of the form $L(A)$, we obtain the following

Theorem 10.7.4 *Let $f : A \to B$ be a homomorphism of commutative rings with unit A and B. If the continuous map $\mathrm{Spec}(f) : \mathrm{Spec}(B) \to \mathrm{Spec}(A)$ restricts to a (continuous) map $\mathrm{Max}(B) \to \mathrm{Max}(A)$ then for any elements c and $\{c_i \mid i \in I\}$ of A such that for every $i \in I$ a power of c_i is a multiple of c and for any finite set of elements $d_1, \ldots, d_n \in A$ such that $(c, d_1, \ldots, d_n) = A$ there exists a finite subset $J \subseteq I$ such that the ideal generated by the d_j (for $j = 1, \ldots, n$) and the c_i (for $i \in J$) is the whole of A, if $(f(c), d'_1, \ldots, d'_m) = B$ for some elements $d'_1, \ldots, d'_m \in B$ then there exists a finite subset $K \subseteq I$ such that the ideal generated by the d'_k (for $k = 1, \ldots, m$) and the $f(c_i)$ (for $i \in K$) is the whole of B. The converse implication holds if $\mathrm{Max}(A)$ is sober.* □

Notice that the above development of a natural analogue, for the maximal spectrum of a commutative ring with unit, of the different points of view available on the Zariski spectrum, has been achieved in a purely canonical way following an implementation of the 'bridge' technique; no choices of essentially arbitrary nature have been made.

10.8 Compactness conditions for geometric theories

In this section we shall identify the essential syntactic properties which characterize the geometric theories belonging to a proper fragment of geometric logic such as cartesian, regular or coherent logic.

We shall say that an object A of a Grothendieck topos $\mathcal{E}$ is *stably irreducible* if it is irreducible and, for any arrows $f, g : B \to A$ in $\mathcal{E}$, where B is an irreducible object, the domain of the pullback of f along g is irreducible.

Given a small site $(\mathcal{C}, J)$, we denote by $l_J^\mathcal{C} : \mathcal{C} \to \mathbf{Sh}(\mathcal{C}, J)$ the composite of the Yoneda embedding $y : \mathcal{C} \to [\mathcal{C}^{\mathrm{op}}, \mathbf{Set}]$ with the associated sheaf functor $a_J : [\mathcal{C}^{\mathrm{op}}, \mathbf{Set}] \to \mathbf{Sh}(\mathcal{C}, J)$.

Recall that a small site $(\mathcal{C}, J)$ is said to be *regular* (resp. *coherent*) if $\mathcal{C}$ is cartesian and J is generated by sieves generated by single arrows (resp. by finite families of arrows).

Proposition 10.8.1 *Let $(\mathcal{C}, J)$ be a small site.*

  (i) *If $\mathcal{C}$ is cartesian and J is the trivial topology on it then, for every $c \in \mathcal{C}$, $l_J^\mathcal{C}(c)$ is a stably irreducible object of $\mathbf{Sh}(\mathcal{C}, J)$.*
  (ii) *If $(\mathcal{C}, J)$ is regular then, for every $c \in \mathcal{C}$, $l_J^\mathcal{C}(c)$ is a supercoherent object of $\mathbf{Sh}(\mathcal{C}, J)$.*
  (iii) *If $(\mathcal{C}, J)$ is coherent then, for every $c \in \mathcal{C}$, $l_J^\mathcal{C}(c)$ is a coherent object of $\mathbf{Sh}(\mathcal{C}, J)$.*
  (iv) *If every J-covering sieve on an object c of $\mathcal{C}$ contains a finite family generating a J-covering sieve, then $l_J^\mathcal{C}(c)$ is a compact object of $\mathbf{Sh}(\mathcal{C}, J)$.*

Proof A proof of parts (ii) and (iii) can be found in [55] (cf. Theorem D3.3.7 and Remark D3.3.10 therein), while part (iv) was proved in [20] (Proposition 2.6(i) therein). So it remains to prove part (i).

The fact that $l_J^\mathcal{C}(c) = y(c)$ is irreducible follows from Proposition 10.3.3. In fact, the category $\mathcal{C}$ being cartesian and hence Cauchy-complete, in the topos $[\mathcal{C}^{\mathrm{op}}, \mathbf{Set}]$ the irreducible objects are precisely, up to isomorphism, those of the form $y(c)$ for some c. The fact that $y(c)$ is stably irreducible thus follows from the fact that the Yoneda embedding y is full and preserves pullbacks. □

Thanks to Proposition 10.8.1, we can characterize the classifying toposes of theories lying in fragments of geometric logic in an invariant way:

Theorem 10.8.2 *Let $\mathcal{E}$ be a Grothendieck topos.*

  (i) *$\mathcal{E}$ is the classifying topos of a cartesian theory if and only if it has a separating set of stably irreducible objects which is closed in $\mathcal{E}$ under finite limits.*

(ii) $\mathcal{E}$ is the classifying topos of a regular theory if and only if it has a separating set of supercoherent objects which is closed in $\mathcal{E}$ under finite limits.

(iii) $\mathcal{E}$ is the classifying topos of a coherent theory if and only if it has a separating set of coherent objects which is closed in $\mathcal{E}$ under finite limits.

Proof

(i) The 'only if' direction follows at once from Proposition 10.8.1(i). Conversely, suppose that $\mathcal{E}$ has a separating set $\mathcal{C}$ of stably irreducible objects which is closed in $\mathcal{E}$ under finite limits; then $\mathcal{C}$, regarded as a full subcategory of $\mathcal{E}$, is cartesian and the Comparison Lemma (Theorem 1.1.8) yields an equivalence $\mathcal{E} \simeq [\mathcal{C}^{\mathrm{op}}, \mathbf{Set}]$. Out thesis thus follows from Theorems 1.4.13 and 2.1.8.

(ii) and (iii) The 'only if' direction follows from Theorem D3.3.7 [55] and Remarks D3.3.9-10 [55]. Conversely, suppose that $\mathcal{E}$ has a separating set $\mathcal{G}$ of coherent (resp. supercoherent) objects which is closed in $\mathcal{E}$ under finite limits. Then $\mathcal{G}$ is a cartesian category and, by the Comparison Lemma (Theorem 1.1.8), $\mathcal{E}$ is equivalent to $\mathbf{Sh}(\mathcal{G}, J_{\mathcal{E}}|_{\mathcal{G}})$; now, the site $(\mathcal{G}, J_{\mathcal{E}}|_{\mathcal{G}})$ is clearly coherent (resp. regular) and hence $\mathcal{E}$ is the classifying topos of a regular (resp. coherent) theory (cf. the proof of Theorem 2.1.11).

□

Remark 10.8.3 From the proof of Theorem 10.8.2 it is clear that one can equivalently replace 'stably irreducible' (resp. 'supercoherent', 'coherent') with 'irreducible' (resp. 'supercompact', 'compact') in its statement.

Let us now interpret the notions of compact and supercompact object of a Grothendieck topos in the context of the geometric syntactic site of a geometric theory; together with the concept of irreducible object, these notions will be the key ingredients for establishing our characterizations of the geometric theories which are cartesian, regular or coherent over their signature.

Definition 10.8.4 Let $\mathbb{T}$ be a geometric theory over a signature Σ and $\phi(\vec{x})$ a geometric formula-in-context over Σ.

(a) We say that $\phi(\vec{x})$ is $\mathbb{T}$-compact if, for any family $\{\psi_i(\vec{x}) \mid i \in I\}$ of geometric formulae over Σ in a given context $\vec{x}$, $(\phi \vdash_{\vec{x}} \bigvee_{i \in I} \psi_i)$ provable in $\mathbb{T}$ implies $(\phi \vdash_{\vec{x}} \bigvee_{i \in I'} \psi_i)$ provable in $\mathbb{T}$ for some finite subset I' of I.

(b) We say that $\phi(\vec{x})$ is $\mathbb{T}$-supercompact if, for any family $\{\psi_i(\vec{x}) \mid i \in I\}$ of geometric formulae over Σ in a given context $\vec{x}$, $(\phi \vdash_{\vec{x}} \bigvee_{i \in I} \psi_i)$ provable in $\mathbb{T}$ implies $(\phi \vdash_{\vec{x}} \psi_i)$ provable in $\mathbb{T}$ for some $i \in I$.

Lemma 10.8.5 (cf. Lemma 3.2 [20]). *Let $\mathbb{T}$ be a geometric theory over a signature Σ, $\phi(\vec{x})$ a geometric formula-in-context over Σ and $y_{\mathbb{T}} : \mathcal{C}_{\mathbb{T}} \hookrightarrow \mathbf{Sh}(\mathcal{C}_{\mathbb{T}}, J_{\mathbb{T}})$ the Yoneda embedding. Then*

(i) $\phi(\vec{x})$ is $\mathbb{T}$-irreducible (in the sense of Definition 6.1.9) if and only if $y_{\mathbb{T}}(\{\vec{x}.\phi\})$ is an irreducible object of $\mathbf{Sh}(\mathcal{C}_{\mathbb{T}}, J_{\mathbb{T}})$.

(ii) $\phi(\vec{x})$ is $\mathbb{T}$-compact if and only if $y_{\mathbb{T}}(\{\vec{x}.\phi\})$ is a compact object of $\mathbf{Sh}(\mathcal{C}_{\mathbb{T}}, J_{\mathbb{T}})$.

(iii) $\phi(\vec{x})$ is $\mathbb{T}$-supercompact if and only if $y_{\mathbb{T}}(\{\vec{x}.\phi\})$ is a supercompact object of $\mathbf{Sh}(\mathcal{C}_{\mathbb{T}}, J_{\mathbb{T}})$.

The following theorem, whose part (ii) extends Lemma D3.3.11 [55], is obtained from Proposition 10.8.1 and Lemma 10.8.5 by applying the 'bridge' technique (cf. section 2.2).

Theorem 10.8.6 *Let $\mathbb{T}$ be a geometric theory over a signature Σ and $\phi(\vec{x})$ a geometric formula-in-context over Σ.*

(i) *If $\mathbb{T}$ is cartesian and $\phi(\vec{x})$ is $\mathbb{T}$-cartesian then $\phi(\vec{x})$ is $\mathbb{T}$-irreducible.*

(ii) *If $\mathbb{T}$ is regular and $\phi(\vec{x})$ is regular then $\phi(\vec{x})$ is $\mathbb{T}$-supercompact.*

(iii) *If $\mathbb{T}$ is coherent and $\phi(\vec{x})$ is coherent then $\phi(\vec{x})$ is $\mathbb{T}$-compact.*

Proof The theorem arises from the consideration of the following 'bridges':

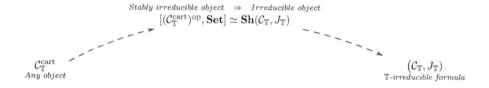

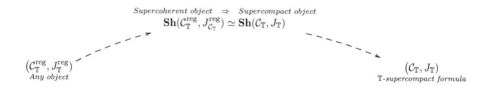

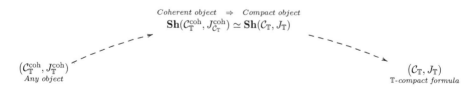

Specifically, if $\mathbb{T}$ is a cartesian (resp. regular, coherent) theory then for any $\mathbb{T}$-cartesian (resp. regular, coherent) formula $\phi(\vec{x})$ over Σ we have that $y_{\mathbb{T}}^{\mathrm{cart}}(\{\vec{x}.\phi\})$

(resp. $y_{\mathbb{T}}^{\text{reg}}(\{\vec{x}\,.\,\phi\})$, $y_{\mathbb{T}}^{\text{coh}}(\{\vec{x}\,.\,\phi\})$) and $y_{\mathbb{T}}(\{\vec{x}\,.\,\phi\})$ correspond to each other under the equivalence
$$[(\mathcal{C}_{\mathbb{T}}^{\text{cart}})^{\text{op}}, \mathbf{Set}] \simeq \mathbf{Sh}(\mathcal{C}_{\mathbb{T}}, J_{\mathbb{T}})$$
(resp.
$$\mathbf{Sh}(\mathcal{C}_{\mathbb{T}}^{\text{reg}}, J_{\mathcal{C}_{\mathbb{T}}}^{\text{reg}}) \simeq \mathbf{Sh}(\mathcal{C}_{\mathbb{T}}, J_{\mathbb{T}}),$$
$$\mathbf{Sh}(\mathcal{C}_{\mathbb{T}}^{\text{reg}}, J_{\mathcal{C}_{\mathbb{T}}}^{\text{coh}}) \simeq \mathbf{Sh}(\mathcal{C}_{\mathbb{T}}, J_{\mathbb{T}})),$$
where $y_{\mathbb{T}}^{\text{cart}}: \mathcal{C}_{\mathbb{T}}^{\text{cart}} \hookrightarrow [(\mathcal{C}_{\mathbb{T}}^{\text{cart}})^{\text{op}}, \mathbf{Set}]$ (resp. $y_{\mathbb{T}}^{\text{reg}}: \mathcal{C}_{\mathbb{T}}^{\text{reg}} \hookrightarrow \mathbf{Sh}(\mathcal{C}_{\mathbb{T}}^{\text{reg}}, J_{\mathbb{T}}^{\text{reg}})$, $y_{\mathbb{T}}^{\text{coh}}: \mathcal{C}_{\mathbb{T}}^{\text{coh}} \hookrightarrow \mathbf{Sh}(\mathcal{C}_{\mathbb{T}}^{\text{coh}}, J_{\mathbb{T}}^{\text{coh}}))$ is the Yoneda embedding.

Now, by Proposition 10.8.1, $y_{\mathbb{T}}^{\text{cart}}(\{\vec{x}\,.\,\phi\})$ (resp. $y_{\mathbb{T}}^{\text{reg}}(\{\vec{x}\,.\,\phi\})$, $y_{\mathbb{T}}^{\text{coh}}(\{\vec{x}\,.\,\phi\})$) is a stably irreducible and hence irreducible (resp. supercoherent and hence supercompact, coherent and hence compact) object of the topos $[(\mathcal{C}_{\mathbb{T}}^{\text{cart}})^{\text{op}}, \mathbf{Set}]$ (resp. $\mathbf{Sh}(\mathcal{C}_{\mathbb{T}}^{\text{reg}}, J_{\mathbb{T}}^{\text{reg}})$, $\mathbf{Sh}(\mathcal{C}_{\mathbb{T}}^{\text{coh}}, J_{\mathbb{T}}^{\text{coh}}))$. Our thesis thus follows from Lemma 10.8.5. □

Theorem 10.8.7 (cf. Theorems 3.3-4-5 [20]). *Let $\mathbb{T}$ be a geometric theory over a signature Σ. Then $\mathbb{T}$ is syntactically equivalent to a cartesian (resp. regular, coherent) theory over Σ if and only if every $\mathbb{T}$-cartesian (resp. regular, coherent) formula over Σ is $\mathbb{T}$-irreducible (resp. $\mathbb{T}$-supercompact, $\mathbb{T}$-compact).*

Proof The 'only if' part of the theorem follows from Theorem 10.8.6. To prove the converse direction we observe that, by Lemma D1.3.8 [55], the full subcategory of the geometric syntactic category $\mathcal{C}_{\mathbb{T}}$ of $\mathbb{T}$ on the Horn formulae is $J_{\mathbb{T}}$-dense. So if every $\mathbb{T}$-cartesian formula over Σ is $\mathbb{T}$-irreducible the Comparison Lemma (Theorem 1.1.8) yields an equivalence
$$\mathbf{Sh}(\mathcal{C}_{\mathbb{T}}, J_{\mathbb{T}}) \simeq [(\mathcal{C}_{\mathbb{T}'}^{\text{cart}})^{\text{op}}, \mathbf{Set}],$$
where $\mathbb{T}'$ is the cartesianization of $\mathbb{T}$ (i.e. the cartesian theory consisting of all the $\mathbb{T}$-cartesian sequents which are provable in $\mathbb{T}$), which shows that $\mathbb{T}$ is syntactically equivalent to $\mathbb{T}'$, as required. Similar arguments work for the regular and coherent cases. □

BIBLIOGRAPHY

[1] Adámek, J., and Rosický, J., *Locally Presentable and Accessible Categories*, Cambridge University Press (1994).
[2] Antonius, W., *Théories cohérentes et prétopos*, Thèse de Maitrise (Mathématiques), Université de Montréal (1975).
[3] Artin, M., Grothendieck, A., and Verdier, J. L., *Théorie des topos et cohomologie étale des schémas*, Séminaire de Géométrie Algébrique du Bois-Marie, année 1963-64; second edition published as Lecture Notes in Mathematics, vols. 269, 270 and 305, Springer-Verlag (1972).
[4] Barr, M., and Diaconescu, R., Atomic Toposes, *Journal of Pure and Applied Algebra* 17 (1980), 1-24.
[5] Beke, T., Theories of presheaf type, *Journal of Symbolic Logic* 69 (3), 923-934 (2004).
[6] Beke, T., Karazeris, P., and Rosický, J., When is flatness coherent? *Communications in Algebra* 33 (6), 1903-1912 (2005).
[7] Blass, A., and Ščedrov, A., Classifying topoi and finite forcing, *Journal of Pure and Applied Algebra* 28 (2), 111-140 (1983).
[8] Borceux, F., *Handbook of categorical algebra*, vol. 1, Cambridge University Press (1994).
[9] Borceux, F., *Handbook of categorical algebra*, vol. 3, Cambridge University Press (1994).
[10] Bunge, M., *Categories of Set-valued functors*, Ph.D. thesis, University of Pennsylvania (1966).
[11] Butz, C., and Moerdijk, I., Representing topoi by topological groupoids, *Journal of Pure and Applied Algebra*, 130 (3), 223-235 (1998).
[12] Caramello, O., *The Duality between Grothendieck Toposes and Geometric Theories*, Ph.D. thesis, University of Cambridge (2009).
[13] Caramello, O., De Morgan classifying toposes, *Advances in Mathematics* 222 (6), 2117-2144 (2009).
[14] Caramello, O., The unification of Mathematics via Topos Theory, *arXiv:math.CT/1006.3930* (2010).
[15] Caramello, O., A characterization theorem for geometric logic, *Annals of Pure and Applied Logic* 162 (4), 318-321 (2011).
[16] Caramello, O., A topos-theoretic approach to Stone-type dualities, *arXiv:math.CT/1006.3930* (2011).
[17] Caramello, O., Yoneda representations of flat functors and classifying toposes, *Theory and Applications of Categories* 26 (21), 538-553 (2012).

[18] Caramello, O., Atomic toposes and countable categoricity, *Applied Categorical Structures* 20 (4), 379-391 (2012).
[19] Caramello, O., Universal models and definability, *Mathematical Proceedings of the Cambridge Philosophical Society* 152 (2), 279-302 (2012).
[20] Caramello, O., Syntactic characterizations of properties of classifying toposes, *Theory and Applications of Categories* 26 (6), 176-193 (2012).
[21] Caramello, O., Site characterizations for geometric invariants of toposes, *Theory and Applications of Categories* 26 (225), 710-728 (2012).
[22] Caramello, O., Gelfand spectra and Wallman compactifications, *arxiv:math.CT/1204.3244* (2012).
[23] Caramello, O., Fraïssé's construction from a topos-theoretic perspective, *Logica Universalis* 8 (2), 261-281 (2014).
[24] Caramello, O., A general method for building reflections, *Applied Categorical Structures* 22 (1), 99-118 (2014).
[25] Caramello, O., Topologies for intermediate logics, *Mathematical Logic Quarterly* 60 (4-5), 335-347 (2014).
[26] Caramello, O., Priestley-type dualities for partially ordered structures, *Annals of Pure and Applied Logic* 167 (9), 820-849 (2016).
[27] Caramello, O., Topological Galois Theory, *Advances in Mathematics* 291, 646-695 (2016).
[28] Caramello, O., *Grothendieck toposes as unifying 'bridges' in Mathematics*, Mémoire d'habilitation à diriger des recherches, Université Paris 7 (2016), available at `http://www.oliviacaramello.com/Unification/HDROliviaCaramello.pdf`.
[29] Caramello, O., and Johnstone, P. T., De Morgan's law and the theory of fields, *Advances in Mathematics* 222 (6), 2145-2152 (2009).
[30] Caramello, O., and Russo, A. C., The Morita-equivalence between MV-algebras and abelian ℓ-groups with strong unit, *Journal of Algebra* 422, 752-787 (2015).
[31] Caramello, O., and Russo, A. C., Lattice-ordered abelian groups and perfect MV-algebras: a topos-theoretic perspective, *Bulletin of Symbolic Logic* 22 (2), 170-214 (2016).
[32] Caramello, O., and Russo, A. C., On the geometric theory of local MV-algebras, *Journal of Algebra* 479, 263-313 (2017).
[33] Caramello, O., and Wentzlaff, N., Cyclic theories, *Applied Categorical Structures* 25 (1), 105-126 (2017).
[34] Cole, J., The bicategory of topoi and spectra, *Reprints in Theory and Applications of Categories* 25, 1-16 (2016).
[35] Connes, A., Cohomologie cyclique et foncteur Ext^n, *C. R. Acad. Sci. Paris, Sér. I Math.* 296, 953-958 (1983).
[36] Connes, A., and Consani, C., The Arithmetic Site, *C. R. Acad. Sci. Paris, Sér. I Math.* 352, 971-975 (2014).
[37] Connes, A., and Consani, C., The Cyclic and Epicyclic Sites, *Rendiconti del Seminario Matematico dell'Università di Padova* 134, 197-238 (2015).

Bibliography

[38] Coste, M., Localisation, spectra and sheaf representation, in *Applications of Sheaves*, Lecture Notes in Mathematics vol. 753 (Springer-Verlag, 1979), 212-238.

[39] Coste, M., and Coste, M.-F., Théories cohérentes et topos cohérents, *Séminaire Bénabou*, Paris (1975).

[40] Coste, M., Lombardi, H., and Roy, M.-F., Dynamical method in algebra: effective Nullstellensätze, *Annals of Pure and Applied Logic*, 111 (3), 203-256 (2001).

[41] Coumans, D., Generalising canonical extension to the categorical setting, *Annals of Pure and Applied Logic*, 163 (12), 1940-1961 (2012).

[42] Giraud, J., Analysis situs, *Séminaire Bourbaki*, exposé 256 (1963).

[43] Elkins, B., and Zilber, J. A., Categories of actions and Morita equivalence, *Rocky Mountain Journal of Mathematics* 6 (2), 199-226 (1976).

[44] Funk, J., Lawson, M. V., and Steinberg B., Characterizations of Morita equivalent inverse semigroups, *Journal of Pure and Applied Algebra* 215 (9) 2262-2279 (2011).

[45] Funk, J., and Hofstra, P., Topos theoretic aspects of semigroup actions, *Theory and Applications of Categories* 24 (6), 117-147 (2010).

[46] Goldblatt, R., *Topoi: The Categorial Analysis of Logic*, Dover, Reprint of the Elsevier North-Holland Publishing Co., New York, 1983 edition.

[47] Grothendieck, A., *Revêtements étales et groupe fondamental (SGA 1)*, Séminaire de Géométrie Algébrique du Bois-Marie, année 1960-61; published as Lecture Notes in Mathematics, vol. 224 (Springer-Verlag, 1971).

[48] Hakim, M., *Topos annelés et schémas relatifs*, Springer-Verlag (1972).

[49] Hodges, W., *Model theory*, Encyclopedia of Mathematics and its Applications vol. 42, Cambridge University Press (1993).

[50] Johnstone, P. T., *Topos theory*, Academic Press (1977).

[51] Johnstone, P. T., A syntactic approach to Diers' localizable categories, in *Applications of sheaves*, Lecture Notes in Mathematics vol. 753 (Springer-Verlag, 1979), 466-478.

[52] Johnstone, P. T., Conditions related to De Morgan's law, in *Applications of Sheaves*, Lecture Notes in Mathematics vol. 753 (Springer-Verlag, 1979), 479-491.

[53] Johnstone, P. T., *Stone spaces*, Cambridge Studies in Advanced Mathematics No. 3, Cambridge University Press (1982).

[54] Johnstone, P. T., Almost maximal ideals, *Fundamenta Mathematicae* 123, 197-209 (1984).

[55] Johnstone, P. T., *Sketches of an Elephant: a topos theory compendium. Vols. 1 and 2*, Oxford University Press (2002).

[56] Johnstone, P. T., and Wraith, G., Algebraic theories in toposes, in *Indexed categories and their applications*, Lecture Notes in Mathematics vol. 661 (Springer-Verlag, 1978), 141-242.

[57] Joyal, A., and Tierney, M., An extension of the Galois theory of Grothendieck, *Memoirs of the American Mathematical Society* 51, no. 309, vii + 71 pp. (1984).

[58] Joyal, A., and Wraith, G., Eilenberg-MacLane toposes and cohomology, *Mathematical applications of category theory*, Contemporary Mathematics, vol. 30 (American Mathematical Society, 1984), 117-131.

[59] Lawvere, W., Functorial Semantics of Algebraic Theories (1963), republished in *Reprints in Theory and Applications of Categories* 5, 1-121 (2004).

[60] Lawvere, W., Intrinsic Co-Heyting Boundaries and the Leibniz Rule in Certain Toposes, in *Category Theory*, Lecture Notes in Mathematics vol. 1488 (Springer-Verlag, 1991), 279-281.

[61] Lawvere, W., Open problems in Topos Theory, 4 April 2009, manuscript available at http://cheng.staff.shef.ac.uk/pssl88/lawvere.pdf.

[62] Mac Lane, S., *Categories for the Working Mathematician*, Graduate Texts in Mathematics, Vol. 5, 2nd ed., Springer-Verlag (1978).

[63] Mac Lane, S., and Moerdijk, I., *Sheaves in geometry and logic: a first introduction to topos theory*, Springer-Verlag (1992).

[64] Makkai, M., and Reyes, G., *First-order categorical logic*, Lecture Notes in Mathematics vol. 611, Springer-Verlag (1977).

[65] Moerdijk, I., Morita equivalence for continuous groups, *Mathematical Proceedings of the Cambridge Philosophical Society* 103 (1), 97-115 (1988).

[66] Moerdijk, I., Cyclic sets as a classifying topos (Utrecht, 1996), available at http://ncatlab.org/nlab/files/MoerdijkCyclic.pdf.

[67] Paré, R., and Schumacher D., Abstract families and the Adjoint Functor Theorems, in *Indexed categories and their applications*, Lecture Notes in Mathematics vol. 661 (Springer-Verlag, 1978), 1-125.

[68] Rajani, R., and Prest, M., Model-theoretic imaginaries and coherent sheaves, *Applied Categorical Structures* 17, 517-559 (2009).

[69] Schreier, O., Die Untergruppen der freien Gruppen, *Abh. Math. Sem. Univ. Hamburg* vol. 5, 161-183 (1927).

[70] Steinberg, S. A., *Lattice-ordered Rings and Modules*, Springer-Verlag (2010).

[71] Talwar, S., Morita equivalence for semigroups, *Journal of the Australian Mathematical Society (Series A)* 59, 81-111 (1995).

[72] Tierney, M., Forcing Topologies and Classifying Topoi, in *Algebra, topology and category theory. A collection of papers in honor of Samuel Eilenberg* (Academic Press, 1976), 211-219.

[73] Vickers, S., Strongly algebraic = SFP (topically), *Mathematical Structures in Computer Science* 11 (6), 717-742 (2001).

[74] Vickers, S., Locales and Toposes as Spaces, in *Handbook of Spatial Logics* (Springer, 2007), 429-496.

INDEX

J-ideal, 20
∞-pretopos, 35

Amalgamation property, 332
Arity (of a function symbol), 23
Associated sheaf functor, 16
Atom, 137
   of a topos, 328
Axiom (of a theory), 25

Bi-interpretation, 49
Booleanization
   of a theory, 131, 338
   of a topos, 130
Boundary
   of a subtopos, 326
Bridge (topos-theoretic)
   arch of, 74
   deck of, 70
   technique, 69

Cartesianization
   of a geometric theory, 26
Categorical equivalence, 73
Category
   of elements, 20
   Boolean coherent, 33
   cartesian, 10
   cartesian closed, 15
   Cauchy-complete, 18
   coherent, 32
   effective regular, 34
   essentially small, 89, 198
   finitely accessible, 260
   geometric, 32
   Heyting, 33
   indexed, 140
   internal to a topos, 140
   positive coherent, 35
   regular, 31
   well-powered, 32
Cauchy completion, 195

Cell
   0-cell, 83
   1-cell, 83
   2-cell, 83
Classifying arrow (of a subobject), 16
Classifying topos, 55
   for a cartesian theory, 57, 64
   for a coherent theory, 57
   for a geometric theory, 55, 57
   for a propositional theory, 63
   for a regular theory, 57
   for a universal Horn theory, 57
   for Boolean algebras, 65
   for commutative rings with unit, 65
   for integral domains, 66
   for intervals, 67
   for local rings, 65
Closure
   of a sieve, 84
Closure operation, 41
   universal, 41
Coherent category
   Boolean, 33
   positive, 35
Colimit, 15
   as tensor product, 146
   of an indexed functor, 142
   of an internal diagram, 142
Comparison Lemma, 14
Completeness
   classical (for coherent logic), 52
   strong, 51
Completeness theorem
   for coherent logic, 52
Cone
   universal, 14
Congruence, 268
Constant symbol, 23
Context, 25
   canonical, 25
   suitable, 25
Continuous G-set, 13
Continuous map, 18

De Morgan's law, 131
Deduction system, 27
    for Grothendieck topologies, 94
Definability theorem (for theories of presheaf type), 201
DeMorganization
    of a theory, 131
    of a topos, 131
Derivation, 27
Diaconescu's equivalence, 22
Diagram
    internal to a topos, 140
Direct image, 18
Discrete fibration, 141
Discrete opfibration, 141
Double negation
    topology, 11
    universal closure operation, 41
Duality, 73

Effectivization (of a regular category), 34
Elementary morphism, 38
Epimorphic family, 12
Equivalence
    categorical, 73
    Morita (of theories), 70
    proof-theoretic, 94
    syntactic (of theories), 89
Expansion
    fully faithful, 267
    hyperconnected, 242
    localic, 242
    of a geometric theory, 242
Exponential, 15

Filter
    prime, 65
Formula
    $\mathbb{T}$-Boolean, 202
    $\mathbb{T}$-complete, 329
    $\mathbb{T}$-consistent, 134
    $\mathbb{T}$-indecomposable, 329
    $\mathbb{T}$-irreducible, 203, 329
    $\mathbb{T}$-stably consistent, 131
    $\mathbb{T}$-compact, 355
    $\mathbb{T}$-supercompact, 355
    atomic, 25
    cartesian, 26
    coherent, 25
    first-order, 25
    geometric, 25
    Horn, 25
    in context, 25
    infinitary, 25
    provably equivalent, 46
    regular, 25
Fraïssé's theorem, 331
Frame, 11
    homomorphism, 11
Frobenius axiom, 28
Function symbol, 23
Functor
    cartesian, 10, 38
    coherent, 38
    conservative, 38
    continuous (for a Grothendieck topology), 22
    cover-preserving, 10
    filtering, 21
    final, 151
    flat, 21
    full, 140
    geometric, 38
    global sections, 18
    Heyting, 38
    indexed, 140
    indexed category of elements of, 144
    internal, 141
    logical, 39
    regular, 38

Galois theory, 340
Geometric category, 32
Geometric inclusion
    dense-closed factorization, 132
    skeletal, 133
Geometric morphism, 17
    coherent, 62
    essential, 18
    hyperconnected, 242
    inclusion, 19
    localic, 242
    regular, 62
    sujection, 19
    surjection-inclusion factorization of, 134
Geometric theory, 26, 53
    countably categorical, 336
    expansion of, 242
    presheaf completion of, 247
    propositional, 26
    quotient of, 89
    subtheory of, 215
    syntactically equivalent to, 89
Geometric transformation, 18
Giraud's theorem, 14
Grothendieck topology, 10, 85
    associated $\mathbb{T}$-quotient of, 93
    atomic, 11
    canonical (on a locale), 11

coherent, 48
dense, 11
double negation, 11
generated by a family of (pre)sieves, 84
generators for, 86
geometric, 48
Heyting implication of Grothendieck topologies, 111
induced, 14
of finite type, 289
operations on Grothendieck topologies, 108
proof-theoretic interpretation, 94
pseudocomplement of, 111
regular, 48
rigid, 202, 295
subcanonical, 13, 339, 344
trivial, 11
Grothendieck topos, 12
Groupoid
 localic, 13
 topological, 13, 39

Heyting algebra, 10
 complete, 10
 internal to a Grothendieck topos, 108
 internal to a topos, 40
Heyting category, 33
Heyting implication
 of Grothendieck topologies, 111
 of theories, 115

Ideal
 for a Grothendieck topology, 20
 maximal, 347
 prime, 66, 347
Image, 31
 direct, 18
 inverse, 18
 of an arrow, 31
Inclusion (geometric), 19
Indexed category, 140
 $\mathcal{E}$-full subcategory of, 151
 filtered, 144
 final subcategory of, 151
 having small homs at 1, 143
 localization of, 143
 locally small, 143
 strictly final subcategory of, 151
Indexed functor
 colimit of, 142
 final, 151
 full, 140
Inductive completion, 199, 284
Inference rule, 27

Infinitary positivization (of a geometric category), 35
Internal category
 externalization of, 142
Internal diagram
 colimit of, 142
 indexation of, 145
Internal hom, 143, 210
Internal language, 34
Interpretation, 49
 cartesian, 60
 coherent, 60
 faithful, 251
 geometric, 60
 regular, 60
Interval, 66
Invariant (topos-theoretic), 69
Inverse image, 18

Joint embedding property, 332

Kan extension, 18
Kripke-Joyal semantics, 44

Language
 first-order, 23
 internal, 34
Lattice, 10
 distributive, 350
Law of excluded middle, 28
Lawvere-Tierney topology, 41
Limit, 14
 finite fc, 340
 inductive, 15
 projective, 14
Lindenbaum-Tarski algebra, 64
Local operator, 41
 closed, 128
 dense, 125
 double negation, 41
 open, 124
 order between local operators, 115
 quasi-closed, 124
 relativization of, 121
Locale, 11
Localic groupoid, 13
Localization
 of an indexed category, 143
Logic, 28
 algebraic, 29
 classical first-order, 29
 coherent, 29
 fragment of, 77
 fragments of, 28
 geometric, 29
 Horn, 29

intuitionistic first-order, 29
regular, 29

Model (of a theory), 37
  J-homogeneous, 276
  S-homogeneous, 275
  finite, 294, 305
  finitely generated, 259, 301
  finitely presentable, 205
  finitely presented, 64, 205
  strongly finitely presented, 213
  universal (in a classifying topos), 56
  universal (in a syntactic category), 47
Monomorphism
  strict, 343
Morita-equivalence, 59, 70
  for associative rings with unit, 72
  for geometric theories, 59
  for inverse semigroups, 73
  for small categories, 72
  for topological groups, 72
Morleyization, 54
Morphism
  of sites, 10
  geometric, 17
  of structures, 30

Negation (of a subobject), 33

Object
  $\mathcal{C}$-homogeneous, 332
  $\mathcal{C}$-ultrahomogeneous, 332
  $\mathcal{C}$-universal, 333
  coherent, 61, 354
  compact, 61
  decidable, 255, 256
  indecomposable, 328
  irreducible, 61, 206, 328
  of homomorphisms, 210
  of morphisms, 143
  stably irreducible, 354
  subterminal, 1, 13
  supercoherent, 61, 354
  supercompact, 61
Order
  between Grothendieck topologies, 108
  between local operators, 115
  between theories, 113
Ore condition
  right, 11

Point (of a topos), 19
Positivization (of a coherent category), 35
Presheaf, 12
Presheaf completion (of a geometric theory), 247

Presieve, 84
Pretopology, 85
Pretopos, 35
Proof, 27
Pseudocomplement (of a subobject), 33

Quotient
  associated $\mathbb{T}$-topology of, 93
  induced $\mathbb{T}$-topology of, 280
  of a geometric theory, 89
  of a model, 259

Reflector, 42
Regular category, 31
  effective, 34
Relation symbol, 23
Relativization of a local operator, 121
Right Ore condition, 11
Rule of inference, 27

Sentence, 24
Separating family
  of objects, 12
  of points, 19
Sequent, 25
  cartesian, 26
  coherent, 25
  derivable, 29
  equivalent to, 89
  first-order, 25
  geometric, 25
  Horn, 25
  provable, 29
  regular, 25
  satisfied in M, 37
Sequent calculus, 27
Sheaf, 12
  for a closure operation, 41
  for a Grothendieck topology, 12
  on a locale, 13
  on a site, 12
  on a space, 13
Sierpiński space, 66
Sierpiński topos, 66
Sieve, 10
  J-closed, 17
  J-closure of, 84
  J-covering, 10
  composite, 85
  generated by a presieve, 84
  principal, 86
  stably non-empty, 11
Signature
  canonical, 34
  first-order, 23

Site, 10
    atomic, 276, 330
    coherent, 354
    locally connected, 330
    morphism of sites, 10
    of definition, 12
    regular, 354
    syntactic, 48
Site characterization, 69, 74
Sort, 23
Soundness theorem, 51
Space
    extrememally disconnected, 75
    sober, 20
Spectrum, 73
    Cole, 73
    Gelfand, 352
    maximal, 347
    prime, 347
    Zariski, 346
Strict monomorphism, 343
Structure
    finitely generated, 259
Structure (for a signature), 30
    homomorphism, 30
    tautological, 34
Subcategory
    J-dense, 13
Subobject, 15
    classifying arrow of, 16
    complemented, 33
    dense, 42
    negation of, 33
    pseudocomplement of, 33
Subobject classifier, 16
Subterminal object, 1
Subterminal topology, 20
Subtopos, 19
    closed, 128
    dense, 125
    of points of a topos, 138
    open, 126
    quasi-closed, 129
Sujection (geometric), 19
Surjection-inclusion factorization, 19
Syntactic category, 45
    algebraic (for a universal Horn theory), 46
    classical (for a first-order theory), 46
    for a cartesian theory, 46
    for a coherent theory, 46
    for a first-order theory, 46
    for a geometric theory, 45, 46
    for a regular theory, 46
Syntactic equivalence (of theories), 89

Tensor product, 146
Term (in formal language), 24
Theory, 25
    finite-limit part of, 26
    algebraic, 25, 26
    atomic, 137
    Boolean, 136
    Booleanization of, 131, 338
    cartesian, 26
    cartesianization of, 26
    coherent, 26
    complete, 55, 137, 335
    completions of, 138, 336
    contradictory, 113, 137
    countably categorical, 336
    DeMorganization of, 131
    first-order, 25
    geometric, 26
    having enough set-based models, 51
    Horn, 26
    Morita-equivalent, 59
    Morleyization of, 54
    of a structure, 134
    of abelian ℓ-groups with strong unit, 322
    of abstract circles, 307
    of algebraic extensions of a field, 317
    of decidable objects, 306
    of Diers fields, 312
    of fields, 312
    of fields of finite characteristic which are algebraic over their prime field, 337
    of finite sets, 310
    of groups (and its injectivization), 318
    of locally finite groups, 320
    of MV-algebras, 323
    of presheaf type, 4, 197
    of separable extensions of a field, 318
    of small categories, 26
    of vector spaces over a field, 321
    presentation of, 89
    presheaf completion of, 247
    propositional, 26, 63
    regular, 26
    subtheory of, 215
    universal Horn, 26
    with enough set-based models, 138
Topology
    atomic, 11
    canonical (on a locale), 11
    dense, 11
    double negation, 11
    Grothendieck, 10
    Lawvere-Tierney, 41

syntactic, 48
trivial, 11
Topos
  as a 'bridge', 69
  atomic, 137, 329, 334
  Boolean, 136
  Booleanization of, 130
  classifying, 55
  coherent, 59
  Connes', 307
  degenerate, 135
  DeMorganization of, 131
  elementary, 39
  epicyclic, 310
  equivalent to a presheaf topos, 329
  Grothendieck, 12
  having enough points, 19
  localic, 13, 63
  locally connected, 329
  of cyclic sets, 307
  of simplicial sets, 66
  point of, 19
  Schanuel, 337
  Sierpiński, 66
  trivial, 135
  two-valued, 135, 137, 335
  with enough points, 138
  Zariski, 65
Type (of a function symbol), 23

Universal closure operation, 41

Variable, 23
  bound, 23
  free, 23

Yoneda lemma
  2-dimensional, 83